Metal Oxide-Based Nanofibers and Their Applications

The Metal Oxides Book Series Edited by Ghenadii Korotcenkov

Forthcoming Titles

- *Palladium Oxides Material Properties, Synthesis and Processing Methods, and Applications*, Alexander M. Samoylov, Vasily N. Popov, 9780128192238
- *Metal Oxides for Non-volatile Memory*, Panagiotis Dimitrakis, Ilia Valov, Stefan Tappertzhofen, 9780128146293
- *Metal Oxide Nanostructured Phosphors*, H. Nagabhushana, Daruka Prasad, S.C. Sharma, 9780128118528
- *Nanostructured Zinc Oxide*, Kamlendra Awasthi, 9780128189009
- *Multifunctional Piezoelectric Oxide Nanostructures*, Sang-Jae Kim, Nagamalleswara Rao Alluri, Yuvasree Purusothaman, 9780128193327
- *Transparent Conductive Oxides*, Mirela Petruta Suchea, Petronela Pascariu, Emmanouel Koudoumas, 9780128206317
- *Metal oxide-based nanofibers and their applications*, Vincenzo Esposito, Debora Marani, 9780128206294
- *Metal-oxides for Biomedical and Biosensor Applications*, Kunal Mondal, 9780128230336
- *Metal Oxide-Carbon Hybrid Materials*, Muhammad Akram, Rafaqat Hussain, Faheem K Butt, 9780128226940
- *Metal Oxide-based heterostructures*, Naveen Kumar, Bernabe Mari Soucase, 9780323852418
- *Metal Oxides and Related Solids for Electrocatalytic Water Splitting*, Junlei Qi, 9780323857352
- *Advances in Metal Oxides and Their Composites for Emerging Applications*, Sagar Delekar, 9780323857055
- *Metallic Glasses and Their Oxidation*, Xinyun Wang, Mao Zhang, 9780323909976
- *Solution Methods for Metal Oxide Nanostructures*, Rajaram S. Mane, Vijaykumar Jadhav, Abdullah M. Al-Enizi, 9780128243534
- *Metal Oxide Defects*, Vijay Kumar, Sudipta Som, Vishal Sharma, Hendrik Swart, 9780323855884
- *Renewable Polymers and Polymer-Metal Oxide Composites*, Sajjad Haider, Adnan Haider, 9780323851558
- *Metal Oxides for Optoelectronics and Optics-based Medical Applications*, Suresh Sagadevan, Jiban Podder, Faruq Mohammad, 9780323858243
- *Graphene Oxide-Metal Oxide and Other Graphene Oxide-Based Composites in Photocatalysis and Electrocatalysis*, Jiaguo Yu, Liuyang Zhang, Panyong Kuang, 9780128245262

Published Titles

- *Metal Oxides in Nanocomposite-Based Electrochemical Sensors for Toxic Chemicals*, Alagarsamy Pandikumar, Perumal Rameshkumar, 9780128207277
- *Metal Oxide-Based Nanostructured Electrocatalysts for Fuel Cells*, Electrolyzers, and Metal-Air Batteries, Teko Napporn, Yaovi Holade, 9780128184967
- *Titanium Dioxide (TiO2) and Its Applications, Leonardo Palmisano*, Francesco Parrino, 9780128199602
- *Solution Processed Metal Oxide Thin Films for Electronic Applications*, Zheng Cui, 9780128149300
- *Metal Oxide Powder Technologies*, Yarub Al-Douri, 9780128175057
- *Colloidal Metal Oxide Nanoparticles*, Sabu Thomas, Anu Tresa Sunny, Prajitha V, 9780128133576
- *Cerium Oxide, Salvatore Scire*, Leonardo Palmisano, 9780128156612
- *Tin Oxide Materials*, Marcelo Ornaghi Orlandi, 9780128159248
- *Metal Oxide Glass Nanocomposites*, Sanjib Bhattacharya, 9780128174586
- *Gas Sensors Based on Conducting Metal Oxides*, Nicolae Barsan, Klaus Schierbaum, 9780128112243
- *Metal Oxides in Energy Technologies*, Yuping Wu, 9780128111673
- *Metal Oxide Nanostructures*, Daniela Nunes, Lidia Santos, Ana Pimentel, Pedro Barquinha, Luis Pereira, Elvira Fortunato, Rodrigo Martins, 9780128115121
- *Gallium Oxide*, Stephen Pearton, Fan Ren, Michael Mastro, 9780128145210
- *Metal Oxide-Based Photocatalysis*, Adriana Zaleska-Medynska, 9780128116340
- *Metal Oxides in Heterogeneous Catalysis*, Jacques C. Vedrine, 9780128116319
- *Magnetic, Ferroelectric, and Multiferroic Metal Oxides*, Biljana Stojanovic, 9780128111802
- *Iron Oxide Nanoparticles for Biomedical Applications*, Sophie Laurent, Morteza Mahmoudi, 9780081019252
- *The Future of Semiconductor Oxides in Next-Generation Solar Cells*, Monica Lira-Cantu, 9780128111659
- *Metal Oxide-Based Thin Film Structures*, Nini Pryds, Vincenzo Esposito, 9780128111666
- *Metal Oxides in Supercapacitors*, Deepak Dubal, Pedro Gomez-Romero, 9780128111697
- *Transition Metal Oxide Thin Film-Based Chromogenics and Devices*, Pandurang Ashrit, 9780081018996

Metal Oxides Series

Metal Oxide-Based Nanofibers and Their Applications

Edited by

Vincenzo Esposito
Department of Energy Conversion and Storage, Technical University of Denmark, Electrovej, Lyngby, Denmark

Debora Marani
Engineering, Modelling and Applied Social Sciences Center (CECS), Federal University of ABC (UFABC), Brazil

Series Editor
Ghenadii Korotcenkov
Department of Physics and Engineering, Moldova State University, Chişinău, Moldova

Elsevier
Radarweg 29, PO Box 211, 1000 AE Amsterdam, Netherlands
The Boulevard, Langford Lane, Kidlington, Oxford OX5 1GB, United Kingdom
50 Hampshire Street, 5th Floor, Cambridge, MA 02139, United States

Copyright © 2022 Elsevier Inc. All rights reserved.

No part of this publication may be reproduced or transmitted in any form or by any means, electronic or mechanical, including photocopying, recording, or any information storage and retrieval system, without permission in writing from the publisher. Details on how to seek permission, further information about the Publisher's permissions policies and our arrangements with organizations such as the Copyright Clearance Center and the Copyright Licensing Agency, can be found at our website: www.elsevier.com/permissions.

This book and the individual contributions contained in it are protected under copyright by the Publisher (other than as may be noted herein).

Notices

Knowledge and best practice in this field are constantly changing. As new research and experience broaden our understanding, changes in research methods, professional practices, or medical treatment may become necessary.

Practitioners and researchers must always rely on their own experience and knowledge in evaluating and using any information, methods, compounds, or experiments described herein. In using such information or methods they should be mindful of their own safety and the safety of others, including parties for whom they have a professional responsibility.

To the fullest extent of the law, neither the Publisher nor the authors, contributors, or editors, assume any liability for any injury and/or damage to persons or property as a matter of products liability, negligence or otherwise, or from any use or operation of any methods, products, instructions, or ideas contained in the material herein.

British Library Cataloguing-in-Publication Data
A catalogue record for this book is available from the British Library

Library of Congress Cataloging-in-Publication Data
A catalog record for this book is available from the Library of Congress

ISBN: 978-0-12-820629-4

For Information on all Elsevier publications
visit our website at https://www.elsevier.com/books-and-journals

Publisher: Matthew Deans
Acquisitions Editor: Kayla Dos Santos
Editorial Project Manager: Rafael G. Trombaco
Production Project Manager: Vijayaraj Purushothaman
Cover Designer: Miles Hitchen

Working together to grow libraries in developing countries

www.elsevier.com • www.bookaid.org

Typeset by MPS Limited, Chennai, India

Contents

List of contributors	**xiii**
About the series editor	**xvii**
About the volume editors	**xix**
Preface to the series	**xxi**
Preface to the volume	**xxv**

Section 1

1   Fundamentals of electrospinning and safety　　　　　　　　**3**
Bussarin Ksapabutr and Manop Panapoy

1.1	Introduction	3
1.2	Electrospinning process for metal oxide nanofibers	6
	1.2.1　Precursor solution for sol−gel electrospinning process toward metal oxide nanofibers	8
	1.2.2　Precursor solution for colloidal method toward metal oxide nanofibers	12
	1.2.3　Electrospinning process parameters	13
1.3	Large-scale production	16
1.4	Electrospinning safety	20
	Acknowledgments	21
	References	22

2   Special techniques and advanced structures　　　　　　　　**31**
Mingyu Tang, Suting Liu, Zhihui Li, Xiaodi Zhang, Zhao Wang,
Yunqian Dai, Yueming Sun, Liqun Zhang and Jiajia Xue

2.1	Introduction	31
2.2	Electrospinning for producing metal oxide nanofibers	33
	2.2.1　Directly electrospinning of the precursor solution	34
	2.2.2　Selectively removing the polymer component in the composite nanofibers	35
	2.2.3　Converting amorphous to crystalline structure	38
	2.2.4　Physical and chemical modifications	39
2.3	Advanced structures of metal oxide nanofibers	40
	2.3.1　Control of core-sheath, hollow, or side-by-side morphology	40
	2.3.2　Control of in-fiber and interfiber porosity	43
	2.3.3　Control of hierarchical surface structures	46
	2.3.4　Control of alignment and patterns	49

	2.3.5	Welding of nanofibers at their cross points	52
	2.3.6	Three-dimensional fibrous aerogels	53
	2.3.7	Mass production of metal oxide nanofibers	54
	Acknowledgments		55
	References		55

3 Nonelectrospun metal oxide nanofibers **65**
Alsiad Ahmed Almetwally

3.1	Introduction		65
3.2	Nonelectrospinning techniques		65
	3.2.1	Solution blow spinning technique	66
	3.2.2	Plasma-induced technique	67
	3.2.3	Drawing technique	67
	3.2.4	CO_2 laser supersonic drawing	68
	3.2.5	Template synthesis	68
	3.2.6	Centrifugal spinning	69
3.3	Synthesis of metal oxide nanofibers		71
	3.3.1	Tin oxide nanofibers	71
	3.3.2	Silica nanofibers	71
	3.3.3	Barium titanate nanofibers	72
	3.3.4	Copper oxide nanofibers	73
	3.3.5	Tungsten oxide nanofibers	74
	3.3.6	Ferric oxide nanofibers	75
	3.3.7	NiO, CeO_2, and NiO-CeO_2 composite nanofibers	77
	3.3.8	Titanium dioxide and zinc oxide nanofibers	78
3.4	Conclusion		79
	References		80

4 Polymer−metal oxide composite nanofibers **89**
Zainab Ibrahim Elkahlout, Abdulrahman Mohmmed AlAhzm,
Maan Omar Alejli, Fatima Zayed AlMaadeed,
Deepalekshmi Ponnamma and Mariam Al Ali Al-Maadeed

4.1	Introduction	89
4.2	Electroactive polymers and their metal oxide composites	90
4.3	Elastomer−metal oxide nanocomposite fibers	93
4.4	Biopolymer/metal oxide nanocomposite fibers	97
4.5	Conclusion	105
	Acknowledgments	105
	References	105

Section 2

5 Metal oxide nanofibers and their applications for biosensing **113**
Kunal Mondal, Raj Kumar, Blesson Isaac and Gorakh Pawar

5.1	Introduction	113

5.2	Synthesis strategies for MONFs		114
	5.2.1	Physicochemical route for MONF fabrication	115
	5.2.2	Spinning technique for fabrication of MONFs	116
	5.2.3	Advanced microfabrication and nanofabrication strategies for MONFs	120
5.3	Biosensing applications of MONFs		121
	5.3.1	Titanium dioxide nanofibers for biosensing	123
	5.3.2	ZnO nanofibers for biosensing	123
	5.3.3	Others MONFs for biosensing	126
5.4	Recent computational advances		128
5.5	Conclusions, challenges, and future scope		130
Acknowledgments			131
References			132

6   Metal oxide-based nanofibers and their gas-sensing applications   139

Ali Mirzaei, Sanjit Manohar Majhi, Hyoun Woo Kim
and Sang Sub Kim

6.1	Introduction		139
6.2	Gas-sensing applications of metal oxide nanofibers		143
	6.2.1	Importance of gas-sensing and gas sensors	143
	6.2.2	Pristine metal oxide nanofibers	144
	6.2.3	Metal oxide composite nanofibers	145
	6.2.4	Loaded or doped metal oxide nanofibers	150
6.3	Conclusions and outlook		153
References			153

7   Metal oxide nanofibers for flexible organic electronics and sensors   159

Roohollah Bagherzadeh and Nikoo Saveh Shemshaki

7.1	Incorporation of nanofibers into electronic devices	159
7.2	Conductive and transparent nanofibrous networks as a futuristic approach toward the flexible displays	161
7.3	Recent progress in electrospun metal oxide nanofibers	163
7.4	Summary and future trends	164
References		167

8   Role of metal oxide nanofibers in water purification   173

Ali A. El-Samak, Hammadur Rahman, Deepalekshmi Ponnamma,
Mohammad K. Hassan, Syed Javaid Zaidi
and Mariam Al Ali Al-Maadeed

8.1	Introduction	173
8.2	Metal oxide as water purifiers	175
8.3	Polymer—metal oxide composite fibers for water treatment	179
8.4	Conclusions	186
Acknowledgment		187
References		187

viii Contents

9 Metal oxide nanofiber for air remediation via filtration, catalysis, and photocatalysis 191

Chin-Shuo Kang, Edward A. Evans and George G. Chase

9.1 Introduction 191
9.2 Filtration for air pollutants 192
    9.2.1 Filtration mechanism 192
    9.2.2 Characterization of filter 194
    9.2.3 Nanofibrous particulate matter filter 198
    9.2.4 Gas filter 200
9.3 Conclusion 205
References 206

Section 3

10 Piezoelectric application of metal oxide nanofibers 215

Tutu Sebastian and Frank Clemens

10.1 Introduction 215
10.2 Inorganic piezoelectric materials and their properties 216
10.3 Synthesis of one-dimensional nanostructures 220
10.4 Hydrothermal synthesis 220
10.5 Electrospinning 223
10.6 Molten salt synthesis 228
10.7 Sol−gel template synthesis 228
10.8 Material and structural characterizations 229
10.9 X-ray diffraction 229
10.10 Raman spectroscopy 229
10.11 Atomic force microscopy 230
10.12 Potential applications 231
    10.12.1 Nanogenerators 231
    10.12.2 High-energy-density storage devices 236
    10.12.3 Structural health monitoring 236
10.13 Summary and outlook 237
References 237

11 Memristive applications of metal oxide nanofibers 247

Shangradhanva E. Vasisth, Parker L. Kotlarz, Elizabeth J. Gager and Juan C. Nino

11.1 Introduction 247
11.2 Recent trends 249
11.3 Memristors and resistive switching 250
11.4 Resistive switching in metal oxide nanofibers 254
    11.4.1 NiO 254
    11.4.2 TiO_2 257
    11.4.3 CuO 258

		11.4.4	ZnO	258
		11.4.5	Nb_2O_5	261
		11.4.6	VO_2	261
		11.4.7	WO_3	261
		11.4.8	Complex Oxide Nanofibers	261
	11.5	Core—shell nanowires	262
	11.6	Perspective and outlook	270
	References		271

12 Metal oxide nanofibers in solar cells   **277**

JinKiong Ling and Rajan Jose

12.1	Introduction: role of nanofibers in various types of solar cells		277
12.2	Photoconversion mechanism in sensitized photovoltaic cells		279
		12.2.1	Excitation of electrons	280
		12.2.2	Generation of photovoltage	281
		12.2.3	Charge extraction and transport	283
		12.2.4	Generation of photocurrent	284
12.3	Metal oxide nanofibers as photoanode in dye-sensitized solar cells		284
12.4	Reducing energy trap states		286
		12.4.1	Improving crystallinity through high sintering	287
		12.4.2	Raising the Fermi energy level	288
		12.4.3	N-type doping induced diffusion coefficient improvement	289
		12.4.4	P-type doping induced Schottky-barrier	291
		12.4.5	Homovalent ion substitution	292
		12.4.6	Composite fibers	293
12.5	Challenges		294
		12.5.1	Conclusion and outlook	294
References				295

13 Metal oxide nanofiber-based electrodes in solid oxide fuel cells   **301**

Paola Costamagna, Peter Holtappels and Caterina Sanna

13.1	Introduction		301
		13.1.1	State-of-the-art architectures and materials for solid oxide fuel cell electrodes	303
13.2	Nanofiber solid oxide fuel cell electrode preparation through electrospinning		307
		13.2.1	Electrode preparation	307
		13.2.2	Typical electrode structures	310
13.3	Overview of electrochemical performance of nanofiber versus conventional solid oxide fuel cell electrodes		311
		13.3.1	Strontium-doped lanthanum manganite	313
		13.3.2	Cobalt-based metal oxides	315
		13.3.3	Cobalt-free metal oxides	318

13.4	Understanding the structure-performance relationship in nanofiber solid oxide fuel cell electrodes: experimental characterization and numerical modeling	319
	13.4.1 Electrochemical impedance spectroscopy: experimental characterization and equivalent circuit modeling	319
	13.4.2 One-dimensional pseudohomogeneous model of infiltrated mixed ionic-electronic conductor nanofiber electrodes	324
13.5	Summary	327
References		327

14 Synthesis of one-dimensional metal oxide−based crystals as energy storage materials
333

Andrea La Monaca, Daniele Campanella and Andrea Paolella

14.1	Introduction	333
14.2	Aluminum oxide	334
14.3	Copper oxide	336
14.4	Iron oxide	336
14.5	Manganese oxide	338
14.6	Nickel oxide	340
14.7	Silicon oxide and silicates	343
14.8	Tin oxide	344
14.9	Titanium oxides and titanates	344
14.10	Tungsten oxide and tungstates	348
14.11	Vanadium oxide	349
14.12	Zinc oxide	352
14.13	Zirconate fibers	354
References		356

15 Supercapacitors based on electrospun metal oxide nanofibers
361

Di Tian, Ce Wang and Xiaofeng Lu

15.1	Introduction	361
15.2	Electrospun metal oxide nanofibers	362
	15.2.1 Single metal oxides	362
	15.2.2 Bimetallic or polymetallic oxides	368
15.3	Electrospun metal oxide nanofiber−based composites	370
	15.3.1 Metal oxide/metal oxide (metal hydroxide, metal) composites	371
	15.3.2 Metal oxide/carbon-based composites	372
	15.3.3 Metal oxide (metal)/carbon nanofibers/conducting polymer composites	381
	15.3.4 Other composites	383
15.4	Conclusion	383
References		384

| | | Contents | | xi |

16 Thermoelectrics based on metal oxide nanofibers   **395**
Yong X. Gan
16.1   Introduction   395
16.2   Thermoelectric metal oxide nanofiber processing technology   396
     16.2.1   Electrospinning   397
     16.2.2   Chemical bath deposition   402
     16.2.3   Template-assisted deposition   406
     16.2.4   Chemical spray pyrolysis   409
     16.2.5   Microlithography and nanolithography   411
     16.2.6   Electrochemical oxidation   414
     16.2.7   Glass-annealing method   416
16.3   Thermoelectric metal oxide nanofiber device concept and characterization   417
16.4   Perspectives and conclusions   420
References   420

Index   **425**

List of contributors

Abdulrahman Mohmmed AlAhzm Department of Chemistry, College of Arts and Science, Qatar University, Doha, Qatar

Maan Omar Alejli Department of Chemistry, College of Arts and Science, Qatar University, Doha, Qatar

Fatima Zayed AlMaadeed Department of Environmental Sciences, Qatar University, Doha, Qatar

Mariam Al Ali Al-Maadeed Materials Science and Technology Program (MATS), College of Arts & Sciences, Qatar University, Doha, Qatar; Center for Advanced Materials, Qatar University, Doha, Qatar

Alsiad Ahmed Almetwally Textile Engineering Department, Textile Research Division, National Research Center, Dokki, Cairo, Egypt

Roohollah Bagherzadeh Advanced Fibrous Materials Lab, Institute for Advanced Textile Materials and Technologies, School of Advanced Materials and Processes, Amirkabir University of Technology (Tehran Polytechnic), Tehran, Iran

Daniele Campanella Centre d'Excellence en Électrification Des Transports et Stockage d'Énergie, Hydro-Québec, Varennes, QC, Canada

George G. Chase Department of Chemical, Biomolecular and Corrosion Engineering, The University of Akron, Akron, OH, United States

Frank Clemens Laboratory for High Performance Ceramics, Empa, Swiss Federal Laboratories for Materials Science and Technology, Dübendorf, Switzerland

Paola Costamagna DCCI, Department of Chemistry and Industrial Chemistry, University of Genoa, Genoa, Italy

Yunqian Dai School of Chemistry and Chemical Engineering, Southeast University, Nanjing, P.R. China; Center for Flexible RF Technology, Southeast University, Purple Mountain Laboratory, Nanjing, P.R. China

Zainab Ibrahim Elkahlout Department of Chemical Engineering, College of Engineering, Qatar University, Doha, Qatar

Ali A. El-Samak Materials Science & Technology Program (MATS), College of Arts & Sciences, Qatar University, Doha, Qatar; Center for Advanced Materials, Qatar University, Doha, Qatar

Edward A. Evans Department of Chemical, Biomolecular and Corrosion Engineering, The University of Akron, Akron, OH, United States

Elizabeth J. Gager University of Florida, Gainesville, FL, United States

Yong X. Gan Department of Mechanical Engineering, College of Engineering, California State Polytechnic University Pomona, Pomona, CA, United States

Mohammad K. Hassan Center for Advanced Materials, Qatar University, Doha, Qatar

Peter Holtappels DTU Energy, Department of Energy Conversion and Storage, Technical University of Denmark, Electrovej, Lyngby, Denmark

Blesson Isaac Chemical and Radiation Measurement Department, Idaho National Laboratory, Idaho Falls, ID, United States

Rajan Jose Nanostructured Renewable Energy Material Laboratory, Faculty of Industrial Sciences & Technology, University Malaysia Pahang, Pahang, Malaysia

Chin-Shuo Kang Department of Chemical, Biomolecular and Corrosion Engineering, The University of Akron, Akron, OH, United States

Hyoun Woo Kim The Research Institute of Industrial Science, Hanyang University, Seoul, Republic of Korea; Division of Materials Science and Engineering, Hanyang University, Seoul, Republic of Korea

Sang Sub Kim Department of Materials Science and Engineering, Inha University, Incheon, Republic of Korea

Parker L. Kotlarz University of Florida, Gainesville, FL, United States

Bussarin Ksapabutr Department of Materials Science and Engineering, Faculty of Engineering and Industrial Technology, Silpakorn University, Nakhon Pathom, Thailand; Center of Excellence on Petrochemical and Materials Technology, Chulalongkorn University, Bangkok, Thailand

Raj Kumar Faculty of Engineering, Bar Ilan Institute of Nanotechnology and Advanced Materials (BINA), Bar Ilan University, Ramat Gan, Israel

List of contributors

Andrea La Monaca Centre d'Excellence en Électrification Des Transports et Stockage d'Énergie, Hydro-Québec, Varennes, QC, Canada

Zhihui Li School of Chemistry and Chemical Engineering, Southeast University, Nanjing, P.R. China

JinKiong Ling Nanostructured Renewable Energy Material Laboratory, Faculty of Industrial Sciences & Technology, University Malaysia Pahang, Pahang, Malaysia

Suting Liu School of Chemistry and Chemical Engineering, Southeast University, Nanjing, P.R. China

Xiaofeng Lu Alan G. MacDiarmid Institute, College of Chemistry, Jilin University, Changchun, P.R. China

Sanjit Manohar Majhi The Research Institute of Industrial Science, Hanyang University, Seoul, Republic of Korea; Division of Materials Science and Engineering, Hanyang University, Seoul, Republic of Korea

Ali Mirzaei Department of Materials Science and Engineering, Shiraz University of Technology, Shiraz, Iran

Kunal Mondal Materials Science and Engineering Department, Idaho National Laboratory, Idaho Falls, ID, United States

Juan C. Nino University of Florida, Gainesville, FL, United States

Manop Panapoy Department of Materials Science and Engineering, Faculty of Engineering and Industrial Technology, Silpakorn University, Nakhon Pathom, Thailand; Center of Excellence on Petrochemical and Materials Technology, Chulalongkorn University, Bangkok, Thailand

Andrea Paolella Centre d'Excellence en Électrification Des Transports et Stockage d'Énergie, Hydro-Québec, Varennes, QC, Canada

Gorakh Pawar Materials Science and Engineering Department, Idaho National Laboratory, Idaho Falls, ID, United States

Deepalekshmi Ponnamma Center for Advanced Materials, Qatar University, Doha, Qatar

Hammadur Rahman Center for Advanced Materials, Qatar University, Doha, Qatar

Caterina Sanna DCCI, Department of Chemistry and Industrial Chemistry, University of Genoa, Genoa, Italy

Nikoo Saveh Shemshaki Department of Biomedical Engineering, University of Connecticut, Storrs, CT, United States

Tutu Sebastian Laboratory for High Performance Ceramics, Empa, Swiss Federal Laboratories for Materials Science and Technology, Dübendorf, Switzerland

Yueming Sun School of Chemistry and Chemical Engineering, Southeast University, Nanjing, P.R. China

Mingyu Tang School of Chemistry and Chemical Engineering, Southeast University, Nanjing, P.R. China

Di Tian Alan G. MacDiarmid Institute, College of Chemistry, Jilin University, Changchun, P.R. China

Shangradhanva E. Vasisth University of Florida, Gainesville, FL, United States

Ce Wang Alan G. MacDiarmid Institute, College of Chemistry, Jilin University, Changchun, P.R. China

Zhao Wang Beijing Laboratory of Biomedical Materials, Beijing University of Chemical Technology, Beijing, P.R. China; Center of Advanced Elastomer Materials, Beijing University of Chemical Technology, Beijing, P.R. China

Jiajia Xue Beijing Laboratory of Biomedical Materials, Beijing University of Chemical Technology, Beijing, P.R. China; Center for Flexible RF Technology, Southeast University, Purple Mountain Laboratory, Nanjing, P.R. China

Syed Javaid Zaidi Center for Advanced Materials, Qatar University, Doha, Qatar

Liqun Zhang The Key Laboratory of Beijing City on Preparation and Processing of Novel Polymer Materials, Beijing University of Chemical Technology, Beijing, P.R. China

Xiaodi Zhang Beijing Laboratory of Biomedical Materials, Beijing University of Chemical Technology, Beijing, P.R. China; Center of Advanced Elastomer Materials, Beijing University of Chemical Technology, Beijing, P.R. China

About the series editor

Ghenadii Korotcenkov received PhD in Physics and Technology of Semiconductor Materials and Devices in 1976 and Dr. Sci. degree (Doc. Hab.) in Physics of Semiconductors and Dielectrics in 1990. He has more than 45 years of experience as a teacher and scientific researcher. He has been a leader of a gas sensor group and manager of various national and international scientific and engineering projects carried out in the Laboratory of Micro-Optoelectronics, Technical University of Moldova, Chisinau, Moldova. In 2007−08 he was an invited scientist at the Korea Institute of Energy Research (Daejeon), after which, until 2017, Dr. Korotcenkov was a research professor at the School of Materials Science and Engineering at Gwangju Institute of Science and Technology (GIST) in Korea. Currently, Dr. Korotcenkov is a chief scientific researcher at the Moldova State University, Chisinau, Moldova. His present scientific interests, starting from 1995, include material sciences, focusing on metal oxide film deposition and characterization, surface science, thermoelectric conversion, and design of physical and chemical sensors, including thin-film gas sensors.

Dr. Korotcenkov is the author or editor of 39 books and special issues, including the 11-volume *Chemical Sensors* series published by Momentum Press; the 15-volume *Chemical Sensors* series published by Harbin Institute of Technology Press, China; the 3-volume *Porous Silicon: From Formation to Application* issue published by CRC Press; the 2-volume *Handbook of Gas Sensor Materials* published by Springer; and the 3-volume *Handbook of Humidity Measurements* published by CRC Press. Currently, he is the series editor of the *Metal Oxides* book series published by Elsevier.

Dr. Korotcenkov is the author or coauthor of more than 650 scientific publications, including 31 review papers, 38 book chapters, and more than 200 peer-reviewed articles published in scientific journals [h-index = 41 (Web of Science), h-index = 42 (Scopus), and h-index = 56 (Google Scholar Citations)]. He is a holder of 18 patents. He has presented more than 250 reports at national and international conferences, including 17 invited talks. As a cochair or a member of program, scientific, and steering committees, Dr. Korotcenkov has participated in the organization of more than 30 international scientific conferences. He is a member of the editorial boards of five scientific international journals. His name and activities have been listed in many biographical publications, including *Who's Who*. His research activities have been honored by, among other awards, the Honorary

Diploma of the Government of the Republic of Moldova (2020); an Award of the Academy of Sciences of Moldova (2019); an Award of the Supreme Council of Science and Advanced Technology of the Republic of Moldova (2004); the Prize of the Presidents of the Ukrainian, Belarus and Moldovan Academies of Sciences (2003); the Senior Research Excellence Award of Technical University of Moldova (2001, 2003, 2005); and the National Youth Prize of the Republic of Moldova in the field of science and technology (1980). Dr. Korotcenkov also received fellowships from the International Research Exchange Board (IREX, United States, 1998), the Brain Korea 21 Program (2008−12), and the BrainPool Program (Korea, 2015−17).

About the volume editors

Vincenzo Esposito is a full professor in ceramics science and engineering and technology coordinator at the Department of Energy Conversion and Storage, Technical University of Denmark. He developed his career at Risø DTU National Laboratory for Sustainable Energy, University of Rome "Tor Vergata," the University of Florida, and the Instituto de Pesquisas Energéticas e Nucleares (IPEN), Brazil. His research interest is primarily in functional inorganic nanomaterials and processing for emerging technologies in energy, catalysis, electromechanical, electronics, and electrochemical systems. His research profile lies at the frontier between nanoionics, solid-state chemistry, and advanced materials processing.

Debora Marani is an independent scientist in materials science affiliated with the Engineering, Modelling and Applied Social Sciences Center (CECS), Federal University of ABC (UFABC), Brazil. She obtained a master's degree in chemistry from the University of Rome "La Sapienza" in 2002 and a "da Vinci Italian-French PhD" in materials science from the University of Rome "Tor Vergata" and the Université Aix-Marseille in 2006. Over the years, she has been working on various topics in materials science in several academic environments in Italy, France, Japan, Brazil, and Denmark, often in close collaboration with the industry. Her scientific interest is in the development of innovative materials (ceramics, hybrids, and polymers) and their energy and environmental technologies applications.

Preface to the series

The field of synthesis, study, and application of metal oxides is one of the most rapidly progressing areas of science and technology. Metal oxides are among the most ubiquitous compound groups on earth, having a large variety of chemical compositions, atomic structures, and crystalline shapes. In addition, metal oxides are known to possess unique functionalities that are absent or inferior in other solid materials. In particular, metal oxides represent an assorted and appealing class of materials, properties of which exhibit a full spectrum of electronic properties from insulating to semiconducting, metallic, and superconducting. Moreover, almost all the known effects, including superconductivity, thermoelectric effects, photoelectrical effects, luminescence, and magnetism, can be observed in metal oxides. Therefore metal oxides have emerged as an important class of multifunctional materials with a rich collection of properties, which have great potential for numerous device applications. Specific properties of the metal oxides, such as the wide variety of materials with different electrophysical, optical, and chemical characteristics, their high thermal and temporal stability, and their ability to function in harsh environments, make metal oxides very suitable materials for designing transparent electrodes, high-mobility transistors, gas sensors, actuators, acoustical transducers, photovoltaic and photonic devices, photocatalysts and heterogeneous catalysts, solid-state coolers, high-frequency and micromechanical devices, energy-harvesting and storage devices, nonvolatile memories, and many other applications in the electronics, energy, and health sectors. In these devices, metal oxides can be successfully used as sensing or active layers, substrates, electrodes, promoters, structure modifiers, membranes, and fibers, that is, they can be used as active and passive components.

Among other advantages of metal, oxides are the low fabrication cost and robustness in practical applications. Furthermore, metal oxides can be prepared in various forms, such as ceramics, thick film, and thin film. For thin film deposition, deposition techniques can be used that are compatible with standard microelectronic technology. The last factor is very important for large-scale production because the microelectronic approach promotes low cost for mass production, offers the possibility of manufacturing devices on a chip, and guarantees good reproducibility. Various metal oxide nanostructures, including nanowires, nanotubes, nanofibers, core−shell structures, and hollow nanostructures also can be synthesized. As is known, the field of metal oxide nanostructured morphologies (e.g., nanowires, nanorods, and nanotubes) has become one of the most active research areas in the nanoscience community.

The ability to create a variety of metal oxide−based composites and the ability to synthesize various multicomponent compounds significantly expand the range of properties that metal oxide—based materials can have, making metal oxides truly versatile multifunctional materials for widespread use. Small changes in their chemical composition and atomic structure can be accompanied by spectacular variations in properties and behavior. Even now, advances in synthesizing and characterizing techniques are revealing numerous new functions of metal oxides.

Taking into account the importance of metal oxides for progress in microelectronics, optoelectronics, photonics, energy conversion, sensors, and catalysis, a large number of various books devoted to this class of materials have been published. However, one should note that some books from this list are too general, some books are collections of various original works without any generalizations, and s were published many years ago. However, during the past decade, great progress has been made in the synthesis of metal oxides as well as their structural, physical, and chemical characterization and application in various devices, and a large number of papers have been published on metal oxides. In addition, until now, many important topics related to metal oxides study and application have not been discussed. To remedy the situation in this area, we decided to generalize and systematize the results of research in this direction and to publish a series of books devoted to metal oxides.

The proposed book series, *Metal Oxides*, is the first one to be devoted solely to the consideration of metal oxides. We believe that combining books on metal oxides in a series could help readers in finding required information on the subject. In particular, we plan that the books in our series, which will have clear specialization by content, will provide interdisciplinary discussion for various oxide materials with a wide range of topics, from material synthesis and deposition to characterizations, processing, and device fabrications and applications. This book series is being prepared by a team of highly qualified experts, which guarantees its high quality.

I hope that our books will be useful and easy to use. I hope that readers will consider this book series to be like an encyclopedia of metal oxides that will enable readers to understand the present status of metal oxides, to evaluate the role of multifunctional metal oxides in the design of advanced devices, and then, on the basis of observed knowledge, to formulate new goals for further research.

The intended audience of the present book series is scientists and researchers who are working or planning to work in the field of materials related to metal oxides, that is, scientists and researchers whose activities are related to electronics, optoelectronics, energy, catalysis, sensors, electrical engineering, ceramics, biomedical designs, and so on. I believe that this *Metal Oxides* book series will also be interesting for practicing engineers or project managers in industries and national laboratories who would like to design metal oxide—based devices but don't know how to do it or how to select the optimal metal oxides for specific applications. With many references to the vast resource of recently published literature on the subject, this book series will serve as a significant and insightful source of valuable information, providing scientists and engineers with new insights for understanding and improving existing metal oxide—based devices and for designing new metal oxide—based materials with new and unexpected properties.

Preface to the series

I believe that this *Metal Oxides* book series will be very helpful for university students, postdoctorate scholars, and professors. The structure of these books offers a basis for courses in the field of material sciences, chemical engineering, electronics, electrical engineering, optoelectronics, energy technologies, environmental control, and many others. Graduate students could also find the book series to be very useful in their research, for understanding features of metal oxide synthesis, and for study and applications of this multifunctional material in various devices. We are sure that all of them will find the information useful for their activity.

Finally, I thank all contributing authors and book editors who have been involved in the creation of these books. I am thankful that they agreed to participate in this project and for their efforts in the preparation of these books. Without their participation, this project would have not been possible. I also express my gratitude to Elsevier for giving us the opportunity to publish this series. I especially thank the team at the editorial office at Elsevier for their patience during the development of this project and for encouraging us during the various stages of preparation.

Ghenadii Korotcenkov

Preface to the volume

In the last few decades the scientific community has progressively shifted its research interest toward the infinitely small: the nanoworld. The trend has opened the unprecedented opportunity to have a detailed picture of the world at the atom level and the unrivaled prospect of tuning and controlling the properties of the materials as never before. At the nanoscale, atoms interact and combine to determine the specific properties of materials (e.g., chemical, physical, electronic, magnetic, optical).

The targeted manipulation of materials at this scale makes it possible to significantly enhance existing properties while new exotic ones can be induced. The behavior of materials can be controlled by controlling the chemical composition (atom by atom) and by confining one or more dimensions into the nanoscale range. Several advanced compositions shaped into novel nanostructures have been engineered and proven to possess superior emerging properties compared to their bulk counterparts. At the same time, a significant number of structures with one or more dimensions in the nanoscale range have emerged in many fields with proven enhanced properties or even entirely new properties. The list includes superior nanomaterials in the 10- to 100-nm scale. This range is above the atom-to-atom size domain. However, these materials still have exceptional properties and unique compositional and microstructural features. Some prominent examples are nanoparticles, nanosheets, nanocages, nanotubes, nanowires, and nanofibers. Many others are expected to be introduced with various shapes and heterostructures. Heterostructures combine two or more nanomaterials with complementary functionalities to enable nanodevices.

Nanofibers are a class of quasi-one-dimensional materials with cross-sectional diameters ranging from tens to hundreds of nanometers and a few tens of micrometers in length. These peculiar dimensional characteristics offer the merits of an extremely high aspect ratio and surface-area-to-volume ratio. In addition, nanofibers are typically arranged into thick membranes or substrates with a wide-open and well-interconnected porous network that offer further potential to significantly affect performance. The relationship between structure and composition is an essential aspect for efficient tuning of properties and performances of the materials. As one of the main strengths, nanofibers are easily fabricated with a sizable and simple spinning system that is flexible for different materials processing, such as ceramic, polymers, metals, and even hybrid materials.

Among the materials, metal oxides exhibit a rich spectrum of properties that can potentially solve many technological challenges, from lower energy consumption to renewable energy sources. Their success is mainly due to compositional diversity, easy tunability of the chemical and physical properties, facile synthesis, high stability, low cost, and environmental friendliness.

When metal oxides are shaped into nanofibers, devices with enhanced performances and emerging behaviors are designed to meet the specific requirements of many different applications. The tuning of specific properties can be attained by manipulating either the chemistry and the structural parameters of the nanofibers (e.g., diameters and crystallite size). In the last decades, many metal oxides have been synthesized in the shape of nanofibers and explored for their applications in diverse fields, such as purification of liquids and gases; remediation of water and air pollutants; sensors; and energy generation, conversion, and storage.

This book provides an overview of the current state of developments, challenges, and perspectives of metal oxide nanofibers. It comprises three main sections covering the synthesis with its related critical aspects and applications in topical areas.

The first section deals with the theoretical and experimental aspects of synthesis and methodologies. The focus is on the control of the microstructure, composition, and shape of nanofibrous metal oxides. The section includes electrospinning and other methods to synthesize nanofibers in either random or allied fashion. The section also includes the safety aspects associated with the fabrication and handling of nanofibers.

The second and third sections emphasize applications of metal oxide nanofibers in diverse technologies, focusing on the relationship between the peculiar structural, morphological, and compositional features of the nanofibers with the performances in specific fields of applications. Specifically, the second section deals with the applications of metal oxide nanofibers in sensors (e.g., biosensing, gas, and vapor sensors) and water and air purification (e.g., catalyst and filter for the abatement of pollutants). The third section discusses energy generation and storage technologies (e.g., piezoelectric, solar cells, solid oxide fuel cells, lithium-ion batteries, supercapacitors, and hydrogen storage).

The book results from the contributions of several eminent worldwide experts along with their collaborators. The editors highly appreciate their time and effort, without which the book would not have been possible. The book will be relevant for students, young research scientists working with metal oxide nanofibers, engineers, and technology developers. The editors have done their best to cover all the areas of great interest in the field of metal oxides. The intent is to transfer the current enthusiasm about metal oxide nanofibers to as large a readership as possible. The editors hope that the book will serve this purpose. Finally, the editors would like to thank the editorial team at Elsevier for the opportunity to publish this book and for the valuable support, patience, and pursuance, which were essential for finalizing it.

Debora Marani and Vincenzo Esposito

Section 1

Fundamentals of electrospinning and safety

Bussarin Ksapabutr[1,2] and Manop Panapoy[1,2]
[1]Department of Materials Science and Engineering, Faculty of Engineering and Industrial Technology, Silpakorn University, Nakhon Pathom, Thailand, [2]Center of Excellence on Petrochemical and Materials Technology, Chulalongkorn University, Bangkok, Thailand

1.1 Introduction

Nanofibers have been considered among the most interesting materials from both the academic and modern industrial point of views [1]. Ultralong nanofibers with dimensional and morphological control are widely used in many fields, such as antivirus face masks, microelectronics, biosensors, batteries, composites, and separation [2−8]. Nanofibers offer great opportunities to develop functional products with new properties and applications by various physical and chemical modification strategies [1]. In particular, metal oxide nanofibers have sparked great attention for diverse potential applications in supercapacitors [9] and batteries for energy storage devices [10], chemical catalysis [11], tissue engineering [12], biomedical devices [13], biosensors [4], gas sensors [14], electronics [15], and wastewater and environmental treatments [16]. Various nanofabrication techniques have been utilized for the fabrication of metal oxide nanofibers, such as plasma-induced synthesis, centrifugal jet spinning [17], vapor-liquid-solid growth, nanocarving, and electrospinning [13,18]. However, research is still ongoing on a technically reliable and economical process for the large-scale production of high-quality one-dimensional (1D) nanofibers [19]. Among those techniques, electrospinning possesses unique advantages, either by modifying the electrospinning setup, by adjusting the spinning parameters, or by selecting more alternative precursor materials to design and fabricate metal oxide nanofibers with special features to meet some special requirements [13].

Electrospinning is one of the most common, efficient, and scalable techniques for the production of continuous 1Dnanostructured fibers of metal oxides, metals, polymers, and composites with diameters ranging from a few tens of nanometers to a few micrometers in the form of continuous nonwoven nanofiber mats with tunable structures and properties [20,21]. The nanofibers that are produced from electrospinning have large specific surface area, high aspect ratio, excellent pore interconnectivity, and other useful properties, which are favorable for many applications. Furthermore, electrospinning offers a simple way to construct 3D porous nanoarchitectures made of nanofibers [22]. Electrospinning evolved from electrospray technology based on high voltage, referring to the phenomenon that charged droplets can be emitted at high speed under a high-voltage electric field [23]. The

---

Metal Oxide-Based Nanofibers and Their Applications. DOI: https://doi.org/10.1016/B978-0-12-820629-4.00004-7
© 2022 Elsevier Inc. All rights reserved.

mechanism of the electrospinning process relies on the ejection and elongation of an organic polymer solution or melt (for fabricating polymer nanofibers) and a mixed inorganic precursor and organic polymer solution (for fabricating composite nanofibers, metal, or ceramic-based nanofibers) under a high-voltage electric field [21]. A typical electrospinning setup consists of four main components, as shown in Fig. 1.1: (1) a high-voltage power supply (usually in the kilovolt range); (2) a liquid supply device (usually syringe pump), which feeds the precursor solution or melt with the help of a syringe; (3) a syringe connected to metallic needle (spinneret); and (4) a metal collector on which the nanofibers are deposited (usually aluminum foil). The syringe containing the precursor solution or melt with a metallic needle is generally connected as a positive electrode [24]. The spinneret can be of any cross-sectional shape. Various spinneret configurations for electrospinning of neat and composite microfibers and nanofibers are shown in Fig. 1.1. The use of a multiple-needle syringe gives an opportunity to scale up and increase throughput of the fiber production. The coaxial needle contains a small inner capillary fitting inside the larger outer capillary. The used spinnerets share an axis, leading to the injection of one solution into the other at the needle tip with the core solution getting drawn within the outer one. This configuration can easily encapsulate a smaller fiber within a larger fiber to produce nanofibers with a core-shell structure using two or more different fiber materials [25,26]. The electrostatic force created by the high-voltage power supply is applied to a precursor solution or melt, which is dispensed through the spinneret at a controlled rate using a syringe pump [4]. The collector can either be grounded or charged opposite to the spinneret. Various collector geometries have been used to meet the specific product requirements. Besides using a static collector plate, rotating drum collector, rotating wire drum, rotating disc,

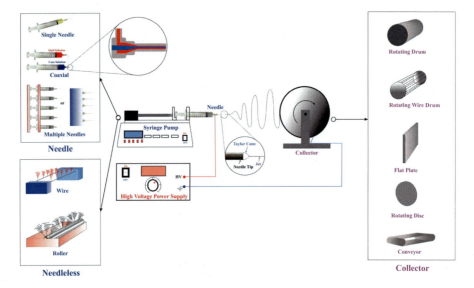

Figure 1.1 Schematic diagram of a typical electrospinning setup.

and conveyor are also commonly employed to collect aligned nanofibers [27], as shown in Fig. 1.1.

In the electrospinning technique a high-voltage electric field is used to create a charged jet of organic polymer solution or inorganic precursor solution. A Taylor cone is required in the electrospinning technique because it defines the onset of subtle velocity gradients for the fiber-forming process. Fig. 1.2 illustrates the formation of a Taylor cone. When a high voltage, usually in the kilovolt range, is applied to the tip of metallic needle containing a liquid droplet of precursor solution held by its own surface tension, the liquid droplet of precursor solution becomes electrostatically charged, and the induced charges are evenly distributed over the surface. The surface tension of the liquid droplet generally results in a hemisphere at equilibrium, but it is deformed in the electric field because the charges within the liquid droplet migrate to the surface facing the collector. The accumulation of the charges around the liquid droplet causes electrostatic repulsion opposite to the surface tension, deforming the liquid droplet into a conical shape known as a Taylor cone [28]. When the applied voltage increases, the repulsive electrostatic force overcomes the surface tension of the liquid droplet, and a charged jet of precursor solution is ejected from the apex of the Taylor cone to an electrode of opposite polarity (or electrical ground) when a critical voltage value (V_c) is obtained. The charged jet then undergoes a stretching and whipping process (a series of connected loops), resulting in the formation of a long, thin fiber. Additionally, the solvent evaporates from the fine fibers during the ejection and elongation process, thereby reducing

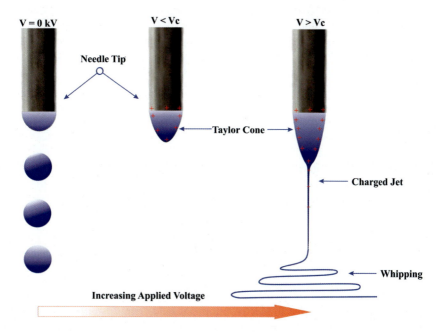

Figure 1.2 Schematic illustration of Taylor cone formation.

the fiber diameter from micrometers to nanometers. Finally, the nanofibrous material is attracted by a metal collector and collected in randomly oriented, nonwoven mats. For each application the nanofibers need to possess certain structural characteristics, such as fiber diameter, morphology, and membrane pore features, and they can easily be controlled by tuning the solution parameters and process parameters. Several factors influence the electrospinning process and the structure of the obtained nanofibers, such as the solvent, precursor type, concentration, viscosity, surface tension, flow rate, applied electric field, distance between tip and collector, and needle diameter.

For metal oxide nanofibers the electrospinning process has been used in the preparation of many types of ceramic nanofibers from mixed inorganic precursor and organic polymer solution or mixed ceramic particles with a particular size and organic polymer solution. Inorganic precursor solution is reacted in the syringe or during electrospinning to yield a chemical synthesis, followed by a thermal treatment process to achieve the desired ceramic phase. Meanwhile, nano-sized ceramic particles must be dispersed in organic polymer matrix to obtain suitable viscosity for electrospinning [29]. Several factors affect the final crystal structure of the ceramic nanofibers. For instance, the amount of ethyl alcohol needed to dissolve polyvinyl pyrrolidone (PVP) and the calcination temperature have a significant effect on the morphology and crystallite size of final zirconia nanofibers [30]. The molecular weight of the polymer and the amount of ethyl alcohol to dissolve PVP should be appropriate to achieve sufficient viscosity to produce continuous fibers. After the electrospinning process, a high-temperature calcination step is often applied to the as-spun polymer/inorganic precursor composite nanofibers to degrade the polymer and to crystallize the ceramic phases [30,31]. As the calcination temperature increases, metal oxide nanocrystals begin to nucleate and grow, whereas the polymer template is burned off and removed completely at a certain temperature. Several effective examples have been reported in the literatures for the fabrication of metal oxide nanofibers by the calcination of appropriate precursors synthesized either via the electrospinning process of ex situ formed colloidal dispersions or via the combination of sol−gel chemistry and electrospinning process [13,32]. Freestanding and flexible porous ceramic fiber membranes produced from the electrospinning technique have been widely investigated as sensor, insulation, heat-resistant filter, catalyst, self-cleaning, supercapacitor, and battery [33]. Recently, numerous materials and experimental designs have been developed by using electrospinning technology (see Table 1.1).

1.2 Electrospinning process for metal oxide nanofibers

Metal oxides cannot be directly electrospun to obtain continuous and uniform nanofibers. However, it is feasible to electrospin metal oxide nanofibers from their melts at excessively high temperatures. Therefore the fabrication of metal oxide nanofibers by electrospinning needs to rely on using a spinnable precursor solution or

Table 1.1 Some metal oxide and composite nanofibers obtained from electrospinning.

Materials	Thermal treatment	Precursor/polymer processing aids	References
YMn_2O_5 nanofibers	Calcination at 800°C	$Mn(CH_3COO)_2$/Y $(NO_3)_3$/PVP	[34]
ZrO_2−carbon nanofibers	Pyrolysis at 500°C, 600°C, and 900°C	UiO-66/PAN	[35]
Au−ZnO composite fibers	Calcination at 650°C	$HAuCl_4$/$Zn(Ac)_2$/ PAN/PVP	[36]
SnO_2−CuO hollow nanofibers	Calcination at 600°C	$SnCl_4$/Cu$(Cu_3COO)_2$/ PVP	[37]
$SrNb_2O_6$ nanofibers	Calcination at 1000°C	$SrNb_2(O^iPr)12$ (HO^iPr)/PVP	[38]
WO_3/SnO_2 hollow nanofibers	Calcination at 600°C	$(NH_4)_{10}W_{12}O_{41}$/ $SnCl_2$/PVP	[39]
MgO−SiO_2 nanofibers	Calcination at 400°C	MgO/TEOS/PVA	[40]
Fe_3O_4 nanofibers	Annealing at 200°C−300°C and carbonization at 500°C−700°C	$Fe(NO_3)_3$/PEO	[41]
Al_2O_3 nanofibers	Calcination at 900°C	$Al(NO_3)_3$/ $C_9H_{21}AlO_3$/PVA or PVB or PVP	[42]
α-Fe_2O_3 hollow spheres and nanofibers	Calcination at 650°C	$Fe(NO_3)_3$/PVP	[43]
Fe_2O_3 nanofibers	Calcination at 600°C	Ferratrane/PVP	[44]
SiC nanofibers	Pyrolysis at 1200°C−1500°C	Polycarbosilane/Ni $(acac)_2$/PS	[45]
ZnO nanoparticles/ Co_3O_4 nanofibers	Calcination at 600°C	$Co(OAc)_2$/ZnO/PVA	[46]
ZrO_2 nanofibers	Calcination at 500°C−800°C	$ZrOCl_2$/PVP	[47]
Ag−TiO_2 nanofibers	Calcination at 500°C	$AgNO_3$/Titanium (IV) butoxide/PVP	[48]

mixing of inorganic colloidal particles with organic polymer solution. There are two different ways to obtain metal oxide nanofibers via electrospinning: the sol−gel process and the colloidal method. Furthermore, some extra steps are required for preparing metal oxide nanofibers, which are usually not essential for organic polymers. The complete procedure for the fabrication of metal oxide nanofibers by electrospinning consists of three major steps:

1. Preparation of a spinning solution in the form of a precursor solution containing a metal alkoxide or metal salt with an organic polymer solution (for the sol−gel process) or in the form of inorganic colloidal particles dispersed in an organic polymer solution (for the

colloidal method). Prior to preparing the spinning solution, the solubility of metal alkoxide, metal salt, and organic polymer in solvent or the dispersion of inorganic colloidal particles in polymer solution should be investigated to obtain the desired viscosity. The most widely used organic polymers in preparation of metal oxide nanofibers are polyvinyl alcohol (PVA), polystyrene (PS), polyvinyl butyral (PVB), PVP, polyethylene oxide (PEO), and polyacrylonitrile (PAN) with proper rheological properties required for electrospinning (Table 1.1).

2. Electrospinning of the prepared spinning solution to generate the composite nanofibers that make up the organic polymer and inorganic components. The electrospinning process for metal oxide nanofibers is typically performed in a controlled environment at room temperature or slightly elevated temperature. Moreover, by using some inorganic precursors such as metal alkoxides or metal salts, a chemical conversion can take place in the syringe or during electrospinning.

3. Calcination or sintering of the electrospun composite nanofibers at elevated temperatures to remove all organic components from the composite nanofibers and to form crystalline metal oxide nanofibers. After calcination, shrinkage in fiber diameter is commonly observed, owing to the decomposition of organic polymer template during calcination.

1.2.1 Precursor solution for sol−gel electrospinning process toward metal oxide nanofibers

Sol−gel electrospinning process is the most feasible approach for the production of high-aspect-ratio metal oxide nanofibers at a large scale. Generally, sol−gel process uses a metal alkoxide or metal salt, which is dissolved in the spinning solution and reacted with or incorporated into the as-spun fiber mats by electrospinning. Sol−gel electrospinning of metal oxides is based on the electrospinning of a solution of metal−organic precursor (alkoxides, chlorides, nitrates, acetates, or acetylacetonates) in a volatile solvent (alcohol, acetone, or water). Additionally, organic polymers (PVA, PVP, PEO, or PAN) and organic acids (acetic acid or citric acid) are commonly employed as spinning aids and catalysts, respectively [49].

Sol−gel chemistry involves a wet chemical method for the preparation of ceramics or inorganic polymers from a chemical solution by transformation from liquid precursors to a sol and finally to a network structure known as a gel [50]. The sol−gel technique has been extensively explored for producing metal oxides such as silica [51], alumina [52], titania [53], zirconia [30,47], and yttria [54]. The unique benefit of the sol−gel process is the excellent control of stoichiometric composition, microstructure, and morphology of products [30,55]. The basic chemical reactions of sol−gel formation are the hydrolysis and condensation of the metal alkoxide precursors, resulting in the formation of $M - O - M$ linkages at low temperature (Fig. 1.3). The properties of the sol−gel-based materials are affected by pH, precursor concentration, water content, temperature, and reaction time. For instance, the hydrolysis reaction contains three steps: nucleophilic addition, proton transfer, and departure of the protonated species (ROH) [56,57]. In considering the inorganic network, the growth of gels with an acid-catalyzed sol−gel process leads to a structure of linear chains, whereas a base-catalyzed process causes the formation of a material with randomly branched structures [58,59]. Simple metal alkoxides (M

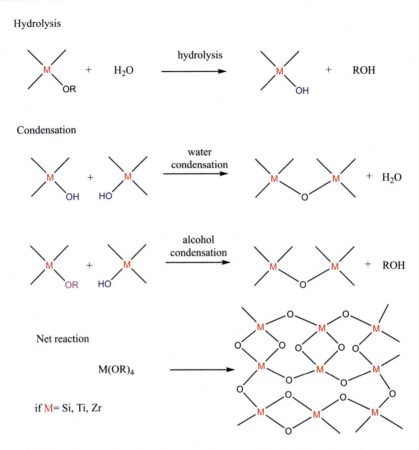

Figure 1.3 Hydrolysis and condensation reactions of metal alkoxide in the sol–gel process.

(OR)$_n$) with usual ligands (OR) are commercially available in a wide range of metals. However, their use is limited because of the relatively high cost, very high reactivity to humidity and water, and high hydrolytic instability. The required careful handling of the water-sensitive metal alkoxides to avoid partial hydrolysis before the hydrolysis reaction also limits the practicability of the sol–gel electrospinning process for metal oxide nanofibers [50].

To overcome this obstacle, the most general approach is to modify simple alkoxide precursors to reduce hydrolytic reactivity and prevent precipitate formation. The substitution of one or more OR groups in metal alkoxides by complexing ligands such as carboxylic acids and beta-diketones [60], which are less hydrolyzable and additionally block coordination sites at the metal, has many structural and chemical consequences for sol–gel technology [50,61]. The degree of crosslinking in the gel network is lower, owing to the decreased portion of hydrolyzable OR groups. The replacement of monodentate alkoxy group by bidentate or multidentate ligands decreases the connectivity of the molecular building blocks, which favors the formation of gels instead of crystalline precipitates. Triisopropanolamine (N(CH$_2$CHCH$_3$OH)$_3$) and especially triethanolamine (N(CH$_2$CH$_2$OH)$_3$) have also been used for the modification of metal alkoxides [61]. However, the modification of simple metal alkoxides is a rather complicated and expensive process. The synthesis of new metal alkoxides with unique structures and properties has thus emerged as an increasingly important and flexible approach for the study of sol−gel processes and the evolution of metal alkoxide chemistry [50]. Metallatrane or atrane compounds are metal complexes formed by complexation of a metal or semimetal with trialkanolamine such as triethanolamine (TEA) or triisopropanolamine (TIS) [30,44,62−66]. TEA or TIS acts as a chelating ligand with the three alkoxy groups coordinating to the central atom and incorporation of a transannular N−M bond (Fig. 1.4).

Generally, the atrane complexes are synthesized by means of a transesterification reaction, using metal alkoxide derivatives in nonaqueous dried solvent under an inert atmosphere. The procedure is still quite complicated, and metal alkoxide starting materials used for the synthesis of atrane complexes are commercially expensive [67]. Therefore the basis of using metal alkoxide starting materials will limit the scale-up of the synthesis procedures and pose a constraint on the application of metal oxide nanofibers in fields in which large amounts of relatively cheap materials are required [68]. To overcome these problems, our group has developed a synthesis route to metal alkoxide precursors directly from the oxides themselves. Reactions between metal oxides or hydroxides and dialkylcarbonates, polyols, or aminoalcohols provide an effective alternative way to synthesize new metal alkoxides of some elements [50]. This route is then extended to a one-step synthesis known as the oxide one-pot synthesis (OOPS) process. The OOPS process offers an economical, simple, and straightforward process to scale up the synthesis procedures of alkoxides for many metals and the preparation of highly moisture stable and pure metal alkoxide precursors using inexpensive and readily available starting materials [67]. For example, we synthesized atrane compounds to further produce metal oxide nanofibers, namely, zirconatrane and ferratrane, which are

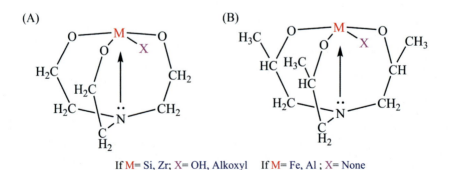

Figure 1.4 General structure of metallatranes. (A) TEA ligands. (B) TIS ligands.

atrane complexes with aminoalkoxide derivatives of zirconium and iron metals, respectively (Fig. 1.5). The presence of triethanolamine ligands was found to be hydrolytically stable in air, thus yielding more controllable chemistry and minimizing special handling requirement in the sol−gel electrospinning process [64]. Both zirconia and hematite nanofibers were fabricated by using the sol−gel electrospinning process using zirconatrane and ferratrane, respectively, as precursor [44,64]. In these works, PVP was used as the template polymer, and ethyl alcohol was used as the solvent for preparing the spinning solution of atrane complexes and PVP solution.

The spinning solution was pumped at a flow rate of 1.0 mL/h. The nozzle tip-to-collector distance and applied voltage were fixed at 15 cm and 20 kV, respectively. The nanofiber membrane was collected after electrospinning for 3 h on the heated collector at 300°C. Fig. 1.6A and B reveal the microstructure of the Fe_2O_3/PVP and ZrO_2/PVP composite nanofibers, respectively.

Some works have reported on the fabrication of metal oxide nanofibers via one-step electrospinning in combination with a citric acid sol−gel method using metal salt precursors. Citric acid plays a multifunctional role in the preparation of spinning solution, including a catalyst in the metal alkoxide hydrolytic process, a chelating agent, and a pore-forming agent in the heat treatment stage [40]. For the synthesis of spinel ferrites, the citric acid and polyacrylic acid act as the chelating agents, which allow the mixing of metal cations at the molecular level in a sol−gel process [69]. Citric acid contains three carboxyl groups (−COOH) and one hydroxyl group (−OH), which provide four lone pair electrons to form a tetradentate ligand. Nevertheless, one metal ion cannot chelate with more than two of the carboxyl groups in a citric acid molecule, and another carboxyl group can be chelated only by other metal ions [70] (Fig. 1.7).

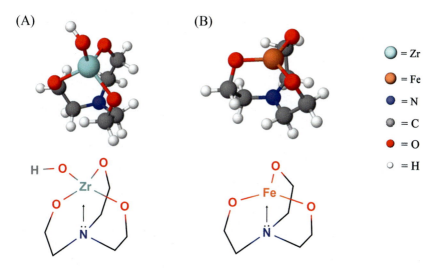

Figure 1.5 Structures of (A) zirconatrane and (B) ferratrane.

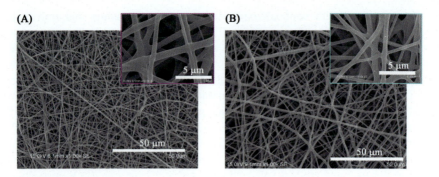

Figure 1.6 SEM images of (A) as-spun Fe$_2$O$_3$/PVP and (B) ZrO$_2$/PVP composite nanofibers.

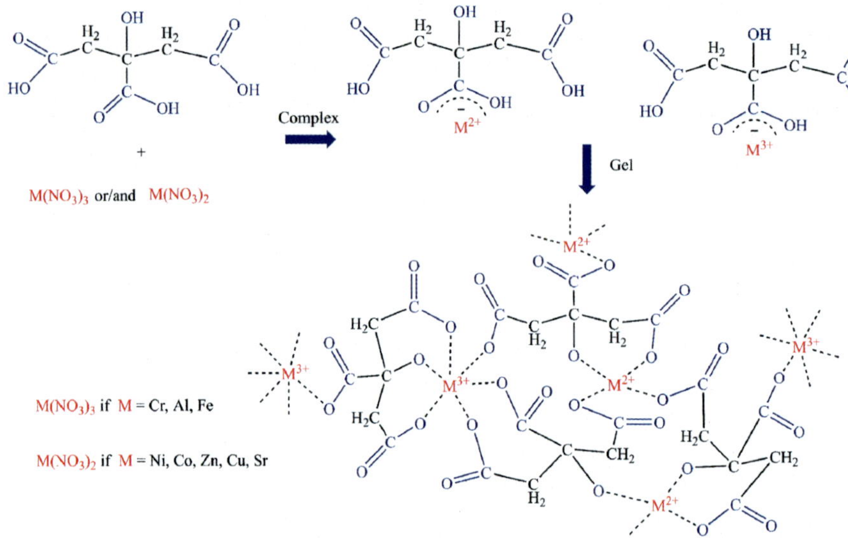

Figure 1.7 Chemical reactions in citric acid-based sol−gel process.

1.2.2 Precursor solution for colloidal method toward metal oxide nanofibers

The electrospinning of colloids, also known as colloid electrospinning, is a new technique that incorporates one or numerous kinds of colloids into the spinning solution. This method offers the possibility to elaborate multicompartment nanomaterials and to obtain fibers with more complicated structures and multiple functionalities [71]. The spinning solution is generally composed of solid nanoparticles and polymer matrix; the solid materials have been chosen to be insoluble in the solution [72]. For example, silica [18], titania [73], and zinc oxide [74] nanoparticles were immobilized in polymer fibers by electrospinning a mixture of polymer solution and these corresponding nanoparticles. The simplest and most commonly used way to combine metal oxide nanoparticles with electrospun nanofibers is the dispersion of metal oxide nanoparticles directly into the polymer solution. The incorporation of a small amount of some metal oxide nanoparticles into the polymer solution can enhance the mechanical, thermal, and antimicrobial properties of the polymer matrix [75]. In colloid electrospinning, the bead formation may result in aggregation of colloids in bead structures with high particle contents. In this case, the particle aggregation in fibers is often observed, especially if the diameter of the fiber is smaller than the size of the particles [76]. Meanwhile, the particle embedding in a certain content range with no bead formation forms green pepper−like structured fibers [77]. Nanoparticle aggregation takes place when attractive interaction forces, commonly van der Waals forces, outweigh other repulsive interaction forces between the particles, such as steric interactions and electrostatic forces [76]. Moreover, the location of nanoparticles in the fibers can be controlled by the composition of the mixture of solvent used to dissolve the polymer matrix, and the presence of a surfactant in the electrospinning solution is essential to disperse the nanoparticles well in solvents [77]. For colloidal precursors the colloid aggregation can induce the gelation reaction, yielding particulate gel. The van der Waals forces are the driving force for gel formation. However, the colloidal precursors cannot form an irreversible permanent chemical gel because of reversible particle agglomeration and gelation by weak interaction forces. Therefore it is difficult to ensure the structural reliability of metal oxide nanofibers. By contrast, metal alkoxide precursors can form polymeric gels by covalent bond between precursors, showing the properties of a permanent chemical gel [57].

1.2.3 Electrospinning process parameters

The properties of the spinning solution for producing metal oxide nanofibers (e.g., solvent and precursor type, molecular weight of polymer matrix, solution concentration and viscosity, surface tension, and conductivity), the process parameters (e.g., applied voltage, solution flow rate, distance between the tip and collector, needle diameter, and collector temperature), and the environmental parameters (e.g., ambient temperature and humidity) have an influence on the morphology of the fiber during electrospinning [23]. Polymers that have low molecular weight tend to form beads during the electrospinning process, whereas the increase in molecular weight of polymers can decrease the number of beads and provides fibers with uniform diameter. When the molecular weight of the polymer is determined, the solution concentration becomes a significant factor influencing the entanglement of the polymer chain in the solution. The viscosity of the polymer solution increases when the concentration of polymer solution increases. Moreover, the viscosity of the spinning solution containing the mixture of organic polymer solution and sol−gel precursor or colloid precursor is also affected by the chemistry of the precursor solution. Metal oxide nanofibers can be formed by controlling the viscosity of the spinning solution within a certain optimal range during electrospinning. The viscosity of the spinning solution has an effect on the morphological features of the electrospun fibers. The surface tension of the spinning solution also plays a critical role in the electrospinning process. The electrostatic repulsive force on the surface of the charged spinning solution must overcome its surface tension. The surface tension of the solution can be adjusted by adding a surfactant to the solution or by changing the solvent composition. The electrical conductivity of the spinning solution involves the electrification ability of the polymer solution, which can be adjusted by adding salt or polyelectrolyte or by changing the composition of the solvent; thus the morphology of the fiber can be changed [23,78].

For our research, we provided results from a study of the effect of polymer concentration on the fiber morphology of freestanding zirconia nanofibers obtained by the sol−gel electrospinning process. The zirconatrane precursor was synthesized by using the OOPS process as previously described by our group [79]. PVP (molecular weight: 1,300,000) was first dissolved in ethyl alcohol (95%). The zirconatrane precursor was then added to the PVP polymer solution under stirring for 30 minutes. The spinning solution was composed of PVP, ethyl alcohol, and zirconatrane in two different weight ratios (1.0:18.0:12.0 and 2.5:18.0:12.0). The spinning solution was pumped at a flow rate of 1.0 mL/h. The nozzle tip-to-collector distance and applied voltage were fixed at 15 cm and 20 kV, respectively. The nanofiber membrane was collected after electrospinning for 3 h on the heated collector at 300°C. A part of the as-spun sample was then calcined at 450°C for 2 h in air. Fig. 1.8A and B show the SEM images of the as-spun nanofibers (collector temperature of 300°C) at different PVP contents. The viscosity of the spinning solution increased when the PVP content increased. It was found that the diameter of the resulting zirconia nanofibers increased with an increase in PVP content in the spinning solution. However, the electrospun metal oxide nanofibers required the calcination step after electrospinning to burn off the polymer matrix and to crystallize the metal oxides. The SEM images of the as-spun nanofibers at various PVP contents calcined at a furnace temperature of 450°C are shown in Fig. 1.8C and D. The shrinkage in fiber diameter of

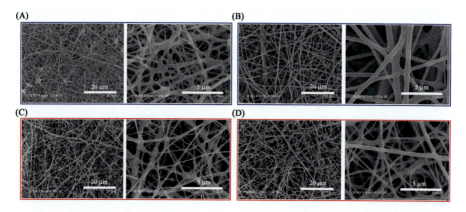

Figure 1.8 SEM images of as-spun zirconia nanofibers at PVP:ethyl alcohol:zirconatrane weight ratios of (A) 1.0:18.0:12.0 and (B) 2.5:18.0:12.0. SEM images of zirconia nanofibers calcined at 450°C at PVP:ethyl alcohol:zirconatrane weight ratios of (C) 1.0:18.0:12.0 and (D) 2.5:18.0:12.0.

the calcined zirconia nanofibers was observed because of the decomposition of PVP during calcination. The ZrO_2 nanofibers have a long length without discontinuity and are well interconnected, being welded together at the fiber junction points to form an interconnected fibrous network [80,81]. Zirconatrane precursor can generate polymeric gels by covalent bond between precursors and fiber junction points via sol−gel reaction [57]. The underlying layers of nanofibers are surmounted by the upper layers of nanofibers [82].

The effect of different precursor types on the morphology of hematite nanofibers prepared by the sol−gel electrospinning process was also investigated. The free-standing hematite (α-Fe_2O_3) nanofibers were prepared by using the same conditions as were used for electrospinning the as-spun zirconia nanofibers. The ferratrane precursor was synthesized via the OOPS process as previously reported [44,67]. The weight ratio of the spinning solution was 1.0:4.0:1.9 (PVP:ethyl alcohol:ferratrane). To identify different types of precursors for preparing hematite nanofiber, another hematite nanofiber was fabricated by using commercial ferric nitrate precursor (Fe $(NO_3)_3$) at a weight ratio of 1.0:4.0:1.5 (PVP:ethyl alcohol:$Fe(NO_3)_3$) under the same electrospinning condition as described above. The as-spun nanofiber membrane was divided into two batches, and one batch was calcined at 600°C in air for 2 h. Fig. 1.9A shows the morphology of the as-spun hematite nanofibers at a collector temperature of 300°C obtained from using ferric nitrate precursor. After calcination at 600°C the continuous inorganic nanofibers were converted into short nanofibers or nanoparticles. It was seen that the calcined hematite nanofibers that were obtained from using ferric nitrate precursor had ruptures and discontinuity in the fiber, as shown in Fig. 1.9B. On the other hand, by using ferratrane precursor, the hematite nanofibers were still continuous after calcination at 600°C owing to the greater stability of the hematite nanofibers (Figs. 1.6A and 1.9C), as a result of the sol−gel process from atrane precursor and the interconnection of hematite nanofibers. It was also confirmed by TEM measurement as shown in Fig. 1.9D. Our previous studies for zirconia nanofibers using zirconatrane precursor also determined that the morphology of zirconia nanofibers changed from interconnected nanofibers (electrospun at 300°C) to thornlike to flower-like architecture only by varying the calcination temperature from 500°C to 700°C [30]. The 3D thornlike zirconia nanostructure of nanofibers was more likely to form at lower calcination temperature, whereas the 3D flower-like zirconia nanostructure was more likely to form at higher calcination temperature [30].

In the preparation of metallic nanofibers, metal oxide nanofibers can be transformed into metallic phases in a subsequent calcination step in a reductive atmosphere or in hydrogen gas atmosphere [49,83]. However, after calcination at high temperature the nanoparticles are segregated, and the discontinuity and rupture of fibers are unavoidable [84]. Unlike high-temperature calcination in air, the carbon originating from organic polymer template at low-temperature calcination or in a reductive atmosphere can help to generate continuous nanofibers and to enhance the reduction to metal. For example, we observed the fabrication of nitrogen-doped carbon coated Ni nanofibers via electrospinning in one step. The spinning solution was prepared by dissolving nickel nitrate ($Ni(NO_3)_2$) and PAN (molecular weight: 150,000) in N,N-dimethylformamide (DMF) at a weight ratio of 1.0:9.0:0.5 (PAN:

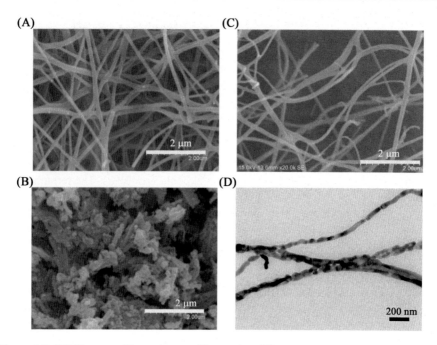

Figure 1.9 SEM images of hematite nanofibers using different types of precursors. (A) As-spun nanofibers (300°C) obtained from ferric nitrate precursor. (B) After calcination at 600°C. (C) The calcined nanofibers (600°C) obtained from ferratrane precursor. (D) TEM image of the calcined nanofibers (600°C) using ferratrane precursor.

DMF:Ni(NO$_3$)$_2$). In this work, PAN acted as both a carbon and nitrogen source and spinning aids. The spinning solution was pumped through a metal nozzle at a flow rate of 1.0 mL/h. The applied voltage, spinning time, and deposition distance were set at 20 kV, 3 h, and 15 cm, respectively. The electrospinning was carried out at room temperature. After electrospinning, the as-spun composite nanofibers were annealed under a nitrogen atmosphere at 600°C for 2 h. Fig. 1.10A and B present the SEM and TEM images of nitrogen-doped carbon-coated Ni nanofibers. In the process of annealing, the nickel nitrate is converted into nickel in the reduction atmosphere generated during PAN pyrolysis. The formed Ni nanoparticles are randomly dispersed in nitrogen-doped carbon nanofibers during the annealing process (inset in Fig. 1.10B) and mostly wrapped by graphitic layers, implying good physical and chemical stability owing to carbon protection [85].

1.3 Large-scale production

Even though electrospinning is considered a powerful tool in a wide range of potential applications, the low production efficiency of this process is a major limitation in applying the technique to large-scale production. Several attempts have been

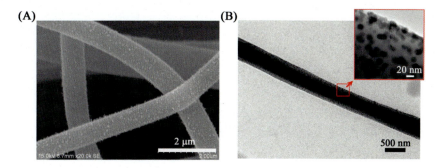

Figure 1.10 (A) SEM and (B) TEM images of nitrogen-doped carbon-coated Ni nanofibers prepared by one-step electrospinning.

made to speed up and develop the process by using multiple needles, needleless spinning, and gas-assisted systems [86–88]. The common needleless electrospinning technologies include metal roller spinneret and wire spinneret, as shown in Fig. 1.1. For needleless spinning technology, efficient fabrication of nanofibers has been obtained for some materials. However, the development of needleless electrospinning technology is still restricted by the product quality of the nanofibers. Because the jet initiation in needleless electrospinning is a self-organized process of liquid that happens on a free liquid surface, the spinning process is difficult to control (e.g., motion of multiple jets, jet point, jet time, jet size) [89–91]. Multineedle technology was developed to scale up and increase throughput of the nanofiber production by increasing the number of needles used. The main advantages of multineedle electrospinning are the ability to mix different precursor solutions at desired ratios, the adaptability of materials, and the evenness of the resulting nanofibers [89,92,93]. Multineedle electrospinning technology is based on a conventional single-needle electrospinning method, but multiple needles are used as spinnerets that contain one or different types of precursor solutions (Fig. 1.11). The precursor solution is forced through multiple needles connected to a high-voltage power supply. A syringe pump is used to pump one or different spinning solutions to the spinneret setup. The use of multiple needles can lead to the formation of multiple Taylor cones during the electrospinning process. The Taylor cones may also undergo splitting into more jets under the electric field and thus substantially enhance the production of nanofibers. The drawbacks of this method include solution clocking at the needle tips and cleaning of multiple needles. Moreover, in multineedle electrospinning, when the distance between needle tips is relatively small, the mutual interferences of the electric field between the adjacent solution jets will be larger. This may be the cause of a nonuniform electric field on each needle tip of the spinneret, leading to instability of the jet flow, which influences the nanofiber quality [88,94]. For example, we tested a multineedle electrospinning device that contained three needles with a linear arrangement to obtain PAN nanofiber membrane (Fig. 1.11). The increase in the number of needles could greatly enhance the nanofiber yield under the same experimental conditions as those used in the

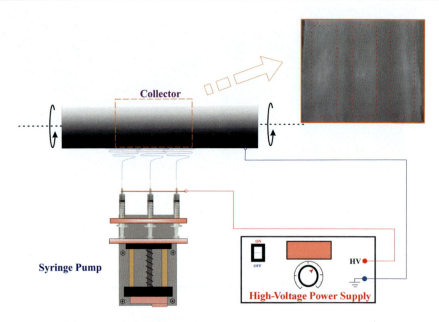

Figure 1.11 Schematic of a multineedle electrospinning device with three needles.

single-needle electrospinning process. However, the electric field interference between multiple needles became stronger, and the unevenness of the nanofibers increased. The fiber mat that was collected on the rotating drum collector with three-needle systems in linear alignment is shown in the inset in Fig. 1.11. The jets that were ejected from needle in the middle position were electrostatically repelled by the adjacent sides of the jets in a three-needle electrospinning device [94,95].

To overcome these limitations, the needles are set at a proper distance, indicating a requirement for larger working space and more complex arrangement of multiple needles. Up to now, developing electrospinning spinnerets with continuous stability, high throughput, and nanofibers produced with precisely regulated diameter remains a great challenge. Modifications of multineedle electrospinning process in this work are based on both phenomenological and numerical approaches to solve the above-mentioned problems in the process. We developed multineedle electrospinning device that contained three needles arranged at a needle spacing of 65 mm in linear alignment system. The device uses a rotating drum roller as the collector, and the linearly arranged three needles mounted on the laterally moving slider crank, as shown in Fig. 1.12A. For the multinozzle electrospinning system with a linear arrangement, when three needles are connected to the slider crank by moving left and right (the moving speed of about 10 mm/s), the electric field distribution between the nozzles becomes more uniform (Fig. 1.12B). It is clear that the distance between the deposition zones of each jet was reduced and disappeared. Therefore multineedles mounted on the laterally moving slider crank can effectively reduce the electric field interference between the needles and give a better spinning

Fundamentals of electrospinning and safety 19

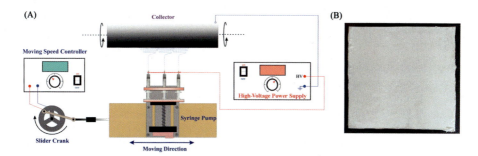

Figure 1.12 (A) Schematic illustration of the three-needle electrospinning setup using the laterally moving slider crank. (B) Fiber mat collected on the laterally moving slider crank.

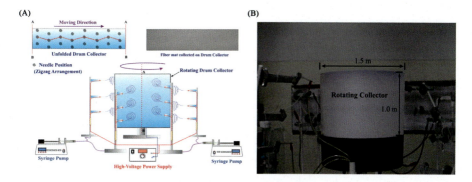

Figure 1.13 (A) Schematic diagram of the 24-needle electrospinning system with the zigzag arrangement. (B) Photograph of massive nanofiber membrane.

effect. However, there is some limitation in increasing the number of needles, owing to the size limitation of the electrospinning device. To further improve the above setup, we designed an electrospinning device not only to reduce the electric field interference between the needles, but also to enhance the nanofiber yield by increasing the number of needles and changing the arrangement of multiple needles. The 24-needle electrospinning setup uses a larger rotating drum collector (1.5 m diameter × 1.0 m width), and the 24 needles are designed to be arranged in the zigzag array at the needle spacing in horizontal and vertical directions of 500 and 200 mm, respectively, as shown in Fig. 1.13A and B. The fiber mat collected from this modified multineedle electrospinning is shown in the inset of Fig. 1.13A. For the 24-needle electrospinning system with the zigzag arrangement, when the spacing between the needles increased, the electric field interference between the nozzles could be ignored. Additionally, this setup modification can produce massive nanofiber fabric with the area of approximate 4.7 m^2. This indicates that the modified multineedle electrospinning system with the zigzag arrangement not only overcomes the mutual electric field interference between needles, but also achieves the goal of increasing the mass production of nanofibers.

1.4 Electrospinning safety

Electrospinning is an effective technique for the construction of microfibrous and nanofibrous structures with considerable potential in applications. Moreover, the large number of reports on various setups have shown significant improvements in the scale-up production of electrospun nanofibers. However, other key aspects, including process safety, environmental impact, and energy consumption, should be carefully considered in the setting of both research and commercialization. The major safety concern arises from the use of high applied voltage for the spinning process (generally more than 10 kV) and volatile, often flammable compounds used in the spinning solution [96−98]. The evaporation of the solvent into the environment during the solution electrospinning will cause environmental, health, and safety effects as well as waste of chemicals. Most of the organic solvents that are used to prepare electrospinning solutions (e.g., methyl alcohol, ethyl alcohol, DMF, chloroform) are combustible, often highly volatile, and extremely flammable and pose long-term health hazards, and some are known to be carcinogenic. Therefore they should always be used and handled with care [99]. Furthermore, the residual solvents can also be retained in the nanofibers as harmful impurities, hence posing a safety concern, especially for products used in pharmaceutical and biomedical applications. When high voltage is being considered, the use of a high voltage may cause danger to the workers, and it is not well applicable to some electric-sensitive materials, especially biomolecules. The charge particle space charges exist in space, owing to the constant direct current (DC) voltage and corona. They move in same direction, causing the ion flow in space under a DC electric field. The field near DC high voltage is a combined field with DC and ion flow fields. The International Commission on Non-Ionizing Radiation Protection has presented guidelines for limited exposure to extremely low frequencies [100]. The reference levels of electric field intensity for safety is lower than 5 kV/m, in accordance with these guidelines [101,102]. The influence of a DC electric shock is examined by the amplitude of the current passing through a human body and the duration of the shock. For a human body the DC current amplitude of 300 mA is considered a safety limit. The DC voltage depends on the resistance of a human body [103]. However, the maximum current allowed to pass through a human body is 25 mA for the precondition with no damage to the human body. In normal conditions, the impedance of the human body is very high. But when the human body touches a high voltage, the impedance of the human body will rapidly decrease [104]. Generally, the voltage applied to the nozzle is very high, in the range of $10-20$ kV, and the distance between the tip of the nozzle and the collector is in the range of approximately $10-15$ cm. The electric field intensity is therefore larger than 5 kV/m, which presents danger of touching a high-voltage part.

When the electric field emitted from the conductor exceeds the initial ionization level (typically $20-30$ kV/cm), molecules in the air will be ionized. The ozone smell, a discharge sound, and sometimes a blue-purple-colored corona can be seen in the dark, related to the emission of heat, sound, and light [105]. The major components of air are nitrogen and oxygen, whose ionized and neutralized molecules emit UV radiation (200 − 400 nm). Preferentially, the electric field intensity for the electrospinning process is lower than 20 − 30 kV/cm, not exceeding the initial ionization level.

The environmental, health, and safety concerns about nanotechnology have attracted extensive attention. Furthermore, the safety of nanofibers has been brought into question again, with some reports suggesting potential adverse impact on the lungs upon inhalation exposure to nanofibers [106,107]. The concern has been expressed that new kinds of nanofibers being produced by nanotechnology industries and their products might pose a risk to human health because of their similar shapes to asbestos fibers. Asbestos fibers may cause lung and pleural disease because of the inhalation of airborne respirable fibers, which enter the peripheral parts of the lungs, causing pulmonary fibrosis and lung cancer, and approach the pleura, resulting in fibrosis and mesothelioma [108,109]. Generally, there are three main routes of human exposure to fibers: by inhalation, by skin penetration, or by ingestion. As inhalation is considered to be the primary entry route and a health concern has been raised that high-aspect-ratio nanomaterials such as nanofibers and nanotubes may cause unintended health consequences, such as asbestos-like lung cancer and mesothelioma upon chronic inhalation [109−111]. Nanofibers, which can be made from a range of materials, including carbon, organic polymer, ceramic, and metal, are approximately 1000 times smaller than the diameter of a human hair and can reach the lung cavity from inhalation exposure [112]. In considering the potential impacts of electrospun nanofibers on the health of humans and other species, the composition, length, and diameter of the nanofibers need to be considered. The most reported harmful effects of nanofibers are based on ceramic, metal, and carbon materials [106−109,111,113]. Schinwald et al. reported that silver nanowires of 5−20 μm in length were lodged in the lungs of mice and resulted in respiratory problems after silver nanowires were injected into the lungs of mice [109]. Some research has also suggested harmful impacts of ceramic nanofibers that were relatively short, with diameters in the range of tens of nanometers or smaller. These short nanofibers can be easily inhaled and may cause a pulmonary inflammatory response in the lungs [96,114]. Meanwhile, electrospun nanofibers are generally organic polymers with lengths mostly longer than a few centimeters and diameters usually in range of hundreds of nanometers. The opportunity for electrospun nanofibers at such a length scale to reach the lungs is quite low [96]. However, there is still no definitive conclusion because only a very limited amount of research has been carried out so far addressing the inhalation safety of electrospun nanofibers.

Acknowledgments

The authors would like to thank the Department of Materials Science and Engineering, Faculty of Engineering and Industrial Technology, Silpakorn University, and the Center of Excellence on Petrochemical and Materials Technology, Chulalongkorn University.

References

[1] A. Barhoum, R. Rasouli, M. Yousefzadeh, H. Rahier, M. Bechelany, Nanofiber technology: history and developments, 2018. https://doi.org/10.1007/978-3-319-42789-8_54-1.

[2] M. Tebyetekerwa, Z. Xu, S. Yang, S. Ramakrishna, Electrospun nanofibers-based face masks, Adv. Fiber Materials (2020). Available from: https://doi.org/10.1007/s42765-020-00049-5.

[3] J.L. Skinner, J.M. Andriolo, J.P. Murphy, B.M. Ross, Electrospinning for nano-to mesoscale photonic structures, Nanophotonics 6 (2017) 765−787. Available from: https://doi.org/10.1515/nanoph-2016-0142.

[4] C. Cleeton, A. Keirouz, X. Chen, N. Radacsi, Electrospun nanofibers for drug delivery and biosensing, ACS Biomater. Sci. Eng. 5 (2019) 4183−4205. Available from: https://doi.org/10.1021/acsbiomaterials.9b00853.

[5] J.P. Yapor, A. Alharby, C. Gentry-Weeks, M.M. Reynolds, A.K.M.M. Alam, Y.V. Li, Polydiacetylene nanofiber composites as a colorimetric sensor responding to *Escherichia coli* and pH, ACS Omega 2 (2017) 7334−7342. Available from: https://doi.org/10.1021/acsomega.7b01136.

[6] M. Liu, P. Zhang, Z. Qu, Y. Yan, C. Lai, T. Liu, et al., Conductive carbon nanofiber interpenetrated graphene architecture for ultra-stable sodium ion battery, Nat. Commun. 10 (2019). Available from: https://doi.org/10.1038/s41467-019-11925-z.

[7] S. Jiang, Y. Chen, G. Duan, C. Mei, A. Greiner, S. Agarwal, Electrospun nanofiber reinforced composites: a review, Polym. Chem. 9 (2018) 2685−2720. Available from: https://doi.org/10.1039/c8py00378e.

[8] S. Zhang, H. Liu, N. Tang, J. Ge, J. Yu, B. Ding, Direct electronetting of high-performance membranes based on self-assembled 2D nanoarchitectured networks, Nat. Commun. 10 (2019) 1−11. Available from: https://doi.org/10.1038/s41467-019-09444-y.

[9] M.A.A. Mohd Abdah, N.H.N. Azman, S. Kulandaivalu, Y. Sulaiman, Asymmetric supercapacitor of functionalised electrospun carbon fibers/poly(3,4-ethylenedioxythiophene)/manganese oxide//activated carbon with superior electrochemical performance, Sci. Rep. 9 (2019) 1−9. Available from: https://doi.org/10.1038/s41598-019-53421-w.

[10] K.N. Jung, S.M. Hwang, M.S. Park, K.J. Kim, J.G. Kim, S.X. Dou, et al., One-dimensional manganese-cobalt oxide nanofibres as bi-functional cathode catalysts for rechargeable metal-air batteries, Sci. Rep. 5 (2015) 1−10. Available from: https://doi.org/10.1038/srep07665.

[11] G. Yanalak, A. Aljabour, E. Aslan, F. Ozel, I.H. Patir, A systematic comparative study of the efficient co-catalyst-free photocatalytic hydrogen evolution by transition metal oxide nanofibers, Int. J. Hydrog. Energy 43 (2018) 17185−17194. Available from: https://doi.org/10.1016/j.ijhydene.2018.07.113.

[12] C.J. Mortimer, C.J. Wright, The fabrication of iron oxide nanoparticle-nanofiber composites by electrospinning and their applications in tissue engineering, Biotechnol. J. 12 (2017) 1−10. Available from: https://doi.org/10.1002/biot.201600693.

[13] K. Mondal, A. Sharma, Recent advances in electrospun metal-oxide nanofiber based interfaces for electrochemical biosensing, RSC Adv. 6 (2016) 94595−94616. Available from: https://doi.org/10.1039/c6ra21477k.

[14] F. Li, T. Zhang, X. Gao, R. Wang, B. Li, Coaxial electrospinning heterojunction SnO_2/Au-doped In_2O_3 core-shell nanofibers for acetone gas sensor, Sens. Actuators B Chem. 252 (2017) 822−830. Available from: https://doi.org/10.1016/j.snb.2017.06.077.

[15] M. Madadi Masouleh, J. Koohsorkhi, R. Askari Moghadam, Direct writing of individual micro/nanofiber patterns suitable for flexible electronics using MEMS-based microneedle, Microelectron. Eng. 229 (2020) 111345. Available from: https://doi.org/10.1016/j.mee.2020.111345.

[16] Z. Duan, Y. Huang, D. Zhang, S. Chen, Electrospinning fabricating Au/TiO2 network-like nanofibers as visible light activated photocatalyst, Sci. Rep. 9 (2019) 1–8. Available from: https://doi.org/10.1038/s41598-019-44422-w.

[17] Kenry, C.T. Lim, Nanofiber technology: current status and emerging developments, Prog. Polym. Sci. 70 (2017) 1–17. Available from: https://doi.org/10.1016/j.progpolymsci.2017.03.002.

[18] N. Horzum, R. Muñoz-Espí, G. Glasser, M.M. Demir, K. Landfester, D. Crespy, Hierarchically structured metal oxide/silica nanofibers by colloid electrospinning, ACS Appl. Mater. Interfaces 4 (2012) 6338–6345. Available from: https://doi.org/10.1021/am301969w.

[19] R. Sahay, P. Suresh Kumar, V. Aravindan, J. Sundaramurthy, W. Chui Ling, S.G. Mhaisalkar, et al., High aspect ratio electrospun CuO nanofibers as anode material for lithium-ion batteries with superior cycleability, J. Phys. Chem. C 116 (2012) 18087–18092. Available from: https://doi.org/10.1021/jp3053949.

[20] M.J. Chang, W.N. Cui, X.J. Chai, J. Liu, K. Wang, L. Qiu, Fabrication of flexible MIL-100(Fe) supported SiO_2 nanofibrous membrane for visible light photocatalysis, J. Mater. Sci. Mater. Electron. 30 (2019) 1009–1016. Available from: https://doi.org/10.1007/s10854-018-0370-9.

[21] Y. Dou, W. Zhang, A. Kaiser, Electrospinning of metal−organic frameworks for energy and environmental applications, Adv. Sci. 7 (2020). Available from: https://doi.org/10.1002/advs.201902590.

[22] P. Lu, S. Murray, M. Zhu, Chapter 23 - Electrospun nanofibers for catalysts, Editor(s): B. Ding, X. Wang, J. Yu, In micro and nano technologies, electrospinning: nanofabrication and applications, William Andrew Publishing, 2019; 695-717, Available from: https://doi.org/10.1016/B978-0-323-51270-1.00023-6.

[23] Y. Sun, S. Cheng, W. Lu, Y. Wang, P. Zhang, Q. Yao, Electrospun fibers and their application in drug controlled release, biological dressings, tissue repair, and enzyme immobilization, RSC Adv. 9 (2019) 25712–25729. Available from: https://doi.org/10.1039/c9ra05012d.

[24] M.K. Selatile, S.S. Ray, V. Ojijo, R. Sadiku, Recent developments in polymeric electrospun nanofibrous membranes for seawater desalination, RSC Adv. 8 (2018) 37915–37938. Available from: https://doi.org/10.1039/C8RA07489E.

[25] R.S. Bhattarai, R.D. Bachu, S.H.S. Boddu, S. Bhaduri, Biomedical applications of electrospun nanofibers: drug and nanoparticle delivery, Pharmaceutics 11 (2019). Available from: https://doi.org/10.3390/pharmaceutics11010005.

[26] P. Zhou, J. Wang, A.L.B. Maçon, A. Obata, J.R. Jones, T. Kasuga, Tailoring the delivery of therapeutic ions from bioactive scaffolds while inhibiting their apatite nucleation: a coaxial electrospinning strategy for soft tissue regeneration, RSC Adv. 7 (2017) 3992–3999. Available from: https://doi.org/10.1039/C6RA25645G.

[27] G. George, Z. Luo, A review on electrospun luminescent nanofibers: photoluminescence characteristics and potential applications, Curr Nanosci. 16 (2019) 321–362. Available from: https://doi.org/10.2174/1573413715666190112121113.

[28] Y. Li, Q. Li, Z. Tan, A review of electrospun nanofiber-based separators for rechargeable lithium-ion batteries, J. Power Sources 443 (2019) 227262. Available from: https://doi.org/10.1016/j.jpowsour.2019.227262.

[29] T. Sebastian, T.R. Preisker, L. Gorjan, T. Graule, C.G. Aneziris, F.J. Clemens, Synthesis of hydroxyapatite fibers using electrospinning: a study of phase evolution based on polymer matrix, J. Eur. Ceram. Soc. 40 (2020) 2489−2496. Available from: https://doi.org/10.1016/j.jeurceramsoc.2020.01.070.

[30] B. Ksapabutr, M. Panapoy, Fabrication of ceramic nanofibers using atrane precursor, Nanofibers (2010). Available from: https://doi.org/10.5772/8163.

[31] M.J. Bauer, X. Wen, P. Tiwari, D.P. Arnold, J.S. Andrew, Magnetic field sensors using arrays of electrospun magnetoelectric Janus nanowires, Microsyst. Nanoeng. 4 (2018). Available from: https://doi.org/10.1038/s41378-018-0038-x.

[32] Z.U. Abideen, J.H. Kim, J.H. Lee, J.Y. Kim, A. Mirzaei, H.W. Kim, et al., Electrospun metal oxide composite nanofibers gas sensors: a review, J. Korean Ceram. Soc. 54 (2017) 366−379. Available from: https://doi.org/10.4191/kcers.2017.54.5.12.

[33] C. Shi, J. Zhu, X. Shen, F. Chen, F. Ning, H. Zhang, et al., Flexible inorganic membranes used as a high thermal safety separator for the lithium-ion battery, RSC Adv. 8 (2018) 4072−4077. Available from: https://doi.org/10.1039/c7ra13058a.

[34] T. Zhang, H. Li, Z. Yang, F. Cao, L. Li, H. Chen, et al., Electrospun YMn_2O_5 nanofibers: a highly catalytic activity for NO oxidation, Appl. Catal. B: Environ. 247 (2019) 133−141. Available from: https://doi.org/10.1016/j.apcatb.2019.02.005.

[35] S. Zhou, M. Hu, X. Huang, N. Zhou, Z. Zhang, M. Wang, et al., Electrospun zirconium oxide embedded in graphene-like nanofiber for aptamer-based impedimetric bioassay toward osteopontin determination, Microchimica Acta 187 (2020). Available from: https://doi.org/10.1007/s00604-020-4187-x.

[36] X. Li, G. Zhu, J. Dou, J. Yang, Y. Ge, J. Liu, Electrospun Au nanoparticle-containing ZnO nanofiber for non-enzyme H_2O_2 sensor, Ionics 25 (2019) 5527−5536. Available from: https://doi.org/10.1007/s11581-019-03118-x.

[37] K.R. Park, H.B. Cho, J. Lee, Y. Song, W.B. Kim, Y.H. Choa, Design of highly porous SnO_2-CuO nanotubes for enhancing H2S gas sensor performance, Sens. Actuators B Chem. 302 (2020) 127179. Available from: https://doi.org/10.1016/j.snb.2019.127179.

[38] D. Graf, A. Queraltó, A. Lepcha, L. Appel, M. Frank, S. Mathur, Electrospun SrNb2O6 photoanodes from single-source precursors for photoelectrochemical water splitting, Sol. Energy Mater. Sol. Cells 210 (2020). Available from: https://doi.org/10.1016/j.solmat.2020.110485.

[39] H. Shao, M. Huang, H. Fu, S. Wang, L. Wang, J. Lu, et al., Hollow WO_3/SnO_2 heteronanofibers: controlled synthesis and high efficiency of acetone vapor detection, Front. Chem. 7 (2019) 1−10. Available from: https://doi.org/10.3389/fchem.2019.00785.

[40] C. Xu, S. Shi, Q. Dong, S. Zhu, Y. Wang, H. Zhou, et al., Citric-acid-assisted sol-gel synthesis of mesoporous silicon-magnesium oxide ceramic fibers and their adsorption characteristics, Ceram. Int. 46 (2020) 10105−10114. Available from: https://doi.org/10.1016/j.ceramint.2019.12.279.

[41] S. Shi, C. Xu, X. Wang, Y. Xie, Y. Wang, Q. Dong, et al., Electrospinning fabrication of flexible Fe_3O_4 fibers by sol-gel method with high saturation magnetization for heavy metal adsorption, Mater. Des. 186 (2020). Available from: https://doi.org/10.1016/j.matdes.2019.108298.

[42] X. Song, K. Zhang, Y. Song, Z. Duan, Q. Liu, Y. Liu, Morphology, microstructure and mechanical properties of electrospun alumina nanofibers prepared using different polymer templates: a comparative study, J. Alloys Compd. 829 (2020) 154502. Available from: https://doi.org/10.1016/j.jallcom.2020.154502.

[43] J. Huang, X. Liu, G. Chen, N. Zhang, R. Ma, G. Qiu, Selective fabrication of porous iron oxides hollow spheres and nanofibers by electrospinning for photocatalytic water

purification, Solid State Sci. 82 (2018) 24−28. Available from: https://doi.org/10.1016/j.solidstatesciences.2018.05.014.

[44] I.O. Sittirug, B. Ksapabutr, M. Panapoy, Freestanding hematite nanofiber membrane for visible-light-responsive photocatalyst, Ceram. Int. 42 (2016) 3864−3875. Available from: https://doi.org/10.1016/j.ceramint.2015.11.051.

[45] N. Wu, B. Wang, C. Han, Q. Tian, C. Wu, X. Zhang, et al., Pt-decorated hierarchical SiC nanofibers constructed by intertwined SiC nanorods for high-temperature ammonia gas sensing, J. Mater. Chem. C 7 (2019) 7299−7307. Available from: https://doi.org/10.1039/c9tc01330j.

[46] M.A. Kanjwal, F.A. Sheikh, N.A.M. Barakat, I.S. Chronakis, H.Y. Kim, Co_3O_4 -ZnO hierarchical nanostructures by electrospinning and hydrothermal methods, Appl. Surf. Sci. 257 (2011) 7975−7981. Available from: https://doi.org/10.1016/j.apsusc.2011.04.034.

[47] S. Khalili, H.M. Chenari, Successful electrospinning fabrication of ZrO_2 nanofibers: a detailed physical—chemical characterization study, J. Alloys Compd. 828 (2020) 154414. Available from: https://doi.org/10.1016/j.jallcom.2020.154414.

[48] K. Roongraung, S. Chuangchote, N. Laosiripojana, T. Sagawa, Electrospun Ag-TiO_2 nanofibers for photocatalytic glucose conversion to high-value chemicals, ACS Omega. 5 (2020) 5862−5872. Available from: https://doi.org/10.1021/acsomega.9b04076.

[49] T. Ludwig, C. Bohr, A. Queraltó, R. Frohnhoven, T. Fischer, S. Mathur, Inorganic nanofibers by electrospinning techniques and their application in energy conversion and storage systems, Semicond. Semimet. 98 (2018) 1−70. Available from: https://doi.org/10.1016/bs.semsem.2018.04.003.

[50] B. Ksapabutr, E. Gulari, S. Wongkasemjit, One-pot synthesis and characterization of novel sodium tris(glycozirconate) and cerium glycolate precursors and their pyrolysis, Mater. Chem. Phys. 83 (2004) 34−42. Available from: https://doi.org/10.1016/j.matchemphys.2003.08.016.

[51] W. Matysiak, T. Tański, Analysis of the morphology, structure and optical properties of 1D SiO_2 nanostructures obtained with sol-gel and electrospinning methods, Appl. Surf. Sci. 489 (2019) 34−43. Available from: https://doi.org/10.1016/j.apsusc.2019.05.090.

[52] J.H. Roque-Ruiz, N.A. Medellín-Castillo, S.Y. Reyes-López, Fabrication of α-alumina fibers by sol-gel and electrospinning of aluminum nitrate precursor solutions, Results Phys. 12 (2019) 193−204. Available from: https://doi.org/10.1016/j.rinp.2018.11.068.

[53] P. Aghasiloo, M. Yousefzadeh, M. Latifi, R. Jose, Highly porous TiO_2 nanofibers by humid-electrospinning with enhanced photocatalytic properties, J. Alloys Compd. 790 (2019) 257−265. Available from: https://doi.org/10.1016/j.jallcom.2019.03.175.

[54] Y. Xie, L. Wang, B. Liu, L. Zhu, S. Shi, X. Wang, Flexible, controllable, and high-strength near-infrared reflective Y_2O_3 nanofiber membrane by electrospinning a polyacetylacetone-yttrium precursor, Mater. Des. 160 (2018) 918−925. Available from: https://doi.org/10.1016/j.matdes.2018.10.017.

[55] B. Naderi-Beni, A. Alizadeh, Development of a new sol-gel route for the preparation of aluminum oxynitride nano-powders, Ceram. Int. 46 (2020) 913−920. Available from: https://doi.org/10.1016/j.ceramint.2019.09.049.

[56] F.X. Perrin, F. Ziarelli, A. Dupuis, Relation between the corrosion resistance and the chemical structure of hybrid sol-gel coatings with interlinked inorganic-organic network, Prog. Org. Coat. 141 (2020). Available from: https://doi.org/10.1016/j.porgcoat.2019.105532.

[57] H. Song, Y. Ma, D. Ko, S. Jo, D.C. Hyun, C.S. Kim, et al., Influence of humidity for preparing sol-gel ZnO layer: characterization and optimization for optoelectronic device

applications, Appl. Surf. Sci. 512 (2020). Available from: https://doi.org/10.1016/j.apsusc.2020.145660.

[58] H. Ye, X. Zhang, Y. Zhang, L. Ye, B. Xiao, H. Lv, et al., Preparation of antireflective coatings with high transmittance and enhanced abrasion-resistance by a base/acid two-step catalyzed sol-gel process, Sol. Energy Mater. Sol. Cells 95 (2011) 2347−2351. Available from: https://doi.org/10.1016/j.solmat.2011.04.004.

[59] C. Agustín-Sáenz, M. Machado, A. Tercjak, Antireflective mesoporous silica coatings by optimization of water content in acid-catalyzed sol-gel method for application in glass covers of concentrated photovoltaic modules, J. Colloid Interface Sci. 534 (2019) 370−380. Available from: https://doi.org/10.1016/j.jcis.2018.09.043.

[60] S. Ambreen, K. Gupta, S. Singh, D.K. Gupta, S. Daniele, N.D. Pandey, et al., Synthesis and structural characterization of some titanium butoxides modified with chloroacetic acids, Transit. Met. Chem. 38 (2013) 835−841. Available from: https://doi.org/10.1007/s11243-013-9756-y.

[61] U. Schubert, Chemical modification of titanium alkoxides for sol-gel processing, J. Mater. Chem. 15 (2005) 3701−3715. Available from: https://doi.org/10.1039/b504269k.

[62] W. Charoenpinijkarn, M. Suwankruhagsn, B. Ksapabutr, S. Wongkasemjit, A.M. Jamieson, Sol-gel processing of silatranes, Eur. Polym. 37 (2001) 1441−1448. Available from: https://doi.org/10.1016/S0014-3057(00)00255-X.

[63] B. Ksapabutr, E. Gulari, S. Wongkasemjit, Sol-gel transition study and pyrolysis of alumina-based gels prepared from alumatrane precursor, Colloids Surf. A Physicochem. Eng. Asp. 233 (2004) 145−153. Available from: https://doi.org/10.1016/j.colsurfa.2003.11.019.

[64] M. Panapoy, B. Ksapabutr, Fabrication of zirconia nanofibers using zirconatrane synthesized by oxide one-pot process as precursor, Adv. Mat. Res. 55−57 (2008) 605−608. Available from: https://doi.org/10.4028/http://www.scientific.net/AMR.55-57.605.

[65] G. Poungchan, B. Ksapabutr, M. Panapoy, One-step synthesis of flower-like carbon-doped ZrO_2 for visible-light-responsive photocatalyst, Mater. Des. 89 (2016) 137−145. Available from: https://doi.org/10.1016/j.matdes.2015.09.136.

[66] M. Cargnello, T.R. Gordon, C.B. Murray, Solution-phase synthesis of titanium dioxide nanoparticles and nanocrystals, Chem. Rev. 114 (2014) 9319−9345. Available from: https://doi.org/10.1021/cr500170p.

[67] M. Panapoy, C. Duangdee, A. Laobuthee, B. Ksapabutr, Synthesis of a novel aminoalkoxide of iron by oxide one-pot process: Its sol-gel application to iron oxide powder, Songklanakarin, J. Sci. Technol. 31 (2009).

[68] H. Schäfer, S. Brandt, B. Milow, S. Ichilmann, M. Steinhart, L. Ratke, Zirconia-based aerogels via hydrolysis of salts and alkoxides: the influence of the synthesis procedures on the properties of the aerogels, Chem. Asian J. 8 (2013) 2211−2219. Available from: https://doi.org/10.1002/asia.201300488.

[69] J. Li, Y. Wu, M. Yang, Y. Yuan, W. Yin, Q. Peng, et al., Electrospun Fe_2O_3 nanotubes and Fe_3O_4 nanofibers by citric acid sol-gel method, J. Am. Ceram. Soc. 100 (2017) 5460−5470. Available from: https://doi.org/10.1111/jace.15164.

[70] B. Niu, F. Zhang, H. Ping, N. Li, J. Zhou, L. Lei, et al., Sol-gel autocombustion synthesis of nanocrystalline high-entropy alloys, Sci. Rep. 7 (2017) 1−7. Available from: https://doi.org/10.1038/s41598-017-03644-6.

[71] C. Wu, W. Yuan, S.S. Al-Deyab, K.Q. Zhang, Tuning porous silica nanofibers by colloid electrospinning for dye adsorption, Appl. Surf. Sci. 313 (2014) 389−395. Available from: https://doi.org/10.1016/j.apsusc.2014.06.002.

[72] N.A.M. Barakat, M.F. Abadir, F.A. Sheikh, M.A. Kanjwal, S.J. Park, H.Y. Kim, Polymeric nanofibers containing solid nanoparticles prepared by electrospinning and their applications, Chem. Eng. J. 156 (2010) 487−495. Available from: https://doi.org/10.1016/j.cej.2009.11.018.

[73] K. Ghosal, C. Agatemor, Z. Špitálsky, S. Thomas, E. Kny, Electrospinning tissue engineering and wound dressing scaffolds from polymer-titanium dioxide nanocomposites, Chem. Eng. J. 358 (2019) 1262−1278. Available from: https://doi.org/10.1016/j.cej.2018.10.117.

[74] A.D. Sekar, V. Kumar, H. Muthukumar, P. Gopinath, M. Matheswaran, Electrospinning of Fe-doped ZnO nanoparticles incorporated polyvinyl alcohol nanofibers for its antibacterial treatment and cytotoxic studies, Eur. Polym. J. 118 (2019) 27−35. Available from: https://doi.org/10.1016/j.eurpolymj.2019.05.038.

[75] Z. Zhang, Y. Wu, Z. Wang, X. Zou, Y. Zhao, L. Sun, Fabrication of silver nanoparticles embedded into polyvinyl alcohol (Ag/PVA) composite nanofibrous films through electrospinning for antibacterial and surface-enhanced Raman scattering (SERS) activities, Mater. Sci. Eng. C 69 (2016) 462−469. Available from: https://doi.org/10.1016/j.msec.2016.07.015.

[76] S. Jiang, W. He, K. Landfester, D. Crespy, S.E. Mylon, The structure of fibers produced by colloid-electrospinning depends on the aggregation state of particles in the electrospinning feed, Polymer 127 (2017) 101−105. Available from: https://doi.org/10.1016/j.polymer.2017.08.061.

[77] D. Crespy, K. Friedemann, A.M. Popa, Colloid-electrospinning: fabrication of multi-compartment nanofibers by the electrospinning of organic or/and inorganic dispersions and emulsions, Macromol. Rapid Commun. 33 (2012) 1978−1995. Available from: https://doi.org/10.1002/marc.201200549.

[78] R. Hongtong, P. Thanwisai, R. Yensano, J. Nash, S. Srilomsak, N. Meethong, Data on effect of electrospinning conditions on morphology and effect of heat-treatment temperature on the cycle and rate properties of core-shell LiFePO$_4$/FeS/C composite fibers for use as cathodes in Li-ion batteries, Data Brief 26 (2019) 1−5. Available from: https://doi.org/10.1016/j.dib.2019.104364.

[79] A. Dankeaw, G. Poungchan, M. Panapoy, B. Ksapabutr, In-situ one-step method for fabricating three-dimensional grass-like carbon-doped ZrO$_2$ films for room temperature alcohol and acetone sensors, Sens. Actuators B Chem. 242 (2017) 202−214. Available from: https://doi.org/10.1016/j.snb.2016.11.055.

[80] W. Wang, H. Wang, H. Wang, X. Jin, J. Li, Z. Zhu, Electrospinning preparation of a large surface area, hierarchically porous, and interconnected carbon nanofibrous network using polysulfone as a sacrificial polymer for high performance supercapacitors, RSC Adv. 8 (2018) 28480−28486. Available from: https://doi.org/10.1039/c8ra05957h.

[81] M. Tanveer, A. Habib, M.B. Khan, Structural and optical properties of electrospun ZnO nanofibres applied to P3HT:PCBM organic photovoltaic devices, J. Exp. Nanosci. 10 (2015) 640−650. Available from: https://doi.org/10.1080/17458080.2013.869841.

[82] S. Lee, J. Ha, J. Choi, T. Song, J.W. Lee, U. Paik, 3D cross-linked nanoweb architecture of binder-free TiO$_2$ electrodes for lithium ion batteries, ACS Appl. Mater. Interfaces 5 (2013) 11525−11529. Available from: https://doi.org/10.1021/am404082h.

[83] H. Wu, L. Hu, M.W. Rowell, D. Kong, J.J. Cha, J.R. McDonough, et al., Electrospun metal nanofiber webs as high-performance transparent electrode, Nano Lett. 10 (2010) 4242−4248. Available from: https://doi.org/10.1021/nl102725k.

[84] A. Choi, J. Park, J. Kang, O. Jonas, D. woo Kim, H. Kim, et al., Surface characterization and investigation on antibacterial activity of CuZn nanofibers prepared by

electrospinning, Appl. Surf. Sci. 508 (2020). Available from: https://doi.org/10.1016/j.apsusc.2019.144883.

[85] Y. Shen, Y. Wei, J. Ma, Q. Li, J. Li, W. Shao, et al., Tunable microwave absorption properties of nickel-carbon nanofibers prepared by electrospinning, Ceram. Int. 45 (2019) 3313–3324. Available from: https://doi.org/10.1016/j.ceramint.2018.10.242.

[86] T.K.-C. Savva Ioanna, Encroachment of traditional electrospinning, Roy. Soc. Chem. (2014). Available from: https://doi.org/10.1039/9781788011006-00024.

[87] M. Wojasiński, J. Goławski, T. Ciach, Blow-assisted multi-jet electrospinning of poly-L-lactic acid nanofibers, J. Polym. Res. 24 (2017). Available from: https://doi.org/10.1007/s10965-017-1233-4.

[88] X. Li, Z. Li, L. Wang, G. Ma, F. Meng, R.H. Pritchard, et al., Low-voltage continuous electrospinning patterning, ACS Appl. Mater. Interfaces 8 (2016) 32120–32131. Available from: https://doi.org/10.1021/acsami.6b07797.

[89] H. Niu, T. Lin, Fiber generators in needleless electrospinning, J. Nanomater. 2012 (2012). Available from: https://doi.org/10.1155/2012/725950.

[90] J. Xiong, Y. Liu, A. Li, L. Wei, L. Wang, X. Qin, et al., Mass production of high-quality nanofibers via constructing pre-Taylor cones with high curvature on needleless electrospinning, Mater. Des. 197 (2021) 109247. Available from: https://doi.org/10.1016/j.matdes.20200.109247.

[91] L. Wei, R. Sun, C. Liu, J. Xiong, X. Qin, Mass production of nanofibers from needle-less electrospinning by a novel annular spinneret, Mater. Des. 179 (2019). Available from: https://doi.org/10.1016/j.matdes.2019.107885.

[92] Z. Zhu, P. Wu, Z. Wang, G. Xu, H. Wang, X. Chen, et al., Optimization of electric field uniformity of multi-needle electrospinning nozzle, AIP Adv. 9 (2019). Available from: https://doi.org/10.1063/1.5111936.

[93] G. Xu, X. Chen, Z. Zhu, P. Wu, H. Wang, X. Chen, et al., Pulse gas-assisted multi-needle electrospinning of nanofibers, Adv. Compos. Hybrid Mater. 3 (2020) 98–113. Available from: https://doi.org/10.1007/s42114-019-00129-0.

[94] S. Xie, Y. Zeng, Effects of electric field on multineedle electrospinning: experiment and simulation study, Ind. Eng. Chem. Res. 51 (2012) 5336–5345. Available from: https://doi.org/10.1021/ie2020763.

[95] L. Tian, C. Zhao, J. li, Z. Pan, Multi-needle, electrospun, nanofiber filaments: effects of the needle arrangement on the nanofiber alignment degree and electrostatic field distribution, Text. Res. J. 85 (2015) 621–631. Available from: https://doi.org/10.1177/0040517514549990.

[96] J. Xue, T. Wu, Y. Dai, Y. Xia, Electrospinning and electrospun nanofibers: methods, materials, and applications, Chem. Rev. 119 (2019) 5298–5415. Available from: https://doi.org/10.1021/acs.chemrev.8b00593.

[97] X. Yan, M. Yu, S. Ramakrishna, S.J. Russell, Y.Z. Long, Advances in portable electrospinning devices for: in situ delivery of personalized wound care, Nanoscale 11 (2019) 19166–19178. Available from: https://doi.org/10.1039/c9nr02802a.

[98] C.J. Luo, S.D. Stoyanov, E. Stride, E. Pelan, M. Edirisinghe, Electrospinning vs fibre production methods: from specifics to technological convergence, Chem. Soc. Rev. 41 (2012) 4708–4735. Available from: https://doi.org/10.1039/c2cs35083a.

[99] M. Wortmann, N. Frese, L. Sabantina, R. Petkau, F. Kinzel, A. Gölzhäuser, et al., New polymers for needleless electrospinning from low-toxic solvents, Nanomaterials 9 (2019). Available from: https://doi.org/10.3390/nano9010052.

[100] R. Matthes, J.H. Bernhardt, A.F. McKinlay, International Commission on Non-Ionizing Radiation Protection, Guidelines on limiting exposure to non-ionizing radiation: a reference book based on the guidelines on limiting exposure to non-ionizing radiation and statements on special applications, 1999.

[101] A.S. Safigianni, A.I. Spyridopoulos, V.L. Kanas, Electric and magnetic field measurements in a high voltage center, 2011, in: 10th International Conference on Environment and Electrical Engineering, EEEIC. EU 2011—Conference Proceedings, 2011, pp. 1−4. https://doi.org/10.1109/EEEIC.2011.5874771.

[102] C. Kocatepe, O. Arikan, C.F. Kumru, A. Erduman, N. Umurkan, Electric field measurement and analysis around a line model at different voltage levels, in: ICHVE 2012−2012 International Conference on High Voltage Engineering and Application, 2012, pp. 39−42. https://doi.org/10.1109/ICHVE.2012.6357005.

[103] F. Hormot, J. Bacmaga, A. Baric, Infrared protection system for high-voltage testing of SiC and GaN FETs used in DC-DC converters, in: 2016 39th International Convention on Information and Communication Technology, Electronics and Microelectronics, MIPRO 2016-Proceedings, 2016, pp. 72−75. https://doi.org/10.1109/MIPRO.2016.7522113.

[104] L. Bi, X. Wei, Z. Sun, A high-voltage safety protection method for electric vehicle based on FPGA, in: 2006 IEEE International Conference on Vehicular Electronics and Safety, ICVES, 2006, pp. 26−31. https://doi.org/10.1109/ICVES.2006.371547.

[105] C. Zang, Z. Xinjie, H. Shuang, H. Lei, J. Zhenglong, Y. Huisheng, et al., Research on mechanism and ultraviolet imaging of corona discharge of electric device faults, in: Conference Record of IEEE International Symposium on Electrical Insulation, vol. 3, 2008, pp. 690−693. https://doi.org/10.1109/ELINSL.2008.4570424.

[106] H.U. Shin, A.B. Stefaniak, N. Stojilovic, G.G. Chase, Comparative dissolution of electrospun Al_2O_3 nanofibres in artificial human lung fluids, Environ. Sci. Nano. 2 (2015) 251−261. Available from: https://doi.org/10.1039/c5en00033e.

[107] G. Oberdörster, V. Castranova, B. Asgharian, P. Sayre, Inhalation exposure to carbon nanotubes (CNT) and carbon nanofibers (CNF): methodology and dosimetry, J. Toxicol. Environ. Health B Crit. Rev. 18 (2015) 121−212. Available from: https://doi.org/10.1080/10937404.2015.1051611.

[108] M. Kucki, J.-P. Kaiser, M.J.D. Clift, B. Rothen-Rutishauser, A. Petri-Fink, P. Wick, The role of the protein corona in fiber structure-activity relationships, Fibers 2 (2014) 187−210. Available from: https://doi.org/10.3390/fib2030187.

[109] A. Schinwald, F.A. Murphy, A. Prina-Mello, C.A. Poland, F. Byrne, D. Movia, et al., The threshold length for fiber-induced acute pleural inflammation: shedding light on the early events in asbestos-induced mesothelioma, Toxicol. Sci. 128 (2012) 461−470. Available from: https://doi.org/10.1093/toxsci/kfs171.

[110] F. Mustafa, S. Andreescu, Nanotechnology-based approaches for food sensing and packaging applications, RSC Adv. 10 (2020) 19309−19336. Available from: https://doi.org/10.1039/d0ra01084g.

[111] A. Genaidy, T. Tolaymat, R. Sequeira, M. Rinder, D. Dionysiou, Health effects of exposure to carbon nanofibers: systematic review, critical appraisal, meta analysis and research to practice perspectives, Sci. Total Environ. 407 (2009) 3686−3701. Available from: https://doi.org/10.1016/j.scitotenv.2008.12.025.

[112] A. Barhoum, K. Pal, H. Rahier, H. Uludag, I.S. Kim, M. Bechelany, Nanofibers as new-generation materials: from spinning and nano-spinning fabrication techniques to emerging applications, Appl. Mater. Today 17 (2019) 1−35. Available from: https://doi.org/10.1016/j.apmt.2019.06.015.

[113] S. Luanpitpong, L. Wang, D.C. Davidson, H. Riedel, Y. Rojanasakul, Carcinogenic potential of high aspect ratio carbon nanomaterials, Environ. Sci. Nano. 3 (2016) 483−493. Available from: https://doi.org/10.1039/c5en00238a.

[114] A. Maynard, R. Aitken, T. Butz, V. Colvin, K. Donaldson, G. Oberdörster, et al., Safe handling of nanotechnology, Nature 444 (2006) 267−269. Available from: https://doi.org/10.1038/444267a.

Special techniques and advanced structures

2

Mingyu Tang[1], Suting Liu[1], Zhihui Li[1], Xiaodi Zhang[2,3], Zhao Wang[2,3], Yunqian Dai[1,4], Yueming Sun[1], Liqun Zhang[5] and Jiajia Xue[2,4]

[1]School of Chemistry and Chemical Engineering, Southeast University, Nanjing, P.R. China, [2]Beijing Laboratory of Biomedical Materials, Beijing University of Chemical Technology, Beijing, P.R. China, [3]Center of Advanced Elastomer Materials, Beijing University of Chemical Technology, Beijing, P.R. China, [4]Center for Flexible RF Technology, Southeast University, Purple Mountain Laboratory, Nanjing, P.R. China, [5]The Key Laboratory of Beijing City on Preparation and Processing of Novel Polymer Materials, Beijing University of Chemical Technology, Beijing, P.R. China

2.1 Introduction

Metal oxide materials with excellent properties are promising materials for various applications in extreme environments because of their high melting point, good stability, and excellent mechanical properties at elevated temperatures [1−3]. By constructing metal oxide materials into nanofibers, they can be applied in more abundant ways. Characterized by the inoxidizability, high temperature stability, high tensile strength, corrosion resistance, and electrical insulation, metal oxide nanofibers can be used as protective and insulative materials for exploration in metallurgy, machinery, aerospace, petroleum, the chemical industry, and others [4].

One or more dimensions in size reduction have been considered an effective means to obtain materials with unique characteristics when compared with their bulk materials, such as the low density of defects and size effect [5,6]. In general, the nanofiber has distinctive characteristics, such as a controllable structure, a large specific surface area, high mechanical strength, and high porosities. With the fast development of nanotechnology, electrospinning has been considered one of the most direct and facile technologies for generating nanostructures. As a versatile and viable technique to fabricate nanomaterials with diameters varying from tens of nanometers to micrometers, electrospinning has been remarkably developed to generate and engineer nanofibers for use in various applications [7]. The remarkable advantages of electrospinning are involved with the facile process, high adaptability, and worth in mass production, making it attractive in diverse application fields [8−10].

The principle of electrospinning is to induce electrostatic charges on the molecules of the electrospinning solution in an electric field. The self-repulsion force causes the solution to stretch into fibers. If no fracture occurs in the ejected solution, the solvent tends to evaporate, leading to the formation of a single continuous

Metal Oxide-Based Nanofibers and Their Applications. DOI: https://doi.org/10.1016/B978-0-12-820629-4.00016-3
© 2022 Elsevier Inc. All rights reserved.

fiber [11,12]. Over the past decades, much attention has been put into the research of electrospun nanofibers, especially for regulating their surface and internal structure. A rich variety of materials have been utilized to fabricate electrospun nanofibers, including organic polymers, colloidal particles, small molecules, and composites. Additionally, by doping nanoscale components with varying dimensions and/or morphologies (e.g., nanoparticles, nanowires, nanorods, nanotubes, nanosheets) into polymer solutions, the resultant mixtures are also available for electrospinning [13]. In addition, electrospun nanofibers have been endowed with unique properties, and the scalable manufacturing of nanofibers is important for future applications.

Metal oxide with one-dimensional nanoarchitectures can provide a good material system to investigate the dependence of electrical, optical, thermal, and mechanical properties on dimensionality and size reduction. The development and application of metal oxide nanofibers have been among the most interesting fields in nanoscience. The production equipment and conditions are not particularly strict to fabricate metal oxide nanofibers via electrospinning. Numerous metal oxide nanofibers, such as nanofibers made of NiO, Co_3O_4, and a mixture of ZnO and Al_2O_3, have all been processed via the sol-gel method during electrospinning [14,15]. The morphologies and diameters of the nanofibers can be simply modified by adjusting the processing parameters. By regulating the raw materials and techniques, the structure of the nanofibers can be prepared with a special function [16]. The nanofibers can be engineered with different structures, such as core-sheath and hollow fibers, or with hierarchical and side-by-side morphologies. These structures result in the enhancement of the specific surface area of the nanofibers, resulting in the improvement of the properties. They have a good prospect in a wide range of fields, including medical, catalyst, and textiles. Apart from controlling the macroscopic structure to gain traditional smooth surface solid nanofibers, nanofibers with complex structures can also be permitted during the electrospinning process [17]. The properties and application fields of the electrospun metal oxide nanofibers can be effectively improved and broadened by combining them with other modification pathways.

Further functionalization of the internal and/or external surfaces of the nanofibers can be achieved by pretreatment or aftertreatment of the precursors before or after the electrospinning process. In addition, diverse secondary structures, such as core-sheath, hollow, and side-by-side morphologies, have been endowed into the metal oxide nanofibers [18]. The electrospun metal oxide nanofibers can also be fitted into ordered arrays by operating the patterning and/or alignment process [19,20]. Introducing pores into the nanofibers makes a lot of sense because of the effective enhancement in the specific surface area of the sample, thus improving the performance of electrospun nanofibers in various applications [21−23]. It is also possible to provide strong connections for the nonwoven mat through effective welding among the nanofibers [24]. Meanwhile, because of the excellent structural flexibility and high porosity of the monolithic structure, a three-dimensional (3D) scaffold based on the electrospun metal oxide nanofibers has aroused great interest [20,25]. Mass production of electrospun metal oxide nanofibers can be reached with the development of electrospinning facilities [26].

2.2 Electrospinning for producing metal oxide nanofibers

To date, with their high porosity, large surface area, and superb thermal stability, metal oxides with a fibrous morphology (i.e., metal oxide nanofibers) have been widely used as chemical catalysts, [27] supercapacitors, [28] fuel cells, [29] energy storage devices [30], and so on. Several bottom-up and top-down approaches are utilized to synthesize and fabricate metal oxide nanofibers, for instance, vapor-liquid-solid, [31] nanocarving [32], and electrospinning. Among them, electrospinning can be considered one of the most adaptable techniques because it is accessible to easily control the composition, diameter, morphology, and orientation of the resultant metal oxide nanofibers. Electrospinning is an electrodynamic process in which an electrospinning solution is charged to produce a jet and then stretched and elongated to produce thin fibers. A setup for electrospinning is illustrated in Fig. 2.1A. The basic components consist of a high-voltage power supply, a spinneret, a syringe pump, and a conductive collector [35−37]. Metal oxide nanofibers can be fabricated by directly electrospinning from the precursor and by selectively

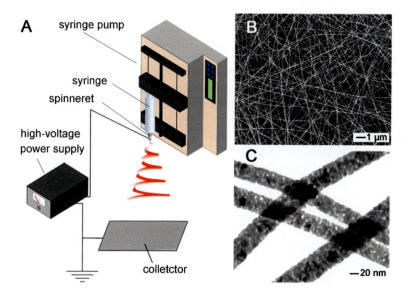

Figure 2.1 (A) Schematic illustration of a typical setup for electrospinning. (B) SEM image of TiO$_2$/PVP composite nanofibers that were electrospun from an ethanol solution containing Ti(OiPr)$_4$ and PVP. (C) TEM image of the TiO$_2$/PVP composite nanofibers after calcined in air at 500°C for 3 h.
Source: Reprinted with permission from (A) J. Xue, et al., Electrospun nanofibers: new concepts, materials, and applications, Acc. Chem. Res. 50 (8) (2017) 1976−1987 [33]. Copyright 2017 American Chemical Society; (B and C) D. Li, Y. Xia, Fabrication of titania nanofibers by electrospinning, Nano Lett., 2003. 3(4): p. 555−560 [34]. Copyright 2003 American Chemical Society.

removing the sacrificial polymer components from the suitable composite fibers. The amorphous structure can also be converted to the crystalline structure through modification methods [33].

2.2.1 Directly electrospinning of the precursor solution

Metal oxide nanofibers can be easily prepared by directly electrospinning the sol-gel precursors composed of an aged inorganic precursor (metal salts or metal alkoxides), a polymer, and a solvent. Metal alkoxides or metal salts are the commonly used inorganic precursors. However, they are easy to hydrolyze and condense, which can result in an unstable electrospinning process by blocking the spinneret. Therefore in preparing a sol-gel precursor solution, it is necessary to avoid the occurrence of sol-gel reactions in the stock solution, including hydrolysis, condensation, and gelation of the precursor. Instead, these reactions should be initiated in the jet by contacting with the surrounding air during the electrospinning process [34,38].

The electrospinnability of the prepared solution depends on the nature of the sol-gel precursor and carrier polymer and on the conductivity and viscosity of the solution [13]. The carrier polymer should be spinnable and have a high molecular weight or a considerable degree of chain entanglement [39]. In a typical example, polyvinyl pyrrolidone (PVP) is one of the most fashionable polymers acted as a matrix featured with high solubility in ethanol and water and good compatibility with metal alkoxides or metal salts. Other polyvinyl alcohol (PVA), polyacrylonitrile (PAN), polymethyl methacrylate, polyvinyl acetate (PVAc), and polyacrylic acid have also been widely used as the carrier polymer for the fabrication of metal oxide nanofibers [34,38,40].

Besides introducing a polymer into the solution as a matrix to adjust the rheological properties, additive such as a catalyst (e.g., acetic acid, hydrochloric acid, propionic acid), can also be added to the solution to facilitate the electrospinning process by stabilizing the precursor and enhancing the electrospinnability [41,42]. The additive can be used as a catalyst for mitigating both the hydrolysis and gelation rates to prevent the blocking of the spinneret for improving the stability of the electrospinning process. As for the solvent, a relatively volatile solvent is often used, such as ethanol, isopropanol, water, dimethylformamide, or chloroform, to enhance the electrospinnability.

During the electrospinning process, as long as the spinneret ejects a liquid jet, solvent evaporation, hydrolysis, condensation, and gelation of the precursor will occur simultaneously in the air. Taking TiO_2 nanofibers as an example, PVP and titanium tetraisopropoxide ($Ti(OiPr)_4$, a precursor to TiO_2) were dissolved in ethanol to obtain the electrospinning precursor solution [34]. Upon electrospinning, the as-spun TiO_2/PVP composite nanofibers displayed a uniform morphology with a smooth surface and exhibited a high length-to-diameter ratio, as shown in Fig. 2.1B [34]. By carefully engineering the electrospinning parameters (e.g., the viscosity of the precursor solution, the voltage value, the distance between spinneret and collector, and the spinning rate), which can largely affect the formation and the diameters of the nanofibers, the desired inorganic/polymer composite nanofibers can be obtained [38]. Adding

Special techniques and advanced structures

salts, for example, sodium chloride or tetramethylammonium chloride, into the electrospinning solution is an available strategy to raise the charge density on the liquid jet and eliminate the formation of beads. Good control of the environment around the jet is also very important. In general, an atmosphere with low relative humidity and/or solvent vapor saturation can significantly reduce the rate of hydrolysis and gelation, thereby promoting a continuous electrospinning process [43].

2.2.2 Selectively removing the polymer component in the composite nanofibers

Metal oxide nanofibers can be derived by removing the polymeric component in composite nanofibers composed of a polymer and an inorganic precursor through calcination. The as-spun inorganic/polymer composite nanofibers undergo calcination or solvent extraction to selectively remove the polymer component. The calcination process is based on two aspects: (1) the removal of the polymeric phase by burning at high temperature in the presence of oxygen and (2) the oxidative conversion of the precursor component to produce the metal oxide by high-temperature nucleation and growth. During the calcination process, the polymer component, such as PVP (as a polymer matrix), was selectively burned out while the nanofibers evolved into polycrystalline anatase (or rutile) (when $Ti(OiPr)_4$ was used as the precursor of TiO_2) with a relatively rough surface and intriguing porous structure [32]. Owing to the removal of the polymer matrix and sintering of the metal oxide phase, the diameter of the nanofibers is often reduced. Taking the fabrication of Al_2O_3 nanofibers as an example, the as-spun composite nanofibers exhibited a smooth surface with a high length-to-diameter ratio [44]. After calcination at 350°C for 2 h in air, the average diameter of the nanofibers was reduced to about 290 nm, and the surface of the nanofibers became porous. Additionally, the metal oxide nanofibers consisting of tiny nanocrystals generally suffer from severe sintering (growth into larger nanocrystals), leaving a bamboo-like morphology, which is considered as one of the major causes of the brittleness of metal oxide nanofibers [45]. Subsequently, as a result of calcination, the cross section of the nanofibers becomes thinner, and the shrinkage and densification of the nanocrystals generate thermal stress and internal mechanical stress; thus the nanofibers usually become more brittle. For example, as-spun TiO_2/PVP composite nanofibers were converted into TiO_2 nanofibers by calcining in air at 500°C for 3 h [34]. The resultant TiO_2 nanofibers have an average diameter of 53 ± 8 nm, which could be ascribed to the loss of PVP and the densification of TiO_2 (Fig. 2.1C). The resultant anatase nanofibers are composed of fused TiO_2 nanoparticles around 10 nm in diameter and include voids among adjacent nanoparticles.

The diameter of the metal oxide nanofibers is highly dependent on the diameter and composition of the composite nanofibers, which can be controlled by adjusting the electrospinning parameters and the calcination conditions. As exemplified by TiO_2 nanofibers, the resultant nanofibers with thinner diameter were prepared from composite nanofibers with smaller diameters [21]. Adding a low percentage of alkoxide precursor and using a higher calcination temperature also lead to the formation of thinner metal oxide nanofibers because of the removal of a large amount of polymer [34]. Note that the composition phase and surface roughness of the as-obtained metal oxide nanofibers can be controlled by adjusting the temperature and/or calcination time. Except for calcination temperature and time, the calcination atmosphere is also of great importance. Calcination in air usually results in forming oxide nanofibers, while thermal treatment under N_2 or NH_3 can produce nitride nanofibers, owing to the nitridation of the metal oxide particles. Therefore the atmosphere needs to be carefully selected and controlled to ensure the formation of metal oxide nanofibers. By carefully selecting the composite nanofibers and engineering the calcination conditions, more than 100 different types of metal oxides (e.g., CeO_2, SnO_2, CuO, Fe_2O_3, SiO_2, V_2O_5, $BaTiO_3$, $ZnCo_2O_4$, $LiNi_{0.5}Mn_{1.5}O_4$) have been successfully prepared as nanofibers.

Interestingly, by carefully engineering the electrospinning parameters and calcination process, the obtained metal oxide nanofibers can exhibit different morphologies. For example, by taking advantage of the diameter-dependent morphology of as-spun composite $Ce(acac)_3$/PVP nanofibers, columnar, celery-like, or beltlike porous CeO_2 nanofibers were generated by simply increasing the diameter of as-spun composite nanofibers (Fig. 2.2A–F) [46]. These CeO_2 nanofibers exhibited good thermal stability in terms of special phase structures and morphologies up to 700°C. In addition to selectively removing the polymer component by calcination, extraction is another approach that is commonly applied to produce hollow nanofibers. For example, ultrathin CeO_2-based nanofibers (Al_2O_3/CeO_2) with the elegant fibril-in-tube structure were generated via single-spinneret-based electrospinning followed by one-step calcination at 500°C (Fig. 2.2G and H) [47,48]. The intriguing fibril-in-tube structure within Al_2O_3/CeO_2 nanofibers was achieved by carefully selecting two different types of metal precursors with different decomposition rates upon calcination. During the calcination process, the use of $Al(acac)_3$ as Al_2O_3 precursor, which could rapidly release gaseous pieces and lead to the growth kinetics varied along the radial direction of nanofibers, was critical to the formation of the fibril-in-tube structure.

Except for removing the polymer component by calcination, extraction is another approach that is commonly applied to obtain hollow nanofibers. A coaxial electrospinning method is often used to generate hollow metal oxide nanofibers with the use of a sacrificial template. The sacrificial component is usually applied as the inner fluid, while the polymer matrix for the hollow fibers serves as the outer fluid to generate core-sheath nanofibers. Followed by the removal of the core component and the organic component from the as-spun nanofibers through calcination, hollow metal oxide nanofibers can thus be produced [44]. In a typical example, mineral oil was applied as the core, owing to its good rheological properties, while a conventional spinnable sol-gel solution is used for the sheath [38] The mineral phase was then removed by extracting with a solvent such as an octane or by burning out at an elevated temperature to generate hollow metal oxide nanofibers.

Metal oxide nanofibers can also be generated by depositing metals on the surface of as-spun polymer nanofibers followed by the selective removal of the polymer

Special techniques and advanced structures 37

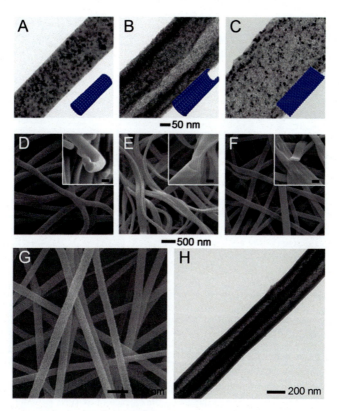

Figure 2.2 (A–F) TEM (*top row*) and SEM (*bottom row*) images of fibrous CeO_2 nanofibers made of (A, D) columnar-like, (B, E) celery-like, and (C, F) beltlike nanofibers, respectively. The insets in A–C are the schematics of nanofibers models, while those in D–F are the corresponding cross-sectional SEM images, respectively. (G) SEM and (H) TEM images of Al_2O_3/CeO_2 nanofibers.
Source: Reprinted with permission from (A–F) Y. Dai, J. Tian, W. Fu, Shape manipulation of porous CeO2 nanofibers: facile fabrication, growth mechanism and catalytic elimination of soot particulates, J. Mater. Sci. 54 (14) (2019) 10141–10152 [46]. Copyright 2019 Springer; (G and H) S. Liu, et al., Constructing fibril-in-tube structures in ultrathin CeO_2-based nanofibers as the ideal support for stabilizing Pt nanoparticles, Mater. Today Chem. 17 (2020) 100333 [47]. Copyright 2020 Elsevier.

template. In one demonstration, as shown in Fig. 2.3A, [49] an as-spun free-standing polymer nanofiber network was coated with a thin layer of metal oxide by using standard thin-film deposition techniques, such as thermal evaporation, electron-beam evaporation, and magnetron sputtering [50]. Subsequently, the polymer templates were removed after dissolving in suitable solvents. The resultant metal oxide nanofibers usually have a hollow-shaped cross section. Fig. 2.3B and C, show the SEM images of a network made of intertwined indium tin oxide nanofibers that were fabricated by using this method [49].

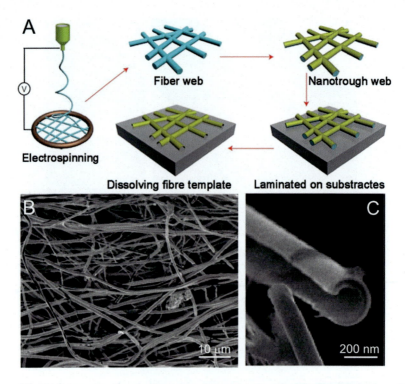

Figure 2.3 (A) Schematic of an indirect, template-assisted method for the fabrication of metal oxide nanofibers by depositing metals on the surface of as-spun polymer nanofibers followed by selective removal of the polymer template. (B, C) SEM images of indium tin oxide nanofibers.
Source: Reprinted with permission from H. Wu, et al., A transparent electrode based on a metal nanotrough network, Nat. Nanotechnol. 8 (6) (2013) 421–425 [49]. Copyright 2013 Springer Nature.

2.2.3 Converting amorphous to crystalline structure

Tailoring the morphological, physical, and chemical properties of metal oxides is a crucial route to optimize their performance for various applications [51]. The crystalline structures of the metal oxide nanofibers are usually affected by the calcination parameters and the type of polymer matrix used for fabricating the composite nanofibers. In an interesting report, the resultant crystalline phases of vanadium oxide samples were entirely different after calcining at different heating rates [52]. The optimized calcination heating rate and temperature enabled modification of the V^{4+}/V^{5+} ratio as well as the crystallite size and crystalline phase of the generated nanofibers. The size of the crystallites was gradually increased with the enhancement of the calcination temperature, guaranteeing an achievement of the similar crystalline phase. In another study, as-prepared ZrO_2 nanofibers after calcination in optimized conditions were found to preserve the cubic ordered porous arrangement after transforming from amorphous to crystalline phases [53]. The residual titanium tetraisopropoxide alkoxy groups stemming from incomplete hydrolysis behaved as structural impurities, which could hinder the TiO_2 crystallization, and the calcination at high temperature could effectively transform TiO_2 from the amorphous to the crystalline phase [54].

The type and concentration of the polymer matrix used for the preparation of metal oxide nanofibers also affect the crystalline structure and thus the mechanical properties of the obtained metal oxide nanofibers [55]. For example, Al_2O_3 nanofibers were fabricated by using aluminum isopropoxide and aluminum nitrate via electrospinning combined with the sol-gel method [56]. The crystallization temperatures of the nanofibers prepared by using PVP, polyvinyl butyral, and PVA as the polymer templates were completely different. In another study, TiO_2 nanofibers prepared by electrospinning of a mixture of PVP and titanium(IV) bis (ammonium lactate) dihydroxide were studied by using different concentrations of the precursors [57]. The as-spun composite fibers were annealed at 600°C for the removal of the PVP components to obtain the TiO_2 nanofibers. The annealing step of titanium (IV) was crucial to achieving the crystalline TiO_2.

2.2.4 Physical and chemical modifications

The structure and property of metal oxide nanofibers can be modified by regulating the parameters during electrospinning and/or by postprocessing modifications of the as-spun nanofibers, mainly through physical or chemical methods [58,59]. For the physical method, direct blending in the electrospinning polymer solution and coaxial electrospinning are the two most often used methods to incorporate functional components into the polymer matrix [60]. Ligand molecules can also be embedded in the bulk material. During the electrospinning process, the addition of ligands can result in greater diameters with the enhancement of the viscosity [61−63]. Chemical modification can be achieved by chemical reactions between selective ligands and the functional groups on the as-spun nanofibers. In this case, targeted ligands can be combined onto the nanofibers, offering multiple possibilities and applications in separation, adsorption, and smart-responsive surface [30,64]. Through chemical treatments, suitable functional groups can be modified onto the surface for ligand binding. Crosslinking is a commonly used method to modify the electrospun nanofibers. The resulting fibers with an acceptable mechanical strength can be easily obtained [65]. The plasma treatment is well known for the physical and chemical modifications of electrospun nanofibers. In one study, the plasma treatment was used to maintain the morphology of polyethylene terephthalate nanofibers [66]. By using plasma treatment, the coaxial fiber was found to preserve two components with a ceramic alumina surface and a polymer component inside. The polymer core can maintain all the advantages of ceramic alumina fibers while keeping the greater flexibility of the coaxial fibers [67].

The morphology of the electrospun metal oxide nanofibers can be significantly influenced by hydrolysis and condensation rates, which can also be affected by the thermal treatment parameters. By regulating the calcination temperatures and durations, various morphologies with different properties can be obtained. For example, the morphology of the nanofibers was changed from a flexible smooth surface by the calcinating temperature at 700°C, got into porous nanograin structure at 900°C, and nonporous coarser nanograins structure at 1100°C. Eventually, elongated submicrometer grains were formed by calcinating at 1300°C−1450°C [68]. The thermal treatment has also been employed for the modification of metal oxide nanofibers. The morphology and the structure of the nanofibers are largely affected by the conditions during the thermal treatment. In one study, TiO_2 nanofibers with an average diameter of 400−500 nm were fabricated by heat treatment at 700°C [69]. The surface of the obtained nanofibers was smooth without obvious defects, and the crystal phase of TiO_2 fibers was anatase. For high-quality TiO_2 nanofibers, by adjusting the molar ratio of acetylacetone to Ti, the precursor of polyacetylacetone titanium with good stability and good spinnability was optimized. The process was also favored for the low thermal conductivity, which guaranteed the high-temperature insulation fields' practical applications. In another study, Al_2O_3-SiO_2 composite nanofibers with a diameter of about 300 nm were obtained by combining the electrospinning with sol-gel process followed by thermally heating the precursor nanofibers at 1300°C [70]. The continuous and uniform structures of precursor nanofibers were obtained by calcinating the PVA and γ-Al_2O_3 phase at 878°C, while a mullite phase was formed at 1322°C upon the thermal treatment of the precursor fibers in an air atmosphere. The homogeneity and diphasic features of nanofibers were confirmed because the Al and Si elements were uniformly distributed in fibers and mixed at nanoscale [11]. In one study, yttrium silicate fiber was successfully prepared and further modified by the titanium silicide to improve the densification degree of the fiber, and different morphologies were obtained by using different proportions and temperatures [71]. By employing suitable chemical reactions, the functional nanofibers can be readily obtained by introducing the functional groups into the polymer matrix during the postprocessing method [72]. As one of the postfunctionalization approaches, amide can be covalently bonding onto the surface of the nanofibers [72,73]. Various methods, such as surface grafting, are used to modify the surface of electrospun metal oxide nanofibers [74].

2.3 Advanced structures of metal oxide nanofibers

2.3.1 Control of core-sheath, hollow, or side-by-side morphology

Electrospinning provides a simple and feasible approach to establish secondary structures in an individual metal oxide nanofiber with high uniformity. Through manipulating the setup for electrospinning, such as the use of a well-designed spinneret, metal oxide nanofibers with various secondary structures in terms of core-sheath, hollow, and side-by-side morphologies can be directly generated or constructed followed by typical postmodification methods.

Metal oxide nanofibers with a core-sheath structure can be fabricated by introducing phase separation between two components, by directly coaxial electrospinning, or by decorating metal oxide on the surface of nanofibers [36]. By introducing phase separation between two components in a homogeneous precursor, a conventional electrospinning setup based on a single spinneret can be used to prepare core-sheath nanofibers [75−77]. As one of the most widely used method, coaxial electrospinning is also applied to produce metal oxide nanofibers with core-sheath structures [78]. Typically, a coaxial spinneret is often used to allow for the introduction of two electrospinning solutions as the outer and inner fluids, respectively, as illustrated in Fig. 2.4A. The basic requirement for a successful coaxial electrospinning process is to guarantee that the inner and outer fluids can form a compound jet and meantime stay together in a concentric manner, as illustrated in Fig. 2.4B [79,84]. In a representative demonstration, highly uniform SnO_2-TiO_2 core-sheath nanofibers with a tunable internal morphology were successfully prepared in one step by coaxial electrospinning [85]. The productivity of core-sheath nanofibers can be enhanced by air blowing−assisted coaxial electrospinning [86]. A core-sheath structure can also be formed by decorating metal oxide on the surface of electrospun nanofibers. For example, TiO_2-$Bi_4Ti_3O_{12}$-MoS_2 three-layered core-sheath nanofibers were constructed by using this method [87]. TiO_2-$Bi_4Ti_3O_{12}$ core-sheath nanofibers were generated by first partially converting TiO_2 to a highly crystallized $Bi_4Ti_3O_{12}$ sheath at $500°C$ via a solid-state reaction using BiOI nanoplates, accompanied by phase transformation of TiO_2 from anatase to rutile to a certain degree. Then several layers of thick MoS_2 nanosheets were evenly modified on the surface of TiO_2-$Bi_4Ti_3O_{12}$ fiber to form three-layered $TiO_2/Bi_4Ti_3O_{12}/MoS_2$ core-shell structures [87].

Hollow nanofibers have gained extensive attention owing to their unique catalytic, electrical and electrochemical properties. Nowadays, there are two main strategies used in fabricating hollow metal oxide nanofibers [88]. The first method is the template-assistant approach, which uses electrospun nanofibers as a sacrificial template to coat the required precursors. Additional removal of polymer cores upon high temperatures is required to generate a representative hollow structure. Multifluid coaxial electrospinning technology is commonly used to prepare ultrathin hollow nanofibers with unique nanotubular or microtubular structures, as demonstrated in Fig. 2.4C [80]. For example, a spinneret containing three coaxial capillaries was used, in which a chemically inert fluid acted as a buffer between the internal and external polymer solutions. Then the liquid in the buffer was removed, resulting in a void between the internal solid nanofibers and the external solid microtubules. Hollow nanofibers based on metal oxide can also be fabricated by heating as-spun nanofibers under appropriate conditions. In this approach, the concentrations of precursor salts located on the surface and inside of the nanofibers are usually different. As the temperature of the heat treatment increases, the concentration gradient drives the precursor salt migrating from the interior of the nanofibers to the surface, ultimately leading to the formation of hollow nanofibers. For instance, by directly calcinating as-spun composite nanofibers composed of a blend of PVP and $Fe(NO_3)_3$, Fe_2O_3 hollow nanofibers were successfully prepared [89].

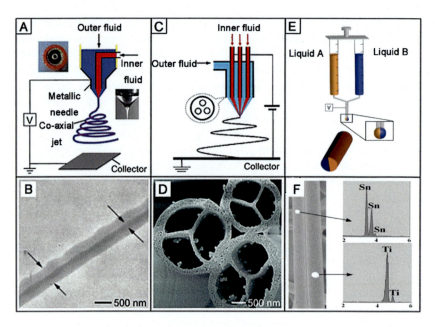

Figure 2.4 (A) Schematic illustration of a coaxial electrospinning process. (B) TEM image of the PANi@PVA nanofibers in a concentric manner. (C) Schematic illustration of the electrospinning process with the use of a spinneret composed of three metallic needles inserted in an outer needle in the pattern of an equilateral triangle. (D) SEM image of the ZnO nanofibers in which the channels were divided into three independent flabellate parts by a Y-shaped inner ridge. (E) Schematic diagram of the experimental setup used for electrospinning bicomponent nanofibers with side-by-side dual spinnerets. (F) Typical EDS microanalysis on selected areas of a single nanofiber, showing that the single nanofiber was indeed composed of two small TiO_2 and SnO_2 nanofibers that were bound together.
Source: (A) Adapted with permission from J.-J. Li, et al., Fast dissolving drug delivery membrane based on the ultra-thin shell of electrospun core-shell nanofibers, Eur. J. Pharm. Sci. 122 (2018) 195–204 [78]. Copyright 2018 Elsevier; Reprinted with permission from (B) A.K. Moghe, B.S. Gupta, Co-axial electrospinning for nanofiber structures: preparation and applications, Polym. Rev. 48 (2) (2008) 353–377 [79]. Copyright 2008 Wiley-VCH; (C) Y. Zhao, X. Cao, L. Jiang, Bio-mimic multichannel microtubes by a facile method, J. Am. Chem. Soc. 129 (4) (2007) 764–765 [80]. Copyright 2007 American Chemical Society; (D) S.-H. Choi, et al., Hollow ZnO nanofibers fabricated using electrospun polymer templates and their electronic transport properties, ACS Nano 3 (9) (2009) 2623–2631 [81]. Copyright 2009 American Chemical Society; (E) Z. Liu, et al., An efficient bicomponent TiO_2/SnO_2 nanofiber photocatalyst fabricated by electrospinning with a side-by-side dual spinneret method, Nano Lett. 7 (4) (2007) 1081–1085 [82]. Copyright 2007 American Ceramic Society; (F) J.D. Starr, M.A.K. Budi, J.S. Andrew, Processing-property relationships in electrospun Janus-type biphasic ceramic nanofibers, J. Am. Ceram. Soc. 98 (1) (2015) 12–19 [83]. Copyright 2015 American Chemical Society.

In one demonstration, hollow ZnO nanofibers were successfully fabricated by utilizing PVAc nanofibers as the sacrificial template (Fig. 2.4D) [81]. Another common method is by coaxial electrospinning of two immiscible liquids with the use of a coaxial spinneret followed by selectively removing the core material by extraction or combustion. In one representative demonstration, mineral oil was used as the inner fluid to form the core because of its good rheological properties, while the spinnable sol-gel solution is used as the outer fluid to form the sheath. For example, it has been reported that when PVP blended with a titanium precursor was used as the sheath material and the mineral oil as the core material, hollow TiO_2 nanofibers could be generated after removal of the mineral oil core and decomposition of PVP components by calcination [77].

Metal oxide nanofibers with a side-by-side morphology enable the control of their composition and surface anisotropy and allow the access to both phases while retaining a reasonably large contact area [90]. Fig. 2.4E shows the schematic of a setup used for generating nanofibers, in which dual spinnerets in a side-by-side format is adopted. During electrospinning, the two jetting liquids simultaneously experience an electric field formed between the liquid tip and the collector. Bicomponent side-by-side TiO_2/SnO_2 nanofibers toward photocatalysis have been produced via this side-by-side electrospinning method [82]. Two oxides with different crystal structures, such as barium titanate (perovskite) and cobalt ferrite (spinel), have been successfully electrospun into a Janus-type nanofiber using a dual-channel syringe [91]. In another demonstration, flexible $Fe_3O_4/NaYF_4$ Janus nanofibers were successfully electrospun by using a parallel spinneret and provided with the feature of simultaneously improved magnetic-photoluminescent bifunction [92]. Principally, it is possible to control the size of the resultant biphasic composite nanofibers by adjusting the properties of the solutions and the electrospinning parameters, such as the conductivity and viscosity of the precursor solutions, as well as the value of the high voltage. It is also promising to produce side-by-side nanofibers made of several compositions by independently changing the ion concentration of each phase [83]. Typical EDS microanalysis on selected areas of a single nanofiber showed that the single nanofiber was indeed composed of two small TiO_2 and SnO_2 nanofibers that were bound together (Fig. 2.4F).

2.3.2 Control of in-fiber and interfiber porosity

Introducing pores into the nanofibers can endow them with superior attributes because the pores can effectively increase the specific surface area of the whole nanofibers and lead to enhanced physiochemical performance for many applications. To this end, previous works have attempted to reliably fabricate porous nanofibers with an easy method and at a low cost. The hierarchical pores of metal oxide nanofibers refer to porous structures in-fiber (i.e., within the individual nanofiber) and interfiber (i.e., among the stacked nanofibers). The in-fiber pores broadly include nanopores and opening structures within individual nanofibers. Resulting from the random arrangement of nanofibers during the electrospinning process, the interfiber pore structures mainly have a disordered porous morphology. These naturally formed pores are commonly macropores. Hence hierarchical pores with different diameters and shapes can form within each individual nanofiber among neighboring nanofibers in two-dimensional (2D) fibrous mats and 3D aerogel or sponge.

Two typical approaches are commonly used for generating pores within individual nanofibers: elegantly inducing phase separation during electrospinning, and selectively removing a component from nanofibers, such as combusting matrices. The in-fiber porosity can be tuned by controlling the shape, size, density, and pore distribution [93]. Inducing phase separation during the electrospinning process by rapidly cooling the nanofibers before they are completely solidified is a feasible approach to generate pores. Cooling can be operated by rapidly evaporating highly volatile solvents or collecting nanofibers in cryogenic liquids [94]. Phase separation, including steam-induced or liquid-induced phase separation, can also be introduced between the polymer and the nonsolvent to produce a porous structure. This pore generation strategy is mostly used for polymer nanofibers and now is a promising approach to construct pores in metal oxide nanofibers by simply adding one or more metal oxide precursors into the traditional polymer solution.

Additionally, porous structures can be introduced by selectively removing a sacrificial phase from the nanofibers, including small molecules and polymers (Fig. 2.5A), which can be easily removed by a leaching or calcination process. For example, porous TiO_2 nanofibers were successfully fabricated by selectively combusting the PVP matrix through a simple calcination process (Fig. 2.5B) [34]. In another report, gradient electrospinning followed by controlled pyrolysis methodology was demonstrated to synthesize various types of mesoporous nanotubes and pealike nanotubes [101]. Some representative examples include multielement oxides, binary-metal oxides, and single metal oxides. The template method is also a conventional and general method to produce porous materials with superb structural characteristics, which can be divided into hard and soft templates according to the characteristics of the template. Ceramic particles and colloidal particles (Fig. 2.5C) often act as hard templates to be incorporated into polymers for electrospinning. Porous fibers composed of particles with gaps between the particles can be obtained after calcination. In one study, 50-nm SiO_2 particles and 237-nm polystyrene (PS) particles were added to an aqueous solution of polyoxyethylene or polyacrylamide (PAM) for electrospinning. The stand-alone porous SiO_2 nanofibers were successfully generated after the removal of PS particles and PAM components by calcination [95]. Soft templates specialize in synthesizing different materials with various morphologies and commonly occur in relatively mild experimental conditions. Typically, macroemulsions and microemulsions, micelles, and some polymers are often used as soft templates in electrospinning. As shown in Fig. 2.5D and E, a special multiscale porous TiO_2 structure was formed by using low-cost paraffin oil microemulsion droplets as the soft template [96]. Nowadays, the porous structures and tailored compositions of metal-organic frameworks (MOFs) make them outstanding templates for creating porous nanofibers (Fig. 2.5F) [97]. In a study, hierarchical fibrous structures composed of ZnO quantum dots, amorphous carbons, and carbon nanofibers are synthesized via a single carbonization process of electrospun ZIF-8/PVA nanofibers, as shown in Fig. 2.5G and H [98].

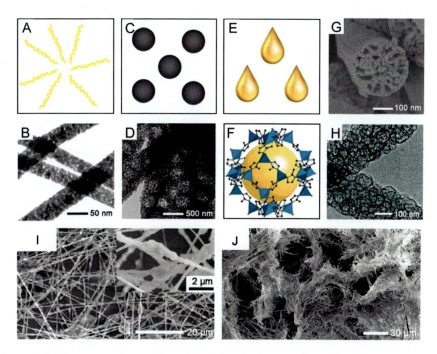

Figure 2.5 The top row shows schematics of the templates. (A) Polymer (PVP). (B) TEM image of the TiO$_2$ fibers calcined in 500°C. (C) Hard template (PS colloid). (D) TEM image of electrospun composite nanofibers of PAM and 50-nm silica particles with 237-nm polystyrene particles after calcination. (E) Soft template (oil microemulsion droplets). (F) MOF material (ZIF-8). (G) An individual TiO$_2$ nanorod showing the hierarchical pore structure. (H) TEM image of ZnO QDs-decorated CNF. (I) SEM images of PAN and SiO$_2$ aerogel/PAN composite nanofiber membranes. (J) SEM image of alumina/silica nanofibrous aerogel.

Source: Reprinted with permission from (A, C, and E) J.-M. Lim, et al., Superhydrophobic films of electrospun fibers with multiple-scale surface morphology, Langmuir 23 (15) (2007) 7981–7989 [95]. Copyright 2007 American Chemical Society; (B) D. Li, Y. Xia, Fabrication of titania nanofibers by electrospinning, Nano Lett., 2003. 3(4): p. 555–560 [34]. Copyright 2003 American Chemical Society; (D) H.-Y. Chen, et al., Electrospun hierarchical TiO$_2$ nanorods with high porosity for efficient dye-sensitized solar cells, ACS Appl. Mater. Interfaces 5 (18) (2013) 9205–9211 [96]. Copyright 2013 American Chemical Society; (F) K.S. Park, et al., Exceptional chemical and thermal stability of zeolitic imidazolate frameworks, Proc. Natl. Acad. Sci. 103 (27) (2006) 10186 [97]. Copyright 2006 National Academy of Sciences; (G and H) G. Lee, Y.D. Seo, J. Jang, ZnO quantum dot-decorated carbon nanofibers derived from electrospun ZIF-8/PVA nanofibers for high-performance energy storage electrodes, Chem. Commun. 53 (83) (2017) 11441–11444 [98]. Copyright 2017 Royal Society of Chemistry; (I) Y. Yu, et al., Electrospun SiO$_2$ aerogel/polyacrylonitrile composited nanofibers with enhanced adsorption performance of volatile organic compounds, Appl. Surf. Sci. 512, (2020) 145697 [99]. Copyright 2020 Elsevier B.V.; (J) R. Liu, et al., Ultralight, thermal insulating, and high-temperature-resistant mullite-based nanofibrous aerogels, Chem. Eng. J. 360 (2019) 464–472 [100]. Copyright 2019 Elsevier B.V.

In addition to the porous structure within an individual nanofiber, the control of interfiber porosity is equally important. The electrospun nanofibers not only can form highly porous 2D mats with controlled pore sizes but also can be functionalized to play an important role in various fields, such as filtration, energy field, environmental engineering, and protective clothing applications [102,103]. Three-dimensional macroporous architectures constructed by electrospinning have attracted considerable attention, owing to their excellent physicochemical properties and wide range of applications [104]. Sponges or aerogels based on frozen cast nanofibers are endowed with a hierarchical porous structure. Recently, nanofibrous aerogel has been fabricated with nanofibers as the main starting materials [105]. Different from the porous structure of traditional aerogel based on nanoparticles, the 3D network structure of nanofiber aerogel is formed by multiple nanofibers intertwined with each other. In particular, metal oxide nanofibrous aerogel can exhibit both low thermal conductivity and superior high-temperature stability. Recently, CeO_2 nanofibers were prepared with uniform 3D structure and large pore size (> 50 nm, macroporous structure) via electrospinning [106]. Additionally, the specific surface area of metal oxide nanofibers can be further increased to promote its adsorption efficiency by adding porous material [84,107]. As shown in Fig. 2.5I, SiO_2 aerogel and PAN were successfully fabricated into a flexible nanofiber mat with a high surface area via electrospinning [99]. SiO_2 aerogel with honeycomb porous structure was assembled on the surface of PAN nanofibers as the solvent evaporated, which greatly improved the specific surface area of the resultant membranes. In another study, fabrication of mullite-based nanofibrous aerogels via the gel casting and freeze drying of electrospun nanofibers with different alumina/silica molar ratios as the matrix and silica sols as the high-temperature binders was reported, as shown in Fig. 2.5J [100].

2.3.3 Control of hierarchical surface structures

Metal oxide nanofibers with hierarchical structures are of practical significance for enriching the morphology of the material and thus expanding their applications [108]. The nanofibers can be obtained through one-step electrospinning by incorporating a variety of materials such as metal alkoxides or metal salts into the electrospinning solution when the mixture can be thermally decomposed and be shaped into interesting hierarchical structures. In general, hierarchical nanostructures such as nanoplates or nanorods can be incorporated into metal oxide nanofibers. The hierarchical nanostructure can provide a high surface-to-volume ratio, which is essential for applying the metal oxides as catalysts, antibacterial agents, piezoelectric materials, optoelectronic devices, and so on [109]. In addition, feasible structures can be designed for free-standing hierarchical structure for the realization of different features with multifunctional properties and the advantages of combining with the functional features and nanofiber morphology [110].

Different types of hierarchical structures, such as other types of nanostructures, including nanorods, nanoplates, and nanoflowers, can be endowed into the metal oxide nanofibers through various methods. A hierarchical structure can be generated in the nanofibers during electrospinning by controlling the electrospinning parameters.

For example, through a coaxial electrospinning process, during which SiO_2 nanoparticle dispersion was used as the outer fluid and polytetrafluoroethylene as the inner fluid, hierarchical nanofibers were obtained followed by in situ surface grafting of perfluorooctyltrimethoxysilane with amphiphilic properties [111]. Then hierarchical $ZnO/CoNiO_2$ nanofibers were achieved after an annealing treatment, which may not only combine the advantages of the two components but also exhibit novel properties on account of the synergistic effects between building blocks [112]. In another typical example, by carefully selecting the precursors and controlling the growth conditions, TiO_2 nanofibers with nanorods uniformly embedded on the surface were obtained through a hierarchically growing procedure, as shown in Fig. 2.6A [113].

Metal oxide nanofibers with hierarchical structures can also be fabricated by the posttreatment of the as-spun nanofibers. In one study, Al_2O_3/La_2O_3 nanofibers decorated with perovskite crystals were fabricated through the self-assembly of $CsPbBr_3/Cs_4PbBr_6$ crystals on the surface of the as-spun nanofibers [117]. Herein, nanofiber mats with excellent mechanical flexibility were obtained under the repeatable various bending radii and mechanical bending. By controlling the assembly of the graphene oxide onto the nanofibers, the hierarchical structure membranes can be developed between the reaction of the groups of nanofibers and the carboxyl or epoxy groups of GO. Similarly to graphene and nanoparticles of silica, TiO_2, Al_2O_3 have been embedded in electrospun nanofibers for the development of hierarchical membranes. In one system as shown in Fig. 2.6B, through adjusting the interaction between the oxygenated functional groups on GO and Al_2O_3/TiO_2 nanofibers, appropriate GO content could regulate the hierarchical growth of Al_2O_3 heterojunctions outside TiO_2 nanofibers [114]. As provided by the controllable hierarchically nano/microbead surface morphology, the membrane was underoiled superhydrophobic and underwater superoleophobic. By employing the electrospinning process, the polyethylene naphthalene 2,6 dicarboxylate/ZnO hierarchical nanostructures were fabricated following the hydrothermal growth [118]. Typically, the "water lily" and "caterpillar" like hierarchical microstructures and nanostructures are formed in fabricating the organic/inorganic hierarchical nanostructures. The growth of the two shapes of hierarchical microstructures and nanostructures is noticeable, depending on the hydrothermal growth period [109]. Two shapes of hierarchical structures of ZnO-deposited electrospun polyamide 6 nanofibers were formed. After 50 cycles of atomic layer deposition (ALD) of ZnO onto nanofibers, ZnO nanoparticles were grown into the cluster morphology, water lily−like nanorods with sharp tips. In the 100 and 150 cycles of ALD ZnO coating onto polyamide 6 nanofibers, caterpillar-like hierarchical nanostructures were formed.

Another method is aimed at processing through the posttreatment of the as-spun nanofibers. The hierarchically structured porous Co_3O_4@carbon fibers were reported by direct electrospinning of ZIF-67 nanoparticles followed by a thermal treatment. Also, TiO_2 nanofibers with hierarchical distinctive microstructures were successfully fabricated by a microemulsion electrospinning process followed by pyrolysis [115]. As illustrated in Fig. 2.6C, the $Ag_2C_2O_4$ nanoplates were grown onto the TiO_2 nanofibers for the fabrication of hierarchical $Ag_2C_2O_4/TiO_2$ nanofibrous membranes through electrospinning technique followed by successive ionic

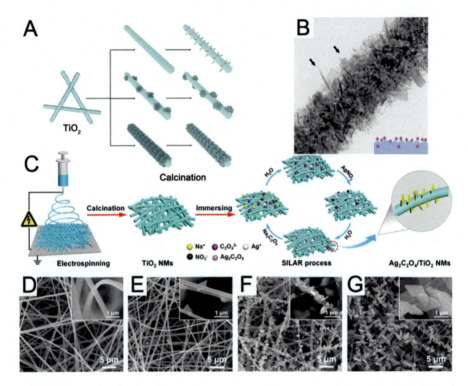

Figure 2.6 (A) Schematic of the synthesis process of nanorod on rough surface of TiO$_2$ nanofibers. (B) TEM images of nanofibers after coupling with GO during hydrothermal reaction. (C) Fabrication process of Ag$_2$C$_2$O$_4$/TiO$_2$ NMs. (D–G) SEM images of Ag$_2$C$_2$O$_4$/TiO$_2$ nanofibers with different circulation times.
Source: Reprinted with permission from (A) W. Fu, et al., Stabilizing 3 nm-Pt nanoparticles in close proximity on rutile nanorods-decorated-TiO$_2$ nanofibers by improving support uniformity for catalytic reactions, Chem. Eng. J. (2020) 126013 [113]. Copyright 2020 Elsevier; (B) Q. Zhan, et al., Graphene-based modulation on the hierarchical growth of Al$_2$O$_3$ heterojunctions outside TiO$_2$ nanofibers via a surfactant-free approach, Compos. Commun. 21 (2020) 100394 [114]. Copyright 2020 Elsevier; (C) C.-L. Zhang, et al., Hierarchically structured Co$_3$O$_4$@carbon porous fibers derived from electrospun ZIF-67/PAN nanofibers as anodes for lithium ion batteries, J. Mater. Chem. A 6 (27) (2018) 12962–12968 [115]. Copyright 2017 Royal Society of Chemistry; (D–G) X. Wu, et al., Thorn-like flexible Ag$_2$C$_2$O$_4$/TiO$_2$ nanofibers as hierarchical heterojunction photocatalysts for efficient visible-light-driven bacteria-killing, J. Colloid Interface Sci. 560 (2020) 681–689 [116]. Copyright 2020 Elsevier.

layer adsorption and reaction process. The TiO$_2$ nanofibers depicted were randomly assembled into a 3D network with an average value of approximately 300 nm (Fig. 2.6D), and the average fiber diameter was maintained almost unchanged for Ag$_2$C$_2$O$_4$/TiO$_2$ nanofibers with different circulation times (Fig. 2.6E–G) [116].

For preparing a fiber sponge scaffold, the main determining parameter is the freezing speed, which determines the interaction of the filler with the fiber matrix, including the spatial arrangement and orientation effects of the materials [119]. This was presented in a report in which a hierarchical homoassembled PAN nanofibrous mat decorated with ZnO nanostructures was constructed by electrospinning. The buildup of layer-by-layer self-assembly multilayers is driven by the electrostatic attractions between the oppositely charged constituents, compared with other surface modification techniques [16].

2.3.4 Control of alignment and patterns

Aligned arrays of metal oxide nanofibers have attracted growing attention and are playing an important role in the relevant applications because of the remarkable anisotropy and enhanced mechanical properties relative to randomly deposited nanofibers. Owing to their high surface-to-volume ratio, remarkable anisotropy, and enhanced mechanical properties, aligned nanofibers have attracted growing attention and have played an important role in the relevant applications of organic and inorganic materials. To optimize the orientation of the nanofiber to produce aligned products, several common setups are necessary with modifications by electrospinning [3]. Until now, several effective methods have been applied to fabricate aligned fibers via electrospinning, including regulation of the electrical field, introducing a rotating mandrel (a disk collector and a drum collector), and magnetic electrospinning. By shortening the distance between the parallel electrodes, the arrangement of optical fibers can be significantly improved [120]. The desired aligned fibers can be obtained in three types of rotating collectors of blade, drum (smooth), and grid [121]. In addition to orientation, the aligned electrospun nanofibers are easier to apply for the production of fiber-reinforced composites with high strength and high toughness [122]. In general, by introducing a rotating mandrel and changing the rotation speed of the mandrel, the alignment of the nanofibers can be effectively controlled when there is a linear velocity match between the rotating mandrel and the jet deposition [123]. As Fig. 2.7A depicts, randomly oriented nanofibers were collected when a static plate collector was used, while partially aligned nanofibers were obtained when a rotating collector was adopted [2,124,126]. During the jet acceleration toward the grounded mandrel, the solvent evaporates, and $Cu(CH_3COO)_2$/PVA nanofibers are collected on an aluminum foil substrate. CuO nanofibers with aligned orientation were fabricated, followed by the calcination process as conductive and high transparent layers [127]. The rotating mandrel was utilized as a grounded electrode with a flow rate of $1 \, mL \, h^{-1}$ for the fabrication of ceramic yttria-stabilized zirconia. The influence on alignment was quantified with the varied rotating speed of the mandrel. The results showed that the alignment is better at a higher speed. This alignment process is not adversely influencing by the formation of thicker layers at longer deposition times and higher flow rates [128].

In the process of electrospinning, the droplet with a certain static voltage is usually in an electric field. Therefore when the jet moves from the end of the metallic needle to the collector device, the acceleration will lead to the stretching of the jet between the two electrodes and the formation of metal oxide nanofibers [129,130]. In one typical study, by applying a collector composed of two parallel copper

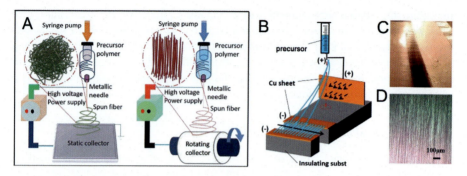

Figure 2.7 (A) Electrospinning setup for randomly oriented nanofibers and rotating mandrel for partially align nanofibers. (B) Schematic of the electrospinning system made of two copper electrodes in a gap and an individual assistant electrode. (C) The image recorded by a camera showing the oriented fibers between two electrodes after 5 min collection. (D) Optical micrograph of PVP/TiO$_2$ fiber; the image in the upper left corner shows the PVP/TiO$_2$ fibers on the glass substrate.
Source: Reprinted with permission from (A) Q.P. Pham, U. Sharma, A.G. Mikos, Electrospun poly(ε-caprolactone) microfiber and multilayer nanofiber/microfiber scaffolds: characterization of scaffolds and measurement of cellular infiltration, Biomacromolecules 7 (10) (2006) 2796−2805 [124]. Copyright 2006 Royal Society of Chemistry; (B−D) J. Wang, et al., Fabrication of a well-aligned TiO2 nanofibrous membrane by modified parallel electrode configuration with enhanced photocatalytic performance, RSC Adv. 6 (37) (2016) 31476−31483 [125]. Copyright 2016 Royal Society of Chemistry.

electrodes and one additional assistant electrode (Fig. 2.7B), uniaxially aligned electrospun PVP/TiO$_2$ fibers were continuously collected [125]. After calcination of the composite nanofibers to decompose the PVP component, highly aligned TiO$_2$ nanofibers were generated, as shown in Fig. 2.7C and D. How to efficiently utilize electrical energy to produce electrospun nanofibers in an orderly manner free from an expensive and time-consuming process has also been investigated [131]. Well-aligned arrays of nanowires made of ZnO have been fabricated by electrospinning with zinc acetate used as the precursor [132]. The two connected parallel collector plates employed with a separating gap resulted in a very high degree of nanowire alignment. By adjusting the size of the gap between the electrodes and deposition time, the spacing of the nanowires could be controlled. The aligned ZnO nanowire arrays could be used to fabricate large-area devices at low cost, with control over the nanowire density, which can be extended to other functional oxides (e.g., MgO, CuO, SnO$_2$, and Fe$_2$O$_3$) [132].

The magnetic field−assisted method is another effective way for the preparation of ordered electrospun metal oxide nanofibers [133]. In a typical report, this magnetic field−assisted method was applied to separately fabricate aligned TiO$_2$ nanofibers and nanotubes, both of which indicated outstanding photocatalytic performances [134]. A simple magnetic-electrospinning method was employed for the preparation of aligned Fe$_2$O$_3$ nanofibers [135]. By introducing the extra magnetic field force on Fe$_2$O$_3$ nanoparticles within composite fibers, the critical voltage for spinning has been reduced, along with decreased fiber diameters. The nanofibers showed increased strength for the magnetic field alignment of the micromagnets, and the attraction between them assisted the increase in fiber strength. A piece of flat aluminum foil used as the collector was stuck to the surface of a permanent magnet. One of three permanent magnets was used by itself each time to generate a magnetic field with different surface magnetic field strengths. The spinning distance was controlled by fixing the permanent magnet and the grounded aluminum foil together on a distance adjuster.

In addition to aligned arrays, the patterned structure can also be endowed on the surface of electrospun metal oxide nanofiber mat, representing another type of special morphology. The incorporation of the patterned structure is beneficial for the enhancement of the specific surface area and certain roughness due to the micro- or nanoscale structures [136]. In general, metal oxide nanofiber mats with a patterned structure can be generated by a pretreatment method or a posttreatment method. In a pretreatment strategy, by controlling operation parameters for electrospinning, the patterned architectures of metal oxide nanofibers can be readily generated during the electrospinning process. It has been demonstrated in various reports to have the ability to accomplish the patterning; among them the application of patterned metal electrodes is the most popular approach to pattern nanofibers on the electrodes [137−139]. Through the use of a properly designed solid conductive collector, the electric field between the metallic needle and the collector can be modulated, guiding the selective deposition of the nanofibers onto the specific region of the collector and thus generating a patterned nanofiber mat [140,141]. A homogeneous collector implemented with geometrical features such as protrusions is capable of producing patterned nanofiber mat [142]. As the Coulomb force and the square of the distance between the two static electric charges are in inverse proportion, the protrusion part is preferentially attracted by the highly charged electrospun nanofibers in comparison to the lower part. Therefore the patterned structure is readily formed on the collector. A metal mesh was chosen as the collector to obtain a uniform pattern of the indium tin oxide nanofiber films with a uniform fiber size. The indium tin oxide nanofibers with a crisscross pattern were prepared by the electrospinning of a precursor solution onto a metal mesh template [143]. In contrast to the reported randomly distributed nanofibers, these patterned nanofiber films are likely to be more suitable for applications such as sensors, solar cells, and electromagnetic field filters, owing to the higher electrical and optical performance [143].

In a typical posttreatment program, patterned structures can be generated on the as-spun metal oxide nanofibers mat through various posttreatment methods, such as calcination and printing [144]. By the calcination process applied to the aligned precursor fibers fabricated via near field electrospinning, precise patterning of ZnO, GaO TiO, GaN, and TiN nanofibers with good repeatability were realized [145]. The shape transformation photolithography method demonstrates the possibility of preparing arbitrarily patterned electrospun nanofibers [146]. A microscale patterning process of electrospun metal oxide nanofibers was implemented through the electrohydrodynamic printing method [147]. Various microscale patterns of electrospun nanofibers could be successfully fabricated by the printing of nanofiber fragment solutions. Several types of electrospun metal oxide nanofibers, such as SnO_2, In_2O_3, WO_3, and NiO nanofibers, were fragmented into smaller pieces by ultrasonication, dispersed in organic solvents, and then utilized as inks for the electrohydrodynamic printing process. The pattern shapes were controlled by using a programmable $X-Y$ stage motion controller with a speed of 1 mm s^{-1}.

2.3.5 Welding of nanofibers at their cross points

Electrospun nanofibers are usually physically stacked layer by layer, thus only a weak interaction exists at the cross points of neighboring nanofibers. When applied as nanofiber-based electrical devices, the inferior interfacial adhesion between the nanofibers and the substrate, and the high contact resistance between the nanofibers usually lead to inferior electrical performance. When a tensile force is applied during a stretching, the stacked nanofibers will slide and be easily deformed. Therefore an extra processing step is expected to improve the connections between the nanofibers and enhance the interfacial adhesion properties. Welding at the intersection of nanofibers is seen as an effective method to improve the mechanical strength and electrical properties of the nanofibers [24]. Recently, three related welding methods involving solvent (or vapor) exposure, covalent crosslinking, and thermal treatment have been developed to offer the interfiber connection.

Solvent vapor treatment of the nanofibers is one of the representative methods for welding the nanofibers, during which the vapor pressure is a determining factor and makes a critical difference in the mechanical properties of the nanofibers. Some of the solvents may reside on the nonwoven structure after electrospinning, which can promote the bonding at junction points [148]. The fusion of the nanofibers can occur at their cross-points solvent or vapor when swelling and partial dissolution phenomena arise. The time period for welding is mainly affected by the type of solvent and the composition of the nanofibers, and the degree of welding is dependent on the vapor pressure and the exposure time [149]. The in situ crosslinking process for welding the nanofibers has several advantages compared to the conventional solvent vapor welding process or hot-pressing process, such as a simpler process, more efficiency, and lack of the requirement for expensive facilities or reagents. Note that the crosslinking welding process is controlled by the spontaneous chemical reaction rather than by high-energy radiation. Thus it is not restricted to the limited area of the substrate. In one report, amine-hardened epoxy resin system served as an adhesion agent to weld the In_2O_3 nanofiber junctions [150]. With the enhancement of interfiber connections and interfacial adhesion, the welded In_2O_3 nanofibers exhibited superior mechanical properties and better electrical performance.

The specific operating temperature is usually related to the melting point of the polymer during the welding process. One-dimensional metal oxide nanofiber networks were successfully prepared via the most powerful approaches. The In_2O_3 nanofiber networks were fabricated by electrospinning with the precursor solution containing polymethyl methacrylate and $In(NO_3)_3 \cdot xH_2O$. Benefiting from the thermoplasticity, the composite In_2O_3 nanofiber networks with impact stacking were automatically welded at a relatively low annealing temperature [151]. In another improvement, upon application of a welding process, the mechanical stability of the as-spun polyvinylidene fluoride (PVDF) nanofiber mats was dramatically improved when the thermal treatment was performed near the melting temperature of PVDF ($\approx 160°C$) [152]. By simple thermal stabilization operation, it is feasible to tailor the mechanical properties of reinforced lignin nanofiber with nanocrystalline cellulose [153]. Localized heat energy could be generated at the junctions of nanofibers, resulting in ultrafast and completely welded fibers [154]. Under moderate crosslinking conditions, without significant change of fiber morphology, the cross points of the fibers can be welded together. Until now, the crosslinking approaches were proposed to the nanofibers to guarantee the integrity and elasticity of the nanofibers [155].

2.3.6 Three-dimensional fibrous aerogels

Although electrospinning nanofibers can provide superior 2D fibrous morphology, there are still limitations regarding the porosity and specific surface area [156]. To overcome the obstacle, various means have been used to fabricate 3D scaffolds based on the electrospun nanofibers [104]. Owing to the extremely excellent structural flexibility, functional performance, as well as highly porous nature, the 3D structures assembled from electrospun nanofiber mat or membranes have become an emerging research topic [157–159].

One of the most common methods of fabricating 3D scaffolds based on electrospun nanofibers is by freeze drying a slurry of short nanofibers [99]. By dispersing and homogenizing solid nanofiber membrane in a solution, a slurry of short nanofibers can be obtained. Afterwards, by carefully processing a freeze-drying procedure, the resultant short electrospun scaffolds can be constructed into aerogels. In one study, electrospun composite nanofibers were converted into porous, hierarchical Al_2O_3/TiO_2 nanofibers by selectively removing PVP matrices upon calcination in air at 600°C. The Al_2O_3/TiO_2 nanofibers, well dispersed in water in different concentrations, were added into the GO aqueous solution to obtain the reaction precursor. The uniformly mixed solution was then hydrothermally reacted at 180°C in the 25-mL autoclave for 8 h. The resulting hydrogels were obtained under controlled temperatures followed by a freeze-drying process. The SEM image of the aerogel is shown in Fig. 2.8A, indicating the uniform fibrous frameworks exciting hierarchical pore structures in aerogels with ultralight weights [160]. This freeze-drying strategy has also been reported to create lamellar-structured metal oxide nanofibrous aerogels with tunable densities and desired shapes. In one typical example, hierarchical cellular structured ceramic nanofibrous aerogels with ultralow thermal conductivity were created by intertwining SiO_2 nanoparticle aerogels and SiO_2 nanofibers [161]. The SiO_2 nanoparticle aerogels and SiO_2 sol were dispersed in polyethylene oxide solution to form a homogeneous dispersion. During freezing by liquid nitrogen the silica nanofibers were rejected from the moving solidification ice front and piled up between the growing cellular crystals; simultaneously, the polyethylene oxide together with SiO_2 nanofibers and SiO_2 sol was also gradually entrapped between the ice crystals and coated on the surface of SiO_2 nanofibers by strong hydrogen bonding. After the sample had completely frozen, they became

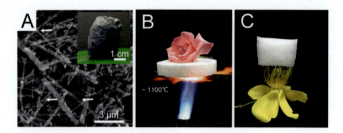

Figure 2.8 (A) SEM image of aerogel fabricated with the fibers concentration of 80 wt.%, showing uniform fibrous frameworks and thin RGO films. The insets show the corresponding optical images of the lightweight aerogels. (B) A ceramic nanofibrous aerogel heated by a butane blowtorch without any damage. (C) The ultralight ceramic nanofibrous aerogels could stand on the top of a flower without bending its stamens.
Source: Reprinted with permission from (A) X. Meng, et al., Coupling of hierarchical Al2O3/TiO2 nanofibers into 3D photothermal aerogels toward simultaneous water evaporation and purification, Adv. Fiber Mater. 2 (2) (2020) 93–104 [160]. Copyright 2020 Springer; (B and C) L. Dou, et al., Hierarchical cellular structured ceramic nanofibrous aerogels with temperature-invariant superelasticity for thermal insulation, ACS Appl. Mater. Interfaces 11 (32) (2019) 29056–29064 [161]. Copyright 2019 American Chemical Society.

tangled and locked into a 3D solid network. Subsequently, the ice crystal was replaced by air during the freeze-drying process via sublimation. Since no capillary effect existing on sublimation, cellular pores could be formed by the removal of ice, whose geometric forms were directly replicated with the pristine ice crystals. The resultant ceramic nanofibrous aerogels were demonstrated to have low thermal conductivity and ultralow density as well as excellent flexibility at 1100°C. As shown in Fig. 2.8B, the metal oxide nanofibrous aerogels were not damaged after being heated by a butane blowtorch. With ultralow density the metal oxide nanofibrous aerogel could stand on the tip of stamens, highlighting the ultralight characteristic (Fig. 2.8C) [162].

2.3.7 *Mass production of metal oxide nanofibers*

Over the past 100 years, the nanofibers that have been obtained have excellent properties have been extensively studied and successfully implemented in a variety of applications. But the electrospinning mass production on a large scale is still a problem worth discussing and pondering. In addition, mass-produced nanofibers are combined with advanced technologies, and the metal oxides are mixed with polymers. The significant advancements and noticeable growth in instrumentation and operating parameters are both beneficial for producing nanofibers on a large scale. In a recent study, the use of multiple needles represents a great potential for large-scale fabrication of nanofibers [163]. Although the yield can be improved by the multinozzle electrospinning method, the equipment structure is still involved in the complex equipment structure. The electric field will seriously interfere with the electrostatic spinning jets during the process. Another needleless electrospinning method can raise productivity by orders of magnitude without any use of electrodes [164]. The solvents are not limited by the dielectric constant, since the electric field is dispensable in this process. When the electric field intensity exceeds the critical value, a large number of jets are formed directly from the open free liquid level in the needleless method, which avoids the defect of needle blockage and can greatly improve the spinning efficiency and fiber yield. Regarding the operation parameters, in addition to traditional thermal treatment methods such as the use of muffle furnace, the use of a picosecond laser to provide the high temperature for the calcination process is also considered as an appropriate method [165]. Although these technologies can overcome the laboratory production scale and have great potential for industrial mass production, they still cannot fill the huge gap between the laboratory and industry. In this regard, deeper investigations into the material properties and fabrication processes are still required to overcome these challenges, which can be adopted to meet the requirements for further industrial manufacture.

Acknowledgments

This work was financially supported by the National Natural Science Foundation of China (Grant NO. 52073014 and 21975042), the Project of Six Talents Climax Foundation of Jiangsu (XCL-082), Young Talent Lifting Project of Jiangsu Science and Technology Associate, National Key R&D Program of China (No. 2020YFC1511902), and the Fundamental Research Funds for the Central Universities. This work was also supported by the Key Program of Beijing Natural Science Foundation (Z200025). We also thank the startup funding support of the Beijing University of Chemical Technology (J. Xue).

References

[1] Y. Khaksarfard, H. Ziyadi, A. Heydari, Preparation of ceramic nanofibers of iron vanadate using electrospinning method, Mater. Sci. Poland 37 (4) (2019) 645−651.
[2] K. Mondal, A. Sharma, Recent advances in electrospun metal-oxide nanofiber based interfaces for electrochemical biosensing, RSC Adv. 6 (97) (2016) 94595−94616.
[3] Y. Huang, et al., Scalable manufacturing and applications of nanofibers, Mater. Today 28 (2019) 98−113.
[4] X. Song, et al., Highly aligned continuous mullite nanofibers: conjugate electrospinning fabrication, microstructure and mechanical properties, Mater. Lett. 212 (2018) 20−24.
[5] H. Esfahani, R. Jose, S. Ramakrishna, Electrospun ceramic nanofiber mats today: synthesis, properties, and applications, Materials (Basel) 10 (11) (2017).
[6] P. Pascariu, M. Homocianu, ZnO-based ceramic nanofibers: preparation, properties and applications, Ceram. Int. 45 (9) (2019) 11158−11173.
[7] M. Rahmati, et al., Electrospinning for tissue engineering applications, Prog. Mater. Sci. (2020) 100721.
[8] B. Zhang, et al., Recent advances in electrospun carbon nanofibers and their application in electrochemical energy storage, Prog. Mater. Sci. 76 (2016) 319−380.

[9] C. Huang, N.L. Thomas, Fabrication of porous fibers via electrospinning: strategies and applications, Polym. Rev. (2019) 1−53.

[10] M. Al-Hashem, S. Akbar, P. Morris, Role of oxygen vacancies in nanostructured metaloxide gas sensors: a review, Sens. Actuators B: Chem. 301 (2019) 126845.

[11] Z. Wen, et al., Electrospinning preparation and microstructure characterization of homogeneous diphasic mullite ceramic nanofibers, Ceram. Int. 46 (8) (2020) 12172−12179.

[12] S. Homaeigohar, et al., The electrospun ceramic hollow nanofibers, Nanomaterials (Basel) 7 (11) (2017).

[13] J. Xue, et al., Electrospinning and electrospun nanofibers: methods, materials, and applications, Chem. Rev. 119 (8) (2019) 5298−5415.

[14] D. Malwal, P. Gopinath, Efficient adsorption and antibacterial properties of electrospun CuO-ZnO composite nanofibers for water remediation, J. Hazard. Mater. 321 (2017) 611−621.

[15] B. Coşkuner Filiz, A.K. Figen, Fabrication of electrospun nanofiber catalysts and ammonia borane hydrogen release efficiency, Int. J. Hydrog. Energy 41 (34) (2016) 15433−15442.

[16] X. Gao, et al., Progress in electrospun composite nanofibers: composition, performance and applications for tissue engineering, J. Mater. Chem. B 7 (45) (2019) 7075−7089.

[17] D. Wang, et al., Electrospun polyimide nonwovens with enhanced mechanical and thermal properties by addition of trace plasticizer, J. Mater. Sci. 55 (13) (2020) 5667−5679.

[18] G. Duan, A. Greiner, Air-blowing-assisted coaxial electrospinning toward high productivity of core/sheath and hollowfibers, Macromol. Mater. Eng. 304 (5) (2019) 1800669.

[19] G. Duan, et al., High-performance polyamide-imide films and electrospun aligned nanofibers from an amide-containing diamine, J. Mater. Sci. 54 (8) (2019) 6719−6727.

[20] G. Duan, et al., Microstructures and mechanical properties of aligned electrospun carbon nanofibers from binary composites of polyacrylonitrile and polyamic acid, J. Mater. Sci. 53 (21) (2018) 15096−15106.

[21] S. Zhan, D. Chen, X. Jiao, *Co-electrospun SiO_2 hollow nanostructured fibers with hierarchical walls.*, J. Colloid Interface Sci. 318 (2) (2008) 331−336.

[22] W. Zheng, et al., *A highly sensitive and fast-responding sensor based on electrospun In_2O3 nanofibers.*, Sens. Actuators B: Chem. 142 (1) (2009) 61−65.

[23] X. Lu, C. Wang, Y. Wei, One-dimensional composite nanomaterials: synthesis by electrospinning and their applications, Small 5 (21) (2009) 2349−2370.

[24] H.-J. Lee, et al., Spontaneous and selective nanowelding of silver nanowires by electrochemical ostwald ripening and high electrostatic potential at the junctions for highperformance stretchable transparent electrodes, ACS Appl. Mater. Interfaces 10 (16) (2018) 14124−14131.

[25] Z. Li, et al., Porous ceramic nanofibers as new catalysts toward heterogeneous reactions, Compos. Commun. 15 (2019) 168−178.

[26] G. Jin, et al., Electrospun three-dimensional aligned nanofibrous scaffolds for tissue engineering, Mater. Sci. Eng. C 92 (2018) 995−1005.

[27] E. Ghasemi, et al., Iron oxide nanofibers: a new magnetic catalyst for azo dyes degradation in aqueous solution, Chem. Eng. J. 264 (2015) 146−151.

[28] J. Cai, et al., High-Performance Supercapacitor Electrode materials from cellulosederived carbon nanofibers, ACS Appl. Mater. Interfaces 7 (27) (2015) 14946−14953.

[29] Z. Shao, et al., A thermally self-sustained micro solid-oxide fuel-cell stack with high power density, Nature 435 (7043) (2005) 795−798.

[30] Shilpa, et al., Electrospun hollow glassy carbon—reduced graphene oxide nanofibers with encapsulated ZnO nanoparticles: a free standing anode for Li-ion batteries, J. Mater. Chem. A 3 (10) (2015) 5344−5351.

[31] Y. Shen, et al., Epitaxy-nnabled vapor—liquid—solid growth of Tin-doped indium oxide nanowires with controlled orientations, Nano Lett. 14 (8) (2014) 4342−4351.

[32] S. Yoo, S.A. Akbar, K.H. Sandhage, Nanocarving of titania (TiO2): a novel approach for fabricating chemical sensing platform, Ceram. Int. 30 (7) (2004) 1121−1126.

[33] J. Xue, et al., Electrospun nanofibers: new concepts, materials, and applications, Acc. Chem. Res. 50 (8) (2017) 1976−1987.

[34] D. Li, Y. Xia, Fabrication of titania nanofibers by electrospinning, Nano Lett. 3 (4) (2003) 555−560.

[35] D. Li, Y. Xia, Electrospinning nanofibers: reinventing wheel? Adv. Mater. 16 (14) (2004) 1151−1170.

[36] B. Sun, et al., Advances in three-dimensional nanofibrous macrostructures via electrospinning, Prog. Polym. Sci. 39 (5) (2014) 862−890.

[37] Y. Liao, et al., Progress in electrospun polymeric nanofibrous membranes for water treatment: fabrication, modification and applications, Prog. Polym. Sci. 77 (2018) 69−94.

[38] Y. Dai, et al., Ceramic nanofibers fabricated by electrospinning and their applications in catalysis, environmental science, and energy technology, Polym. Adv. Technol. 22 (3) (2011) 326−338.

[39] C.D. Saquing, et al., Alginate—polyethylene oxide blend nanofibers and the role of the carrier polymer in electrospinning, Ind. Eng. Chem. Res. 52 (26) (2013) 8692−8704.

[40] D. Malwal, P. Gopinath, Fabrication and characterization of poly(ethylene oxide) templated nickel oxide nanofibers for dye degradation, Environ. Sci. Nano 2 (1) (2015) 78−85.

[41] P. Viswanathamurthi, et al., *GeO$_2$ fibers: preparation, morphology and photoluminescence property*, J. Chem. Phys. 121 (1) (2004) 441−445.

[42] S.-S. Choi, et al., Silica nanofibers from electrospinning/sol-gel process, J. Mater. Sci. Lett. 22 (12) (2003) 891−893.

[43] D. Li, et al., Electrospinning: a simple and versatile technique for producing ceramic nanofibers and nanotubes, J. Am. Ceram. Soc. 89 (6) (2006) 1861−1869.

[44] W. Fu, et al., Unusual hollow Al2O3 nanofibers with loofah-like skins: intriguing catalyst supports for thermal stabilization of Pt nanocrystals, ACS Appl. Mater. Interfaces 9 (25) (2017) 21258−21266.

[45] L. Yao, et al., Stabilizing nanocrystalline oxide nanofibers at elevated temperatures by coating nanoscale surface amorphous films, Nano Lett. 18 (1) (2018) 130−136.

[46] Y. Dai, J. Tian, W. Fu, *Shape manipulation of porous CeO$_2$ nanofibers: facile fabrication, growth mechanism and catalytic elimination of soot particulates.*, J. Mater. Sci. 54 (14) (2019) 10141−10152.

[47] S. Liu, et al., Constructing fibril-in-tube structures in ultrathin CeO2-based nanofibers as the ideal support for stabilizing Pt nanoparticles, Mater. Today Chem. 17 (2020) 100333.

[48] D. Li, Y. Xia, Direct fabrication of composite and ceramic hollow nanofibers by electrospinning, Nano Lett. 4 (5) (2004) 933−938.

[49] H. Wu, et al., A transparent electrode based on a metal nanotrough network, Nat. Nanotechnol. 8 (6) (2013) 421−425.

[50] P.J. Kelly, R.D. Arnell, Magnetron sputtering: a review of recent developments and applications, Vacuum 56 (3) (2000) 159−172.

[51] J. Kim, et al., Dual-gate crystalline oxide-nanowire field-effect transistors utilizing ion-gel gate dielectric, Appl. Surf. Sci. 515 (2020) 145988.

[52] R. Berenguer, et al., *Synthesis of vanadium oxide nanofibers with variable crystallinity and $V^{(5+)}/V^{(4+)}$ ratios.*, ACS Omega 2 (11) (2017) 7739−7745.

[53] S. Chattopadhyay, et al., *Electrospun ZrO_2 nanofibers: precursor controlled mesopore ordering and evolution of garland-like nanocrystal arrays.*, Dalton Trans. 47 (16) (2018) 5789−5800.

[54] Y. Park, E. Ford, Titanium oxide sol-gel induced wrinkling of electrospun nanofibers, Macromol. Chem. Phys. 219 (13) (2018) 1800028.

[55] O. Kéri, et al., Thermal properties of electrospun polyvinylpyrrolidone/titanium tetraisopropoxide composite nanofibers, J. Therm. Anal. Calorim. 137 (4) (2019) 1249−1254.

[56] X. Song, et al., Morphology, microstructure and mechanical properties of electrospun alumina nanofibers prepared using different polymer templates: a comparative study, J. Alloy. Compd. 829 (2020) 154502.

[57] O.V. Otieno, et al., Synthesis of TiO2 nanofibers by electrospinning using water-soluble Ti-precursor, J. Therm. Anal. Calorim. 139 (1) (2019) 57−66.

[58] O. Pereao, et al., Morphology, modification and characterisation of electrospun polymer nanofiber adsorbent material used in metal ion removal, J. Polym. Environ. 27 (9) (2019) 1843−1860.

[59] H. Shao, et al., *Hollow WO_3/SnO_2 hetero-nanofibers: controlled synthesis and high efficiency of acetone vapor detection*, Front. Chem. 7 (2019).

[60] H.S. Koh, et al., Enhancement of neurite outgrowth using nano-structured scaffolds coupled with laminin, Biomaterials 29 (26) (2008) 3574−3582.

[61] A.C.B. Vicente, et al., Influence of process variables on the yield and diameter of zein-poly(N-isopropylacrylamide) fiber blends obtained by electrospinning, J. Mol. Liq. 292 (2019) 109971.

[62] A.-F. Che, X.-J. Huang, Z.-K. Xu, Polyacrylonitrile-based nanofibrous membrane with glycosylated surface for lectin affinity adsorption, J. Membr. Sci. 366 (1) (2011) 272−277.

[63] K.H. Na, et al., *Microstructure of $Ni_{0.5}Zn_{0.5}Fe_2O_4$ nanofiber with metal nitrates in electrospinning precursor*, Nanomaterials (Basel) 10 (7) (2020).

[64] Y. Wang, et al., Electrospun flexible self-standing γ-alumina fibrous membranes and their potential as high-efficiency fine particulate filtration media, J. Mater. Chem. A 2 (36) (2014) 15124−15131.

[65] O. Sandoval, et al., Morphological, electrical, and chemical characteristics of poly (sodium 4-styrenesulfonate) coated PVDF ultrafiltration membranes after plasma treatment, Polymers 11 (10) (2019) 1689.

[66] Z. Ma, et al., Surface engineering of electrospun polyethylene terephthalate (PET) nanofibers towards development of a new material for blood vessel engineering, Biomaterials 26 (15) (2005) 2527−2536.

[67] E. Mudra, et al., *Development of Al_2O_3 electrospun fibers prepared by conventional sintering method or plasma assisted surface calcination.*, Appl. Surf. Sci. 415 (2017) 90−98.

[68] K. Castkova, et al., Electrospinning and thermal treatment of yttria doped zirconia fibres, Ceram. Int. 43 (10) (2017) 7581−7587.

[69] L. Wang, et al., *Flexible TiO_2 ceramic fibers near-infrared reflective membrane fabricated by electrospinning.*, Ceram. Int. 45 (6) (2019) 6959−6965.

[70] C. Li, et al., *Preparation and characterization of Al_2O_3/SiO_2 composite nanofibers by using electrostatic spinning method.*, Inorg. Nano-Metal Chem. 47 (9) (2016) 1275−1278.

[71] Q.K. Li, et al., Chemical modification of electrospun yttrium silicate fiber with self-healing properties, J. Sol-Gel Sci. Technol. 93 (1) (2019) 142−148.

[72] P.S. Curti, et al., Surface modification of polystyrene and poly(ethylene terephtalate) by grafting poly(N-isopropylacrylamide), J. Mater. Sci. Mater. Med. 13 (12) (2002) 1175−1180.

[73] C. Zhang, et al., Electrospinning of nanofibers: potentials and perspectives for active food packaging, Compr. Rev. Food Sci. Food Saf. 19 (2) (2020) 479−502.

[74] N.H. Ladizesky, I.M. Ward, A review of plasma treatment and the clinical application of polyethylene fibres to reinforcement of acrylic resins, J. Mater. Sci. Mater. Med. 6 (9) (1995) 497−504.

[75] S. Li, et al., Core-sheath zirconia/silica microfibers for dental composites reinforcement, Mater. Lett. 142 (2015) 204−206.

[76] J. Rajala, et al., Core−shell electrospun hollow aluminum oxide ceramic fibers, Fibers 3 (4) (2015) 450−462.

[77] J.T. McCann, D. Li, Y. Xia, Electrospinning of nanofibers with core-sheath, hollow, or porous structures, J. Mater. Chem. 15 (7) (2005) 735.

[78] J.-J. Li, et al., Fast dissolving drug delivery membrane based on the ultra-thin shell of electrospun core-shell nanofibers, Eur. J. Pharm. Sci. 122 (2018) 195−204.

[79] A.K. Moghe, B.S. Gupta, Co-axial electrospinning for nanofiber structures: preparation and applications, Polym. Rev. 48 (2) (2008) 353−377.

[80] Y. Zhao, X. Cao, L. Jiang, Bio-mimic multichannel microtubes by a facile method, J. Am. Chem. Soc. 129 (4) (2007) 764−765.

[81] S.-H. Choi, et al., Hollow ZnO nanofibers fabricated using electrospun polymer templates and their electronic transport properties, ACS Nano 3 (9) (2009) 2623−2631.

[82] Z. Liu, et al., *An efficient bicomponent TiO_2/SnO_2 nanofiber photocatalyst fabricated by electrospinning with a side-by-side dual spinneret method.*, Nano Lett. 7 (4) (2007) 1081−1085.

[83] J.D. Starr, M.A.K. Budi, J.S. Andrew, Processing-property relationships in electrospun Janus-type biphasic ceramic nanofibers, J. Am. Ceram. Soc. 98 (1) (2015) 12−19.

[84] W. Zhang, et al., Structural design and environmental applications of electrospun nanofibers, Compos. Part A: Appl. Sci. Manuf. 137 (2020) 106009.

[85] X. Peng, et al., Fabrication and enhanced photocatalytic activity of inorganic core−shell nanofibers produced by coaxial electrospinning, Chem. Sci. 3 (4) (2012) 1262−1272.

[86] G. Duan, A. Greiner, Air-blowing-assisted coaxial electrospinning toward high productivity of core/sheath and hollow fibers, Macromol. Mater. Eng. 304 (5) (2019) 1800669.

[87] M.-J. Chang, et al., *Construction of novel $TiO_2/Bi_4Ti_3O_{12}/MoS_2$ core/shell nanofibers for enhanced visible light photocatalysis.*, J. Mater. Sci. Technol. 36 (2020) 97−105.

[88] Y. Dai, et al., Ceramic nanofibers fabricated by electrospinning and their applications in catalysis, environmental science, and energy technology, Polym. Adv. Technol. 22 (3) (2011) 326−338.

[89] Y. Cheng, et al., Formation mechanism of Fe2O3 hollow fibers by direct annealing of the electrospun composite fibers and their magnetic, electrochemical properties, CrystEngComm 13 (8) (2011) 2863−2870.

[90] K.-H. Roh, D.C. Martin, J. Lahann, Biphasic Janus particles with nanoscale anisotropy, Nat. Mater. 4 (10) (2005) 759–763.

[91] J.D. Starr, J.S. Andrew, Janus-type bi-phasic functional nanofibers, Chem. Commun. 49 (39) (2013) 4151–4153.

[92] X. Xi, et al., Flexible Janus nanofiber: a new tactics to realize tunable and enhanced magnetic-luminescent bifunction, Chem. Eng. J. 254 (2014) 259–267.

[93] J. Xue, et al., Electrospinning and electrospun nanofibers: methods, materials, and applications, Chem. Rev. 119 (8) (2019) 5298–5415.

[94] Y. Si, et al., Ultralight nanofibre-assembled cellular aerogels with superelasticity and multifunctionality, Nat. Commun. 5 (1) (2014) 5802.

[95] J.-M. Lim, et al., Superhydrophobic films of electrospun fibers with multiple-scale surface morphology, Langmuir 23 (15) (2007) 7981–7989.

[96] H.-Y. Chen, et al., *Electrospun hierarchical TiO2 nanorods with high porosity for efficient dye-sensitized solar cells.*, ACS Appl. Mater. Interfaces 5 (18) (2013) 9205–9211.

[97] K.S. Park, et al., Exceptional chemical and thermal stability of zeolitic imidazolate frameworks, Proc. Natl. Acad. Sci. 103 (27) (2006) 10186.

[98] G. Lee, Y.D. Seo, J. Jang, ZnO quantum dot-decorated carbon nanofibers derived from electrospun ZIF-8/PVA nanofibers for high-performance energy storage electrodes, Chem. Commun. 53 (83) (2017) 11441–11444.

[99] Y. Yu, et al., *Electrospun SiO2 aerogel/polyacrylonitrile composited nanofibers with enhanced adsorption performance of volatile organic compounds.*, Appl. Surf. Sci. 512 (2020) 145697.

[100] R. Liu, et al., Ultralight, thermal insulating, and high-temperature-resistant mullite-based nanofibrous aerogels, Chem. Eng. J. 360 (2019) 464–472.

[101] C. Niu, et al., General synthesis of complex nanotubes by gradient electrospinning and controlled pyrolysis, Nat. Commun. 6 (1) (2015) 7402.

[102] X. Wang, B.S. Hsiao, Electrospun nanofiber membranes, Curr. Opin. Chem. Eng. 12 (2016) 62–81.

[103] X. Lu, C. Wang, Y. Wei, One-dimensional composite nanomaterials: synthesis by electrospinning and their applications, Small 5 (21) (2009) 2349–2370.

[104] Y. Wang, et al., Self-assembly of ultralight and compressible inorganic sponges with hierarchical porosity by electrospinning, Ceram. Int. 46 (1) (2020) 768–774.

[105] H.-Y. Mi, et al., Fabrication of fibrous silica sponges by self-assembly electrospinning and their application in tissue engineering for three-dimensional tissue regeneration, Chem. Eng. J. 331 (2018) 652–662.

[106] C. Lee, et al., *Ag supported on electrospun macro-structure CeO2 fibrous mats for diesel soot oxidation.*, Appl. Catal. B: Environ. 174–175 (2015) 185–192.

[107] B. Robert, G. Nallathambi, *A concise review on electrospun nanofibres/nanonets for filtration of gaseous and solid constituents (PM2.5) from polluted air*, Colloid Interface Sci. Commun. 37 (2020) 100275.

[108] F. Lu, et al., Uniform deposition of Ag nanoparticles on ZnO nanorod arrays grown on polyimide/Ag nanofibers by electrospinning, hydrothermal, and photoreduction processes, Mater. Des. 181 (2019) 108069.

[109] Z. Wang, et al., The antibacterial polyamide 6-ZnO hierarchical nanofibers fabricated by atomic layer deposition and hydrothermal growth, Nanoscale Res. Lett. 12 (1) (2017).

[110] F. Lu, et al., *3D hierarchical carbon nanofibers/TiO2@MoS2 core-shell heterostructures by electrospinning, hydrothermal and in-situ growth for flexible electrode materials.*, Mater. Des. 189 (2020) 108503.

[111] X. Zhu, et al., *Perfluorinated superhydrophobic and oleophobic SiO₂@PTFE nanofiber membrane with hierarchical nanostructures for oily fume purification.*, J. Membr. Sci. 594 (2020) 117473.

[112] K.T. Alali, et al., *Preparation and characterization of ZnO/CoNiO₂ hollow nanofibers by electrospinning method with enhanced gas sensing properties.*, J. Alloy. Compd. 702 (2017) 20−30.

[113] W. Fu, et al., *Stabilizing 3 nm-Pt nanoparticles in close proximity on rutile nanorods-decorated-TiO₂ nanofibers by improving support uniformity for catalytic reactions*, Chem. Eng. J. (2020) 126013.

[114] Q. Zhan, et al., *Graphene-based modulation on the hierarchical growth of Al₂O₃ heterojunctions outside TiO₂ nanofibers via a surfactant-free approach*, Compos. Commun. 21 (2020) 100394.

[115] C.-L. Zhang, et al., Hierarchically structured Co3O4@carbon porous fibers derived from electrospun ZIF-67/PAN nanofibers as anodes for lithium ion batteries, J. Mater. Chem. A 6 (27) (2018) 12962−12968.

[116] X. Wu, et al., *Thorn-like flexible Ag₂C₂O₄/TiO₂ nanofibers as hierarchical heterojunction photocatalysts for efficient visible-light-driven bacteria-killing.*, J. Colloid Interface Sci. 560 (2020) 681−689.

[117] W. Han, et al., *Self-assembly of perovskite crystals anchored Al₂O₃-La₂O₃ nanofibrous membranes with robust flexibility and luminescence*, Small 14 (45) (2018) e1801963.

[118] F. Kayaci, et al., Enhanced photocatalytic activity of homoassembled ZnO nanostructures on electrospun polymeric nanofibers: a combination of atomic layer deposition and hydrothermal growth, Appl. Catal. B: Environ. 156−157 (2014) 173−183.

[119] S. Camarero-Espinosa, I. Stefani, J. Cooper-White, Hierarchical "as-electrospun" self-assembled fibrous scaffolds deconvolute impacts of chemically defined extracellular matrix- and cell adhesion-type interactions on stem cell haptokinesis, ACS Macro Lett. 6 (12) (2017) 1420−1425.

[120] R. Rega, et al., Maskless arrayed nanofiber mats by bipolar pyroelectrospinning, ACS Appl. Mater. Interfaces 11 (3) (2019) 3382−3387.

[121] I. Borisova, et al., Modulating the mechanical properties of electrospun PHB/PCL materials by using different types of collectors and heat sealing, Polymers (Basel) 12 (3) (2020).

[122] H. Yang, et al., Molecular orientation in aligned electrospun polyimide nanofibers by polarized FT-IR spectroscopy, Spectrochim. Acta Part A: Mol. Biomol. Spectrosc. 200 (2018) 339−344.

[123] C. Yang, et al., Electrospun pH-sensitive core−shell polymer nanocomposites fabricated using a tri-axial process, Acta Biomater. 35 (2016) 77−86.

[124] Q.P. Pham, U. Sharma, A.G. Mikos, Electrospun poly(ε-caprolactone) microfiber and multilayer nanofiber/microfiber scaffolds: characterization of scaffolds and measurement of cellular infiltration, Biomacromolecules 7 (10) (2006) 2796−2805.

[125] J. Wang, et al., *Fabrication of a well-aligned TiO₂ nanofibrous membrane by modified parallel electrode configuration with enhanced photocatalytic performance.*, RSC Adv. 6 (37) (2016) 31476−31483.

[126] T. Krishnamoorthy, et al., *A facile route to vertically aligned electrospun SnO₂ nanowires on a transparent conducting oxide substrate for dye-sensitized solar cells.*, J. Mater. Chem. 22 (5) (2012) 2166−2172.

[127] N. Saveh-Shemshaki, R. Bagherzadeh, M. Latifi, Electrospun metal oxide nanofibrous mat as a transparent conductive layer, Org. Electron. 70 (2019) 131−139.

[128] G. Cadafalch Gazquez, et al., Influence of solution properties and process parameters on the formation and morphology of YSZ and NiO ceramic nanofibers by electrospinning, Nanomaterials (Basel) 7 (1) (2017).

[129] D. Li, Y. Wang, Y. Xia, Electrospinning of polymeric and ceramic nanofibers as uniaxially zligned arrays, Nano Lett. 3 (8) (2003) 1167−1171.

[130] M. Pokorny, K. Niedoba, V. Velebny, Transversal electrostatic strength of patterned collector affecting alignment of electrospun nanofibers, Appl. Phys. Lett. 96 (19) (2010) 193111.

[131] Y.-H. Wu, et al., Effective utilization of the electrostatic repulsion for improved alignment of electrospun nanofibers, J. Nanomater. 2016 (2016) 1−8.

[132] G. Cadafalch Gazquez, et al., Low-cost, large-area, facile, and rapid fabrication of aligned ZnO nanowire device arrays, ACS Appl. Mater. Interfaces 8 (21) (2016) 13466−13471.

[133] D. Yang, et al., Fabrication of aligned fibrous arrays by magnetic electrospinning, Adv. Mater. 19 (21) (2007) 3702−3706.

[134] T. Wang, et al., Preparation of ordered TiO_2 nanofibers/nanotubes by magnetic field assisted electrospinning and the study of their photocatalytic properties, Ceram. Int. 45 (11) (2019) 14404−14410.

[135] J. Zheng, et al., Magnetic-electrospinning synthesis of γ-Fe_2O_3 nanoparticle−embedded flexible nanofibrous films for electromagnetic dhielding, Polymers 12 (3) (2020) 695.

[136] S.M. Park, et al., Direct fabrication of spatially patterned or aligned electrospun nanofiber mats on dielectric polymer surfaces, Chem. Eng. J. 335 (2018) 712−719.

[137] C. Jia, et al., Patterned electrospun nanofiber matrices via localized dissolution: potential for guided tissue formation, Adv. Mater. 26 (48) (2014) 8192−8197.

[138] D. Zhang, J. Chang, Patterning of electrospun fibers using electroconductive templates, Adv. Mater. 19 (21) (2007) 3664−3667.

[139] Z. Ding, A. Salim, B. Ziaie, Selective nanofiber deposition through field-enhanced electrospinning, Langmuir 25 (17) (2009) 9648−9652.

[140] S.J. Cho, et al., Replicable multilayered nanofibrous patterns on a flexible film, Langmuir 26 (18) (2010) 14395−14399.

[141] D. Li, et al., Collecting electrospun nanofibers with patterned electrodes, Nano Lett. 5 (5) (2005) 913−916.

[142] S. Zhao, et al., Nanofibrous patterns by direct electrospinning of nanofibers onto topographically structured non-conductive substrates, Nanoscale 5 (11) (2013) 4993−5000.

[143] M.M. Munir, et al., Patterned indium tin oxide nanofiber films and their electrical and optical performance, Nanotechnology 19 (37) (2008) 375601.

[144] J. Zheng, et al., Electrospun aligned fibrous arrays and twisted ropes: fabrication, mechanical and electrical properties, and application in strain sensors, Nanoscale Res. Lett. 10 (1) (2015).

[145] S. Hu, et al., *Parallel patterning of SiO2 wafer via near-field electrospinning of metallic salts and polymeric solution mixtures.*, Nanotechnology 28 (41) (2017) 415301.

[146] K. Lim, et al., Metal oxide patterns of one-dimensional nanofibers: on-demand, direct-write fabrication, and application as a novel platform for gas detection, J. Mater. Chem. A 7 (43) (2019) 24919−24928.

[147] K. Kang, et al., Micropatterning of metal oxide nanofibers by electrohydrodynamic (EHD) printing towards highly integrated and multiplexed gas sensor applications, Sens. Actuators B: Chem. 250 (2017) 574−583.

[148] L. Huang, S.S. Manickam, J.R. McCutcheon, Increasing strength of electrospun nanofiber membranes for water filtration using solvent vapor, J. Membr. Sci. 436 (2013) 213–220.

[149] C. Su, et al., Dilute solvent welding: a quick and scalable approach for enhancing the mechanical properties and narrowing the pore size distribution of electrospun nanofibrous membrane, J. Membr. Sci. 595 (2020) 117548.

[150] Y. Cui, et al., High performance electronic devices based on nanofibers via a cross-linking welding process, Nanoscale 10 (41) (2018) 19427–19434.

[151] C. Fu, et al., Self-welding and low-temperature formation of metal oxide nanofiber networks and its application to electronic devices, IEEE Electron. Device Lett. 41 (1) (2020) 62–65.

[152] S. Cheon, et al., High-performance triboelectric nanogenerators based on electrospun polyvinylidene fluoride-silver nanowire composite nanofibers, Adv. Funct. Mater. 28 (2) (2018) 1703778.

[153] M. Cho, et al., Enhancement of the mechanical properties of electrospun lignin-based nanofibers by heat treatment, J. Mater. Sci. 52 (16) (2017) 9602–9614.

[154] J.H. Park, et al., Flash-induced self-limited plasmonic welding of silver nanowire network for transparent flexible energy harvester, Adv. Mater. 29 (5) (2017) 1603473.

[155] W. Chen, et al., Superelastic, superabsorbent and 3D nanofiber-assembled scaffold for tissue engineering, Colloids Surf. B: Biointerfaces 142 (2016) 165–172.

[156] P. Moradipour, et al., Fabrication and characterization of new bulky layer mixed metal oxide ceramic nanofibers through two nozzle electrospinning method, Ceram. Int. 42 (12) (2016) 13449–13458.

[157] T. Xu, et al., Three-dimensional monolithic porous structures assembled from fragmented electrospun nanofiber mats/membranes: methods, properties, and applications, Prog. Mater. Sci. 112 (2020) 100656.

[158] T.-W. Kim, S.-J. Park, Synthesis of reduced graphene oxide/thorn-like titanium dioxide nanofiber aerogels with enhanced electrochemical performance for supercapacitor, J. Colloid Interface Sci. 486 (2017) 287–295.

[159] T. Pirzada, et al., Cellulose silica hybrid nanofiber aerogels: from sol–gel electrospun nanofibers to multifunctional aerogels, Adv. Funct. Mater. 30 (5) (2019) 1907359.

[160] X. Meng, et al., *Coupling of hierarchical Al_2O_3/TiO_2 nanofibers into 3D photothermal aerogels toward simultaneous water evaporation and purification*, Adv. Fiber Mater. 2 (2) (2020) 93–104.

[161] L. Dou, et al., Hierarchical cellular structured ceramic nanofibrous aerogels with temperature-invariant superelasticity for thermal insulation, ACS Appl. Mater. Interfaces 11 (32) (2019) 29056–29064.

[162] Y. Si, et al., Ultralight and fire-resistant ceramic nanofibrous aerogels with temperature-invariant superelasticity, Sci. Adv. 4 (4) (2020) eaas8925.

[163] E.-J. Lee, et al., *Advanced multi-nozzle electrospun functionalized titanium dioxide/ polyvinylidene fluoride-co-hexafluoropropylene (TiO_2/PVDF-HFP) composite membranes for direct contact membrane distillation.*, J. Membr. Sci. 524 (2017) 712–720.

[164] L. Wei, et al., Mass production of nanofibers from needleless electrospinning by a novel annular spinneret, Mater. Des. 179 (2019) 107885.

[165] T.-L. Chang, et al., Direct fabrication of nanofiber scaffolds in pillar-based microfluidic device by using electrospinning and picosecond laser pulses, Microelectron. Eng. 177 (2017) 52–58.

Nonelectrospun metal oxide nanofibers

3

Alsiad Ahmed Almetwally
Textile Engineering Department, Textile Research Division, National Research Center, Dokki, Cairo, Egypt

3.1 Introduction

In general, according to the British Standards Institution, a nanomaterial can be defined as any material whose external dimensions, or its surface or internal structure, are in the nanoscale range (100 nm) [1,2]. In addition, the National Science Foundation defined nanofibers as fibers with a size of less than or equal to 100 nm [3].

Nanofibers have similar dimensions on their x- and y-axes (for their diameter, $x = y$) in the nanoscale range, whereas the third dimension (nanofiber length in the z-axis) is substantially larger. There are many types of nanofibers, which can be classified according to their composition (carbon-based, organic, inorganic-based, and composite nanofibers), natural basis (engineered, human-made, and natural nanofibers), rigidity (stiff or flexible), and structure (hollow, solid, porous, nonporous, and core-shell) [4,5].

Up to now, different approaches for manufacturing of nanofibers have been described in the literature under different rankings and names. For example, nanofiber production techniques can be categorized as top-down and bottom-up. The latter technique to fabricate nanofibers includes chemical vapor deposition, physical vapor deposition, blowing solution [6], drawing methods [7], centrifugal spinning [8], self-assembly [9], template synthesis [10], phase separation [11], interfacial polymerization [12], freeze-drying synthesis [13], and electrospinning [14]. In bottom-up approaches, nanofibers are fabricated from ions, atoms, molecules, and nanoparticles. However, refining, grinding, sequential cutting, or milling of a larger bulk material is the key process to produce nanofibers in the top-down approach [15]. In this chapter, nanofiber production techniques are categorized into electrospinning and nonelectrospinning techniques. The electrospinning method to produce nanofibers depends mainly on an electric voltage that controls fiber morphology, while the nonelectrospinning methods use other forces such as centrifugal force and air pressure to control the nanofiber morphology [16]. This chapter describes nonelectrospinning techniques that are used to fabricate nanofibers, especially the synthesis of metal oxide nanofibers.

3.2 Nonelectrospinning techniques

Although the electrospinning technique is the best-known technique that has been used to fabricate nanofibers since the end of the 20th century, it has some

Metal Oxide-Based Nanofibers and Their Applications. DOI: https://doi.org/10.1016/B978-0-12-820629-4.00019-9
© 2022 Elsevier Inc. All rights reserved.

drawbacks. The disadvantages of the electrospinning method include the low production rate, higher cost compared to that of other conventional methods, and safety and health-related problems owing to the emission of vapors and use of high voltage. Furthermore, the electrospinning technique is not suitable for the mass production of certain materials [17,18]. Owing to the limitations of the electrospinning method and to improve nanofiber production, recently, most researchers have relied on other production techniques to fabricate nanofibers.

These approaches include centrifugal spinning, template synthesis, phase separation, drawing, freeze-drying synthesis, solution blowing, self-assembly, interfacial polymerization, plasma-induced technique, laser spinning, and splitting techniques. Among these, and on the basis of scalability and industrialization, solution blowing and centrifugal spinning are the most advantageous techniques. However, for some polymers and their parameters, other methods (e.g., phase separation, template synthesis, and freeze-drying synthesis) may need to be used.

3.2.1 Solution blow spinning technique

Solution blow spinning (SBS), which was introduced by Medeiros et al. [19], uses a combination of electrospinning, melt spinning, and solution spinning techniques to produce fibers with a diameter of a few tens of nanometers. This technique is developed primarily to overcome the drawbacks of melt spinning and electrospinning techniques [20]. The spinning apparatus for the solution blowing technique consists mainly of a compressed gas source supplied with a pressure valve, a 5-mL hypodermic syringe, an injection pump (KD Scientific, United States) to control the polymer solution flow rate, a concentric, and a collector with an adjustable rotation speed. The most commonly used gases are argon, air, and nitrogen. The working distance between the collector and the nozzle is always constant. The spraying apparatus consists mainly of concentric outer and inner nozzles. The polymer solution is pumped through the inner nozzle, whereas the pressurized high-velocity gas is delivered through the concentric outer nozzle [19].

On the basis of Bernoulli's principle, the velocity of air increases, owing to its rapid decompression. Therefore, according to the viscous interaction between the polymer solution and air, a conical-shaped structure is created at the exit of the inner nozzle. With the aid of viscous forces of the flowing air that overcome the surface tension forces, polymeric nanofibers are dragged from the formed cone. The emerging nanofibers are collected on the controllable rotating drum placed 200 mm away from the nozzle system [21,22].

Although SBS nanofibers are similar to those produced by the electrospinning technique, they are superior to electrospun nanofibers in terms of low cost, ease of implementation, and high production rate [23]. In addition, the blow spinning technique can use both polar and nonpolar solvents. In addition, the industrial models that are used for nanofibers production have been efficiently developed, owing to the simplicity and feasibility of SBS [24,25].

Gas pressure, feeding rate, polymer concentration, solvent volatility, solution viscosity, and polymer type are considered to be the key processing parameters of

SBS. These variables were determined to have a considerable impact on fiber diameter and characteristics [26–29].

3.2.2 Plasma-induced technique

In this technique, nanofibers can be obtained by creating plasma between two metal electrodes placed in water without the need for any inert gases or chemicals. The bombardment of energetic radicals on the electrodes' surface causes the formation of nanofibers. Owing to the bombardment by radicals, the vapor and cloud of atoms are emitted in the area containing plasma. Owing to the considerable difference in the temperature and pressure inside water, the created cloud of atoms is condensed, and the plasma is expanded accordingly. Eventually, oxygen (which is present in water) interacts with the metallic atoms, resulting in the formation of nanofibers. Lately, copper oxide (CuO) nanofibers with diameters in the range of 15–25 nm have been efficiently synthesized by using the plasma-induced water technique [30]. A schematic representation of experimental setup for a plasma-induced method is shown in Fig. 3.1.

3.2.3 Drawing technique

In the drawing method, single nanofibers can be produced differently from electrospinning. This technique utilizes a spinneret with a free rotating micropipette and a substrate to draw polymeric individual fibers on its flat surface. As shown in Fig. 3.1, polymer droplets are formed by using a sharp probe tip. Depending on the polymer type and using a specific rate, individual nanofibers can be formed from

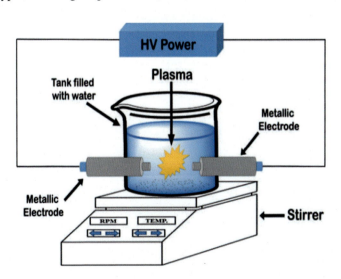

Figure 3.1 Graphical representation of an experimental setup for the plasma-induced technique [30].

the droplets [31]. Owing to the deposition of a polymer solution that is extruded from the micropipette onto the substrate, the movable spinneret of the micropipette pulls the solution to form single nanofibers. The most important processing parameters, which determine the spinnability of this technique, are the polymer solution concentration, polymer type, spinneret rotating movement, polymer solution viscosity, surface tension, and substrate shape. The major advantages of the drawing technique are the minimal demands of the system and the capability of studying nanofibers individually. However, the major disadvantage is that nanofibers can be drawn only for a specific amount of time [7,31−33]. To overcome this problem, it is recommended to use a hollow glass micropipette in which the quantity of the polymer remains constant. This provides more flexibility in drawing speed, in viscosity, and in waiting time before drawing. Thus the diameter of the produced fibers can be better controlled, and fiber drawing can be performed repeatedly [34].

In general, there are two variants of draw spinning: molten draw spinning and solution draw spinning. Basalt fibers or glass fibers with diameters higher than several micrometers are widely used by the molten draw spinning technique, while solution draw spinning is used mainly for producing nanofibers [35,36].

3.2.4 CO_2 laser supersonic drawing

An indefinite length of nanofibers can be produced by using the CO_2 laser supersonic drawing technique, which is commonly referred to as CLSD; this technique relies on a single continuous operation without the help of any additional chemical solvents. The system is supplied with 100- to 200-μm synthetic fibers that are melted and drawn into nanofibers with the help of irradiation by a carbon dioxide laser and the force of air with a supersonic flow. In general, this technique can be applied to drawing a wide range of thermoplastic polymers, such as polylactic acid, polyglycolic acid, and polyethylene terephthalate [37]. Nylon 66 nanofibers, whose melting point can be achieved by a CO_2 laser, were synthesized by using the CLSD method. Compared to electrospinning, this simple technique provides better environmental safety; in addition, the fabricated nanofibers do not scatter in air and have extended chains that improve their mechanical characteristics [38]. A schematic drawing of a CO_2 laser supersonic drawing apparatus to fabricate polyethylene terephthalate nanofibers is shown in Fig. 3.2 [39].

3.2.5 Template synthesis

Template synthesis is another commonly used method for the fabrication of polymeric, metallic, semiconducting, carbon, and ceramic nanofibers using a nonporous metal oxide membrane [40−42]. In this technique, nanofibers are formed by passing polymer solutions through the pores of the membrane with a nanoscale diameter thickness by applying water pressure. Once the fibers come out from the membrane pores, they will be solidified in the solidification solution [43,44]. Long nanofibers cannot be produced by this technique. The main advantage of this method is that nanofibers with different diameters can be produced by using different templates.

Nonelectrospun metal oxide nanofibers

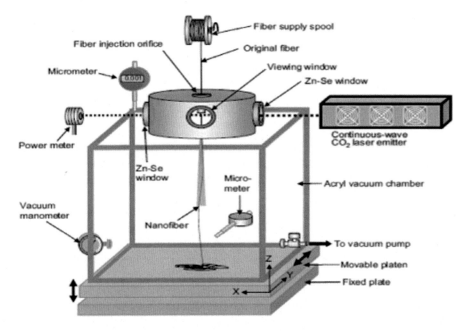

Figure 3.2 Schematic drawing of CO$_2$ laser supersonic drawing [39].

In general, the diameters of the resultant fibers range from a few nanometers to hundreds of nanometers [45].

3.2.6 Centrifugal spinning

Centrifugal spinning, or Forcespinning, is a recently developed nanofiber forming method. This approach has attracted extensive interest, mainly owing to its high production rate, which is 500 times faster than that of traditional electrospinning [46]. In addition, centrifugal spinning allows production of nanofibers from polymers with high concentration, which reduces the amount of used solvent and decreases the production cost [47].

Rather than using an electrostatic force, centrifugal spinning relies on centrifugal force to realize the high-rate production of nanofibers [48]. Centrifugal spinning can be used to fabricate nanofibers from polymer solutions or polymer melts without the dielectric constant restrictions and the involvement of a high-voltage electric field. Carbon, ceramic, and metal fibers can also be fabricated by centrifugal spinning [49,50]. The centrifugal spinning process was initially developed in 1924 by Hooper to produce artificial silk fibers from viscose by applying centrifugal forces to a viscous material [51]. Therefore this method has been used for fiber production since it was established by Hooper. The fiber formation process of centrifugal spinning relies on the competition between the centrifugal force and the Laplace force (arises from surface curvature) [52]. During centrifugal spinning, the nanofiber formation process can be divided into three stages: (1) jet-initiation to force the polymer solution stream through the orifice, (2) jet extension to enhance the surface area of the forced polymer stream, and (3) solvent evaporation to harden and shrink the polymer jet, as shown in Fig. 3.3 [53].

During the initial step, a combination of centrifugal and hydrostatic pressure at the capillary end exceeds the flow-resistant capillary force and forces the liquid polymer through the nozzle capillary as a jet [54]. The external radial centrifugal force stretches the polymer jet when it extends toward the collector wall; however, the jet moves in a warped curve, owing to the rotation-dependent inertia. The stretching of the extruded polymer jet is essential for the jet diameter reduction over the distance from the nozzle to the collector. At the same time, the solvent in the polymer solution evaporates, solidifies, and contracts the jet. The solvent evaporation rate depends on its stability. For a highly volatile solvent, the jets form a thicker fiber because fast evaporation potentiates fast solidification, which hinders the jet extension [55]. Regarding centrifugal spinning, the parameters that affect the structure of the resultant nanofibers and the spinning process include solution surface tension, polymer viscoelasticity, orifice diameter, spinneret angular velocity, solvent evaporation rate, temperature, and the distance between the nozzle and the collector [43,56]. Finally, the main drawback of centrifugal spinning is that productivity and fiber quality are considerably affected by the spinneret configuration and material characteristics [57].

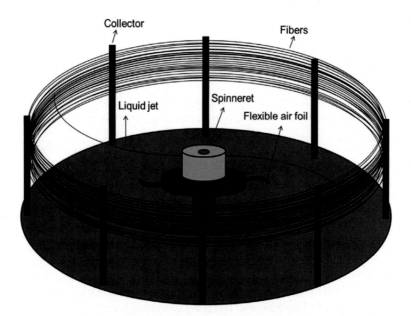

Figure 3.3 Schematic drawing of the centrifugal spinning process [53].

3.3 Synthesis of metal oxide nanofibers

Owing to their distinctive physicochemical and electrical characteristics, metal oxide nanofibers have proven to be excellent candidates for various applications, such as biosensors, lithium-ion batteries, light-emitting diodes, photovoltaic cells, liquid crystal displays, and gas sensors [58].

Metal oxide nanofibers, such as tin oxide (SnO_2), zinc oxide (ZnO), titanium dioxide (TiO_2), nickel oxide (NiO), ferric oxide (Fe_2O_3), CuO, silica (SiO_2), and cerium oxide (CeO_2), can be synthesized by using electrospinning and nonelectrospinning techniques [59−65]. The following subsections shed light on the fabrication of metal oxide nanofibers using nonelectrospinning methods.

3.3.1 Tin oxide nanofibers

Tin oxide nanostructures are widely used in different applications, such as photosensors, gas sensors, transparent electrodes, solar cells, ultraviolet detection, antistatic coating, and medical devices [66,67].

Haolun et al. [68] prepared SnO_2 nanofibers by using the SBS technique. First, they prepared a precursor solution that was similar to the one that was used for electrospinning. A total of 0.6 g of polyvinyl butyral (PVB) and 1.4 g of tin chloride pentahydrate ($SnCl_4 \cdot 5H_2O$) were added to 8 g of ethanol (EtOH). After being stirred for 6 h at room temperature, the precursor solution was transferred into a 1-mL injector. Then the blow spinning setup was operated, using the following parameters: The injection rate was 3 mL h^{-1}; the air flow rate was 10 ms^{-1}; the size of the needle was 30 gauge (outer diameter = 0.31 mm; inner diameter = 0.16 mm); and the distance between the needle tip and the collector was 10 cm. After the setup was run for 10 min, an $SnCl_4$/PVB nanofiber sponge was obtained.

The nanofiber sponge was immediately moved to the muffle. The nanofiber sponge was heated at 500°C in air for 100 min with a heating rate of 2°C min^{-1}; thus SnO_2 nanofibers were obtained. Haolun et al. have also reported that the diameter of SnO_2 nanofibers was approximately 500 nm and the distribution of the fibers and pores was uniform. In addition, scanning electron microscopy (SEM) images revealed that the surface of the SnO_2 nanofibers was porous and rough, which meant that it had a high specific area

3.3.2 Silica nanofibers

Calisir and Kilic [69] prepared SiO_2 nanofibers by two nonelectrospinning routes (centrifugal spinning (CS) and SBS) and compared the obtained materials. After calcination, amorphous SiO_2 was obtained by the two methods. First, a precursor solution was prepared by adding tetraethyl orthosilicate (TEOS) (reagent 98%) to a 10-wt.% solution composed of polyvinylpyrrolidone (PVP) (Mw = 1,300,000) and EtOH (Merck, 95%). The concentration of EtOH in the precursor solution was 15%. The solution was stirred until it became clear. Then distilled water and 1-M hydrochloric acid (HCl) (37 wt.%) as a catalyst was added to the TEOS-PVP-EtOH solution. The molar ratios of H_2O:TEOS and HCL:TEOS were 2.65 and 0.015, respectively. To obtain the final solution with the viscosity of 53.5 mPa·s to be used directly in the spinning devices, the prepared solution should be mixed for 3 h to confirm the condensation reaction and hydrolysis.

The precursor solution was fed into a rotating spinneret and into a concentric nozzle in the CS and SBS devices, respectively. The CS parameters were as follows: The spinneret rotating speed was 7000 rpm, and the feeding rate was 7.5 mL h^{-1}, and the needle size was of gauge 22. The spinning parameters of the SBS were as follows: The air pressure was 0.5 bar, and the inner diameter of the concentric nozzle was 0.4 μm. The feeding rate was the same as that of the CS device. The fibers produces by both spinning systems were calcined for 1 h at 600°C in a muffle furnace and were allowed to cool down to 25°C. The SEM images of both types of SiO_2 fibers before and after calcinations are shown in Fig. 3.4.

In addition, Calisir and Kilic stated that owing to the higher shear forces associated with SBS, the fiber structures made by this technique differed considerably from those spun by CS. These researchers pointed out that CS-SiO_2 nanofiber mats were fluffy, were free from defects, and had a larger diameter than those spun on the SB device. The lower pore size and fiber diameter associated with SB-SiO_2 nanofibers restrict their use to separation and filtration applications. However, the fluffy characteristics of CS-SiO_2 mats made them suitable for thermal insulation applications.

3.3.3 Barium titanate nanofibers

Owing to its strong characteristics in microwave absorption and high-performance capacitors and its capability to increase the density of ferroelectric nonvolatile memories, barium titanate ($BaTiO_3$) has been broadly investigated [70–75]. The high aspect ratio of $BaTiO_3$ nanostructures makes them suitable for use in miniature piezoelectric transducers and actuators, switchable aircraft antenna systems, ultrasonic devices, and medical imaging detectors [76–78].

Liyun and Shiva [79] fabricated $BaTiO_3$ ceramic nanofibers by using centrifugal jet spinning. First, they prepared the spinning solution as a precursor of $BaTiO_3$ using the sol-gel technique. The starting materials include titanium isopropoxide and barium acetate with a 1:1 molar ratio, which were used to obtain stoichiometric $BaTiO_3$. First, a solution composed of 6 mL of acetic acid and 1.275 g of Ba $(CH_3CO)_2$ was prepared by stirring for 10 min. Then, with continuous stirring, 1.475 mL of $[(CH_3)_2CHO]_4Ti$ was added dropwise. Without purification, EtOH (200 proof) and PVP (Mw = 1,300,000 Da; PDI = 2.5) were also used. To prepare 10 wt.% of a homogeneous polymer solution, PVP was dissolved in EtOH and exposed to sound waves for 10 min. Then the obtained PVP-EtOH solution was added continuously to the $BaTiO_3$ sol-gel with strong stirring followed by vortexing.

Nonelectrospun metal oxide nanofibers

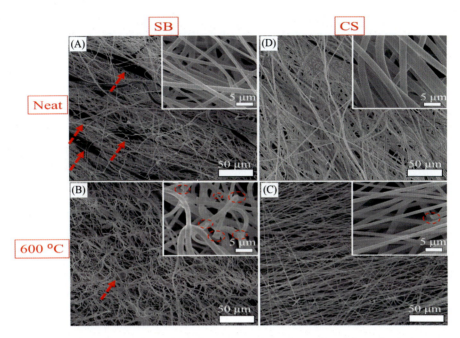

Figure 3.4 SEM images of PVP-TEOS and SiO$_2$ samples produced by centrifugal spinning and solution blow spinning techniques (A, B) before and (C, D) after calcination [69].

To fabricate BaTiO$_3$ by using the centrifugal jet spinning system, the following parameters were used: The rotating hollow chamber of the spinning system was composed of a Teflon tube with inner and outer diameters of 10 and 15 cm, respectively; the system was driven by a 9–18 V hobby motor from Radio Shack. The orifices distributed around the chamber surface had a diameter of 400 μm. The rotational speed of the chamber varied between 900 and 7000 rpm. The spinning solution was fed at a constant rate of 1 mL min^{-1}. The result of the spinning process was PVP-BaTiO$_3$ composite fibers, which were calcined in a furnace for 6 h at 850°C to obtain BaTiO$_3$ nanofibers.

3.3.4 Copper oxide nanofibers

Owing to their superior performance as heterogeneous catalysts, high-critical-temperature superconductors, and p-type semiconductors (band gap = 1.2 eV) and to their complex magnetic phases, CuO nanofibers are extensively used in electrochemical cells, field emission emitters, and gas and biosensors.

Yan et al. [80] prepared CuO nanofibers with 1300-nm length and 30- to 50-nm diameter using dielectric barrier discharge plasma (DBD reactor). This type of plasma is called cold plasma, and it operates under ambient conditions. The DBD reactor is composed of two electrodes, which are made of stainless steel plates with a diameter of 50 mm. Both electrodes are attached to two quartz or ceramic plates

(2-mm thickness and 60-mm diameter). The gap between the two quartz plates is 14 mm. Cu(OH)$_2$ was placed in the gap of the DBD reactor on a quartz plate. The DBD reactor operated for 3 min at a time for three cycles. Thus the total time of decomposition was 9 min. The sample was manually stirred each time. The SEM image of the produced nanofibers is shown in Fig. 3.5.

Xiulan et al. [80] prepared CuO nanofibers by using solution plasma. An electrode made of copper wire with a diameter of 1 mm was used. Pure water (0.0007 wt.%) was also used as a solution medium. To control the conductivity of water, sodium chloride (NaCl) was used to adjust the discharge in water. Two copper wire electrodes were placed in water at a distance of 0.3−0.5 mm from each other. During the discharge period, the distance between the two electrodes was set by using a screw micrometer. To generate the plasma, the two electrodes were connected to a high-voltage power supply (10- to 20-kHz frequency and 1- to 2-μs pulse). When the voltage was supplied, the gas phase began to form, owing to Joule heating. The discharge becomes visible when the voltage increases to break down. Thus water is decomposed into oxygen and hydrogen bubbles in addition to other energetic radical particles such as O, O_2^-, H_2O, H, and OH [81,82]. For this technique, plasma exists at the center and is encircled by the gas phase, which in turn is encompassed by a liquid phase. Therefore the energetic radical particles in the plasma region acquire kinetic energy. As soon as the radical particles reach the region between the gas and liquid medium, they return back into water. During the discharge period, colorless water becomes yellow and finally becomes dark. Using centrifugation, CuO nanofibers were easily collected.

3.3.5 Tungsten oxide nanofibers

Over the last several years, tungsten oxide (WO$_3$) nanofibers have attracted considerable interest, owing to their fascinating chemical and physical properties, which

Figure 3.5 SEM image of CuO nanofibers prepared by cold plasma [80].

have made them a candidate for a wide range of applications, such as photocatalysts, lithium-ion batteries, gas sensors, photochromics, and electrochromics [83–85].

Shixiu et al. [86] synthesized uniform WO_3 nanofibers with a 100-nm diameter and tens of micrometers in length using the one-step hydrothermal method. First, 0.025-M sodium sulfate (Na_2SO_4), 0.1 g of citric acid, and 0.005-M sodium tungstate ($Na_2WO4 \cdot 2H_2O$) were dissolved in 50 mL of deionized water with continual and vigorous stirring. Second, to adjust the pH value, 1.5–2.2 mol L^{-1} HCl was added dropwise. Third, the resulting solution was transferred into a stainless steel autoclave lined with 100 mL of Teflon. Then the prepared solution was heated in an oven at 180°C for one full day. Next, using centrifugation, a white sample was collected. The prepared sample was washed several times with absolute EtOH and water and finally dried in a vacuum oven for 10 h at 60°C. Fig. 3.6 shows the SEM images of WO_3 nanofibers prepared by using the hydrothermal technique.

3.3.6 Ferric oxide nanofibers

Owing to their promising potential applications, the synthesis of Fe_2O_3 nanofibers has gained considerable interest from researchers. This type of metal oxide nanofibers has a wide range of applications, such as in catalysts, supercapacitor electrodes, electrochemical energy, gas sensors, and water treatment [87–98].

Tongyan et al. [99] prepared Fe_2O_3 nanofibers via the hydrothermal technique followed by calcination in ambient air. A clear solution was formed by dissolving

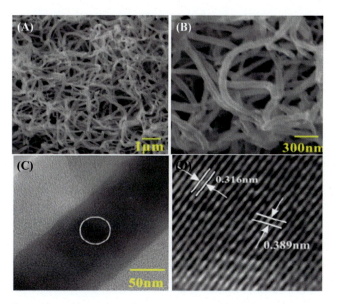

Figure 3.6 SEM (a-b) and TEM (c-d) images of WO3 nanofibers prepared by hydrothermal technique [85].

2 mmol of FeCl$_3$·6H$_2$O in 14 mL of distilled water. The pH value of the solution was adjusted to 13—14 by using an NaOH solution. The prepared solution was vigorously stirred for 30 min; then it was transferred into a stainless steel autoclave lined with 25 mL of Teflon. Next, the solution was heated at 100°C for 8 h. After cooling at room temperature, the resulting brown-yellow powder was collected and washed several times with distilled water. Then the obtained powder was calcined in ambient air at 400°C to form brown-red Fe$_2$O$_3$ nanofibers. Fig. 3.7 shows that the prepared Fe$_2$O$_3$ nanofibers are approximately 40 nm in diameter, and the length ranges between hundreds of nanometers and several micrometers. In addition, abundant pores are observed in individual nanofibers. The pores are clearly observed under high magnification. Fig. 3.4C shows two overlapping porous nanofibers. The pore size ranges between 10 and 25 nm.

In another study, Mandana et al. [100] synthesized multiwall Fe$_2$O$_3$ hollow fibers using the CS technique. Specifically, first, a polymer solution composed of 2.8 g of PVP and 6 g of water was prepared and magnetically stirred for 3 h. Meanwhile, another solution composed of 1.95 g of ferric nitrate (Fe(NO$_3$)0.9H$_2$O) in 2 g of water was also prepared. Then both solutions were mixed at room temperature with an iron precursor/polymer ratio of 0.7 and vigorously stirred for 2 h. A total of 2 mL of the resulting solution was fed into the spinneret of the centrifugal device, which was operating at an angular velocity of 7000—7500 rpm at a relative humidity of 50%—60%.

The fibers were produced in the form of a mat with a shiny yellow color. To eliminate organic compounds and obtain brown-red iron oxide fibers, the collected fibers were calcined at 650°C for 2 h with a heating rate of 3°C min^{-1}. Fig. 3.8

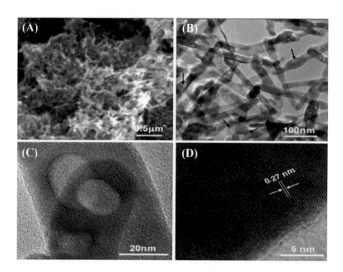

Figure 3.7 (A) SEM image and (B—D) TEM images of Fe$_2$O$_3$ nanofibers prepared by the hydrothermal method [99].

shows the SEM images of PVP/iron nitrate composite fibers, fibers calcined at 650°C, and split fibers and the histogram of composite fiber diameters.

3.3.7 NiO, CeO₂, and NiO-CeO₂ composite nanofibers

Owing to the simple synthesis, low cost, and well-defined redox behavior (Ni^{2+}/Ni^{3+}) in the charge-discharge process, NiO nanofibers have gained considerable attention for use in electrochemical energy storage applications [101,102]. In addition, CeO_2 nanofibers have recently been used as an energy storage material, owing to their redox properties (Ce^{3+}/Ce^{4+}) excellent cyclability, hydrophilicity, and environmentally friendly nature [103,104].

Vinícius et al. [105] prepared NiO, CeO_2, and their composite nanofibers via the SBS technique. Specifically, nickel nitrate hexahydrate ($Ni(NO_3)_2 0.6H_2O$), cerium nitrate ($Ce(NO_3)_3 0.6H_2O$), PVP with Mw = 1,300,000, EtOH, dimethylformamide (DMF), polytetrafluoroethylene, and commercial nickel foam were used as starting materials. The fabrication details of hollow nanofibers were based on the work

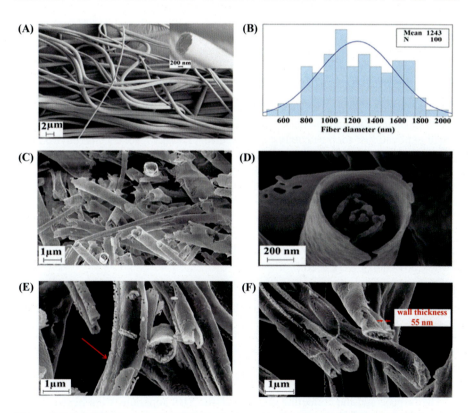

Figure 3.8 (A) SEM images of PVP/iron nitrate composite fibers. (B) Histogram of the diameters of the composite fibers. (C–D) Images of fibers calcined at 650°C shown at different magnifications. (E, F) Image of the longitudinal section of split fibers [100].

reported by Silva et al. [106]. Three different precursor solutions containing 10 mL of the solvent (consisting of EtOH/DMF with a volume ratio of 1:1) were prepared. Then 0.8 each of $Ni(NO_3)_20.6H_2O$ and $Ce(NO_3)_30.6H_2O$ were separately added to the solvent. Next, to adjust the viscosity of each respective precursor solution, 1.2 g of PVP was added with vigorous magnetic stirring for 12 h. To fabricate $NiO-CeO_2$ composite hollow fibers, 0.315 g of $Ce(NO_3)_30.6H_2O$ and 0.496 g of nickel nitrate were dissolved in 10 mL of the solvent to obtain 50 wt.% of each oxide phase. In addition, 1.2 g of PVP was added with strong stirring overnight. The precursor solutions were fed to the concentric nozzle of the SBS setup, which operated at the following experimental parameters: 3-mL h^{-1} injection rate, 0.41-MPa gas pressure, and 60-cm distance between the nozzle and the collector. The obtained nanofibers were collected on aluminum foil wrapped on the SBS collector, which was preheated to 60°C [107]. The obtained nanofiber mats were put into an oven at 80°C for 12 h to eliminate solvents and then subjected to a two-step calcination process to promote the crystallization and formation of $NiO-CeO_2$ hollow nanofibers [108]. The first calcination process was performed at 800°C in air with a heating rate of 2°C min^{-1} for 1 h; the second heating process was implemented at 200°C for 1 h. The EDS mapping and FESEM images of NiO, CeO_2, and their composite nanofibers are shown in Fig. 3.9.

3.3.8 Titanium dioxide and zinc oxide nanofibers

In recent years, TiO_2 and ZnO nanofibers have attracted considerable interest, owing to their enhanced characteristics, which have allowed them to be used in different applications, such as catalysts, gas sensors, batteries, solar cells, environmental protection, and cleaning [109–112].

Costa et al. [113] synthesized nanofibers of TiO_2 and ZnO by SBS. Zinc acetate dihydrate, Zn $(CH_3CO_2)_20.2H_2O$, $Zn(OAc)_2$ and titanium (IV) isopropoxide, Ti $(OCH(CH_3)_2)_4$, TTIP were used as Zn and Ti sources, respectively. For the production of solutions, three solvents and two polymers were also utilized: EtOH, tetrahydrofuran (THF), and DMF; PVP (average MW~ 1,300,000) and poly (vinyl chloride) (PVC) (average MW~ 50,360). To prepare the precursor solutions, the following procedures were used:

1. $Zn(OAc)_2$ (0.5%, w/v; 0.05 g/10 mL) or TTIP (12%, v/v; 1.2 mL/10 mL) was dissolved in THF (10 mL) (100 µL of HCl was added); thereafter, PVC (10%, w/v; 1 g/10 mL) was added to these solutions.
2. TTIP (12%, v/v; 1.2 mL/10 mL) was dissolved in EtOH (10 mL) (50 µL of glacial acetic acid was added); in addition, a solution of EtOH and DMF (10 mL) (80/20, v/v; 8 mL/ 2 mL) was used to dissolve $Zn(OAc)_2$ (5%, w/v; 0.5 g/10 mL) (100 µL of HCl was added); then PVP (10%, w/v; 1 g/10 mL) was also added to these solutions.

The furnace and the collecting chamber of the SBS apparatus were heated during the preparation of PVP-based solutions. For TiO_2 precursor solutions, the injection rates of PVC and PVP solutions were adjusted to 10 mL h^{-1} and 4.4 mL h^{-1}, respectively; the air pressure was kept constant at 483 KPa. For ZnO precursor

Nonelectrospun metal oxide nanofibers

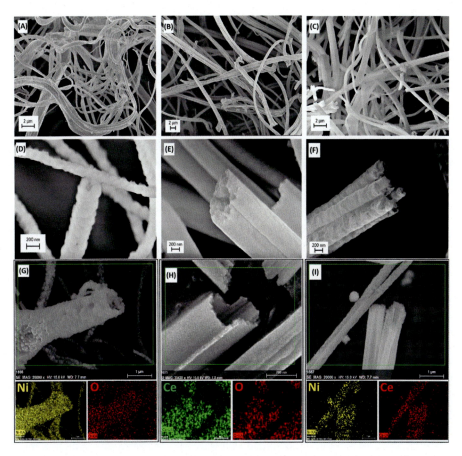

Figure 3.9 FESEM images and EDS mapping of (A, D, G) NiO, (B, E, H) CeO$_2$, and (C, F, I) NiO-CeO$_2$ hollow fibers [105].

solutions, the injection rate and air pressure were adjusted to 7.2 mL h^{-1} and 345 KPa, respectively. For all conditions, the distance between the collector and the concentric nozzle was fixed at 600 mm. The fibers were released and collected at two different temperatures: 600°C and 700°C. The SEM micrographs of TiO$_2$ and ZnO nanofibers are shown in Fig. 3.10.

3.4 Conclusion

Owing to the drawbacks associated with the electrospinning technique, which include a low production rate, a high cost compared to conventional methods, and many safety and related problems such as vapor emitted from it and high voltage rate, researchers have turned to nonelelectrospinning methods to fabricate metal

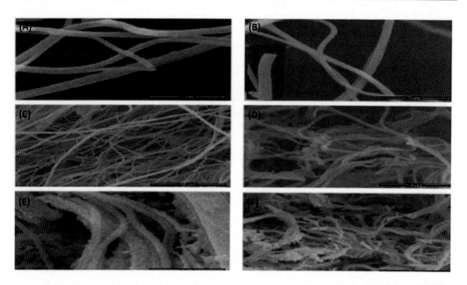

Figure 3.10 SEM micrographs of the titanium oxide and zinc oxide nanofibers. (A) TiO$_2$-PVC fired at 700°C. (B, C) TiO$_2$-PVP fired at 700°C. (D) ZnO-PVC fired at 600°C. (E, F) ZnO-PVP fired at 600°C [113].

oxide nanofibers. The nonelectrospinning techniques of fabricating nanofibers include centrifugal spinning, template synthesis, phase separation, drawing, solution blowing, self assembly, interfacial polymerization, plasma-induced technique, laser spinning, and splitting techniques. In terms of scalability and industrialization, solution blowing and centrifugal spinning might be viewed as the most advantageous techniques. On the other hand, for some polymers and their parameters, other methods such as phase separation, template synthesis, and freeze-drying synthesis might have to be used.

In general, metal oxide nanofibers can be produced from different metals and metal oxides, such as Fe, Cu, Ti, Ba, Si, Sn, Ni, Mo, Sb, CeO$_2$, CuO, TiO$_2$, MnO$_2$, Fe$_2$O$_3$, SnO$_2$, and NiCo$_2$O$_4$. This chapter reviewed fabrication techniques and the synthesis methods of nonelectronspun metal oxides nanofibers. The synthesis and characterization of nanofibers from tin oxide, silica, barium titanate, copper oxide, tungsten oxide, ferric oxide, titanium dioxide, zinc oxide, nickel oxide, and cerium oxide and their mixtures were explained and discussed in detail.

References

[1] B. Jaison Jeevanandam, Y.S. Chan, A. Dufresne, M.K. Danquah, Review on nanoparticles and nanostructured materials: history, sources, toxicity and regulations, Beilstein J. Nanotechnol. 9 (2018) 1050–1074.

[2] A. Barhoum, K. Pal, H. Rahier, H. Uludag, I.S. Kim, M. Bechelany, Nanofibers as new-generation materials: from spinning and nano-spinning fabrication techniques to emerging applications, Appl. Mater. Today 17 (2019) 1−35.

[3] Z. Wang, Z. Li, J. Sun, H. Zhang, W. Wang, W. Zheng, et al., Improved hydrogen monitoring properties based on p-NiO/n-SnO2 heterojunction composite NFs, J. Phys. Chem. C. 114 (2010) 6100−6105.

[4] W.E. Teo, S. Ramakrishna, A review on electrospinning design and nanofibre assemblies, Nanotechnology 17 (14) (2006) R89−R106.

[5] Z. Su, G. Wei, Electrospinning: a facile technique for fabricating polymeric nanofibers doped with carbon nanotubes and metallic nanoparticles for sensor applications, RSC Adv. 4 (94) (2014) 52598−52610.

[6] E.S. Medeiros, et al., Solution blow spinning: a new method to produce micro-and nanofibers from polymer solutions, J. Appl. Polym. Sci. 113 (4) (2009) 2322−2330.

[7] X. Xing, Y. Wang, B. Li, Nanofiber drawing and nanodevice assembly in poly(trimethylene terephthalate), Opt. Express 16 (14) (2008) 10815−10822.

[8] R. Weitz, et al., Polymer nanofibers via nozzle-free centrifugal spinning, Nano Lett. 8 (4) (2008) 1187−1191.

[9] H.-S. Liao, et al., Self-assembly mechanisms of nanofibers from peptide amphiphiles in solution and on substrate surfaces, Nanoscale 8 (31) (2016) 14814−14820.

[10] S.L. Tao, T.A. Desai, Aligned arrays of biodegradable poly(caprolactone) nanowires and nanofibers by template synthesis, Nano Lett. 7 (6) (2007) 1463−1468.

[11] L. He, et al., Fabrication and characterization of poly(l-lactic acid) 3D nanofibrous scaffolds with controlled architecture by liquid−liquid phase separation from a ternary polymer−solvent system, Polymer 50 (16) (2009) 4128−4138.

[12] H. Guan, et al., Polyaniline nanofibers obtained by interfacial polymerization for high-rate supercapacitors, Electrochim. Acta 56 (2) (2010) 964−968.

[13] J. Wu, J.C. Meredith, Assembly of chitin nanofibers into porous biomimetic structures via freeze drying, ACS Macro Lett. 3 (2) (2014) 185−190.

[14] S.S. Ray, et al., A comprehensive review: electrospinning technique for fabrication and surface modification of membranes for water treatment application, RSC Adv. 6 (88) (2016) 85495−85514.

[15] X. Zhang, M. Rolandi, Engineering strategies for chitin nanofibers, J. Mater. Chem. B 5 (14) (2017) 2547−2559.

[16] E. Stojanoversuska, et al., A review on non-electro nanofibre spinning techniques, RSC Adv. 6 (87) (2016) 83783−83801.

[17] E. Stojanoversuska, E. Canbay, E.S. Pampal, M.D. Calisir, O. Agma, et al., A review on non-electro nanofibre spinning, RSC Adv. 6 (2016) 83783.

[18] A.A. Almetwally, M. El-Sakhawy, M.H. Elshakankery, M.H. Kasem, Technology of nano-fibers: production techniques and properties - critical review, J. Text. 78 (1) (2017) 5−14.

[19] E.S. Medeiros, G.M. Glenn, A.P. Klamczynski, W.J. Orts, L.H.C. Mattoso, Solution blow spinning: a new method to produce micro- and nano fibers from polymer solutions, J. Appl. Polym. Sci. 113 (2009) 2322−2330.

[20] A.M. Behrens, B.J. Casey, M.J. Sikorski, K.L. Wu, W. Tutak, A.D. Sandelr, et al., In situ deposition of PLGA nanofibers via solution blow spinning, ACS Macro Lett. 3 (2014) 249−254.

[21] M. Wojasiński, M. Pilarek, T. Ciach, Comparative studies of electrospinning and solution blow spinning processes for the production of nanofi brous poly(L-lactic acid) materials for engineering, Pol. J. Chem. Technol. 16 (2) (2014) 43−50.

[22] X. Zhuang, X. Yang, L. Shi, B. Cheng, K. Guan, W. Kang, Solution blowing of submicron-scale cellulose fibers, Carbohydr. Polym. 90 (2) (2012) 982–987.

[23] H.J. Jin, D.L. Kaplan, Mechanism of silk processing in insects and spiders, Nature 424 (6952) (2003) 1057–1061.

[24] A.M.C. Santos, E.L.G. Medeiros, J.J. Blaker, E.S. Medeiros, Aqueous solution blow-spinning of poly (vinyl alcohol) micro-and nanofibers, Mater. Lett. 176 (2016) 122–126.

[25] A. Kolbasov, S. Sinha-Ray, A. Joijode, M.A. Hassan, D. Brown, B. Maze, et al., Industrial-scale solution blowing of soy proteinnanofibers, Ind. Eng. Chem. Res. 55 (2016) 323–333.

[26] G. Sabbatier, P. Abadie, F. Dieval, B. Durand, G. Laroche, Evaluation of an air spinning process to produce tailored biosynthetic nanofibre scaffolds, Mater. Sci. Eng. C Mater Bio 35 (2) (2014) 347–353.

[27] D.D. da Silva Parize, M.M. Foschini, J.E. de Oliveira, A.P. Klamczynski, G.M. Glenn, et al., Solution blow spinning optimization and effects on the properties of nanofibers from poly(lactic acid)/dimethyl carbonate solutions, J. Mater. Sci. 51 (2016) 4627–4638.

[28] J. González Benito, J. Teno, D. Torres, M. Díaz, Solution blow spinning and obtaining submicrometric fibers of different polymers, Int. J. Nanopart. Nanotechnol. 3 (1) (2017).

[29] J. Li, G. Song, J. Yu, Y. Wang, J. Zhu, Z. Hu, Preparation of solution blown polyamic acid and their imidization into polyimide nanofiber mats, Nanomaterials (Basel) 7 (11) (2017) 395.

[30] X. Hu, X. Zhang, X. Shen, H. Li, O. Takai, N. Saito, Plasma-induced synthesis of CuO nanofibers and ZnO nanoflowers in water, Plasma Chem. Plasma Process. 34 (5) (2014) 1129–1139.

[31] J. Wang, A.S. Nain, Suspended micro/nanofiber hierarchical biologicalscaffolds fabricated using non-electrospinning STEP technique, Langmuir 30 (45) (2014) 13641–13649.

[32] K. Ahmed Barhoum, H. Pal, H. Rahier, I. Uludag, Soo, Kim, et al., Nanofibers as new-generation materials: from spinning and nano-spinning fabrication techniques to emerging applications, Appl. Mater. Today 17 (2019) 1–35.

[33] A.S. Nain, M. Sitti, A. Jacobson, T. Kowalewski, C. Amon, Dry spinning based bpinneret based tunable engineered parameters (STEP) technique for controlled and aligned deposition of polymeric n nanofibers, Macromol. Rapid Commun. 30 (2009) 1406–1412.

[34] A.S. Nain, et al., Drawing suspended polymer micro-/nanofibers using glassmicropipettes, Appl. Phys. Lett. 89 (18) (2006) 183105.

[35] S.D. Savage, C.A. Miller, D. Furniss, A.B. Seddon, Extrusion of chalcogenide glass performs and drawing to multimode optical fibers, J. Non-Crystalline Solids 354 (29) (2008) 3418–3427.

[36] C. Vitale-Brovarone, G. Novajra, D. Milanese, J. Lousteau, J.C. Knowles, Novel phosphate glasses with different amounts of TiO_2 for biomedical applications: dissolution tests and proof of concept of fibre drawing, Mat. Sci. Eng. C Mater. 31 (2) (2011) 434–442.

[37] T. Hasegawa, T. Mikuni, Higher-order structural analysis of nylon-66 nanofibers prepared by carbon dioxide laser supersonic drawing and exhibiting near equilibrium melting temperature, J. Appl. Polym. Sci. 131 (40361) (2014) 1–8.

[38] A. Suzuki, Ethylene tetrafluoroethylene nanofibers prepared by CO2 laser supersonic drawing, Express Polym. Lett. 7 (6) (2013) 519−527.

[39] A. Suzuki, K. Tanizawa, Poly(ethylene terephthalate) nanofibers prepared by CO2 laser supersonic drawing, Polymer 50 (2009) 913−921.

[40] H.F. Yang, Y. Yan, Y. Liu, A Simple melt impregnation method to synthesize ordered mesoporous carbon and carbon nanofiber bundles with graphitized structure from pitches, J. Phys. Chem. B 108 (45) (2004) 17320−17328.

[41] X. Li, S. Tian, Y. Ping, D.H. Kim, W. Knoll, One-step route to the fabrication of highly porous polyaniline nanofiber films by using PSb-PVP diblock copolymers as templates, Langmuir 21 (3) (2005) 9393−9397.

[42] C. Toro, J.M. Buriak, Template synthesis approach to nanomaterials: Charles Martin, Chem. Mater. 26 (17) (2014) 4889−4890.

[43] X. Zhang, Y. Lua, Centrifugal spinning: an alternative approach to fabricate nanofibers at high speed and low cost, Polym. Rev. 54 (2014) 677−701. Available from: https://doi.org/10.1080/15583724.2014.935858.

[44] P. Kumar, Effect of Collector on Electrospinning to Fabricate Aligned Nanofiber, Department of Biotechnology & Medical Engineering National Institute of Technology, Rourkela, 2012.

[45] T.L. Wadea, J.E. Wegrowe, Template synthesis of nanomaterials, Eur. Phys. J. Appl. Phys. 29 (01) (2005) 3−22.

[46] L. Ren, R. Ozisik, S.P. Kotha, Rapid and efficient fabrication of multilevel structured silica micro-/nanofibers by centrifugal jet spinning, J. Colloid Interface Sci. 425 (2014) 136−142.

[47] Y. Zhao, Y. Qiu, H. Wang, C. Yu, S. Jin, S. Chen, Preparation of nanofibers with renewable polymers and their application in wound dressing, Int. J. Polym. Sci. 2016 (2016) 4672839. Available from: https://doi.org/10.1155/2016/4672839.

[48] K. Sarkar, C. Gomez, S. Zambrano, M. Rmirez, E. Hoyos, H. Vasquez, et al., Electrospinning to forcespinning, Mater. Today 3 (11) (2010) 12−14.

[49] S. Padron, A. Fuentes, D. Caruntu, K. Lozano, Experimental study of nanofiber production through forcespinning, J. Appl. Phys. 113 (2013) 1−9.

[50] B. Raghavan, H. Soto, K. Lozano, Fabrication of melt spun polypropylene nanofibers by forcespinning, J. Eng. Fibers Fabr. 8 (1) (2013) 52−60.

[51] J.P. Hooper, Centrifugal Spinneret, in United States patent United States 1500931 A, United States, 1924.

[52] R.T. Weitz, L. Harnau, S. Rauschenbach, M. Burghard, K. Kern, Polymer nanofibers via nozzle-free centrifugal spinning, Nano Lett. 8 (4) (2008) 1187−1191.

[53] Y. Meltem Yanilmaz, Y. Lu, X.Z. Li, SiO2/polyacrylonitrile membranes via centrifugal spinning as a separator for Li-ion batteries, J. Power Sources 273 (2015) 1114e1119.

[54] J. Ducree, S. Haeberle, S. Lutz, S. Pausch, F.V. Stetten, R. Zengerle, The centrifugal microfluidic bio-disk platform, J. Micromech. Microeng. 17 (7) (2007) S103−S115.

[55] Y. Lu, Y. Li, S. Zhang, G. Xu, K. Fu, H. Lee, et al., Parameter study and characterization for polyacrylonitrile nanofibers fabricated via centrifugal spinning process, Eur. Polym. J. 49 (12) (2013) 3834−3845.

[56] S. Padron, 2D Modeling of Forcespinning (TM) Nanofiber Formation With Experimental Study and Validation (M.S. thesis), The University of Texas-Pan American, 2012.

[57] C.J. Luo, S.D. Stoyanov, E. Stride, E. Pelan, M. Edirisinghe, Electrospinning vs fibre production methods: from specifics to technological convergence, R. Soc. Chem. 41 (2012) 4708−4735. Available from: https://doi.org/10.1039/c2cs35083a.

[58] K. Mondal, Recent advances in the synthesis of metal oxide nanofibers and their environmental remediation applications, Inventions 2 (2017) 9. Available from: https://doi.org/10.3390/inventions2020009.

[59] P. Singh, K. Mondal, A. Sharma, Reusable electrospun mesoporous ZnO nanofiber mats for photocatalytic degradation of polycyclic aromatic hydrocarbon dyes in wastewater, J. Colloid Interface Sci. 394 (2013) 208−215.

[60] K. Mondal, S. Bhattacharyya, A. Sharma, Photocatalytic degradation of naphthalene by electrospun mesoporous carbon-doped anatase TiO2 nanofiber mats, Ind. Eng. Chem. Res. 53 (2014) 18900−18909.

[61] K. Mondal, A. Sharma, Recent advances in electrospun metal-oxide nanofiber based interfaces for electrochemical biosensing, RSC Adv. 6 (2016) 94595−94616.

[62] K. Mondal, M.A. Ali, V.V. Agrawal, B.D. Malhotra, A. Sharma, Highly sensitive biofunctionalized mesoporous electrospun TiO_2 nanofiber based interface for biosensing, ACS Appl. Mater. Interfaces 6 (2014) 2516−2527.

[63] M.A. Ali, K. Mondal, C. Singh, B. Dhar Malhotra, A. Sharma, Anti-epidermal growth factor receptor conjugated mesoporous zinc oxide nanofibers for breast cancer diagnostics, Nanoscale 7 (2015) 7234−7245.

[64] S.-H. Park, H.-R. Jung, W.-J. Lee, Hollow activated carbon nanofibers prepared by electrospinning as counter electrodes for dye-sensitized solar cells, Electrochim. Acta 102 (2013) 423−428.

[65] D. Crespy, K. Friedemann, A.-M. Popa, Colloid-electrospinning: fabrication of multi-compartment nanofibers by the electrospinning of organic or/and inorganic dispersions and emulsions, Macromol. Rapid Commun. 33 (2012) 1978−1995.

[66] L. Berry, J. Brunet, Oxygen influence on the interaction mechanisms of ozone on SnO2 sensors, Sens, Actuators B: Chem. 129 (2008) 450−458.

[67] R. Ab Kadir, Z. Li, A.Z. Sadek, R. Abdul Rani, A.S. Zoolfakar, M.R. Field, et al., Electrospun granular hollow SnO_2 nanofibers hydrogen gas sensors operating at low temperatures, J. Phys. Chem. C 118 (2014) 3129−3139.

[68] H. Wang, Y. Huang, S. Liao, H. He, H. Wu, Tin oxide nanofiber and 3D sponge structure by blow spinning. In: IOP Conf, Series: Earth and Environmental Science (2019) 358.−2019.

[69] M.D. Calisir, A. Kilic, A comparative study on SiO2 nanofiber production via two novel non-electrospinning methods: centrifugal spinning vs solution blowing, Mater. Lett. 258 (2020) 126751.

[70] J.Y. Howe, N.E. Hedin, L. Zhang, Y. Zhang, R. Chandrasekar, H. Fong, Fabrication and characterization of electrospun titania nanofibers, J. Mater. Sci. 44 (2009) 1198−1205.

[71] J.H. Yu, S.V. Fridrikh, G.C. Rutledge, Production of submicrometer diameter fibers by two-fluid electrospinning, Adv. Mater. 16 (2004) 1562.

[72] P. Kim, S.C. Jones, P.J. Hotchkiss, J.N. Haddock, B. Kippelen, S.R. Marder, et al., Phosphonic acid-modified barium titanate polymer nanocomposites with high permittivity and dielectric strength, Adv. Mater. 19 (2007) 1001−1005.

[73] J.J. Urban, J.E. Spanier, L. Ouyang, W.S. Yun, H. Park, Single-crystalline barium titanate nanowires, Adv. Mater. 15 (2003) 423−426.

[74] W.S. Yun, J.J. Urban, Q. Gu, H. Park, Ferroelectric properties of individual barium titanate nanowires investigated by scanned probe microscopy, Nano Lett. 2 (5) (2002) 447−450.

[75] Y. He, Heat capacity, thermal conductivity, and thermal expansion of barium titanate-based ceramics, Thermochim. Acta 419 (1−2) (2004) 135−141.

[76] M. Wegmann, R. Brönnimann, F. Clemens, T. Graule, Barium titanate-based PTCR thermistor fibers: processing and properties, Sens. Actuat A-Phys 135 (2) (2007) 394−404.

[77] E. Burcsu, G. Ravichandran, K. Bhattacharya, Large electrostrictive actuation of barium titanate single crystals, J. Mech. Phys. Solids 52 (4) (2004) 823−846.

[78] F. Wang, Y.-W. Mai, D. Wang, R. Ding, W. Sh, High quality barium titanate nanofibers for flexible piezoelectric device applications, Sens. Actuators A 233 (2015) 195−201.

[79] L. Ren, S.P. Koth, Centrifugal jetspinning for highly efficient and large-scale fabrication of barium titanate nanofibers, Mater. Lett. 117 (2014) 153−157.

[80] Y. Li, P. Kuai, P. Huo, C.-J. Liu, Fabrication of CuO nanofibers via the plasma decomposition of Cu(OH)2, Mater. Lett. 63 (2009) 188−190.

[81] C. Miron, M.A. Bratescu, N. Saito, O. Takai, Time-resolved optical emission spectroscopy in water electrical discharges, Plasma Chem. Plasma Process. 30 (5) (2011) 619−631.

[82] E. Acayanka, A. Tiya Djowe, S. Laminsi, C.C. Tchoumkwe, S. Nzali, A. Poupi Mbouopda, et al., Plasma-assisted synthesis of TiO_2 nanorods by gliding arc discharge processing at atmospheric pressure for photocatalytic applications, Plasma Chem. Plasma Process. 33 (4) (2013) 725−735.

[83] W. Zeng, C.N. Dong, B. Miao, H. Zhang, S.B. Xu, X.Z. Ding, et al., Preparation, characterization and gas sensing properties of sub-micron porous WO_3 spheres, Mater. Lett. 117 (2014) 41−44.

[84] X. Gao, C. Yang, F. Xiao, Y. Zhu, J. Wang, X. Su, $WO_3 \cdot 0.33H_2O$ nanoplates: hydrothermal synthesis, photocatalytic and gas-sensing properties, Mater. Lett. 84 (2012) 151−153.

[85] B. Miao, W. Zeng, S. Hussain, Q.P. Mei, S.B. Xu, H. Zhang, et al., Large scale hydrothermal synthesis of monodisperse hexagonal WO_3 nanowire and the growth mechanism, Mate Lett. 147 (2015) 12−15.

[86] S. Cao, C. Zhao, T. Han, L. Peng, Hydrothermal synthesis, characterization and gas sensing properties of the WO3 nanofibers, Mater. Lett. 169 (2016) 17−20.

[87] L. Ji, et al., α-Fe2O3 nanoparticle-loaded carbon nanofibers as stable and highcapacity anodes for rechargeable lithium-ion batteries, ACS Appl. Mater. Interfaces 4 (5) (2012) 2672−2679.

[88] J.S. Cho, Y.J. Hong, Y.C. Kang, Design and synthesis of bubble-nanorod-structured Fe2O3−carbon nanofibers as advanced anode material for Li-ion batteries, ACS Nano 9 (4) (2015) 4026−4035.

[89] S. Chaudhari, M. Srinivasan, 1D hollow α-Fe2O3 electrospun nanofibers as high performance anode material for lithium ion batteries, J. Mater. Chem. 22 (43) (2012) 23049−23056.

[90] J. Ma, et al., FeOx-based materials for electrochemical energy storage, Adv. Sci. 5 (6) (2018) 1−28 (1700986).

[91] J. Guo, et al., Synthesis of α-Fe2O3, Fe3O4 and Fe2N magnetic hollow nanofibers as anode materials for Li-ion batteries, RSC Adv. 6 (112) (2016) 111447−111456.

[92] G. Qiu, et al., Microwave-assisted hydrothermal synthesis of nanosized α-Fe2O3 for catalysts and adsorbents, J. Phys. Chem. C 115 (40) (2011) 19626−19631.

[93] G. Binitha, et al., Electrospun α-Fe2O3 nanostructures for supercapacitor applications, J. Mater. Chem. A 1 (38) (2013) 11698−11704.

[94] C.-Y. Cao, et al., Low-cost synthesis of flowerlike α-Fe2O3 nanostructures for heavy metal ion removal: adsorption property and mechanism, Langmuir 28 (9) (2012) 4573–4579.

[95] J. Chang, et al., Preparation of α-Fe2O3/polyacrylonitrile nanofiber mat as an effective lead adsorbent, Environ. Sci. Nano 3 (4) (2016) 894–901.

[96] Q. Gao, et al., Novel hollow α-Fe2O3 nanofibers via electrospinning for dye adsorption, Nanoscale Res. Lett. 10 (1) (2015) 176.

[97] L. Wang, et al., Ethanol gas detection using a yolk-shell (core-shell) α-Fe2O3 nanospheres as sensing material, ACS Appl. Mater. Interfaces 7 (23) (2015) 13098–13104.

[98] Y. Huang, et al., A high performance hydrogen sulfide gas sensor based on porous α-Fe2O3 operates at room-temperature, Appl. Surf. Sci. 351 (2015) 1025–1033.

[99] T. Ren, P. He, W. Niu, Y. Wu, L. Ai, X. Gou, Synthesis of α-Fe2O3 nanofibers for applications in removal and recovery of Cr(VI) from wastewater, Environ. Sci. Pollut. Res. 20 (2013) 155–162. Available from: https://doi.org/10.1007/s11356-012-0842-z.

[100] M. Akia, K.A. Mkhoyan, K. Lozano, Synthesis of multiwall α-Fe2O3 hollow fibers via a centrifugal spinning technique, Mater. Sci. Eng. C 102 (2019) 552–557.

[101] B. Ren, M. Fan, Q. Liu, J. Wang, D. Song, X. Bai, Hollow NiO nanofibers modified by citric acid and the performances as supercapacitor electrode, Electrochim. Acta 92 (2013) 197e204. Available from: https://doi.org/10.1016/j.electacta.2013.01.009.

[102] J. Wang, Y. Cui, D. Wang, Design of hollow nanostructures for energy storage, conversion and production, Adv. Mater. 1801993 (2018) 1801993. Available from: https://doi.org/10.1002/adma.201801993.

[103] N. Maheswari, G. Muralidharan, Supercapacitor behavior of cerium oxide nanoparticles in neutral aqueous electrolytes, Energy Fuel 29 (2015) 8246e8253. Available from: https://doi.org/10.1021/acs.energyfuels.5b02144.

[104] S. Maiti, A. Pramanik, S. Mahanty, Extraordinarily high pseudocapacitance of metal organic framework derived nanostructured cerium oxide, Chem. Commun. 50 (2014). Available from: https://doi.org/10.1039/C4CC05363J. 11717e11720.

[105] V.D. Silva, L.S. Ferreira, A.J.M. Araújo, T.A. Simoes, J.P.F. Grilo, M. Tahir, et al., Ni and Ce oxide-based hollow fibers as battery-like electrodes, J. Alloy. Compd. 830 (2020) 154633.

[106] V.D. Silva, T.A. Sim~oes, F.J.A. Loureiro, D.P. Fagg, F.M.L. Figueiredo, E.S. Medeiros, et al., Solution blow spun nickel oxide/carbon nanocomposite hollow fibres as an efficient oxygen evolution reaction electrocatalyst, Int. J. Hydrog. Energy 44 (2019). 14877e14888.

[107] V.D. Silva, T.A. Sim~oes, F.J.A. Loureiro, D.P. Fagg, E.S. Medeiros, D.A. Macedo, Electrochemical assessment of Ca3Co4O9 nanofibres obtained by solution blow spinning, Mater. Lett. 221 (2018) 81e84. Available from: https://doi.org/10.1016/j.matlet.2018.03.088.

[108] V.D. Silva, L.S. Ferreira, T.A. Simoes, E.S. Medeiros, D.A. Macedo, 1D hollow MFe2O4 (M = Cu, Co, Ni) fibers by solution blow spinning for oxygen evolution reaction, J. Colloid Interface Sci. 540 (2019) 59e65. Available from: https://doi.org/10.1016/j.jcis.2019.01.003.

[109] M. Imran, S. Haider, K. Ahmad, A. Mahmood, W.A. Al-masry, Fabrication and characterization of zinc oxide nanofibers for renewable energy applications, Arab. J. Chem. 10 (2017) S1067–S1072.

[110] R.G.F. Costa, C. Ribeiro, L.H.C. Mattoso, Study of the effect of rutile/anatase TiO2 nanoparticles synthesized by hydrothermal route in electrospun PVA/TiO2 nanocomposites, J. Appl. Polym. Sci. 127 (6) (2013) 4463−4469.

[111] X.J. Shen, B.Z. Tian, J.L. Zhang, Tailored preparation of titania with controllable phases of anatase and brookite by an alkalescent hydrothermal route, Catal. Today 201 (2013) 151−158.

[112] A. Stafiniak, B. Boratynski, A. Baranowska-Korczyc, K. Fronc, D. Elbaum, M. Tłaczała, Electrical conduction of a single electrospun ZnO nanofiber, J. Am. Ceram. Soc. 97 (2014) 1157−1163.

[113] D.L. Costa, R.S. Leite, G.A. Neves, L.N. de Lima Santana, E.S. Medeiros, R.R. Menezes, Synthesis of TiO2 and ZnO nano and submicrometric fibers by solution blow spinning, Mater. Lett. 183 (2016) 109−113.

Polymer−metal oxide composite nanofibers

4

*Zainab Ibrahim Elkahlout[1], Abdulrahman Mohmmed AlAhzm[2],
Maan Omar Alejli[2], Fatima Zayed AlMaadeed[3],
Deepalekshmi Ponnamma[4] and Mariam Al Ali Al-Maadeed[4,5]*

[1]Department of Chemical Engineering, College of Engineering, Qatar University, Doha, Qatar, [2]Department of Chemistry, College of Arts and Science, Qatar University, Doha, Qatar, [3]Department of Environmental Sciences, Qatar University, Doha, Qatar, [4]Center for Advanced Materials, Qatar University, Doha, Qatar, [5]Materials Science and Technology Program (MATS), College of Arts & Sciences, Qatar University, Doha, Qatar

4.1 Introduction

Polymeric nanocomposites are highly needed materials in science, owing to their significant properties and multiple applications in industrial and technological fields [1−5]. They also possess advanced features, such as morphological diversity, multifunctionalities, and adaptive designs. Polymer-based fibers are of enormous interest, as they can provide good mechanical strength and are soft and flexible in appearance [6−11]. There are several procedures available for fabricating them in microfiber and nanofiber forms, such as melt blowing [2,3], electrospinning [4−6], solution blowing [7,8], self-assembly [9], phase separation [10], drawing [12], template synthesis, wet spinning, dry spinning, and melt spinning [13,14]. These characteristics make the fibers applicable in environmental and biomedical fields at the same time. Though all of the individual fabrication processes have their own benefits, the most common one is electrospinning because of its simplicity, scalability, and wider applicability [4−6].

The electrospinning process produces long, three-dimensional, ultrafine fibers in submicron to nanoscale dimensions using electrostatic forces [5]. The most important feature of this method is the simplicity compared to traditional spinning methods for producing fibers. The electrospinning technology is an up-and-coming technology that can be at the forefront of all manufacturing technologies because it provides various designs and involves many nanomaterials [15]. Electrospinning also helps in fiber alignment and hence reduces the Gibbs free energy. This technology is qualified to alter specific fiber structures and to adjust the morphology [16,17]. The method is effectively adaptable to all types of materials, including organic and inorganic polymers. With the feed polymers in several forms, such as solutions, emulsions, mixtures, or melts, electrospinning produces multifaceted nanofibers structures, such as porous, hollow, core-shell, and other types [1]. The benefit of electrospinning is

Metal Oxide-Based Nanofibers and Their Applications. DOI: https://doi.org/10.1016/B978-0-12-820629-4.00005-9
© 2022 Elsevier Inc. All rights reserved.

that it does not require any process of functionalization; it requires an efficient solvent to disperse nanoparticles (NPs) and dissolve the polymer.

Though electrospinning is an easy process, several major limitations also exist for this technique, related to the polymer solution properties, such as viscosity, concentration, conductivity, and surface tension. Numerous scientists have experimented to extrinsically modify these factors, causing major changes in the total spinning, such as lowering the surface tension by adding surfactants [18]. In addition, NaCl addition specifically to the polymer polyethylene oxide, enhances the polymer spinnability [19]. Furthermore, the voltage application, distance between the needle and collector, relative humidity, temperature, and flow rate are significant factors too. Hartman et al. [20] displayed the relationship between the charged polymer (D) and feed flow rate (Q) as $D \propto Q^{0.48}$, considering the bending volatility of the charged jets. Moreover, the extreme distance between the needle and the collector affects the filaments in three ways: (1) increase in diameter due to polymer chain relaxation [21], (2) reduction in diameter due to fast solvent vaporization before getting to the collector [22], and (3) beaded filaments [23]. Humidity and temperature also affect the vaporization rate and thus are considered the critical factors for uniform filament diameter production [24]. Conversely, some biological materials like enzymes, coflowing within the polymer solution, high temperature, and so on can also be damaging to the fiber uniqueness.

Numerous nanofillers are utilized for generating polymer composites with improved reinforcement strength when compared to micro- and macro-sized fillers. This is achieved mostly with $1-5$ wt.% of the nanofillers, with a strength improvement that can be more than 30% when compared to the unfilled polymer. In electrospinning, the creation of nanofibers is based on the uniaxial stretching of a viscoelastic solution. Compared with conventional methods of spinning fibers, such as dry spinning and melt spinning, electrospinning uses electrostatic forces to stretch the solution as the solvent evaporates. Similar to conventional methods of spinning fibers, drawing the solution continues as long as there is sufficient solution to feed the electrospinning jet [2]. Thus the creation of the fiber should be continuous without any interruption during electrospinning process.

This chapter sheds light on different polymer-based metal oxide nanocomposite fibers made by electrospinning. Three main category of polymers—electroactive polymers, elastomers, and biopolymers—are studied for their different abilities of fiber formation especially with metal oxide NPs and the functional significance. The main characteristic properties of the fibers and applications are discussed along with the electrospinning conditions.

4.2 Electroactive polymers and their metal oxide composites

Electroactive polymers (EAPs) alter their bulk, shape, or volume characteristics in response to a strong electric field [25,26]. EAPs stand out in the area of active materials (e.g., piezoelectrics, thermoelastic polymers, shape retention alloys and polymers, magnetostrictive materials), owing to their strong deformation capacity, fast reaction speed, low density, and increased durability. They are extremely lightweight, cost-effective, tolerant, and compliant to fractures. EAPs can be divided into two main categories according to their working principle: ionic and electronic EAPs. Ionic EAPs are driven by the displacement of ions leading to shape or volume change during electrical stimulation [26]. Their main advantage is that the voltages that can actuate them are as low as $1-2$ V, since the ions are spread within an electrolyte. They are most often used as bending actuators with strong bending potentials but with a rather slow response rate. In electronic EAPs, the electrostatic forces that occur lead to an electromechanical change in the shape of the material. They are typically used as planar actuators because of their large in-plane deformations. Unlike ionic EAPs, they operate under dry conditions, but they require very high activation voltages in the range of many kilovolts. Among the electronic EAPs, soft dielectric EAPs are known as dielectric elastomers (DEs) or artificial muscles with promising functional properties [27]. The central element of DE actuators is a thin elastomeric film (e.g., silicone or acrylic), which is coated with or sandwiched between two compliant electrodes on both sides. Silicones, polyurethanes, and polymers based on acrylic units have emerged as the most appropriate DEs [28]. Besides being used as mechanical actuators, DEs can also perform as sensors, since the serial resistance within the electrodes changes when the charged DE device is stretched or contracted. They can also be used to convert mechanical work into electrical energy and as such can be applied as DE generators, with additional qualities being provided by their softness and flexibility in form and design [29]. Polyaniline (PANI), polypyrrole, and poly(p-phenylene-vinylene) are notable for their special conduction mechanisms, unique electrical properties, controllable chemical and electrochemical properties, and processability.

A typical EAP, PANI possesses controllable electrical conductivity, environmental stability, and fascinating redox properties. It also acts as a possible substrate for gas-sensing applications, since the charge delocalization provides several active locations on its backbone for gas analyte adsorption and desorption [30]. In general, metal oxide semiconductor, are used to fabricate gas sensors, as their conductivity changes through interaction with gas molecules, depending on the operating temperature [31,32]. PANI is not as prone to gas species as metal oxides, and its low solubility in organic solvents restricts its applications, but it is ideal as a matrix for conducting polymer nanocomposite preparation [33]. Hybridization of metal oxide and conductive polymer may enhance the properties of conducting gas sensors, and the material possesses the synergistic impact of both polymers and metal oxide. The nanofibers of polymeric materials have a large aspect ratio, large surface area to volume ratio, and large porosity, and such materials are best suited for sensing different gases, owing to greater analyte gas absorption capacity [30].

PANI and PANI/zinc oxide (ZnO) nanofibers prepared by electrospinning are reported for liquefied petroleum gas (LPG) sensing applications [33]. For this, PANI/emeraldine salt composites are first transformed into emeraldine base type and then protonated with D-camphor-10-sulfonic acid to make the solution feasible

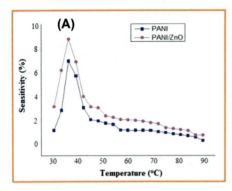 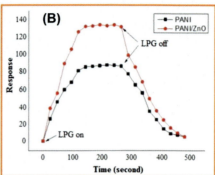

Figure 4.1 (A) Sensitivity and (B) Gas response of PANI and PANI/ZnO nanofibers. *Source*: Reprinted from P.T. Patil, R.S. Anwane, S.B. Kondawar. Development of electrospun polyaniline/ZnO composite nanofibers for LPG sensing. Procedia Mater. Sci. 10 (2015) 195–204 [33] with permission.

for drawing nanofibers. The sensing performance of the PANI/ZnO fibers is demonstrated by the response/recovery times to 1000 ppm LPG (Fig. 4.1). The response time is the time needed to achieve 90% of the final resistance shift when the gas is switched on and off, respectively. As the period of exposure of the gas to the sensor increases, the sensor response rises slowly, and the maximum response is observed at 150 s for both samples. PANI/ZnO nanofibers of 200–300 nm diameter demonstrate quick response and recovery compared to pure PANI nanofibers, suggesting PANI/ZnO as the strongest choice for LPG detection at room temperature [33].

PANI/TiO$_2$ nanocomposite, when mixed with a carrier polymer of 10% polyvinyl alcohol (PVA), also produces electrospun nanofibers. TiO$_2$ possesses unique physical and chemical properties, such as a large energy gap and dielectric constant, which helps PANI to achieve good electrical response to variable concentrations of CO$_2$. The spinning solution is kept in a vertical syringe with a 0.55-mm orifice in a stainless steel needle. The rotating, 30-mm-diameter cylindrical collector is mounted under the ground and the whole electrospinning setup is maintained at room temperature and at 60% relative humidity. Fibers are obtained at 20 kV applied voltage, 1000 rpm rotation speed of the collector and 0.2 mL h^{-1} solution flow rate. Wrapped on a cylindrically revolving tray, the fibers are packed on 0.5 mm thick aluminum foil [34]. These PANI/TiO$_2$ nanofibers on exposure to CO$_2$, formed a p–n junction with increased sensitivity. In fact, a positively charged depletion layer is formed on the TiO$_2$ surface, which could be formed at the heterojunction as a result of the electron migration from TiO$_2$ to PANI. This reduces the activation energy and physisorption enthalpy for CO$_2$. The nanofiber morphology with an average diameter of 250 nm appears to contribute to the sensors' short response time and good reversibility (at 1000 ppm CO$_2$ at low temperature).

Fig. 4.2 illustrates the sensitivity and response of the 1% Al-SnO$_2$/PANI nanofibers of around 200–300 nm diameter to H$_2$ gas. The H$_2$ sensors measure hydrogen easily and accurately over a large spectrum of oxygen concentrations. H$_2$ sensors based on

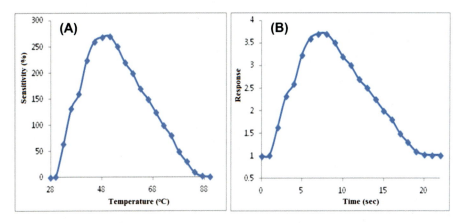

Figure 4.2 (A) Sensitivity and (B) Response of 1% Al-SnO$_2$/PANI nanofibers.
Source: Reprinted from H.J. Sharma, D.V. Jamkar, S.B. Kondawar, Electrospun nanofibers of conducting polyaniline/Al-SnO2 composites for hydrogen sensing applications, Procedia Mater. Sci. 10 (2015) 186–194 [35] with permission.

polymer nanocomposites operating at or closer to room temperature offer microelectronic integrated circuits that are portable, cheap, and easy to implant. While comparing the gas-sensing property to 1000 ppm H$_2$ for the 1% Al-SnO$_2$/PANI and 1% Al-SnO$_2$ nanofibers with temperature, a higher sensing ability is noticed for composites of PANI. In fact, when PANI is mixed with Al-SnO$_2$, the synergistic effect of the two components requires less energy for the electrons to move from valance band to conductive band, that is, increased carrier mobility. Structure of PANI electronically changes in the presence of Al-SnO$_2$ crystallites. The resulting material therefore operates at lower temperatures as compared with pure Al-SnO$_2$. On exposure to H$_2$ gas, 1% Al-SnO$_2$/PANI nanofibers show decreased resistance from its original value by more than an order of magnitude, indicating the film's sensitive nature. Being an n-type surrounded by p-type PANI molecules, Al-SnO$_2$ crystallites make p–n junctions in composite films [35].

4.3 Elastomer–metal oxide nanocomposite fibers

Elastomers are polymers that have viscoelastic characteristics with very weak intermolecular forces. Natural rubber, polyurethanes, polybutadiene, neoprene, and silicone are examples of elastomers. Generally, there are two types of elastomers: saturated and unsaturated elastomers. While the unsaturated elastomers can be cured by vulcanization, the saturated polymers cannot be cured. Elastomers can also be thermoplastic elastomers (polyurethanes, polyamides), thermosets (vulcanized rubbers), and rubber-like solids (neoprene, buna-s). In fact, elastomers play a vital role in industries, owing to their insolubility, elasticity, and flexibility. Because of these features, elastomers are used for manufacturing consumer products, constructions, industrial equipment, wires, and cables.

Elastomer properties vary upon the introduction of different particles or fillers in different concentrations. In general, composites of elastomers are manufactured by solution processing, melt blending, and milling processes, similar to other polymers. However, the tensile properties and flexibility change with the nature and amount of the filler in addition to the vulcanization or crosslinking [36,37]. Electrospinning is another way to generate elastomer composites, in the form of fibers. Polyurethane (PU) solution in dimethylformamide and tetrahydrofuran (50/50, 10 wt.%) containing 10 wt.% CuO particles at 17 kV produces nanofibers when electrospun, as illustrated by Nirmala et al. [38]. Fibers are collected on a rotating iron drum that is kept at a distance of 15 cm from the syringe microtip. During electrospinning, the spout was (internal diameter = 0.51 mm) kept moving and along the side (i.e., back and forward) on its pivot at a distance of 150 mm and a straight speed of 100 mm min^{-1} [39]. As is shown in Fig. 4.3A, the flawless PU nanofibers exhibited a smooth surface and a bead-free nature.

Fig. 4.3 also shows low- and high-magnification SEM images of PU/CuO composite nanofibers with concentrations of 1%, 5%, and 10% of CuO particles [38]. Significant morphological differences between pristine and composite PU nanofibers are observed. Owing to the presence of CuO particles in the PU polymer solution, the electrical conduction of the solution is enhanced, further increasing the acceleration of jetting polymer solution during the electrospinning. The highly conductive polymer solution therefore provides faster electrospinning. As shown by the SEM images, the CuO particles are attached to the periphery of the PU nanofibers. The diameters of the CuO particles are observed to be between 200 and 500 nm. At the same time, the diameters of PU/CuO composite nanofibers are almost similar to those of pristine PU nanofibers.

The EDX spectrum confirms the presence of copper, oxygen, and carbon elements in the PU composite fibers, indicating the successful incorporation of CuO particles. Fig. 4.4 shows the XRD and FTIR pattern of pristine and composite PU/CuO fibers. The XRD pattern of pristine PU nanofibers shows a wide peak at 20.1degrees due to the amorphous nature of the sample. For composite nanofiber mats, the existence of prominent XRD patterns corresponding to the CuO particles in the PU phase is observed as indexed in the figure. Intense diffraction peaks at 35.1, 38.1, 44.1, 48.1, 64.5, and 78.1degrees correspond to (002), (111), (200), (202), (113), and (220) CuO planes and are consistent with JCPDS data (JCPDS file No. 05−661). PU/CuO composite nanofibers with a CuO content of 1% do not show any significant peaks. This can be due to the very low content of CuO particles in composite nanofibers. No significant diffraction peaks for any other phase or impurity can be detected in the composites as well. The bonding configurations and the interaction of PU nanofibers with CuO particles were further identified by using the FT-IR spectroscopy. The characteristic peaks of the PU composite appear in the range of 500−1750 cm. The transmittance at 3240, 2960, 1740, 1530, 1220, 1110, and 777 cm^{-1}, represent the LEB (N−H), LEB (C−H), v (C1/4 O), v (C1/4 C), v (C−C), v (C−O), and v (C−H) modes, respectively [38−40]. A wide transmission band centered at 3500 cm^{-1} corresponding to −OH stretch vibration was

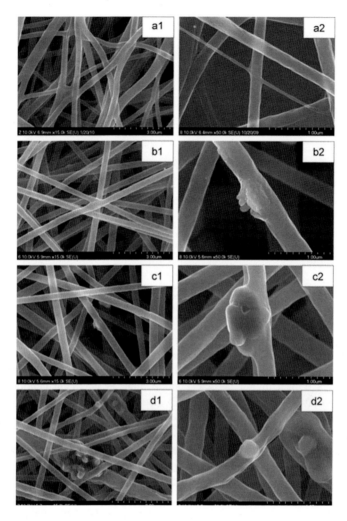

Figure 4.3 SEM images at low (1) and high (2) magnifications of (A) pristine PU nanofibers, (B) PU/1% CuO, (C) PU/5%CuO, and (D) PU/10%CuO.
Source: Reprinted from R. Nirmala, K.S. Jeon, B.H. Lim, R. Navamathavan, H.Y. Kim, Preparation and characterization of copper oxide particles incorporated polyurethane composite nanofibers by electrospinning, Ceram. Int. 39 (8) (2013) 9651–9658 [38] with permission.

observed. At the same time, there is a broad peak at 3200 and 3600 cm^{-1} that is attributed to the increase in −OH stretching. Besides, compared with neat PU nanofibers, there is a transmittance crest at 1600 cm^{-1} for the PU/CuO composite nanofibers due to the isocyanate gather of CuO particles.

In another research work, PU nanofibers decorated with Fe$_3$O$_4$NP were studied [39]. Fig. 4.5 shows a fairly smooth surface of the PU nanofibers and an ultrafine

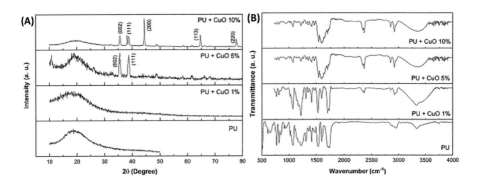

Figure 4.4 (A) XRD patterns of PU and PU-CuO composite nanofibers. (B) FT-IR spectra of PU and PU-CuO composite nanofiber.
Source: Reprinted from R. Nirmala, K.S. Jeon, B.H. Lim, R. Navamathavan, H.Y. Kim, Preparation and characterization of copper oxide particles incorporated polyurethane composite nanofibers by electrospinning, Ceram. Int. 39 (8) (2013) 9651–9658 [38] with permission.

interlocking fiber matrix with submicron size, high porosity, and random morphology. The PU membrane's porous structure is not altered, though smooth surfaces are significantly roughened after the 20-h immersion in Fe_3O_4 NPs. The FE-SEM image shows that the Fe_3O_4 NPs are present not only on PU nanofibers but also below the top layer of nanofibers. The density of Fe_3O_4 is further augmented after the immersion progresses with a higher particle content, that is, magnetic nanofibrous membrane $(MNF)_1$-based membranes (Fig. 4.5C). The surface of the PU nanofiber is also coated entirely with a sheath of tightly packed Fe_3O_4 NPs for membranes of MNF_2 (Fig. 4.5D). The membrane turns from white to pale brown to profound brown after the Fe_3O_4 decoration when the concentration of NPs is doubled, maintaining the same color after various washings with water (insets of Fig. 4.5A–D).

XRD peaks confirm the physical blending of Fe_3O_4 and PU nanofibers without chemical reaction. However, studies of the phase composition and oxidation (valence) state of the test samples are required, since magnetite (Fe_3O_4) and maghemite (c-Fe_2O_3) have an inverse spinel structure with very similar XRD patterns. As is shown in Fig. 4.6, a strong peak at 539 cm^{-1} is assigned to the Fe–O bond vibrations. In addition, the broad and weak peak around 3347 cm^{-1} corresponds to the O–H stretching vibration of H_2O in Fe_3O_4. Peaks for the neat PU nanofibrous at 3314 cm^{-1} (hydrogen bonded NH stretching), 1729 cm^{-1} (free bonded >C=O (amide I band)), 1703 cm^{-1} (H-bonded >C=O (amide I region)), and 1596 cm^{-1} (C=C (benzene ring)) are also observed (Fig. 4.6B). The absorption bands related to asymmetric and symmetric $-CH_2$ stretching are observed at 2944 and 2922 cm^{-1}, respectively, while various modes of $-CH_2$ vibrations are manifested in the 1219–1413 cm^{-1} range. Fig. 4.6C shows where the FT-IR peaks for MNF_2 composites either shifted or appeared. For example, the characteristic peak corresponding to the stretching vibration of FeAO bond not only occurred at 560 cm^{-1} as a new peak, but also is shifted to a higher wavenumber compared to characteristic peak of Fe_3O_4 at 359 cm^{-1}.

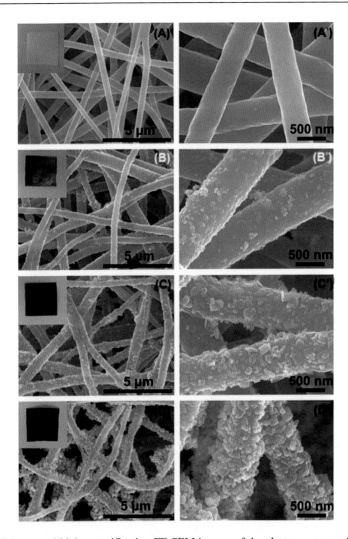

Figure 4.5 Low- and high-magnification FE-SEM images of the electrospun nanofibers. (A and A′) PU; (B and B′) MNF$_{0.5}$; (C and C′) MNF$_1$; (D and D′) MNF$_2$ with 0.5, 1 and 2 mg/ml Fe$_3$O$_4$ NP concentration. Membrane photos are added as insets.
Source: Reprinted from A. Amarjargal, L.D. Tijing, C.H. Park, I.T. Im, C.S. Kim, Controlled assembly of superparamagnetic iron oxide nanoparticles on electrospun PU nanofibrous membrane: a novel heat-generating substrate for magnetic hyperthermia application, Eur. Polym. J. 49 (12) (2013) 3796–3805 [39] with permission.

4.4 Biopolymer/metal oxide nanocomposite fibers

Biopolymers are defined on the basis of the raw materials and biodegradability of the polymer. The various types of developed biopolymer include those made from

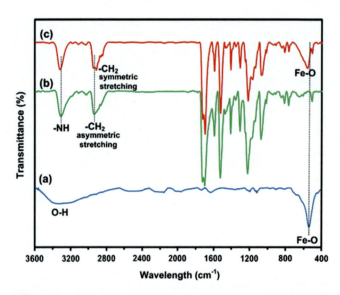

Figure 4.6 FTIR spectra of (A) Fe$_3$O$_4$ NPs, (B) neat PU nanofibers, and (C) MNF$_2$ membranes. *Source*: Reprinted from A. Amarjargal, L.D. Tijing, C.H. Park, I.T. Im, C.S. Kim, Controlled assembly of superparamagnetic iron oxide nanoparticles on electrospun PU nanofibrous membrane: a novel heat-generating substrate for magnetic hyperthermia application, Eur. Polym. J. 49 (12) (2013) 3796–3805 [39] with permission.

Table 4.1 Biodegradable versus bio-based polymer.

Origin	Biodegradable	Nonbiodegradable
Bio-based	CA, CAB, CAP, CN, P3HB, PHBHV, PLA, starch, chitosan	PE (LDPE), PA 11, PA 12, PET, PTT
Partially bio-based	PBS, PBAT, PLA blends, starch blends	PBT, PET, PTT, PVC, SBR, ABS, PU, epoxy resin
Fossil fuel-based	PBS, PBSA, PBSL, PBST, PCL, PGA, PTMAT, PVOH	PE (LDPE, HDPE), PP, PS, PVC, ABS, PBT, PET, PS, PA 6, PA 6.6, PU, epoxy resin, synthetic rubber

renewable raw materials (biodegradable), those made from renewable raw material (not biodegradable), and those synthesized from fossil fuels (biodegradable). Examples of biopolymers include the following: in the chemical synthesis category, polylactic acid (PLA) by lactic acid polymerization, polybutylene succinate (PBS) biopolymer succinic acid); in the manufacture of microorganism category, polyhydroxyalkanoates (PHAs) such as poly(3-hydroxybutyrate) with copolymers such as 3-hydroxybutyrate and 3-hydroxyhexanoate; and in the natural category, cellulose acetate, esterified starch, chitosan-cellulose-starch, and starch-modified PVA [41–43]. The different categories of biopolymers are listed in Table 4.1. The fabrication of microfibers or nanofibers from multiple biopolymers, such as cellulose, chitosan, lignin and other proteins, protein isolated, or their derived materials, is discussed in this section.

Electrospinning of polysaccharides, such as cellulose, or animal-derived biopolymers is extremely complex [41] in contrast to the electrospinning of synthetic polymers [42]. The impartial charged biopolymer encounters long-range electrostatic interface, owing to the occurrence of counterions [43] that prevents the desired electrospinning. Though cellulose is a widely applied biopolymer, its derivatives, such as cellulose acetate (CA), cellulose acetate phthalate (CAP), cellulose acetate butyrate, cellulose acetate trimelitate, hydroxupropylmethyl cellulosephthalate, and the derived ether, such as methyl cellulose (MC), ethyl cellulose, hydroxyethyl cellulose, hydroxypropyl cellulose, hydroxypropylmethyl cellulose, carboxymethyl cellulose, and sodium carboxymethyl cellulose, are exceedingly tough to process [44]. Cellulose is not dissolvable in normal solvents, owing to the tough intramolecular hydrogen bonds, which are considered among the strongest bonds in chemistry. To dissolve cellulose, it must be done via dimethylsulfoxide or paraformaldehyde and sulfur dioxide or others in the same category [45]. The mentioned solvents are not appropriate for electrospinning uses because of the evaporation of solvent in the spinning procedure. Scientists usually utilize ether- or ester-derived cellulose for the method of electrospinning, although this risks the degradation and physical strength of the cellulose [46].

When electrospinning was first developed, in 1934, the cellulose-derived acetate and propionyl cellulose were electrospun with the addition of clean acetone and alcohol mixed with 1 g of softening agents, such as Solactol and Palatinol [47]. Marketable cellulosic filaments were achieved by mixing dissolved cellulose with 50% water solution of N-mthylmorpholine N-oxide (NMMO) solvent, with 1 wt.% antioxidants. The antioxidants were added to effortlessly break down the intramolecular hydrogen bond [45]. The filaments varied from 200 to 400 nm, similar to a noodle-like structure. Later, 3% cellulose with changed solvent mixture of lithium chloride and N,N-dimethylacetamide (DMAc) was electrospun to collect steady and dry nanofibers. The existence of lithium chloride without additional salt is essential to link the electrostatic interface between DMAc and cellulose [48].

PLA is another biopolymer that is notable for its biocompatible and biodegradable properties and the mammalian physiques [49]. Because of the high cost of production (polycondensation process to convert lactic acid into PLA) in addition to the low molecular weight and low mechanical properties, its application is restricted to certain specific areas [50] such as household, pharmaceutical/biomedical, and engineering [50]. Some of the household applications include the textile industry and food-packaging industry. PVA is another significant biopolymer that is widely applied in the biomedical field because of its main property of fiber formation. PVA is also blended with chitosan [51] to make composite membrane through strong hydrogen bonds and to improve tensile strength. Blending was done by mixing a 70−30 ratio of chitosan (in acetic acid) and PVA (distilled water) solutions with sodium hydroxide (NaOH) followed by electrospinning. The fiber's mean diameter with NaOH is 80−150 nm, whereas without NaOH it is 150−300 nm [52]. During electrospinning, the diameter of the nanofibers again decreases from

300 to 125 nm when the PVA concentration is lowered. To increase the average diameter of the nanofibers in electrospinning from 60 to 420 nm, the concentration of PVA/chitosan is increased from 3% to 9%. [53,54]. For any concentration less than 3%, beads developed. Similar results were achieved by Jin et al. [55] while crosslinking PVA/chitosan blend nanofibers with polyethyleneglycol-600-dimethylacrylate and photoinitiator 2-hydroxy-1-[4-(2-hydroxyethoxy) phenyl]-2-methyl-1-propanone.

In previous years, tissue engineering has developed as a major area in biomedical research [56]. Biopolymers alone or blended with polyesters or polycarbonates have been utilized to fabricate several apparatuses, such as catheters, surgery panels, screws, and platforms [57]. However, the nanofibrous and biomimetic supports face significant limitations in mechanical stability and dissimilarity issues in holding extra-cellular matrix (ECM) cell when compared to a natural matrix [58]. Other than the biological applications, technological uses such as water purification have been addressed also by the electrospun metal oxide–based polymer composites. TiO_2 is specifically useful in this regard, owing to its role in advanced oxidation processes for pollutants removal and as photocatalysts. TiO_2 functionalized electrospun gum karaya/polyvinyl alcohol (GK/PVA) has been studied for removing diclofenanc and bisphenol from water [59]. Different ratios of 3 wt.% GK and 10 wt.% PVA solutions, such as 0/100, 40/60, 50/50, 60/40, 70/30, 80/20, 90/10, and 100/0, were tested to determine the spinnability and uniformity in nanofiber distribution. The electro-spinning conditions that were applied were voltage 0–50 kV, substrate speed 0.015–1.95 m min^{-1}, spinning electrode width 500 mm, effective nanofiber layer width 200–500 mm, spinning distance 130–280 mm, and process air flow 20–150 m^3 h^{-1}. Detailed analysis of the membrane revealed the TiO_2 bonding with polymer through the bidentate coordination of two oxygen atoms of −COOH group with Ti^{4+}. In addition, there were hydrogen bonds between the polymeric carbonyl and hydroxyl groups with TiO_2. The influence of plasma and ultraviolet (UV) treatment on the membranes with TiO_2 NPs finely and tightly packed on the surface increased the hydrophobicity and UV resistance. This facilitates the membrane's ability to recover after the micropollutants' treatment. Also, the plasma-treated membrane shows higher thermal stability with no dissolution of TiO_2 particles in water, whereas the UV treated membranes have lower lifetime.

Another biocompatible and biodegradable thermoplastic polyester is the PHAs, with applications in the pharmaceutical, biomedical, and environmental fields. A study by Yu et al. examined the inclusive influence of 1% ZnO NPs on the crystallization behavior of poly(3-hydroxybutyrate-co-3-hydroxyvalerate) (PHBV)/ZnO nanofibers [60]. During the composite preparation, 1% tetrabutylammonium bromide was also added to 20% wt PHBV, since it promotes fiber formation. Electrospun fibers of PHBV-ZnO with 3%–7% ZnO exhibit a higher I_{uv}/I_{vis} ratio by 12.8 time; however, lower ZnO concentrations do not influence the crystallization. The authors propose the application of such biocompatible composites for good antimicrobial activity and biophotonics. Electrospun nanofibers of chitosan have been investigated for healing diabetic wounds because of their significant resemblance to ECM [61]. Chitosan is a natural and polycationic polysaccharide with strong antioxidant activity. Such mechanically strong nanofibers effectively heal wounds and are therapeutically functional, especially when inorganic NPs such as ZnO are embedded. ZnO also enhances the microbicidal and upgraded deposition of collagen in the area of the wound. Electrospinning was performed for a chitosan/PVA solution (1:4) containing ZnO NPs of 40 nm diameter at a flow rate of 0.5 mL h^{-1} at 15 kV applied voltage and at 7 cm needle-collector distance. Collected nanofibers were tested for wound-healing efficiency in rabbits (Fig. 4.7). Though chitosan/PVA and Chitosan/PVA/ZnO have significant antimicrobial potential, the inhibition zones of the latter against *Pseudomonas aeruginosa*, *Escherichia coli*, *Staphylococcus aureus*, and *Bacillus subtilis* are 21.8 ± 1.5, 20.2 ± 1.0, 21.5 ± 0.5, and 15.5 ± 0.8 mm, respectively, in comparison to the former at 15.8 ± 1.0, 14.1 ± 0.8, 5.4 ± 0.5, and 13.0 ± 0.7 mm, respectively.

Toniatto et al. [62] compared the structural compatibility and surface skin tissue engineering properties of PLA-TiO$_2$ nanofibers containing TiO$_2$ (1–5 wt.%). Other than its easiness of processability, biocompatibility, and renewability, PLA represents an energy-saving alternative. Optimal electrospinning conditions for fiber formation were optimized as humidity ∼45%, voltage 12 kV, temperature ∼21°C, gravity-assisted flow, and needle-collector distance 10 cm. As a higher concentration of TiO$_2$ increases the conductivity, both the standard deviations and the lower diameters are preferred for TiO$_2$ to induce a net charge increase in the solutions and thus stable electrospinning jets. Higher TiO$_2$ concentrations cause density enhancement. Compatibility of the PLA/TiO$_2$ nanostructure on a L929 cell line was explored by using a cytotoxicity MTT assay in vitro. Additionally, at 1, 3, and 5 wt.% of TiO$_2$, no significant difference was observed when compared to PLA. This also reflects the viability of the cell at around 100%. PLA/TiO$_2$ fibers showed the capability to promote death of

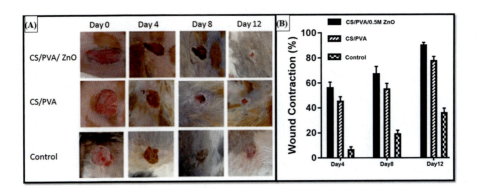

Figure 4.7 (A) Wound closure time in course of 12 days treated with control, CS/PVA, and CS/PVA/ZnO nanofiber mats. (B) Graph showing percent wound contraction over a period of 12 days.
Source: Reprinted from R. Ahmed, M. Tariq, I. Ali, R. Asghar, P.N. Khanam, R. Augustine, et al., Novel electrospun chitosan/polyvinyl alcohol/zinc oxide nanofibrous mats with antibacterial and antioxidant properties for diabetic wound healing, Int. J. Biol. Macromol. 120 (2018) 385–393 [61] with permission.

bacteria at longer incubation times and at higher TiO$_2$ concentrations (5 wt.%). PLA/TiO$_2$ nanofibers are promising as an inert implant coating material that should be useful in a variety of surgical implants (Fig. 4.8).

Electrospinning produces fiber mats of optimum porosity and surface area and contributes to the structural simulation of the extracellular matrix, making it applicable in tissue engineering. Furthermore, the scaffolds' nanofibrous topography has a significant role in controlling the adhesion of cells, differentiation and proliferation [63]. Flexibility of the polymers and mechanical resistance of the ceramic NPs are combined in fabricating biocompatible PLC/CaCO$_3$ and HA/TiO$_2$ nanocomposites for bone regeneration by electrospinning. The high specific surface area of the NPs allows good adhesion with the matrix polymer, thus increasing the chemical properties, flexibility, and mechanical strength. In addition, the TiO$_2$ improves the bioactivity as a result of the presence of —OH groups, which may cause the facilitation of intermolecular interactions and favor the osseointegration or osseoinduction processes. By making a composite with TiO$_2$, the limiting properties of PHBV, such as hydrophobicity, poor mechanical and thermal properties, high crystallinity, and slow

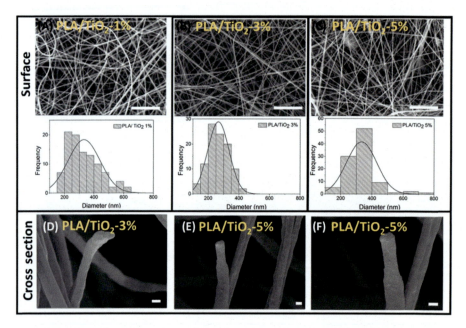

Figure 4.8 FE-SEM micrographs of the surfaces and respective histogram of fiber distribution of (A) PLA/TiO$_2$−1%, (B) PLA/TiO$_2$−3% and (C) PLA/TiO$_2$−5% (scale bar = 10 μm). FE-SEM micrographs from the cross sections of (D) PLA/TiO$_2$−3% and (E, F) PLA/TiO$_2$−5% (scale bar = 300 nm).
Source: Reprinted from T.V. Toniatto, B.V. Rodrigues, T.C. Marsi, R. Ricci, F.R. Marciano, T.J. Webster, et al., Nanostructured poly (lactic acid) electrospun fiber with high loadings of TiO2 nanoparticles: insights into bactericidal activity and cell viability, Mater. Sci. Eng. C 71 (2017) 381−385 [62] with permission.

and uncontrollable degradation, can be omitted, as TiO_2 covalently links with PHBV functional groups. Highly permeable membranes similar in size scale and architecture to natural ECM can be realized for tissue engineering. The natural polymer PHBV itself possesses high cell compatibility and low cytotoxicity properties.

Fiber mats with interconnected structure and porosity allow the transportation of the cell and mimic the ECM. Hydrothermally made TiO_2 particles are used for the thermal destabilization of the PCT gel. The applied voltage of 20 kV, needle tip distance of 10 cm, needle diameter of 0.7 mm, and feed rate of 0.5 mL min^{-1} are applied to get the fibers on a rotating collector at 60 rpm. All fibers are uniform and homogenous, with no porosity and a cylindrical appearance, confirming the efficiency of electrospinning. It has been observed that the homogenous distribution of NPs promote cellular growth. Also, a higher viscosity of the solution provides fibers of larger diameter, since more energy is required to overcome the surface tension during electrospinning. In addition, the overall tension in the fibers is dependent upon the excess charges' self-repulsion on the jet.

As found by Abdalkarim et al., incorporating cellulose nanocrystals (CNC-ZnO) into the matrix of PHBV improves the nanocomposites' hydrophilicity [64]. The water solubility is also improved with the hydrogen bond interaction between CNC-ZnO and PHBV. The nanocomposites are also composed of CNCs containing more hydroxyl groups, helping in the water absorption behavior. It has been found that MEF cells incubated on electrospun mats of PHBV have a good growth rate and good adhesion. However, the mats containing TiO_2 NPs have higher cellular proliferation during the 24-h period. Higher TiO_2 amounts enhance cell viability, since it promotes the adhesion of cells and increases the proliferation rate. The nanocomposite's increased cell viability is a result of both the physicochemical changes and the nanotopography in the polymer induced by the NPs and finally results in PHBV's hydrophilicity.

By maintaining the electrospinning parameters as 1.0 mL h^{-1} (flow rate), 15 cm (tip to collector), and 18 kV (voltage), PLA/TiO_2 nanocomposites containing TiO_2 at 1.25 wt.% can be fabricated by electrospinning. At 1.0 wt.% or higher loading, the smooth surface characteristics of the film change. Incorporation of nano-TiO_2 also creates porous architecture in PLA composites, and the accumulated TiO_2 provides an uneven surface. The films are denser than the nanofibers because of the strength of the intermolecular crosslinking, where the pores are occupied by the PLA molecules. There is also an overall density reduction due to TiO_2 addition in the composites, with the internal space network belonging to the PLA polymer being destroyed. Overall, the nanofibers' density is lower than that of the film. This could be due to the smaller average size of the TiO_2 NPs on the PLA nanofibers. The solubility of the film is also enhanced with TiO_2, and so is its ability to be an effective barrier, and under UV-A radiation the antibacterial activity improves. With the observed functionality, TiO_2 content of 0.75 wt.% for the PLA/TiO_2 nanofibers is the optimum.

Immobilization of Ag NPs on TiO_2 is done [65] by coating the TiO_2 NPs with polydopamine hydrochloride and $AgNO_3$. The hybrid TiO_2/Ag NPs are incorporated into cellulose acetate nanofiber matrix by electrospinning to fabricate CA/TiO_2/Ag NP composite nanofibers (Fig. 4.9). Owing to the propensity of silver leaching from

Figure 4.9 SEM images of the nanofibers (A) CA, (B) CA/TiO$_2$, (C) CA/TiO$_2$/Ag NP$_1$, and (D) CA/TiO$_2$/Ag NP$_2$.
Source: Reprinted from A.W. Jatoi, I.S. Kim, Q.Q Ni, Cellulose acetate nanofibers embedded with AgNPs anchored TiO2 nanoparticles for long term excellent antibacterial applications, Carbohydr. Polym. 207 (2019) 640–649 [65] with permission.

the nanofiber mats being reduced, the antibacterial performance of the Ag NPs anchored on TiO$_2$ NP surfaces and their polymer composites will be reduced. Ag NPs are effective in bacteria colonization inhibition by disrupting the bacterial cell wall and denaturing their DNA by blocking their respiratory systems. The immobilization of Ag NPs with TiO$_2$ and the polymer also prevents the possible excessive releases of silver in human cells. The antibacterial tests for CA/TiO$_2$/Ag NP confirm the growth inhibition against *E. coli* and *S. aureus* as well.

Nanofibers of polylactide blends loaded with TiO$_2$ and blended with PVP are more uniform and are considered smoother than the samples without PVP [66]. For instance, while the fibers containing TiO$_2$ show an absorption band of 200–380 nm in the UV region, neat PLA fibers show a band around 200 nm. Furthermore, the results of the photodegradation process shows that the PVP components are dissolved in the solution of PBS, and the matrix of PLA undergoes degradation as a function of time. After that, the fibers that have been fabricated are used for application of a catalytic system for the epoxidation of sunflower oil (unsaturated) and for the usage as plasticizers or additives, which resulted in the employment of an oxidizing agent that is also a performer acid. The fibers with specifically PVP are able to enhance the oil yield epoxidation with a rate that is slow and of side reactions that are undesirable, breaking the ester bond of triglycerides in order to begin the generation of free fatty acids. The biodegradability and photodegradability are higher in the composites, especially under UV radiation and with an activator present, in comparison to the neat PLA. A solid catalyst that is of typical nature and is based upon Ti-SiO$_2$ that is nanosized revealed an epoxide conversion yield of 100%, higher in triglycerides and fatty acid epoxidation reactions. With cycling times of at least 5, the PLA/PVP/TiO$_2$ composite fiber (PLA/PVP (5:1)/TiO$_2$ 5% and PLA/TiO$_2$ 5%, with PLA weight 0.7 g), exhibits catalytic activity in the SFO epoxidation.

4.5 Conclusion

This chapter has reported some of the general characteristics of the polymer–metal oxide nanofibers, especially focusing on the electroactive polymers, elastomeric polymers, and biopolymers. The polymer categories were selected on the basis of the industrial and technological applications of the materials. For instance, the electrospun fibers of PANI with different metal oxides (ZnO, TiO$_2$, Al-SnO$_2$) are used in fabricating LPG sensors, CO$_2$ sensors, and H$_2$ sensors. Electrospinning also generates various biopolymer composites applicable in different fields, such as the biomedical, environmental, and industrial areas. Several biopolymers, such as chitosan, cellulose, lignin, protein, and silk composites, are also manufactured by electrospinning. The polymers interact with the metal oxides through stronger intermolecular hydrogen bonds, and the electrospun fibers find applications in tissue engineering, household applications, and food packaging.

Acknowledgments

This publication is made possible by UREP grant 24-142-1-032 from the Qatar National Research Fund (a member of Qatar Foundation). The statements made herein are solely the responsibility of the authors.

References

[1] B. Ding, X. Wang, J. Yu, Electrospinning: Nanofabrication and Applications, Elsevier, Amsterdam, Netherlands, 2019.

[2] W. Fang, S. Yang, T.-Q. Yuan, A. Charlton, R.-C. Sun, Effects of various surfactants on alkali lignin electrospinning ability and spun fibers, Ind. Eng. Chem. Res. 56 (34) (2017) 9551–9559.

[3] J.-H. He, Y.-Q. Wan, L. Xu, Nano-effects, quantum-like properties in electrospun nanofibers, Chaos Solitons Fractals 33 (1) (2007) 26–37.

[4] D. Ponnamma, H. Parangusan, A. Tanvir, M.A. AlMa'adeed, Smart and robust electrospun fabrics of piezoelectric polymer nanocomposite for self-powering electronic textiles, Mater. Des. 184 (2019) 108176.

[5] D. Ponnamma, K.K. Sadasivuni, M.A. Al-Maadeed, S. Thomas, Developing polyaniline filled isoprene composite fibers by electrospinning: effect of filler concentration on the morphology and glass transition, Polym. Sci. Ser. A 61 (2) (2019) 194–202.

[6] H. Parangusan, D. Ponnamma, M.K. Hassan, S. Adham, M.A. Al-Maadeed, Designing carbon nanotube-based oil absorbing membranes from gamma irradiated and electrospun polystyrene nanocomposites, Materials 12 (5) (2019) 709.

[7] T. Amna, J. Yang, K.-S. Ryu, I.H. Hwang, Electrospun antimicrobial hybrid mats: innovative packaging material for meat and meat-products, J. Food Sci. Technol. 52 (7) (2014) 4600–4606.

[8] S. Sinha-Ray, S. Khansari, A.L. Yarin, B. Pourdeyhimi, Effect of chemical and physical cross-linking on tensile characteristics of solution-blown soy protein nanofiber mats, Ind. Eng. Chem. Res. 51 (46) (2012) 15109–15121.

[9] P.X. Ma, R. Zhang, Synthetic nano-scale fibrous extracellular matrix, J. Biomed. Mater. Res. 46 (1) (1999) 60–72.

[10] K. Nakata, K. Fujii, Y. Ohkoshi, Y. Gotoh, M. Nagura, M. Numata, et al., Poly(ethylene terephthalate) nanofibers made by Sea–Island-type conjugated melt spinning and laser-heated flow drawing, Macromol. Rapid Commun. 28 (6) (2007) 792–795.

[11] N. Raphael, K. Namratha, B.N. Chandrashekar, K.K. Sadasivuni, D. Ponnamma, A.S. Smitha, et al., Surface modification and grafting of carbon fibers: a route to better interface, Prog. Cryst. Growth Charact. Mater. 64 (3) (2018) 75–101.

[12] T. Ondarçuhu, C. Joachim, Drawing a single nanofibre over hundreds of microns, Europhys. Lett. (EPL) 42 (2) (1998) 215–220.

[13] M.A. Hassan, B.Y. Yeom, A. Wilkie, B. Pourdeyhimi, S.A. Khan, Fabrication of nanofiber meltblown membranes and their filtration properties, J. Membr. Sci. 427 (2013) 336–344.

[14] D. Liu, H. Zhang, P.C.M. Grim, S.D. Feyter, U.-M. Wiesler, A.J. Berresheim, et al., Self-assembly of polyphenylene dendrimers into micrometer long nanofibers: an atomic force microscopy study, Langmuir 18 (6) (2002) 2385–2391.

[15] W.E. Teo, S. Ramakrishna, A review on electrospinning design and nanofibre assemblies, Nanotechnology 17 (14) (2006) R89–R106.

[16] G. Taylor, Electrically driven jets, Proc. R. Soc. A 313 (1515) (1969) 453–475.

[17] A.L. Yarin, S. Koombhongse, D.H. Reneker, Taylor cone and jetting from liquid droplets in electrospinning of nanofibers, J. Appl. Phys. 90 (9) (2001) 4836–4846.

[18] K. Lee, H. Kim, H. Bang, Y. Jung, S. Lee, The change of bead morphology formed on electrospun polystyrene fibers, Polymer 44 (14) (2003) 4029–4034.

[19] W. Fu, Z. Liu, B. Feng, R. Hu, X. He, H. Wang, et al., Electrospun gelatin/PCL and collagen/PLCL scaffolds for vascular tissue engineering, Int. J. Nanomed. 9 (2014) 2335.

[20] R. Hartman, D. Brunner, D. Camelot, J. Marijnissen, B. Scarlett, Jet break-up in electrohydrodynamic atomization in the cone-jet mode, J. Aerosol Sci. 31 (1) (2000) 65–95.

[21] X. Li, H. Liu, J. Wang, H. Cui, X. Zhang, F. Han, Preparation of YAG:Nd nano-sized powder by co-precipitation method, Mater. Sci. Eng. A 379 (1–2) (2004) 347–350.

[22] C. Shin, G. Chase, D. Reneker, Recycled expanded polystyrene nanofibers applied in filter media, Colloids Surf. A: Physicochem. Eng. Asp. 262 (1–3) (2005) 211–215.

[23] M. Jaworska, K. Sakurai, P. Gaudon, E. Guibal, Influence of chitosan characteristics on polymer properties. I: Crystallographic properties, Polym. Int. 52 (2) (2003) 198–205.

[24] H.S. Yoo, T.G. Kim, T.G. Park, Surface-functionalized electrospun nanofibers for tissue engineering and drug delivery, Adv. Drug Deliv. Rev. 61 (12) (2009) 1033–1042.

[25] G.M. Scheutz, J.J. Lessard, M.B. Sims, B.S. Sumerlin, Adaptable crosslinks in polymeric materials: resolving the intersection of thermoplastics and thermosets, J. Am. Chem. Soc. 141 (41) (2019) 16181−16196.

[26] Y. Bar-Cohen, Electroactive polymers: current capabilities and challenges, Smart Structures and Materials 2002: Electroactive Polymer Actuators and Devices (EAPAD), vol. 4695, International Society for Optics and Photonics, 2002, pp. 1−7.

[27] Smart Materials in Architecture, Interior Architecture and Design: Axel Ritter: 9783764373276 https://www.bookdepository.com/Smart-Materials-Architecture-Interior-Architecture-Design-Axel-Ritter/9783764373276 (accessed June 18, 2020).

[28] F. Ko, Y. Gogotsi, A. Ali, N. Naguib, H. Ye, G.L. Yang, et al., Electrospinning of continuous carbon nanotube-filled nanofiber yarns, Adv. Mater. 15 (14) (2003) 1161−1165.

[29] Jordi-2011-Biomimetic_airship_driven_by_dielectric-(Published_version).Pdf.

[30] P. Patil, Chemical vapour sensing properties of electrospun nanofibers of polyaniline/ZnO nanocomposites, Adv. Mater. Lett. 5 (2014) 389−395.

[31] S.M. Shibli, M.A. Sha, B.L. Anisha, D. Ponnamma, K.K. Sadasivuni, Effect of phosphorus on controlling and enhancing electrocatalytic performance of Ni−P−TiO_2−MnO_2 coatings, J. Electroanal. Chem. 826 (2018) 104−116.

[32] D. Ponnamma, S. Goutham, K.K. Sadasivuni, K.V. Rao, J.J. Cabibihan, M.A. Al-Maadeed, Controlling the sensing performance of rGO filled PVDF nanocomposite with the addition of secondary nanofillers, Synth. Met. 243 (2018) 34−43.

[33] P.T. Patil, R.S. Anwane, S.B. Kondawar, Development of electrospun polyaniline/ZnO composite nanofibers for LPG sensing, Procedia Mater. Sci. 10 (2015) 195−204.

[34] S.H. Nimkar, S.P. Agrawal, S.B. Kondawar, Fabrication of electrospun nanofibers of titanium dioxide intercalated polyaniline nanocomposites for CO_2 gas sensor, Procedia Mater. Sci. 10 (2015) 572−579.

[35] H.J. Sharma, D.V. Jamkar, S.B. Kondawar, Electrospun nanofibers of conducting polyaniline/Al-SnO_2 composites for hydrogen sensing applications, Procedia Mater. Sci. 10 (2015) 186−194.

[36] D. Ponnamma, K.K. Sadasivuni, Y. Grohens, Q. Guo, S. Thomas, Carbon nanotube based elastomer composites—an approach towards multifunctional materials, J. Mater. Chem. C 2 (40) (2014) 8446−8485.

[37] K.K. Sadasivuni, D. Ponnamma, S. Thomas, Y. Grohens, Evolution from graphite to graphene elastomer composites, Prog. Polym. Sci. 39 (4) (2014) 749−780.

[38] R. Nirmala, K.S. Jeon, B.H. Lim, R. Navamathavan, H.Y. Kim, Preparation and characterization of copper oxide particles incorporated polyurethane composite nanofibers by electrospinning, Ceram. Int. 39 (8) (2013) 9651−9658.

[39] A. Amarjargal, L.D. Tijing, C.H. Park, I.T. Im, C.S. Kim, Controlled assembly of superparamagnetic iron oxide nanoparticles on electrospun PU nanofibrous membrane: a novel heat-generating substrate for magnetic hyperthermia application, Eur. Polym. J. 49 (12) (2013) 3796−3805.

[40] H. Deka, N. Karak, R.D. Kalita, A.K. Buragohain, Bio-based thermo- stable, biodegradable and biocompatible hyperbranched polyurethane/Ag nanocomposites with antimicrobial activity, Polym. Degrad. Stab. 95 (2010) 1509−1511.

[41] M. Stevens, S. Plimpton, The effect of added salt on polyelectrolyte structure, Eur. Phys. J. B 2 (3) (1998) 341−345.

[42] R.P. Santos, B.V. Rodrigues, D.M. Santos, S.P. Campana-Filho, A.C. Ruvolo-Filho, E. Frollini, Electrospun recycled PET-based mats: tuning the properties by addition of cellulose and/or lignin, Polym. Test. 60 (2017) 422−431.

[43] C.H. Lee, H.S. Lim, J. Kim, J.H. Cho, Counterion-induced reversibly switchable transparency in smart windows, ACS Nano 5 (9) (2011) 7397−7403.

[44] S. Sinha-Ray, S. Sinha-Ray, A.L. Yarin, B. Pourdeyhimi, Theoretical and experimental investigation of physical mechanisms responsible for polymer nanofiber formation in solution blowing, Polymer 56 (2015) 452−463.

[45] P. Kulpinski, Cellulose nanofibers prepared by theN-methylmorpholine-N-oxide method, J. Appl. Polym. Sci. 98 (4) (2005) 1855−1859.

[46] W.K. Son, J.H. Youk, W.H. Park, Preparation of ultrafine oxidized cellulose mats via electrospinning, Biomacromolecules 5 (1) (2004) 197−201.

[47] F. Anton, Richard Schreiber Gastell. Process and Apparatus for Preparing Artificial Threads. United States Patent United States1,975,504, 1934.

[48] X. Zhuang, X. Yang, L. Shi, B. Cheng, K. Guan, W. Kang, Solution blowing of submicron-scale cellulose fibers, Carbohydr. Polym. 90 (2) (2012) 982−987.

[49] D. Ponnamma, K.K. Sadasivuni, M.A. AlMaadeed, Introduction of biopolymer composites: what to do in electronics? Biopolymer Composites in Electronics, Elsevier, 2017, pp. 1−12.

[50] S. Ebnesajjad, Handbook of Biopolymers and Biodegradable Plastics Properties, Processing, and Applications, Elsevier/William Andrew, Amsterdam, 2013.

[51] M. Miya, R. Iwamoto, S. Mima, FT-IR study of intermolecular interactions in polymer blends, J. Polym. Sci. Polym. Phys. (Ed.) 22 (6) (1984) 1149−1151.

[52] Z. Huang, Y. Zhang, M. Kotaki, S. Ramakrishna, A review on polymer nanofibers by electrospinning and their applications in nanocomposites, Compos. Sci. Technol. 63 (15) (2003) 2223−2253.

[53] Y. Jia, J. Gong, X. Gu, H. Kim, J. Dong, X. Shen, Fabrication and characterization of poly (vinyl alcohol)/chitosan blend nanofibers produced by electrospinning method, Carbohydr. Polym. 67 (3) (2007) 403−409.

[54] Z. Jiang, H. Zhang, M. Zhu, D. Lv, J. Yao, R. Xiong, et al., Electrospun soy-protein-based nanofibrous membranes for effective antimicrobial air filtration, J. Appl. Polym. Sci. 135 (8) (2017) 45766.

[55] Y. Jin, D. Yang, Y. Zhou, G. Ma, J. Nie, Photocrosslinked electrospun chitosan-based biocompatible nanofibers, J. Appl. Polym. Sci. 109 (5) (2008) 3337−3343.

[56] S. Sankar, C.S. Sharma, S.N. Rath, S. Ramakrishna, Electrospun nanofibres to mimic natural hierarchical structure of tissues: application in musculoskeletal regeneration, J. Tissue Eng. Regen. Med. 12 (1) (2017) 604−619.

[57] S. Lecommandoux, É. Garanger, Precision polymers with biological activity: design towards self-assembly and bioactivity, Comptes Rendus Chimie 19 (1−2) (2016) 143−147.

[58] X. Li, D. Yu, C. Fu, R. Wang, X. Wang, Ketoprofen/ethyl cellulose nanofibers fabricated using an epoxy-coated Spinneret, Model. Numer. Simul. Mater. Sci. 03 (04) (2013) 6−10.

[59] E. Kudlek, D. Silvestri, S. Wacławek, V.V. Padil, M. Stuchlík, L. Voleský, et al., TiO_2 immobilised on biopolymer nanofibers for the removal of bisphenol A and diclofenac from water, Ecol. Chem. Eng. S 24 (3) (2017) 417−429.

[60] R. Naphade, J. Jog, Electrospinning of PHBV/ZnO membranes: structure and properties, Fibers Polym. 13 (6) (2012) 692−697.

[61] R. Ahmed, M. Tariq, I. Ali, R. Asghar, P.N. Khanam, R. Augustine, et al., Novel electrospun chitosan/polyvinyl alcohol/zinc oxide nanofibrous mats with antibacterial and antioxidant properties for diabetic wound healing, Int. J. Biol. Macromol. 120 (2018) 385−393.

[62] T.V. Toniatto, B.V. Rodrigues, T.C. Marsi, R. Ricci, F.R. Marciano, T.J. Webster, et al., Nanostructured poly (lactic acid) electrospun fiber with high loadings of TiO_2 nanoparticles: insights into bactericidal activity and cell viability, Mater. Sci. Eng. C 71 (2017) 381−385.

[63] N.F. Braga, D.A. Vital, L.M. Guerrini, A.P. Lemes, D.M. Formaggio, D.B. Tada, et al., PHBV-TiO_2 mats prepared by electrospinning technique: physico-chemical properties and cytocompatibility, Biopolymers 109 (5) (2018) e23120.

[64] S. Feng, F. Zhang, S. Ahmed, Y. Liu, Physico-mechanical and antibacterial properties of PLA/TiO_2 composite materials synthesized via electrospinning and solution casting processes, Coatings 9 (8) (2019) 525.

[65] A.W. Jatoi, I.S. Kim, Q.Q. Ni, Cellulose acetate nanofibers embedded with AgNPs anchored TiO_2 nanoparticles for long term excellent antibacterial applications, Carbohydr. Polym. 207 (2019) 640−649.

[66] B. Nim, P. Sreearunothai, A. Petchsuk, P. Opaprakasit, Preparation of TiO_2-loaded electrospun fibers of polylactide/poly (vinylpyrrolidone) blends for use as catalysts in epoxidation of unsaturated oils, J. Nanopart. Res. 20 (4) (2018) 100.

Section 2

Metal oxide nanofibers and their applications for biosensing

Kunal Mondal[1], Raj Kumar[2], Blesson Isaac[3] and Gorakh Pawar[1]
[1]Materials Science and Engineering Department, Idaho National Laboratory, Idaho Falls, ID, United States, [2]Faculty of Engineering, Bar Ilan Institute of Nanotechnology and Advanced Materials (BINA), Bar Ilan University, Ramat Gan, Israel, [3]Chemical and Radiation Measurement Department, Idaho National Laboratory, Idaho Falls, ID, United States

5.1 Introduction

Nanomaterials have had a huge impact globally on various technologically and commercially important sectors, such as energy, the environment, automobiles, space, defense, and health care [1,2]. Nanofibers are one of the most interesting categories of nanomaterials with a typical fiber diameter of 100 nm or less and a significantly higher aspect ratio (i.e., nanofiber length/diameter) [3]. The engineered metal oxide nanofibers (MONFs) and polymer nanofibers are the most promising MONF candidates employed in a wide range of applications, primarily because of their unique optical properties. Further, the MONFs demonstrate a lot of brilliant features, such as high surface area, ability of easy surface functionalization, controllable porosity, superior mechanical performance compared to other nanostructures, and a widespread range of material selection [4,5]. Similarly, the mats of MONFs provide a fibrous configuration similar to native extracellular matrix with high interconnected porosity ($>60\%$), abundant absorbance, stable moisture, and gas permeability that convey a suitable environment to shield the wound from exogenous infection [4]. In addition, the ability to load bioactive compounds into nanofibers offers a great environment together with metal oxides required to treat infections in the wound regions, prohibition of bacterial biofilm formation, prolonging drug release, and decreasing the time of the wound-healing process [6,7]. These significant characteristics along with the unusual adsorptive properties and fast diffusivities, quick and sensitive detection portability, and low cost compared to other conventional systems make the MONFs perfect candidates for a wide range of biomedical applications containing tissue-engineered scaffolds (e.g., cartilage, skin, bone, blood vessel), wound dressings, biomedical devices, and drug and therapeutic delivery systems [8–10]. Finally, the capabilities to immobilize enzymes, antimicrobial peptides, antibiotics, and growth hormones to nanofibers or to be loaded into the core of nanofibers are establishing MONFs as a suitable platform for biosensing applications [11,12]. Fig. 5.1 depicts various application categories based on MONFs.

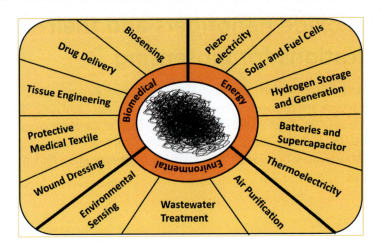

Figure 5.1 Applications of metal oxide nanofibers.

Nanofibers are one of the interesting groups of nanomaterials with two similar external dimensions in the nanoscale (≤ 100 nm) and the third dimension significantly larger. Nanofibers exhibit a lot of wonderful features, such as large surface area, possibilities for surface functionalization, tunable porosity, a wide range of material selection, and MONF-based biosensors have been widely explored in biomolecule detection, owing to their benefits such as low cost, easy fabrication, neat size, and simple electronic counterparts [13]. However, the performance of such sensors is greatly influenced by the morphology and structure of the sensing materials, causing great difficulty for biosensors based on bulk materials to achieve highly sensitivity and other sensor properties [14]. Basing biosensors on nanostructured materials presents great opportunities to improve biosensing properties such as sensitivity, selectivity, and response time. There have been many reports on metal oxide–based biosensors [15], However, major challenges remain to overcome limitations, such as specificity, selectivity, industrialization, cost-effectiveness, flexibility, and accuracy. Moreover, it is still necessary to comprehensively summarize the important nanostructures of metal oxides from the perspective of nanotechnology. In this chapter we provide a comprehensive discussion of fibrous metal oxide nanostructure synthesis procedures; examples of widely studied MONFs and their biosensing mechanisms; properties from the aspects of fiber diameter, morphology, doping, and materials selection; and biosensing of various species ranging from chemical to biological. In the following section we discuss the synthesis strategies of MONFs.

5.2 Synthesis strategies for MONFs

Several nanotechnology-based synthetic strategies have been developed to fabricate MONFs. The most widely used strategies are based on mechanical, chemical,

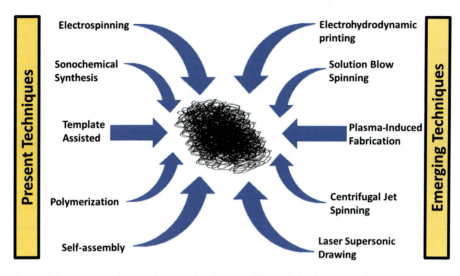

Figure 5.2 Present and emerging metal oxide nanofiber fabrication strategies.

thermal, and electrostatic-based fabrication techniques. As far synthetic strategies are concerned, they are generally classified as top down and bottom up [16]. In the bottom-up approaches, ions, atoms, molecules, or even nanoparticles are used to make the building blocks for the formation of MONFs (e.g., gas phase production of nanofibers). By contrast, top-down approaches involve slicing or successive breaking of a bulk material into small pieces via attrition, grinding, or milling to harvest nanofibers (e.g., production of cellulose nanofibers from wood pulp) [17]. However, large-scale production of nanofibers with high yield and well-defined morphology remains a challenge. Fig. 5.2 depicts the fabrication of metal oxide nanofibers based on traditional synthesis processes and emerging techniques. Fabrication of nanofibers of metal oxides can be commonly classified into two leading categories: physicochemical route and spinning and nonspinning fabrication route. Among various techniques, electrospinning techniques are easy, simple, and cost-effective, and they provide the ability to control size, shape, and porosity. Hence they are the most studied and applied technology for nanofibers of various materials, including metal oxide. In this chapter we discuss them comprehensively, especially electrospinning -based nanofiber fabrication technology. Various physicochemical characteristics, such as nanofiber length and diameter, interfiber spacing, mechanical strength (Young's modulus), and energy of adhesion, are also discussed.

5.2.1 Physicochemical route for MONF fabrication

Physical routes of MONF fabrication involve brute forces such as mechanical pressure, high-energy radiation, electrical energy, or thermal energy that induce material melting, abrasion, evaporation, or condensation and produce nanofibers. Further, the physical vapor deposition, laser ablation techniques are also the most common examples of physical nanofiber fabrication techniques. In case of physical vapor depositions (e.g., thermal evaporation), pulsed laser deposition and plasma sputtering have been expended to make MONFs (or short fibers) and carbon/MONF composites [18]. In these techniques the precursor metal oxide material from a solid or liquid phase transforms to a vapor phase and then recondensed to form nanofibers.

Chemical techniques include a chemical reaction among reacting chemical species to produce nanofibers. The chemical reactions can proceed spontaneously or by an external applied force or energy such as high-energy radiation, thermal energy, electrical energy, or electromagnetic energy to form nanofibers. Usually, catalysts and growth-controlling mediators are added before or during the reaction or immediately after the precipitation to control the growth of initial nanoparticles towards the nanofibers. Chemical vapor deposition (CVD), plasma-enhanced chemical vapor deposition (PECVD), electrochemical deposition, phase separation, microemulsion, the sol-gel method, and hydrothermal synthesis are some of the most commonly used chemical methods for nanofiber synthesis [4]. Most of these methods are modified to fabricate MONFs. For instance, CVD is often used as is or in a modified format for production of metal oxide, carbon, and metal oxide/carbon nanocomposite nanofibers from vapor phase through catalytic reactions in a high-temperature environment inside furnaces in air or inert atmosphere. Liu et al. [19] described a simple vapor deposition method to fabricate large-scale arrays of aligned tungsten oxide submicrofibers using WO_3 powder as the raw material. In the PECVD method, chemical reactions are also involved in the process, which occurs after formation of a plasma of the reacting gases [20]. Hydrothermal treatment [21], microwaves [22], and ultrasound [23] have lately been employed for wet chemistry synthesis of nanowires and nanofibers of MONFs. In addition, soft (e.g., surfactants and polymers, swollen liquid crystals) [24] and hard templates (e.g., porous anodic aluminum oxide membranes) [25] are frequently used together with the chemical synthesis techniques (e.g., sol-gel, electrochemical deposition, CVDs) to yield a range of nanofibers starting from metals, semiconductors, metal oxides, conductive polymers, and carbons.

5.2.2 Spinning technique for fabrication of MONFs

The aforementioned physical and chemical nanofiber fabrication techniques are nonspinning in nature. Typically, the spinning systems use external forces, such as centrifugal force, electric force, or compressed air or gas, to draw filaments of polymer solutions or polymer melts up to fiber with diameters ranging from a few micrometers to hundreds of nanometers [4]. Overall, nanofiber spinning has attracted a huge amount of scientific and industrial attention, owing to its flexibility, cost-effectiveness, option for variety of materials selection, and their possible inclusion in a wide range of practical applications including the environmental remediation, crop protection, and biomedical applications. The nanofiber-spinning techniques can further be categorized into two main types.

5.2.2.1 Electrospinning technique for fabrication of MONFs

Electrospun MONFs have exhibited excellent properties such as a high surface area to volume ratio and tunable porosity [13]. Electrospinning, a top-down approach for nanofabrication, has been acknowledged to be an efficient technique for a low-cost, large-scale production with scalability and easy synthesis of long, continuous fibers [26,27]. By adjusting the properties of those nanofibers through surface functionalization, highly accessible active surface areas can be grasped by easily fine-tuning nanofiber size, with a desired amount of mass and material composition. Consequently, the morphologies and surface textures of electrospun nanofibers can be beneficial in novel applications, such as catalysis and photocatalysis, water purification, clean energy, micro/nanofluidic, drug delivery and therapeutic release, porous membranes and supports, energy storage, photon harvesting, gas sensors, and biosensing [17,28]. A plethora of MONFs, such as zinc oxide (ZnO), titanium dioxide (TiO$_2$), tin oxide (SnO$_2$), iron oxide (Fe$_2$O$_3$), manganese dioxide (MnO$_2$), silicon dioxide (SiO$_2$), vanadium pentoxide (V$_2$O$_5$), and their composites, have been synthesized via electrospinning [29]. Depending upon the system geometry, electrospinning is mainly categorized into two configurations: horizontal electrospinning and vertical electrospinning [30]. A typical horizontal electrospinning device is shown in Fig. 5.3. The diameter of the electrospun nanofibers can be assorted from approximately 10 nm to few micrometers and can be further fine-tuned by simply optimizing various electrospinning parameters, such as polymer concentration, electric field strength, and flow rate of the feed solution [31]. The length and orientation of fibers can be contingent upon these parameters and the geometry and size of the collector [32]. Further requirements, such as fast electron transfer, stable redox potential, and cyclability of the fabricated fibrous assemblies, are crucial, and these can be controlled by selecting

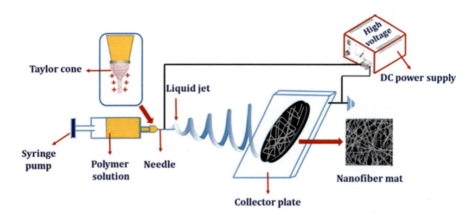

Figure 5.3 A schematic view of horizontal electrospinning technique with a static collector. *Source*: Reproduced with permission from R. Kumar, et al., Advances in nanotechnology based strategies for synthesis of nanoparticles of lignin, in: S. Sharma, A. Kumar (Eds.), Lignin, Springer International Publishing, 2020, pp. 203–229. doi:10.1007/978-3-030-40663-9_7 [28] Copyright: Springer Nature Switzerland AG 2020.

the proper precursor materials during electrospinning. Ambient parameters, such as temperature and humidity, can disturb the fiber diameter and morphology [33].

Coaxial, colloid, melt, and solution electrospinning are the techniques that are most used to harvest MONFs [13,34]. However, the presence of metal oxide colloidal particles as the precursor material along with carrier polymer of the electrospinning feed complicates the spinning. However, recently, many innovative approaches for the electrospinning of metal oxides precursors have drawn much attention for the synthesis of MONFs [13,35]. Fig. 5.4 depicts two approaches for the fabrication of MONF using electrospinning. A metal oxide precursor either could be added into carrier polymer solution before or during the electrospinning or could be commenced after electrospinning process with the spun fibers.

Generally, heating of as-spun MONFs in oxygen is a traditional route for the manufacture of MONFs because it is simple and capable of mass production. The calcination process works in two phases, in which the polymeric part is eliminated from the fibers by burning in air or oxygen at high temperature, and an oxidative conversion

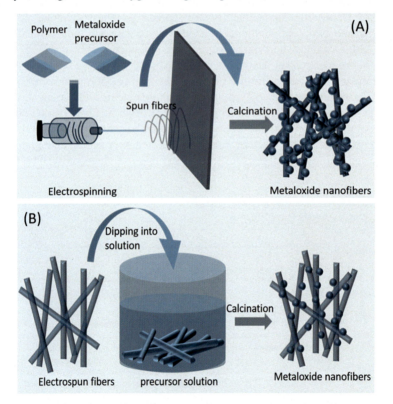

Figure 5.4 Scheme of electrospun metal oxide nanofibers by (A) in situ incorporation of metal oxide precursor and (B) incorporation of metal oxides after electrospinning by dipping spun nanofibers into a solution containing a precursor for metal oxides.
Source: Reproduced with permission from K. Mondal, A. Sharma, Recent advances in electrospun metal-oxide nanofiber based interfaces for electrochemical biosensing, RSC Adv. 6 (2016) 94595–94616 [13]. Copyright 2016, Royal Society of Chemistry.

of the metal oxide precursor yields MONFs by high-temperature nucleation and growth [36].

In the case of the electrospinning of inorganic metal oxide precursor, heating causes shrinkage in the nanofiber, since the carrier polymer is eliminated. The MONFs usually become brittle because of their thinner cross section, so the thermal and internal mechanical stress generated through the shrinkage in size by calcination [37,38]. Thus an accompanying material that is chemically and mechanically unaltered by heating needs to be combined with or impregnated into the electrospun nanofibers. For example, zinc, silica, and TiO_2 nanoparticles were incorporated into polymer nanofibers by electrospinning of a blend of polymer and dispersion of those nanoparticles [39–41]. Crespy and his colleagues fabricated metal oxide/silica nanofibers by using colloid electrospinning technique [37]. Mondal et al. [42] used titanium isopropoxide, a sol-gel precursor of TiO_2, in a carrier polymer polyvinylpyrrolidone (PVP) to create electrospun carbon-doped TiO_2 nanofibers. The as-spun nanofiber mats were dried in air to finish the hydrolysis of TiO_2 precursor and afterwards calcined in air to generate mesoporous TiO_2 fibers. The calcination removes the PVP and produced mesopores in TiO_2 nanofibers, which helps in highly sensitive electrochemical biosensing by allowing more loading of enzymes and biomolecules. TiO_2 nanofibers (shown in Fig. 5.5) can be doped with carbon by just tweaking the calcination temperature duration of heating [38,42].

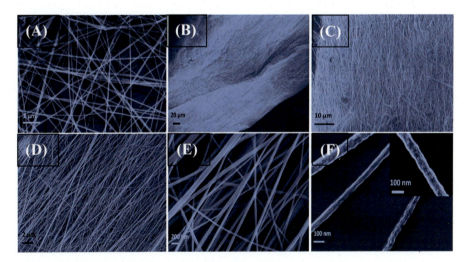

Figure 5.5 Scanning electron microscopic (FE-SEM) micrographs of calcined TiO_2 nanofibers fabricated by electrospinning. (A) Unaligned fibrous web. (B) A free-standing fiber mat. (C) Magnified image of the mat containing aligned nanofibers. Micrographs of TiO_2 nanofibers at (D) low and (E) high resolution. (F) Images of distinct TiO_2 MONFs and (*inset*) a single nanofiber.
Source: Reprinted with permission from K. Mondal, M.A. Ali, V.V. Agrawal, B.D. Malhotra, A. Sharma, Highly sensitive biofunctionalized mesoporous electrospun TiO2 nanofiber based interface for biosensing, ACS Appl. Mater. Interfaces 6 (2014) 2516–2527 [38]. Copyright (2014), American Chemical Society.

Among all strategies, electrospinning is considered the most versatile technique to generate ultrafine nanofibers of metal oxides. However, the electrospinning technique has some limitations [13]. For instance, the selection of polymers is limited when there is a need to produce organic nanofibers by electrospinning. The direct electrospinning of metal oxide is not imaginable, and a carrier polymer is always required, which sometime inhibits the use of this technique for direct application. Although electrospinning is already being used in industries, the yield rate of fibers by electrospinning is a slow process as compared to other spinning techniques. Finally, poor mechanical strength due to brittleness of the calcined nanofibers prevents electrospinning from being used for the direct application of the nanofibrous mats.

5.2.2.2 Nonelectrospinning for fabrication of MONFs

Nonelectrospinning methods use centrifugal force or compressed gases in place of an applied electric field to produce nanofibers. Centrifugal spinning [43], fiber drawing [44], and blowing bubble spinning (gas jet spinning) [45] are the three of the most usual nonelectrospinning MONF-manufacturing techniques. These techniques use less solvent, which increases the productivity and lowers the fabrication cost. The blown bubble spinning technique uses moving air or applied mechanical force to overcome the surface tension of the polymer blend and creates nanofibers [46,47]. Among these nonelectrospinning methods, centrifugal spinning is a high-output nanoscale fiber production method, but it cannot yield high-quality nanofibers. Thus centrifugal spinning has not gained much endorsement for biomedical applications. The unpredictable performance of the fibers may impede the bioactivity of the fibers during their use [48].

Recently, the magnetospinning technique has been invented as a new method for spinning of continuous microfibers and nanofibers with the help of a permanent rotating magnet [49]. The technique utilizes magnetic forces and the hydrodynamic characters of stretched filaments of precursor polymer blend to produce highly loaded, fine magnetic MONFs. The magnetospinning activity is free of the solution dielectric properties and demands no high electric field strengths, in contrast to traditional electrospinning.

5.2.3 Advanced microfabrication and nanofabrication strategies for MONFs

Three-dimensional (3D) printing and additive micro/nano manufacturing are fast-developing fabrication technologies. Currently, additive manufacturing can integrate many polymer, metal, and metal oxide nanostructures and is rapidly becoming an integral part of biomedical developments.

The gravitational fiber drawing technique, which is a fiber drawing method that produces microfibers and nanofibers (100 μm−100 nm in diameter) and enables their assembly in highly ordered 3D arrays, in conjunction with the additive manufacturing technique, collectively offers a unique capability to mix fibers with different diameters with a fine control over fiber spacing and orientation into the scaffold, thus meticulously mimicking the natural extracellular matrices. Overall, such a scaffold offers a promising candidate in configuration of cell differentiation, proliferation, tissue engineering, and biosensing platforms [50].

Recently, a 3D printing technique that uses the fused deposition modeling technology on electrospun nanofiber mats has gained significant attention for fabricating composites with a nanofiber layer [51]. An additional approach has been reported that involved the electrospinning of nanofibers on a translating stage to control the deposition [52]. Therefore it is anticipated that the electrospinning or electrospraying and advanced 3D printing and additive manufacturing can be beneficially combined to garner the attention of the biomedical and biosensors community. This combined manufacturing approach is possibly very powerful, as the number of polymers and metal oxides that have been explored so far can be expanded further [53] to design and fabricate more complex 3D structures.

5.3 Biosensing applications of MONFs

By definition, a biosensor is a device that detects, records, and transmits the data pertinent to the physicochemical change or process of biomolecules [13]. In other words, the biological recognition element responds to a target material, and the transducer then changes the biological reaction to a perceptible signal, which can be measured in terms of optical, acoustical, mechanical, electrochemical, calorimetric, or electronic forms and then be connected with the concentration of the analyte molecules [54–58]. Fig. 5.6 shows a schematic of an electrochemical biosensor that comprises a recognition component that lets the selective reaction to a specific analyte or an assembly of analytes, consequently reducing interference from other sample modules. A typical biosensing process involves the recognition of a bioanalyte followed by transducing the event into detectable signals where the MONFs can serve either or both processes. Fig. 5.6 provides some representative examples to illustrate the role of nanofibers of metal oxides in biosensing.

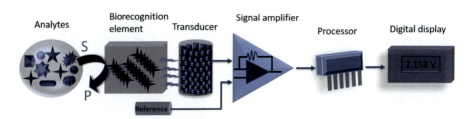

Figure 5.6 Schematic of a biosensor, in which the biocatalyst transforms the substrate (S) into a product (P) phase. The transducer then translates the biological response into an electrical signal. The signal then passes through a power amplifier, followed by transistor processor, and the output is then shown on a computer display.
Source: Reproduced with permission from K. Mondal, A. Sharma, Recent advances in electrospun metal-oxide nanofiber based interfaces for electrochemical biosensing, RSC Adv. 6 (2016) 94595–94616 [13]. Copyright 2016, Royal Society of Chemistry.

The metal oxide−based biosensors that have been reported in the literature rely mainly on electrochemical transduction [15,59], in which the sensing elements that are utilized include enzymes, antibodies, and, more rarely, DNA strands or aptamers. The interface between the metal oxides and biomolecules confirms faster electron transfer, which further improves the biosensing performance [60]. Thus the choice of a proper metal oxide platform for better immobilization of the biomolecules on the nanostructured surface is very important, and nanofibers offer the best selection [13]. Fig. 5.7 describes the mechanism of electrochemical biosensing by immobilization of enzyme on the surface of MONFs. The formation of nanointerfaces and electrochemical performances depends on the physicochemical nature and microenvironment of the metal oxide surfaces. Additional factors, including the overall available active surface area, total amount of charge on the surface, surface energy and adhesion, surface textures, porosity and total pore volume, location of the valence/conduction band in the metal oxides, and functional groups present on the surface, can influence the formation of the bionanointerface [61].

MONFs show their notable potential in chemical sensing and electrochemical biosensing, profiting from their nano-size, high surface to volume ratio, and extraordinary length over diameter ratio [62]. Until now, various metal oxide, metal/metal

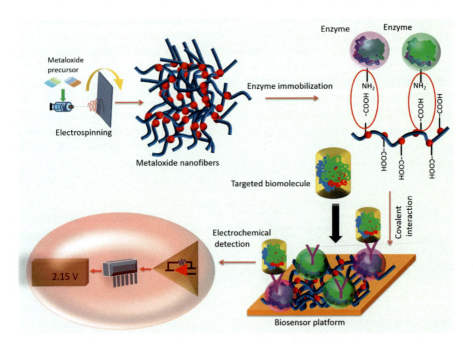

Figure 5.7 Enzyme immobilization onto electrospun nanofiber surfaces for electrochemical biosensing.
Source: Reproduced with permission from K. Mondal, A. Sharma, Recent advances in electrospun metal-oxide nanofiber based interfaces for electrochemical biosensing, RSC Adv. 6 (2016) 94595−94616 [13]. Copyright 2016, Royal Society of Chemistry.

oxide, and carbon and their composite nanofibers have been explored for biosensors applications. Consistent and faster detection of biomolecules based on electrospun nanofibers is one of the most studied area of the clinical diagnostics, biotechnology, and biomedical industries [63−65]. Several biosensors, such as glucose, cholesterol, catechol, urea, low-density lipoproteins (LDLs), triglycerides, dopamine, adenine, and cancer biomarkers, have been reported using nanofibers of TiO_2, ZnO, cobalt tetraoxide (Co_3O_4), MnO_2, copper oxide (CuO), nickel oxide (NiO), and zirconium dioxide (ZrO_2) [15]. Here, we briefly discuss a few of the most commonly used MONFs in biosensing platforms.

5.3.1 Titanium dioxide nanofibers for biosensing

Titanium dioxide nanofibers are one of the distinguishing biocompatible nanostructures that have been widely used in biotechnology and biomedical applications, owing to their strong oxidizing property, chemical inertness, and nontoxicity. TiO_2 nanofibers are biocompatible and environmentally green and have a favorable interface for the enzyme immobilization that helps in biosensing. TiO_2 nanofibers can strongly maintain the biocatalytic activities of the immobilized enzymes as Ti builds covalent bonding with enzyme's amine and carboxyl groups. Nanofibers of TiO_2 offer a higher surface area and a 3D porous morphology, and this helps in enhanced immobilization of enzymes on the fiber surface.

In a recent study, Lee et al. [66] have fabricated a lab-on-a-disc biosensing platform using electrospun TiO_2 nanofibers. It has been shown that this biosensor can perform a stable detection of serum proteins with a wide dynamic range, with merely 10 μL of whole blood within half an hour.

Recently, Mondal et al. [38] used TiO_2 nanofibers for electrochemical sensing of esterified cholesterol using cholesterol esterase (ChEt)/cholesterol oxidase (ChOx) dual enzymes. Fig. 5.8 shows a schematic that describes the cholesterol sensing platform on electrospun TiO_2 nanofibers. The electrospinning parameters were optimized to fabricate TiO_2 nanofibers with uniform diameters of 30−60 nm. The porosity of nanofibers was controlled by using a sacrificial PVP templating polymer in the sol-gel precursor of TiO_2 (titanium isopropoxide). Fig. 5.9A and B show the enzyme-loaded TiO_2 nanofibers along with transmission electron microscopic (TEM) images and selected area electron diffraction (SAED) pattern of the nanofibers (Fig. 5.9C and D). The biosensor displays extraordinary voltammetric and biocatalytic response to esterified cholesterol sensing with a detection limit of 0.49 mM. It was stated that high aspect ratio, mesoporosity, and high enzyme loading on the MONFs afford a fast detection response, within 20 s, and an excellent biosensitivity of 181.6 $\mu A/(mg/dL)/cm^2$.

5.3.2 ZnO nanofibers for biosensing

The biosensor performance depends on the sensing materials that are used. ZnO nanostructures play significant roles as the active sites where biological events happen, which influence the performance by defining the sensitivity and stability of the

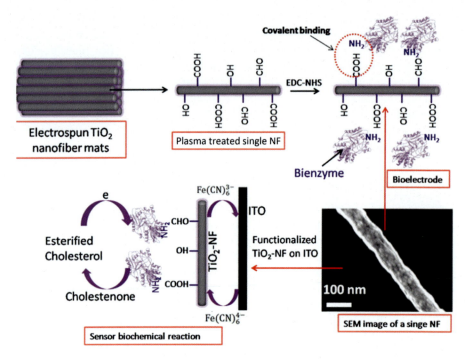

Figure 5.8 Schematic depiction of the TiO$_2$ nanofiber–based biosensing platform for esterified cholesterol sensing.
Source: Reprinted with permission from K. Mondal, M.A. Ali, V.V. Agrawal, B.D. Malhotra, A. Sharma, Highly sensitive biofunctionalized mesoporous electrospun TiO$_2$ nanofiber based interface for biosensing, ACS Appl. Mater. Interfaces 6 (2014) 2516–2527 [38]. Copyright (2014), American Chemical Society.

sensor device. MONFs of ZnO could be of distinctive research focus in the field of biosensors and bioelectronics, as it shows minimal biotoxicity, is transparent, and has a big excitonic binding energy and direct band gap n-type semiconductor. The ZnO nanofibers show interesting characteristics in defining their conductivity and continuous electron transfer through the nanofiber; their exceptional biocompatibility and larger isoelectric point (IEP ∼9.5) make ZnO a suitable candidate for enzymatic electrochemical biosensing [67]. Also, ZnO nanofibers provide a favorable microenvironment for enzyme immobilization on nanofiber surfaces without hampering their bioactivity. As a result, ZnO nanofibers have attracted a great amount of interest for application in the field of biosensing platforms with improved biocatalytic performances. ZnO-based MONFs have been explored for electrochemical biosensing of glucose, cholesterol, H$_2$O$_2$, LDL, uric acid, triglyceride, dopamine, phenol, and so on.

Lately, a uric acid biosensor on carbon paste/ZnO nanofibers and quercetin has been exhibited by Arvand et al. [68]. The fabricated MONF composite bioelectrode was capable of sensing uric acid traces in blood serum and plasma of a leukemia

Metal oxide nanofibers and their applications for biosensing 125

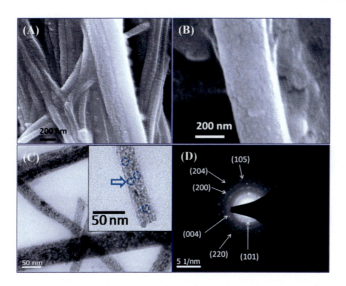

Figure 5.9 FE-SEM micrographs of (A) the ChEt/ChOx dual enzyme loaded nanofibers and (B) its magnified view. (C) The TEM micrographs of TiO₂ nanofibers, with mesopores shown in the inset, and (D) SAED pattern of the nanofibers.
Source: Reprinted with permission from K. Mondal, M.A. Ali, V.V. Agrawal, B.D. Malhotra, A. Sharma, Highly sensitive biofunctionalized mesoporous electrospun TiO₂ nanofiber based interface for biosensing, ACS Appl. Mater. Interfaces 6 (2014) 2516–2527 [38]. Copyright (2014), American Chemical Society.

patient as well as in healthy persons. The reported detection limit was 0.05 μM L^{-1} with a relative standard deviation of 0.2% for uric acid sensing.

In a recent study, Ali et al. [69] reported a point-of-care immunosensor device as shown in Fig. 5.10 (and described a scheme for the working principle of the sensor) for label-free breast cancer biomarker (anti-ErbB2; epidermal growth factor receptor 2) detection that is based on ZnO nanofibers. The nanofibers were fabricated by electrospinning, and the diameters lie in the range of 50–150 nm. It was demonstrated that the biosensor was more sensitive (7.76 kΩ μM^{-1}) than the best found in the literature and much better than that available ELISA (enzyme-linked immunosorbent assay) standard for the detection of breast cancer biomarker. The biosensor was proven to be efficient, highly selective, and reproducible with femtomolar sensitivity to detect an early stage of breast cancer. The ZnO nanofibers were fabricated by electrospinning of a polyacrylonitrile/zinc acetate precursor solution followed by controlled calcination. Fig. 5.11A–C show the scanning electron micrographs of a freestanding ZnO nanofibrous web and the nanofibers after being subjected to heating in air. The bioelectrodes were prepared by ZnO nanofibrous film deposition on an indium tin oxide substrate. Then ErbB2 breast cancer biomarker was immobilized onto the bioelectrodes. Fig. 5.11D–E show the FE-SEM micrographs of anti-ErbB2 immobilized onto ZnO nanofibers, and Fig. 5.11F displays a TEM micrograph of nanofibers and endorses its crystallinity. Ali et al.'s results revealed that their

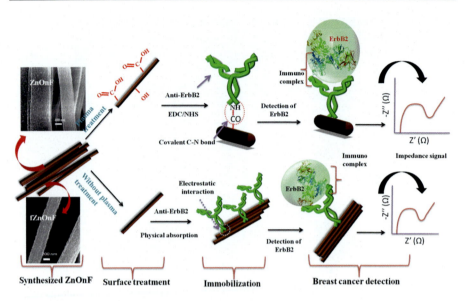

Figure 5.10 Schematic visualization of the fabricated ZnO nanofibers based biosensing platform for a breast cancer biomarker detection.
Source: Reproduced with permission from M.A. Ali, K. Mondal, C. Singh, B. Dhar Malhotra, A. Sharma, Anti-epidermal growth factor receptor conjugated mesoporous zinc oxide nanofibers for breast cancer diagnostics. Nanoscale 7 (2015) 7234–7245 [69]. Copyright 2015, Royal Society of Chemistry.

immunosensor can work in an extensive detection test range of 1.0 fM–0.5 μM and can offer a fast detection capability, such as within 128 s.

5.3.3 Others MONFs for biosensing

Iron oxide nanoparticles, mostly hematite (Fe_2O_3) and magnetite (Fe_3O_4) nanostructures, are imparting a candidate in electrochemical sensing. The oxidation states are Fe^{2+} and Fe^{3+} in magnetite and a single state as Fe^{3+} in hematite. The magnetite shows higher electrical conductivity at room temperature, owing to the electron-hopping activities between the Fe^{2+} and Fe^{3+} states. Iron oxide nanoparticles were extensively used to modify electrodes in the detection of many inorganic analytes, such as H_2O_2, glucose, nitrites, and nitrates; heavy metals (lead, zinc, and cadmium); and many organic targets (urea, bisphenol A, dopamine) [70]. About 10 years ago, Ding et al. [71] reported on hemoglobin (Hb)-carbon nanofibers (CNF) on glassy carbon electrode (Hb-CNFs/GCE) for the amperometric detection of H_2O_2, and the results showed fast response, high sensitivity, excellent reproducibility, good selectivity, and a wide dynamic range with a good limit of detection. It is important to note that Hb-CNF contains Fe species (Fe_2O_3 and/or $Fe-N_4$ moiety).

Nanostructures of cobalt oxides are extremely chemically reactive, easily available, highly stable, low cost, and good thermoelectric material with p-type

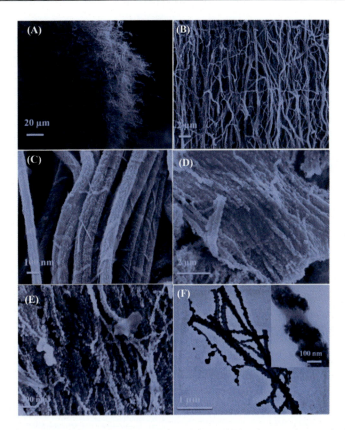

Figure 5.11 FE-SEM images of ZnO nanofibers fabricated by using an electrospinning technique followed by calcination. A free-standing MONFs mat at (A) low magnification, (B) high magnification, and (C) individual ZnO nanofibers at high magnification. FE-SEM micrographs of ErbB2 antibody-loaded ZnO nanofibers at (D) low and (E) high magnification. (F) TEM image of a sole ZnO nanofiber with its distinct grains shown in the inset.
Source: Reproduced with permission from M.A. Ali, K. Mondal, C. Singh, B. Dhar Malhotra, A. Sharma, Anti-epidermal growth factor receptor conjugated mesoporous zinc oxide nanofibers for breast cancer diagnostics. Nanoscale 7 (2015) 7234–7245 [69]. Copyright 2015, Royal Society of Chemistry.

semiconducting property. These nanostructures have obtained a big research thrust in electrochemical sensing and gas sensors. MONFs of Co_3O_4 have wonderful electrocatalytic activity and can be easily deposited on any support structures by electrochemical deposition, which could be advantageous in biomedical and diagnostic devices and biosensing applications. The literature shows that cobalt oxide nanofibers have been used for detection of various biomolecules (e.g., cysteine, glucose, propylamine, methanol, hydroquinone) [72]. For example, Kumar et al. and his colleagues [73] have explained a design of an enzyme-free highly sensitive and selective glucose biosensor on Co_3O_4/NiO composite nanofibrous platform. The cyclic voltammetry and amperometry techniques were used in the sensing of glucose, and the results displayed a high sensitivity of $2477\,\mu\text{A}\,\text{mM}^{-1}\,\text{cm}^{-2}$. A detection limit of $0.17\,\mu\text{M}$ with a broad linear range of $1\,\mu\text{M}-9.055\,\text{mM}$ was also reported for the biosensor.

Numerous other MONFs, such as MnO_2, CuO, ZrO_2, and NiO, including metal, carbon nanomaterials (graphene, reduced graphene oxide, carbon nanotubes, CNF), or even another metal oxide phase, have also been used in biosensors to detect an extensive choice of biomolecules. Hence further studies need to be done to understand the various nanomaterial combination-based biosensing strategies to developed more facile, innovative, mobile sensing technology for better health care in future. A summary of biosensors based on MONFs is described in Table 5.1.

5.4 Recent computational advances

The experimental methods have been a backbone of many early scientific investigations in pursuit of advanced biosensor design and development [92,93]. Unsurprisingly, the computational methods played a somewhat limited role in such early endeavors and were mainly aimed toward the validation of available experimental data and/or to provide a scientific explanation of a very limited ad hoc scientific phenomenon [94,95]. This was primarily because of the limited availability of computational resources and hardware capabilities. However, the landscape has changed significantly in recent years, owing to the continuous and unprecedented advancement in the hardware capabilities, unrestricted availability of the pertinent open source computational tools and techniques, and the explosive growth and utilization of the emerging technologies such as artificial intelligence, data science, and data analytics in mainstream computational investigations. This is evident from the enormous amount of available open scientific literature on the aforementioned advances [96–99]. Nevertheless, it can be safely argued that the recent scientific, computational, and hardware advances are just a beginning of the new technological era in which the true potential of modern hardware and computational tools is still untapped; therefore tremendous opportunities exist to significantly accelerate the design and development of exclusive nanomaterials such as MONFs and consequently enable groundbreaking scientific discoveries in the biomedical sciences.

In the context of the MONFs and their specific application to biosensor design, the computational techniques could provide numerous opportunities to enable the cost-effective and accelerated design and development of biosensors (described in Fig. 5.12). The unparalleled capabilities of multiscale modeling and simulations can be leveraged to shift the paradigm of biosensor design from a traditional validation/verification regime to an advanced computationally guided experimental synthesis, characterization, and optimization regime. This will enable a significant advance in the fundamental material structure characterization and the material structure-property-performance relationship evaluation. With such advancement it will be possible to fine-tune the fundamental biomaterial structures and their interfaces

Table 5.1 Metal oxide nanofiber−based biosensors.

MONF- based electrodes	Fiber diameter (nm)	Analyte	Linear range (μM)	Sensitivity	Limit of detection (μM)	References
RuO_2−TiO_2	100	Ascorbic acid	10−1500	$268.2 \pm 3.7 \, \mu A \, mM^{-1} \, cm^{-2}$	1.8	[74]
TiO_2/Pt	72.6	Hydrazine	up to 1030	$44.42 \, \mu A \, mM^{-1} \, cm^{-2}$	0.142	[75]
TiO_2/graphene oxide/ carbon	70	Adenine	0.1−10	$0.1823 \, \mu A \, \mu M^{-1}$	1.71	[76]
TiO_2/graphite oxide	40−70	Levodopa	0.3−60	$0.0806 \, \mu A \, \mu M^{-1}$	15.94	[77]
CuO/Pd	90−140	Glucose	0.2−2500	$1061.4 \, \mu A \, mM^{-1} \, cm^{-2}$	0.019	[78]
SiO_2/Au	—	H_2O_2	5−1000	—	2	[79]
V_2O_5	58.87	Adenine	0.5−512	$8.5333 \, \mu A \, \mu M^{-1} \, cm^{-2}$	0.013	[80]
NiO/reduced graphene oxide (rGO)	350	Glucose	2−600	$1100 \, \mu A \, mM^{-1} \, cm^{-2}$	0.77	[81]
Ag/CeO_2-Au/carbon	200	Levofloxacin	0.03−10	$1240 \, \mu A \, mM^{-1} \, cm^{-2}$	0.01	[82]
Fe_3O_4	200	Aflatoxin B1	0.000159−0.636	—	0.00636	[83]
$CoFe_2O_4$	200−300	Hydrazine	100−11000	$503 \, \mu A \, mM^{-1} \, cm^{-2}$	1000	[84]
$La_{0.88}Sr_{0.12}MnO_3$/carbon	300−400	Glucose	0.05−100	$1111.11 \, \mu A \, mM^{-1} \, cm^{-2}$	0.0312	[85]
$MnCo_2O_4$/graphene	150−300	Glucose	0.005−800	$1813.8 \, \mu A \, mM^{-1} \, cm^{-2}$	0.001	[86]
ZnO/GOx	100	Glucose	—	$69 \, \mu A \, mM^{-1} \, cm^{-2}$	10	[87]
Polyaniline (PANI)/ MnO_2	130	Adenine	10−100	$1.6 \, \mu A \, cm^{-2} \, \mu M^{-1}$	4.8	[88]
Cu/ZnO	150−200	Plasmodium falciparum histidine-rich protein-2 (HRP2)	$10 \, ag \, mL^{-1}$−$10 \, \mu g \, mL^{-1}$	$28.5 \, k\Omega \, (g/mL)^{-1} \, cm^{-2}$	$6.8 \, ag \, mL^{-1}$	[89]
Graphene/Mn_2O_3	~300	DNA mutation	0.000009−1	—	0.0000008	[90]
CeO_2/Carbon	100−200	L-cysteine (CysH)	up to 200	$120 \, \mu A \, mM^{-1} \, cm^{-2}$	0.02	[91]

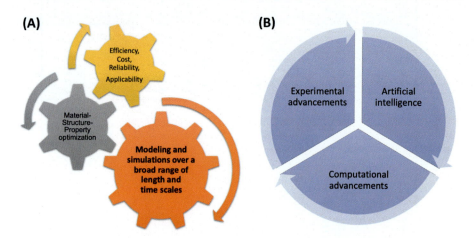

Figure 5.12 (A) An example of the utility of multiscale modeling and simulation techniques that can be used to significantly accelerate the MONF design via material-structure-property optimization to deliver efficient, cost-effective, and reliable biosensors designs. (B) The experimental, computational/hardware, and advancement techniques, such as artificial intelligence, machine learning, and data analytics, could be leveraged to enable computationally driven MOFN material design, synthesis, optimization, and discovery.

with similar or dissimilar biomaterials, which altogether dictates the physicochemical interactions between biomaterials and adsorption and diffusion process at the biomaterial surfaces. Such understanding is critical to ensure the fundamental efficacy and accuracy of biosensors and their stability in a wide range of thermochemical operating environments. The first principle, molecular dynamics, and mesoscopic and macroscopic modeling and simulation techniques could be leveraged to explain such multiscale processes on the pertinent length and time scales via the following multi-pronged strategies:

1. Utilization of computational techniques (in conjunction with the advanced techniques such as artificial intelligence) to provide a rational design and valuable guidance to significantly reduce the experimental synthesis, characterization, and development time of biosensors
2. Synergistic integration of various computational techniques to provide a consistent access to the material-structure-property relationship at multiple length (from the electronic to macroscopic scales) and time (from a few femtoseconds to several days) scales
3. Potentially avoid repetitive experimentation requirements via development of a standard computational-based guidance

5.5 Conclusions, challenges, and future scope

MONFs have proven their potential in biosensing and biomedical devices, owing to their easy and lucrative fabrication from a variety of material options, highly accessible surface area, tunable porosity, and easy surface modifiability. The high sensitivities of MONF-based biosensors are the result of their efficient charge separation and transportation at the surface, high enzyme loading ability by excellent adsorption, entrapment and strong covalent bonding of biomolecules with fibers due to their high aspect ratio, and vast interconnected three-dimensional porosity.

This chapter serve as a platform for MONF-based biosensing applications. We have provided detailed synthetic strategies for MONF fabrication, effects of various parameters, most important MONFs, and biosensing mechanisms, and their applications in biosensing of various molecules, entities, and analytes have been discussed. Finally, we summarized the most promising literature reports and compared the properties and performance of several MONF-based biosensors. The MONFs discussed in this chapter include oxides of titanium, zinc, iron, cobalt, vanadium, copper, nickel, manganese, and cerium. TiO_2, ZnO, iron oxide, cobalt oxide, and manganese oxide nanofibers have been found to be the most applied and efficient metal oxides for biosensors. It was also established that the cooccurrence of MONFs with different nanophase materials is beneficial for enzyme immobilization and biosensing.

Besides having the huge advantages mentioned, these MONFs have some shortcomings. The majority of them have a wide band gap, which makes them as semiconductors or even makes them poor as semiconductors. Other difficulties include poor ion transport kinetics and the pulverization of electrodes resulting from the pronounced volume expansion and contraction during the electrochemical processes. These undesirable complications can be overcome by bandgap modification via doping with metal nanoparticles, carbonaceous materials, polymers, and other nanoparticles.

Challenges remain in enzyme-based biosensing to achieve the efficient immobilization of enzymes into the substrate through the fiber network, poor adhesion between substrate and nanofibers, which need to be addressed. Regarding these issues, engineering of composite MONFs with organometallic perovskite, graphene, and Mxenes (two-dimensional metal chalocogenides) with diverse hierarchical morphologies may be a viable solution. Developing robust electrobiocatalysts with better stability will be an interesting area of imminent research. This chapter emphasizes the significance of MONFs in biosensing. However, applications of these nanofibers are open for other areas too, such as energy generation and storage, environmental remediation, biomedical, and tissue engineering applications.

Acknowledgments

KM, BI, and GP gratefully acknowledge the Energy & Environment Science & Technology (EES&T) at the Idaho National Laboratory (INL), Idaho Falls, Idaho, United States, for the support. GP also would like to thank Idaho National Laboratory's high-performance computational facility for facilitating the computational resources. RK thanks PBC Outstanding Postdoctoral Fellowship by Council of Higher Education, Israel, and Bar Ilan University, Israel.

References

[1] K. Mondal, G. Pawar, M.D. McMurtrey, A. Sharma, Finetuning hierarchical energy material microstructure via high temperature material synthesis route, Mater. Today Chem. 16 (2020) 100269.

[2] R. Kumar, V.B. Kumar, A. Gedanken, Sonochemical synthesis of carbon dots, mechanism, effect of parameters, and catalytic, energy, biomedical and tissue engineering applications, Ultrason. Sonochem. 64 (2020) 105009.

[3] Z.-M. Huang, Y.-Z. Zhang, M. Kotaki, S. Ramakrishna, A review on polymer nanofibers by electrospinning and their applications in nanocomposites, Compos. Sci. Technol. 63 (2003) 2223−2253.

[4] R. Rasouli, A. Barhoum, M. Bechelany, A. Dufresne, Nanofibers for biomedical and healthcare applications, Macromol. Biosci. 19 (2019) 1800256.

[5] K. Mondal, Recent advances in the synthesis of metal oxide nanofibers and their environmental remediation applications, Inventions 2 (2017) 9.

[6] A. Greiner, J.H. Wendorff, Cover picture: electrospinning: a fascinating method for the preparation of ultrathin fibers (Angew. Chem. Int. (Ed.) 30/2007), Angew. Chem. Int. (Ed.) 46 (2007) 5633.

[7] E.A.T. Vargas, N.C. do Vale Baracho, J. de Brito, A.A.A. de Queiroz, Hyperbranched polyglycerol electrospun nanofibers for wound dressing applications, Acta Biomater. 6 (2010) 1069−1078.

[8] S.E. Gilchrist, et al., Fusidic acid and rifampicin co-loaded PLGA nanofibers for the prevention of orthopedic implant associated infections, J. Controlled Rel. 170 (2013) 64−73.

[9] J. Xue, et al., Drug loaded homogeneous electrospun PCL/gelatin hybrid nanofiber structures for anti-infective tissue regeneration membranes, Biomaterials 35 (2014) 9395−9405.

[10] K. Feng, et al., Novel antibacterial nanofibrous PLLA scaffolds, J. Controlled Rel. 146 (2010) 363−369.

[11] M.A. Ali, et al., Microfluidic immuno-biochip for detection of breast cancer biomarkers using hierarchical composite of porous graphene and titanium dioxide nanofibers, ACS Appl. Mater. Interfaces 8 (2016) 20570−20582.

[12] T.G. Kim, D.S. Lee, T.G. Park, Controlled protein release from electrospun biodegradable fiber mesh composed of poly(ε-caprolactone) and poly(ethylene oxide), Int. J. Pharm. 338 (2007) 276−283.

[13] K. Mondal, A. Sharma, Recent advances in electrospun metal-oxide nanofiber based interfaces for electrochemical biosensing, RSC Adv. 6 (2016) 94595−94616.

[14] R. Ford, S. Quinn, R. O'Neill, Characterization of biosensors based on recombinant glutamate oxidase: comparison of crosslinking agents in terms of enzyme loading and efficiency parameters, Sensors 16 (2016) 1565.

[15] P.R. Solanki, A. Kaushik, V.V. Agrawal, B.D. Malhotra, Nanostructured metal oxide-based biosensors, NPG Asia Mater. 3 (2011) 17−24.

[16] S. Dutt, P.F. Siril, S. Remita, Swollen liquid crystals (SLCs): a versatile template for the synthesis of nano structured materials, RSC Adv. 7 (2017) 5733−5750.

[17] Kenry, C.T. Lim, Nanofiber technology: current status and emerging developments, Prog. Polym. Sci. 70 (2017) 1−17.

[18] Handbook of Nanofibers. Springer International Publishing, 2019. Available from: http://doi.org/10.1007/978-3-319-53655-2.

[19] X. Liu, Y. He, S. Wang, Q. Zhang, M. Song, Controllable synthesis and tunable field-emission properties of tungsten oxide sub-micro fibers, Int. J. Refract. Met. Hard Mater. 34 (2012) 47−52.

[20] M. Zhou, J. Zhou, R. Li, E. Xie, Preparation of aligned ultra-long and diameter-controlled silicon oxide nanotubes by plasma enhanced chemical vapor deposition using electrospun PVP nanofiber template, Nanoscale Res. Lett. 5 (2010) 279−285.

[21] Y. Fang, Q. Pang, X. Wen, J. Wang, S. Yang, Synthesis of ultrathin ZnO nanofibers aligned on a zinc substrate, Small 2 (2006) 612−615.

[22] N. Takahashi, Simple and rapid synthesis of ZnO nano-fiber by means of a domestic microwave oven, Mater. Lett. 62 (2008) 1652−1654.

[23] P.K. Nayak, N. Munichandraiah, Rapid sonochemical synthesis of mesoporous MnO_2 for supercapacitor applications, Mater. Sci. Eng. B 177 (2012) 849−854.

[24] D. Li, et al., Soft-template construction of three-dimensionally ordered inverse opal structure from Li_2 $FeSiO_4$/C composite nanofibers for high-rate lithium-ion batteries, Nanoscale 8 (2016) 12202−12214.

[25] W.-S. Chae, S.-W. Lee, Y.-R. Kim, Templating route to mesoporous nanocrystalline titania nanofibers, Chem. Mater. 17 (2005) 3072−3074.

[26] K. Mondal, et al., 110th anniversary: particle size effect on enhanced graphitization and electrical conductivity of suspended gold/carbon composite nanofibers, Ind. Eng. Chem. Res. 59 (2020) 1944−1952.

[27] J. Xue, T. Wu, Y. Dai, Y. Xia, Electrospinning and electrospun nanofibers: methods, materials, and applications, Chem. Rev. 119 (2019) 5298−5415.

[28] R. Kumar, et al., Advances in nanotechnology based strategies for synthesis of nanoparticles of lignin, in: S. Sharma, A. Kumar (Eds.), Lignin, Springer International Publishing, 2020, pp. 203−229. Available from: http://doi.org/10.1007/978-3-030-40663-9_7.

[29] S. Ramakrishna, S. Peng, Electrospun metal oxides for energy applications, in: T. Yao (Ed.), Zero-Carbon Energy Kyoto 2011, Springer, Japan, 2012, pp. 97−108. Available from: http://doi.org/10.1007/978-4-431-54067-0_10.

[30] C. Yang, et al., Comparisons of fibers properties between vertical and horizontal type electrospinning systems, in: 2009 IEEE Conference on Electrical Insulation and Dielectric Phenomena, IEEE, 2009, pp. 204−207. Available from: http://doi.org/10.1109/CEIDP.2009.5377758.

[31] P. Singh, K. Mondal, A. Sharma, Reusable electrospun mesoporous ZnO nanofiber mats for photocatalytic degradation of polycyclic aromatic hydrocarbon dyes in wastewater, J. Colloid Interface Sci. 394 (2013) 208−215.

[32] N. Singh, K. Mondal, M. Misra, A. Sharma, R.K. Gupta, Quantum dot sensitized electrospun mesoporous titanium dioxide hollow nanofibers for photocatalytic applications, RSC Adv. 6 (2016) 48109−48119.

[33] C.L. Casper, J.S. Stephens, N.G. Tassi, D.B. Chase, J.F. Rabolt, Controlling surface morphology of electrospun polystyrene fibers: effect of humidity and molecular weight in the electrospinning process, Macromolecules 37 (2004) 573−578.

[34] D. Crespy, K. Friedemann, A.-M. Popa, Colloid-electrospinning: fabrication of multi-compartment nanofibers by the electrospinning of organic or/and inorganic dispersions and emulsions, Macromol. Rapid Commun. 33 (2012) 1978−1995.

[35] Y. Liao, T. Fukuda, S. Wang, Electrospun metal oxide nanofibers and their energy applications, in: M.M. Rahman, A.M. Asiri (Eds.), Nanofiber Research-Reaching New Heights, InTech, 2016. Available from: http://doi.org/10.5772/63414.

[36] L. Li, W.H. Meyer, G. Wegner, M. Wohlfahrt-Mehrens, Synthesis of submicrometer-sized electrochemically active lithium cobalt oxide via a polymer precursor, Adv. Mater. 17 (2005) 984−988.

[37] N. Horzum, D. Taşçıoglu, S. Okur, M.M. Demir, Humidity sensing properties of ZnO-based fibers by electrospinning, Talanta 85 (2011) 1105−1111.

[38] K. Mondal, M.A. Ali, V.V. Agrawal, B.D. Malhotra, A. Sharma, Highly sensitive bio-functionalized mesoporous electrospun TiO_2 nanofiber based interface for biosensing, ACS Appl. Mater. Interfaces 6 (2014) 2516−2527.

[39] K. Friedemann, T. Corrales, M. Kappl, K. Landfester, D. Crespy, Facile and large-scale fabrication of anisometric particles from fibers synthesized by colloid-electrospinning, Small 8 (2012) 144−153.

[40] J.-M. Lim, J.H. Moon, G.-R. Yi, C.-J. Heo, S.-M. Yang, Fabrication of one-dimensional colloidal assemblies from electrospun nanofibers, Langmuir 22 (2006) 3445−3449.

[41] C. Wessel, R. Ostermann, R. Dersch, B.M. Smarsly, Formation of inorganic nanofibers from preformed TiO_2 nanoparticles via electrospinning, J. Phys. Chem. C. 115 (2011) 362−372.

[42] K. Mondal, S. Bhattacharyya, A. Sharma, Photocatalytic degradation of naphthalene by electrospun mesoporous carbon-doped anatase TiO_2 nanofiber mats, Ind. Eng. Chem. Res. 53 (2014) 18900−18909.

[43] C. Chen, M. Dirican, X. Zhang, Centrifugal spinning—high rate production of nanofibers, Electrospinning: Nanofabrication and Applications, Elsevier, 2019, pp. 321−338. Available from: http://doi.org/10.1016/B978-0-323-51270-1.00010-8.

[44] X. Xing, Y. Wang, B. Li, Nanofibers drawing and nanodevices assembly in poly(tri-methylene terephthalate), Opt. Express 16 (2008) 10815.

[45] H. Dou, B.-Q. Zuo, J.-H. He, Blown bubble-spinning for fabrication of superfine fibers, Therm. Sci. 16 (2012) 1465−1466.

[46] F. Liu, et al., Fabrication of highly oriented nanoporous fibers via airflow bubble-spinning, Appl. Surf. Sci. 421 (2017) 61−67.

[47] Y. Fang, F. Liu, L. Xu, P. Wang, J. He, Preparation of PLGA/MWCNT composite nanofibers by airflow bubble-spinning and their characterization, Polymers 10 (2018) 481.

[48] H. Peng, Y. Liu, S. Ramakrishna, Recent development of centrifugal electrospinning, J. Appl. Polym. Sci. 134 (2017). Available from: http://doi.org/10.1002/APP.44578.

[49] A. Tokarev, O. Trotsenko, I.M. Griffiths, H.A. Stone, S. Minko, Magnetospinning of nano- and microfibers, Adv. Mater. 27 (2015) 3560−3565.

[50] N.S. Yadavalli, et al., Gravity drawing of micro- and nanofibers for additive manufacturing of well-organized 3D-nanostructured scaffolds, Small 16 (2020) 1907422.

[51] T. Kozior, et al., Electrospinning on 3D printed polymers for mechanically stabilized filter composites, Polymers 11 (2019) 2034.

[52] J. Lee, S.Y. Lee, J. Jang, Y.H. Jeong, D.-W. Cho, Fabrication of patterned nanofibrous mats using direct-write electrospinning, Langmuir 28 (2012) 7267−7275.

[53] P.D. Dalton, et al., Electrospinning and additive manufacturing: converging technologies, Biomater. Sci. 1 (2013) 171−185.

[54] M.A. Ali, et al., A surface functionalized nanoporous titania integrated microfluidic biochip, Nanoscale 6 (2014) 13958−13969.

[55] K. Mondal, M.A. Ali, S. Srivastava, B.D. Malhotra, A. Sharma, Electrospun functional micro/nanochannels embedded in porous carbon electrodes for microfluidic biosensing, Sens. Actuators B Chem. 229 (2016) 82−91.

[56] M.A. Ali, et al., Mesoporous few-layer graphene platform for affinity biosensing application, ACS Appl. Mater. Interfaces 8 (2016) 7646−7656.

[57] M.A. Ali, et al., In situ integration of graphene foam−titanium nitride based bioscaffolds and microfluidic structures for soil nutrient sensors, Lab. Chip 17 (2017) 274−285.

[58] K. Mondal, et al., Highly sensitive porous carbon and metal/carbon conducting nanofiber based enzymatic biosensors for triglyceride detection, Sens. Actuators B Chem. 246 (2017) 202−214.

[59] B. Liu, J. Liu, Sensors and biosensors based on metal oxide nanomaterials, TrAC. Trends Anal. Chem. 121 (2019) 115690.

[60] M.J. Limo, et al., Interactions between metal oxides and biomolecules: from fundamental understanding to applications, Chem. Rev. 118 (2018) 11118−11193.

[61] M. Marcus, et al., Interactions of neurons with physical environments, Adv. Healthc. Mater. 6 (2017) 1700267.

[62] M.U. Anu Prathap, B. Kaur, R. Srivastava, Electrochemical sensor platforms based on nanostructured metal oxides, and zeolite-based materials, Chem. Rec. 19 (2019) 883−907.

[63] L. Matlock-Colangelo, A.J. Baeumner, Recent progress in the design of nanofiber-based biosensing devices, Lab. Chip 12 (2012) 2612.

[64] R. Kumar, P.F. Siril, F. Javid, Unusual anti-leukemia activity of nanoformulated naproxen and other non-steroidal anti-inflammatory drugs, Mater. Sci. Eng. C 69 (2016) 1335−1344.

[65] R. Kumar, A. Singh, N. Garg, P.F. Siril, Solid lipid nanoparticles for the controlled delivery of poorly water soluble non-steroidal anti-inflammatory drugs, Ultrason. Sonochem. 40 (2018) 686−696.

[66] W.S. Lee, V. Sunkara, J.-R. Han, Y.-S. Park, Y.-K. Cho, Electrospun TiO_2 nanofiber integrated lab-on-a-disc for ultrasensitive protein detection from whole blood, Lab. Chip 15 (2015) 478−485.

[67] Sribastava et al., Application of ZnO Nanoparticles for Improving the Thermal and pH Stability of Crude Cellulase Obtained from Aspergillus fumigatus AA001″ Front. Microbiol. 18 April 2016. Available from: https://doi.org/10.3389/fmicb.2016.00514.

[68] M. Arvand, S. Tajyani, N. Ghodsi, Electrodeposition of quercetin on the electrospun zinc oxide nanofibers and its application as a sensing platform for uric acid, Mater. Sci. Eng. C 46 (2015) 325−332.

[69] M.A. Ali, K. Mondal, C. Singh, B. Dhar Malhotra, A. Sharma, Anti-epidermal growth factor receptor conjugated mesoporous zinc oxide nanofibers for breast cancer diagnostics, Nanoscale 7 (2015) 7234−7245.

[70] J.M. George, A. Antony, B. Mathew, Metal oxide nanoparticles in electrochemical sensing and biosensing: a review, Microchim. Acta 185 (2018) 358.

[71] Y. Ding, et al., Carbonized hemoglobin nanofibers for enhanced H_2O_2 detection, Electroanalysis 22 (2010) 1911−1917.

[72] Y. Ding, et al., Electrospun Co_3O_4 nanofibers for sensitive and selective glucose detection, Biosens. Bioelectron. 26 (2010) 542−548.

[73] R. Ramasamy, et al., Design and development of Co_3O_4/NiO composite nanofibers for the application of highly sensitive and selective non-enzymatic glucose sensors, RSC Adv. 5 (2015) 76538−76547.

[74] S. Kim, Y.K. Cho, C. Lee, M.H. Kim, Y. Lee, Real-time direct electrochemical sensing of ascorbic acid over rat liver tissues using RuO_2 nanowires on electrospun TiO_2 nanofibers, Biosens. Bioelectron. 77 (2016) 1144−1152.

[75] Y. Ding, et al., Preparation of TiO_2−Pt hybrid nanofibers and their application for sensitive hydrazine detection, Nanoscale 3 (2011) 1149.

[76] M. Arvand, N. Ghodsi, M.A. Zanjanchi, A new microplatform based on titanium dioxide nanofibers/graphene oxide nanosheets nanocomposite modified screen printed carbon electrode for electrochemical determination of adenine in the presence of guanine, Biosens. Bioelectron. 77 (2016) 837−844.

[77] M. Arvand, N. Ghodsi, Electrospun TiO_2 nanofiber/graphite oxide modified electrode for electrochemical detection of l-DOPA in human cerebrospinal fluid, Sens. Actuators B Chem. 204 (2014) 393−401.

[78] W. Wang, et al., Electrospun palladium (IV)-doped copper oxide composite nanofibers for non-enzymatic glucose sensors, Electrochem. Commun. 11 (2009) 1811−1814.

[79] J. Shen, et al., Gold-coated silica-fiber hybrid materials for application in a novel hydrogen peroxide biosensor, Biosens. Bioelectron. 34 (2012) 132−136.

[80] Electroanalysis and Bioelectrochemistry Lab, Department of Chemical Engineering and Biotechnology, National Taipei University of Technology, No. 1, Section 3, Chung-Hsiao East Road, Taipei 106, Taiwan, ROC; M. Annalakshmi, Novel electrochemical sensor for highly sensitive detection of adenine based on vanadium pentoxide nanofibers modified screen printed carbon electrode, Int. J. Electrochem. Sci. (2018) 6218−6228. Available from: http://doi.org/10.20964/2018.07.41.

[81] Y. Zhang, Y. Wang, J. Jia, J. Wang, Nonenzymatic glucose sensor based on graphene oxide and electrospun NiO nanofibers, Sens. Actuators B Chem. 171−172 (2012) 580−587.

[82] L. Tang, et al., Ag nanoparticles and electrospun CeO_2-Au composite nanofibers modified glassy carbon electrode for determination of levofloxacin, Sens. Actuators B Chem. 203 (2014) 95−101.

[83] G. Xu, et al., Magnetic functionalized electrospun nanofibers for magnetically controlled ultrasensitive label-free electrochemiluminescent immune detection of aflatoxin B1, Sens. Actuators B Chem. 222 (2016) 707−713.

[84] J. Liu, J. Shen, M. Li, L.-P. Guo, A high-efficient amperometric hydrazine sensor based on novel electrospun $CoFe_2O_4$ spinel nanofibers, Chin. Chem. Lett. 26 (2015) 1478−1484.

[85] D. Xu, L. Luo, Y. Ding, P. Xu, Sensitive electrochemical detection of glucose based on electrospun La0.88Sr0.12MnO3 naonofibers modified electrode, Anal. Biochem. 489 (2015) 38−43.

[86] Y. Zhang, et al., Electrospun graphene decorated $MnCo_2O_4$ composite nanofibers for glucose biosensing, Biosens. Bioelectron. 66 (2015) 308−315.

[87] J.Y. Huang, M.G. Zhao, Z.Z. Ye, Electrospun porous ZnO nanofibers for glucose biosensors, Adv. Mater. Res. 950 (2014) 3−6.

[88] M.U. Anu Prathap, R. Srivastava, B. Satpati, Simultaneous detection of guanine, adenine, thymine, and cytosine at polyaniline/MnO_2 modified electrode, Electrochim. Acta 114 (2013) 285−295.

[89] K. Brince Paul, et al., A highly sensitive self assembled monolayer modified copper doped zinc oxide nanofiber interface for detection of Plasmodium falciparum histidine-rich protein-2: targeted towards rapid, early diagnosis of malaria, Biosens. Bioelectron. 80 (2016) 39−46.

[90] S. Tripathy, et al., Graphene doped Mn_2O_3 nanofibers as a facile electroanalytical DNA point mutation detection platform for early diagnosis of breast/ovarian cancer, Electroanalysis 30 (2018) 2110−2120.

[91] F. Cao, et al., Sensitive and selective electrochemical determination of L-cysteine based on cerium oxide nanofibers modified screen printed carbon electrode, Electroanalysis 30 (2018) 1133–1139.

[92] S. Zhang, Materials and techniques for electrochemical biosensor design and construction, Biosens. Bioelectron. 15 (2000) 273–282.

[93] J. Castillo, et al., Biosensors for life quality, Sens. Actuators B Chem. 102 (2004) 179–194.

[94] J.K. Bording, B.Q. Li, Y.F. Shi, J.M. Zuo, Size- and shape-dependent energetics of nanocrystal interfaces: experiment and simulation, Phys. Rev. Lett. 90 (2003) 226104.

[95] C. Jang, et al., Molecular dynamics simulations of oxidized vapor-grown carbon nanofiber surface interactions with vinyl ester resin monomers, Carbon 50 (2012) 748–760.

[96] M. Tonezzer, D.T.T. Le, S. Iannotta, N. Van Hieu, Selective discrimination of hazardous gases using one single metal oxide resistive sensor, Sens. Actuators B Chem. 277 (2018) 121–128.

[97] I. Furxhi, F. Murphy, M. Mullins, A. Arvanitis, C.A. Poland, Practices and trends of machine learning application in nanotoxicology, Nanomaterials 10 (2020) 116.

[98] E. Papa, J.P. Doucet, A. Sangion, A. Doucet-Panaye, Investigation of the influence of protein corona composition on gold nanoparticle bioactivity using machine learning approaches, SAR. QSAR Environ. Res. 27 (2016) 521–538.

[99] H. Choi, H. Kang, K.-C. Chung, H. Park, Development and application of a comprehensive machine learning program for predicting molecular biochemical and pharmacological properties, Phys. Chem. Chem. Phys 21 (2019) 5189–5199.

Metal oxide-based nanofibers and their gas-sensing applications

Ali Mirzaei[1], Sanjit Manohar Majhi[2,3], Hyoun Woo Kim[2,3] and Sang Sub Kim[4]
[1]Department of Materials Science and Engineering, Shiraz University of Technology, Shiraz, Iran, [2]The Research Institute of Industrial Science, Hanyang University, Seoul, Republic of Korea, [3]Division of Materials Science and Engineering, Hanyang University, Seoul, Republic of Korea, [4]Department of Materials Science and Engineering, Inha University, Incheon, Republic of Korea

6.1 Introduction

Metal oxide nanofibers (NFs) are currently used in numerous areas, including antimicrobial applications [1], photoelectrochemistry [2], photocatalysis [3], magnetism [4], microwave attenuation [5], solar cells [6], photodetectors [7], light-emitting devices [8], energy harvesting [9], surgical gowns [10], drug delivery/tissue engineering [11], air filtration [12], desalination, supercapacitors [13], lithium-ion batteries [14], battery separators [15], food packaging [16], fuel cells [17], and the textile industry [18]. This is mainly because of their advantageous properties and peculiar morphologies, as they offer high surface areas that are beneficial for different applications.

Nanofibrous materials can be produced by using different techniques, such as electrospinning (ES), phase separation, template synthesis, and drawing [19]. However, the most widely used method for the synthesis of metal oxide NFs is ES, and we will focus on that technique in this chapter. The term electrospinningwas derived from "electrostatic spinning," and the fundamental idea behind it was presented by Formhals 60 years ago in making some novel arrangements to produce yarns made out of electrospun fibers [20]. Since then, ES has gradually become a reliable and highly reproducible method for the production of different materials with fibrous morphologies. Several advantageous attributes are associated with the ES process, which produces NFs with unique features, such as high surface areas with nano-scale diameters and porous NFs with sizes ranging from a few nanometers to a few microns [21,22]. In addition to metal oxides in pristine [23] or composite [24] forms, polymers [25], metals [26], porous carbon [27], and even nanotubes [28] have been synthesized by direct ES or through postprocessing.

The formation of NFs through ES is based on the uniaxial stretching of a viscoelastic solution [29]. The ES setup (Fig. 6.1) consists of four major parts: (1) a glass syringe containing a polymer solution, (2) a metallic needle, (3) a high-voltage

Metal Oxide-Based Nanofibers and Their Applications. DOI: https://doi.org/10.1016/B978-0-12-820629-4.00008-4
© 2022 Elsevier Inc. All rights reserved.

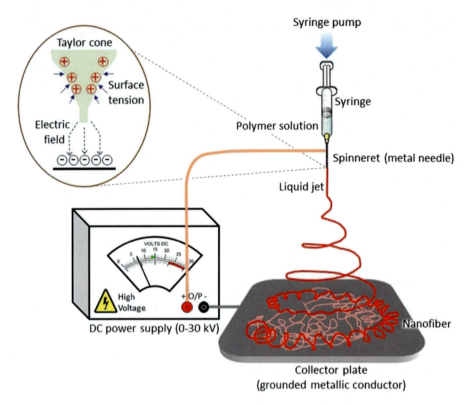

Figure 6.1 Schematic illustration of an electrospinning setup [30].
Source: From K. Ghosal, A. Chandra, G. Praveen, S. Snigdha, S. Roy, C. Agatemor, et al., Electrospinning over solvent casting: tuning of mechanical properties of membranes. Sci. Rep. 8(1) (2018) 1–9.

power source, and (4) a metallic collector (typically in a flat plate or rotating drum configuration) [31]. The quality of electrospun fibers is controlled by the syringe, which connects with a syringe pump to maintain a constant and adjustable feeding rate of the solution. The humidity is controllable as well [32]. Generally, the distance between the spinneret and the collector is 525 cm in most laboratory-scale setups [33].

The use of volatile organic solvents (ethanol, isopropanol, chloroform, and dimethylformamide) is helpful in promoting the feasibility of electrospun NFs. Moreover, they facilitate the expansion of the polymer in the solution as well as fluidic jet drying during the process [34]. A good solvent should not only be able to dissolve polymer but also have a low boiling point (BP). The solvent plays a significant role in the synthesis of electrospun NFs. For example, solvents with lower BPs facilitate the dehydration of the NFs, owing to rapid evaporation, whereas the solvents with high BP result in ribbon-like flat NFs [35]. If the rheological properties are not appropriate and the hydrolysis rates of precursors are very high, then it is very difficult to control the ES process. To overcome this, a polymer can be used in the solution to control the rheological properties. Polyvinyl pyrrolidone, an extremely efficient polymer, is highly soluble in ethanol or water and possesses good compatibility with many precursors. In addition, additives—such as salts and catalysts (hydrochloric acid, propionic acid, and acetic acid) are being added in small amounts to control the hydrolysis and gelation rates of the precursor, thus stabilizing the solution as well as the jet while preventing the solution from blocking the spinneret. In addition, sodium chloride can be used as a salt to enhance the charge density on the liquid jet, thereby eliminating the formation of beads, a common issue in ES [36].

ES initially occurs when an electric field is supplied to a polymer solution via the metallic needle, causing the induction of charges on the polymer droplet. In the meantime a repulsion of charges in the polymer solution produces a type of force that helps the solution to flow in the direction of the electric field. By further increasing the electric field, the spherical droplet is transformed to a conical shape droplet known as a Taylor cone, releasing ultrafine NFs, which are then collected on the metallic collector at a certain time interval. A polymer solution with sufficient cohesive force can form a stable charge jet. The whipping motion of the liquid jet formed through the internal and external charges forces helps the polymer to stretch, forming NFs [37]. Over the long period of time until the solvent evaporates, ultrathin fibers form and are deposited on the collector by electrostatic attraction [32].

The viscosity of the ES solution plays a significant role in defining the nature of the NFs. For instance, the jet will break up at a low viscosity and form polymer droplets, a process known as electrospraying. When the solution has a high viscosity ($1-200$ poise), NFs are produced continuously without any damage to the jet [35]. In general, when the applied voltage is higher than a certain value, the NF diameter first decreases and then increases at a certain point [35].

The shape, size, and porosity of the electrospun NFs depend on the flow rate of the polymer solution; therefore a minimum flow rate is maintained [35]. Furthermore, concentrations of the polymeric solution affect the NF formation. For example, at a low polymer concentration the charged jet produces droplets before reaching the collector. However, with the gradual increase in the concentration of the ES solution, the NFs are formed. Therefore the concentration of ES solution can affect both the viscosity and the surface tension and possibility of NF formation. However, when the viscosity increases by more than a certain value, the polymer solution flow through the capillary is disrupted [35]. An optimized gap between the syringe tip and collector is necessary for the formation of desired NF structures, and a minor deviation would result in either electrospraying or bead formation. A big gap between the tip and collector results in a decrease in the diameter of electrospun NFs, whereas at a smaller distance, NFs with flattened structures are formed, owing to the lack of sufficient time for the evaporation of the solvent [35].

For the synthesis of crystalline metal oxide NFs, as-synthesized NFs, which contain inorganic precursors, should be heat-treated under controlled atmospheres of air or inert gases for carbonization and decomposition of the sacrificial polymer and crystallization of the metal oxides [33]. Owing to the removal of the sacrificial polymer, the size of the NFs is often reduced. By controlling the heat treatment temperature, the morphology of the resulting NFs can be controlled (Fig. 6.2) [36]. Low-temperature calcination is good for obtaining smooth NF surfaces, whereas calcination at high temperatures results in porous NFs. A further increase in temperature will cause the grain to coalesce, resulting in the formation of a totally solid NF [39].

Coaxial ES (Fig. 6.3) involves two coaxial nozzles connected to two reservoirs of precursor [41], which provide separate pathways for the flow of solutions [40]. This method is widely used for synthesis of core−shell (C−S) NFs.

During the process, the shell layer acts as a shield to prevent the evaporation of the core, thus ensuring smooth mass transfer. The concentrations of both phases, which affect the diameter of NFs, should exceed a critical value to obtain the desired morphology. Owing to the C−S structure, the two flow rates should be the same. At a low voltage the electrostatic force is not capable of extracting the entire liquid; thus only the outer shell is handled in the jets. Meanwhile, a high voltage would increase the environmental temperature, leading to poor electrospinnability [42]. The coaxial ES technique is a versatile technique for synthesizing hollow NFs and offers excellent control over the morphology of the fibers, providing greater surface area, which could be increased to approximately 200% relative to

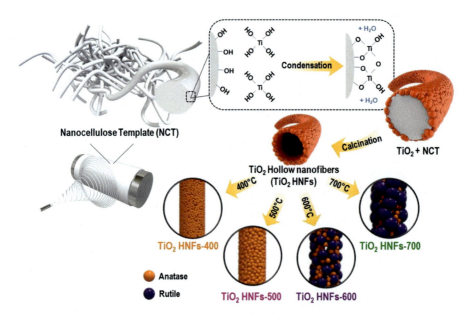

Figure 6.2 Effect of heat treatment temperature on the crystal structures of hollow titanium dioxide (TiO$_2$) nanofibers [38].
Source: From S.-I. Oh, J.-C. Kim, M.A. Dar, D.-W. Kim, Synthesis and characterization of uniform hollow TiO$_2$ nanofibers using electrospun fibrous cellulosic templates for lithium-ion battery electrodes. J. Alloy. Compd., 800 (2019) 483−489.

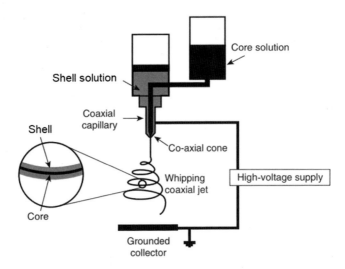

Figure 6.3 A schematic of coaxial electrospinning [40].
Source: From Y. Lu, J. Huang, G. Yu, R. Cardenas, S. Wei, E.K. Wujcik, et al., Coaxial electrospun fibers: Applications in drug delivery and tissue engineering. WIREs Nanomed. Nanobiotechnol 8 (2016) 654−677.

that of C−S NFs [42]. However, a post−heat treatment method is essential to obtaining hollow NFs or even tubelike structures [33].

6.2 Gas-sensing applications of metal oxide nanofibers

6.2.1 Importance of gas-sensing and gas sensors

Today, more than 92% of the people in the world live in a polluted atmosphere. Various toxic gases present in the atmosphere cause air pollution, which is monitored by measuring the concentrations of CO, O_3, SO_2, NO_2, and particulate matter at different regions by using gas-sensing devices [43]. The use of inexpensive gas sensors with high sensitivity can significantly reduce the overall cost of the monitoring of gases. An ideal gas sensor should show high sensitivity to target gases, selectively sense a specific gas, and exhibit high stability, fast response, and fast recovery speed with low fabrication cost [44]. In the literature, various types of gas-sensing devices, including surface acoustic wave [45], electrochemical [46], optical [47], thermoelectric [48], gasochromic [49], and chemiresistive [50] devices have been introduced for the detection of many volatile organic compounds and toxic gases.

Among them, chemiresistive-based metal oxide gas sensors have been the most widely accepted tools for detecting hazardous gases and vapors over the past few decades [51]. Metal oxides can be incorporated into two types of devices: chemiresistive gas sensors and field-effect transistors [39]. This chapter focuses only on chemiresistive gas sensors based on metal oxide NFs. Although metal sulfides have also been used for sensing studies [52] metal oxides are dominant in the fabrication of chemiresistive gas sensors. Among the different morphologies used for gas sensing, NFs are widely employed as gas-sensing materials. The following sections discuss the different types of NF metal oxides used for the realization of gas sensors and their associated sensing mechanisms.

6.2.2 Pristine metal oxide nanofibers

For a gas sensor, because of its special morphology, a high surface area is beneficial; however, the presence of pores can also significantly enhance the sensing performance. Electrospun NFs can be made to be porous with the appropriate heat treatment. For example, porous tin oxide (SnO_2) NFs were produced by using an ES technique and subsequent heat treatment at 600°C. The porous NFs were able to detect very low concentrations of hydrogen sulfide (H_2S) gas. The excellent sensing results of the SnO_2 NFs toward H_2S were attributed to the variation in the potential barriers formed at the SnO_2/SnO_2 interface and the resistance modulation, as shown in Fig. 6.4. In air, the resistance increased, and after H_2S injection, the resistance decreased. Moreover, by injection of H_2S gas to SnO_2 NFs, the oxidation state of Sn^{4+} was changed to Sn^{2+}, which supplied extra electrons to the SnO_2 NFs. Finally, owing to the high surface area and presence of pores on the SnO_2 NFs, the diffusion and reaction of H_2S was facilitated, leading to an enhanced gas response. Good selectivity to H_2S gas was due to the higher reactivity of H_2S and smaller bond energy of H-SH (only 381 kJ mol^{-1}) than the other interfering gases. The H-SH bond was easily broken and interacted with the absorbed oxygen ions (O^-), contributing to the sensor signal. Furthermore, SnO_2 showed high affinity to the strong H_2S reducing gas, which resulted in the good adsorption of H_2S by the SnO_2 surface and conversion to SnS_2 [53].

In addition to SnO_2, zinc oxide (ZnO) has been investigated as a promising sensing material [54]. Some strategies, such as electron beam irradiation, can positively affect the sensing properties, owing to the generation of structural defects [55]. In an attempt, ZnO NFs were synthesized by an ES technique to study the effect of electronbeam doses on the H_2 gas-sensing properties of ZnO NFs, as shown in Fig. 6.5 [56]. The enhanced H_2 responses of the 100- and 150-kGy dose-irradiated ZnO NF gas sensors were ascribed to features such as structural defects, for example, oxygen vacancies and Zn interstitials, which increased the number of electrons in the lattice, thereby leading to greater oxygen adsorption and faster reactions with H_2. Furthermore, owing to a relatively high sensing temperature (300°C), the metallization of ZnO significantly changed the resistance of ZnO. In fact, the surface of ZnO was reduced to metallic Zn when H_2 gas was injected. Thus the resistance of the ZnO sensor changed, thus improving the response.

Not only simple binary metal oxides but also more complex metal oxides, such as perovskites, can also be used for sensing studies [57]. Hollow NFs of $PrFeO_3$ (Fig. 6.6) were successfully synthesized via an ES method followed by calcination at 600°C for 2 h. The sensor exhibited p-type semiconducting characteristics, owing

Metal oxide-based nanofibers and their gas-sensing applications 145

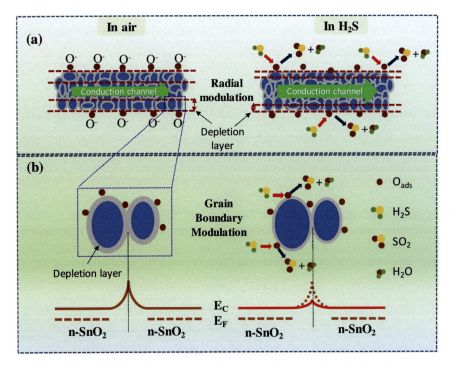

Figure 6.4 Illustration of H$_2$S sensing mechanisms of porous SnO$_2$ NFs. (A) Radial and (B) grain boundary modulation.
Source: From C. Feng, Z. Jiang, J. Wu, B. Chen, G. Lu, C. Huang, Pt-Cr$_2$O$_3$-WO$_3$ composite nanofibers as gas sensors for ultra-high sensitive and selective xylene detection. Sens. Actuators B: Chem., 300 (2019) 127008.

to the ionization of the Pr^{3+} cation vacancies. Acetone gas sensing properties demonstrated that the PrFeO$_3$ hollow NF sensor showed an excellent response with selectivity and good stability at 180°C. The modulation of the resistance of the sensor in the presence of the target gas contributed to the high sensing signal [58].

6.2.3 Metal oxide composite nanofibers

Composite NFs are widely used as sensing materials [59]. Owing to the quite high (typically in the range of 150°–500°C) operating temperature of metal oxide gas sensors, they consume more power, thus often limiting their practical application because of the difficulty in sensor fabrication and degradation in sensor stability. Meanwhile, reduced graphene oxide (rGO), a derivative of graphene, has been employed as a gas-sensing material; however, the weak van der Waals interactions result in poor gas absorption. In addition, the sluggish recovery and poor selectivity characteristics of rGO need to be improved [60]. The combination of rGO and metal oxides can be a good strategy to improve gas-sensing performance. Furthermore,

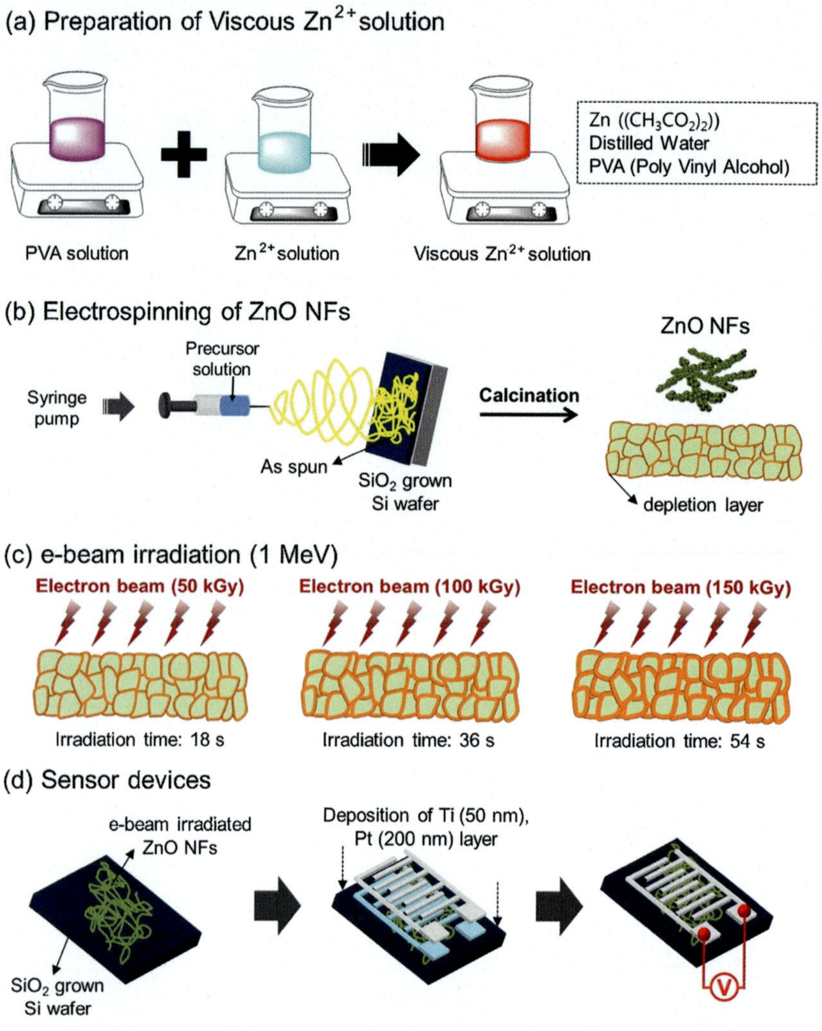

(a) Preparation of Viscous Zn^{2+} solution

Zn $((CH_3CO_2)_2))$
Distilled Water
PVA (Poly Vinyl Alcohol)

PVA solution Zn^{2+} solution   Viscous Zn^{2+} solution

(b) Electrospinning of ZnO NFs

ZnO NFs

Precursor solution

Syringe pump

Calcination

As spun
SiO_2 grown
Si wafer

depletion layer

(c) e-beam irradiation (1 MeV)

Electron beam (50 kGy)   **Electron beam (100 kGy)**   **Electron beam (150 kGy)**

Irradiation time: 18 s   Irradiation time: 36 s   Irradiation time: 54 s

(d) Sensor devices

e-beam irradiated
ZnO NFs

Deposition of Ti (50 nm),
Pt (200 nm) layer

SiO_2 grown
Si wafer

Figure 6.5 Schematic illustration of (A) preparation of Zn^{2+} polymeric solution, (B) electrospinning process, (C) electronbeam irradiation, and (D) sensor fabrication procedure.
Source: From J.-H. Kim, A. Mirzaei, H.W. Kim, P. Wu, S.S. Kim, Design of supersensitive and selective ZnO-nanofiber-based sensors for H_2 gas sensing by electron-beam irradiation. Sens. Actuators B: Chem., 293 (2019) 210−223.

photoexcitation is an effective approach to increasing the carrier density in the conduction band (CB) of metal oxide sensors; this would generate more active absorption sites on the surface of the sensors, thereby leading to improvement in the sensing properties [61]. To improve the NO_2 selectivity using a single sensor, an ultraviolet (UV) light-activated, rGO-functionalized gas sensor with hollow

Metal oxide-based nanofibers and their gas-sensing applications

Figure 6.6 Transmission electron microscopy images of PrFeO$_3$ hollow nanofibers.
Source: From L. Ma, S.Y. Ma, X.F. Shen, T.T. Wang, X.H. Jiang, Q. Chen, et al., PrFeO$_3$ hollow nanofibers as a highly efficient gas sensor for acetone detection. Sens. Actuators B: Chem., 255 (2018) 2546–2554.

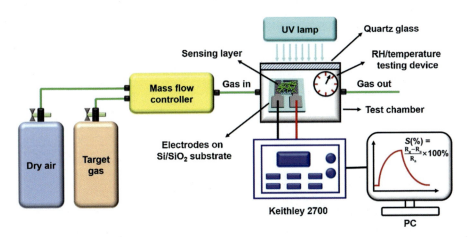

Figure 6.7 Schematic representation of the ultraviolet light-activated gas-sensing layout.
Source: From W. Li, J. Guo, L. Cai, W. Qi, Y. Sun, J.-L. Xu, et al., UV light irradiation enhanced gas sensor selectivity of NO$_2$ and SO$_2$ using rGO functionalized with hollow SnO$_2$ nanofibers. Sens. Actuators B: Chem., 290 (2019) 443–452.

SnO$_2$ NFs (rGO-SnO$_2$ NFs) operating at room temperature was introduced [62]. Fig. 6.7 schematically shows the sensing layout.

The presence of rGO offers the pathway for the conduction of electrons between SnO$_2$ grains, due to the high carrier mobility of rGO. The hybrid hollow SnO$_2$

NF-rGO nanosheet structure provided more contact sites for the adsorption and diffusion of gas molecules. At 97 mW cm^{-2} of UV illumination, the sensor showed excellent response (102%) to NO_2, compared with that in the dark. A p-n heterojunction formed between rGO and SnO_2, owing to the differences in their work functions [rGO ($\sim$4.75 eV) > SnO_2 (4.55 eV)]; hence the electrons flowed from lower to higher energy level. In the UV light-activated rGO/SnO_2 sensor, rGO acted as a photoelectron acceptor, which was conducive for charge transport, whereas SnO_2 behaved as a photoelectron collector and UV absorber. The UV-activated electron hole pairs were generated, and the photo-generated electrons were captured by O_2, thus resulting in photo-generated ionic oxygen species according to the following equations:

$$h\nu \rightarrow h^+ + e^- \tag{6.1}$$

$$O_2 + e^-(h\nu) \rightarrow O_2^-(h\nu) \tag{6.2}$$

The holes (photo-generated) also reacted with the ionic oxygen species to produce molecular oxygen as follows:

$$O_2^-(ads) + h^+(h\nu) \rightarrow O_2(gas) \tag{6.3}$$

Fig. 6.8 displays the gas-sensing mechanism of rGO/SnO_2 nanofibers in the presence of NO_2 under ultraviolet light irradiation and dark. Under dark conditions, as

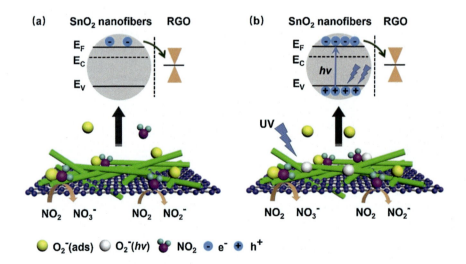

Figure 6.8 Schematic illustration of the sensing mechanism of rGO/SnO_2 composites to NO_2 (A) in dark and (B) under ultraviolet light illumination.
Source: From W. Li, J. Guo, L. Cai, W. Qi, Y. Sun, J.-L. Xu, et al., UV light irradiation enhanced gas sensor selectivity of NO_2 and SO_2 using rGO functionalized with hollow SnO_2 nanofibers. Sens. Actuators B: Chem., 290 (2019) 443–452.

shown in (Fig. 6.8A), a large number of photoelectrons are captured by target gases, leading to the ultraviolet-controlled gas-sensing performance. Because of the coexisting effect of oxygen absorption/desorption arising from ultraviolet irradiation, an appropriate intensity of ultraviolet light can lead to an improvement in the gas-sensing response.

Platinum (Pt)−chromium oxide (Cr_2O_3)−tungsten oxide (WO_3) composite NFs were synthesized by the ES method for the detection of xylene gas, which is an indoor pollutant. It can cause serious health-related problems in human beings, affecting their kidneys, nervous system, liver, skin, and eyes [63]. The response of the $Pt-Cr_2O_3-WO_3$ composite NF sensors was found to be greater ($R_a/R_g = 74.3$) toward xylene compared with that of the WO_3 NF ($R_a/R_g = 2.9$) sensor tested at 325°C. First, the formation of $Cr_2O_3-WO_3$ p−n junctions and the decrease in the energy barrier height of $Cr_2O_3-WO_3$ in the presence of xylene increases the sensing response. Second, the catalytic activity of individual Cr_2O_3 can induce the dissociation of xylene gas molecules into more reactive species. Third, the high surface area and porous architecture of $Pt-Cr_2O_3-WO_3$ composite NFs facilitate the gas diffusion of target gases into numerous pores of the sensing materials. Finally, the electronic and chemical sensitization of Pt contributes to the high gas-sensing performance [53].

Titanium dioxide (TiO_2)-SnO_2 C−S composite NFs were fabricated by a coaxial ES method, which showed a response in the range of 13.7−100 ppm acetone at 280°C, with a fast response time (6.6 s). According to the XPS studies, the oxygen vacancies in TiO_2-SnO_2 C-S NFs were increased relative to pristine sensors, which improved the sensing performance. Furthermore, SnO_2 and TiO_2 as semiconducting oxides with exposed high-energy facets showed improved performance, as they have more dangling bonds associated with high-energy facets [64].

SnO_2−indium oxide (In_2O_3) NF composites were fabricated by using a double-jet-modified ES system. The response of the SnO_2/In_2O_3 hetero-NF gas sensor at 375°C was higher than that of the pristine sensor with respect to formaldehyde gas. The sensing performance was significantly enhanced by controlling grain size and the surface area of sensing materials. In addition, two types of channels existed in the SnO_2/In_2O_3 hetero-NF system, as two types of fibers were constructed with different nanograin sizes. Furthermore, the formation of n-SnO_2/n-In_2O_3 heterojunctions resulted in potential barriers for the movement of electrons. However, in the presence of formaldehyde the height of potential barriers changed, thus contributing to the sensing signal [64].

Cobalt tetraoxide (Co_3O_4)−ZnO C−S NFs were prepared by ES for formaldehyde sensing. The Co_3O_4-ZnO C-S NF sensor showed p-type conductivity, as the ZnO shell was very thin and the change in conductivity of the Co_3O_4 core was the dominant factor. Co_3O_4-ZnO C-S NFs exhibited a high sensitivity to the formaldehyde gas with fast response and recovery times. The electrons of ZnO were captured by oxygen molecules in the air, and potential barriers were established between the ZnO NPs, whereas the holes accumulated on Co_3O_4 with a decrease in the potential barrier. When the p-n junction formed, the holes of Co_3O_4 were transferred to n-type ZnO and vice versa, which led to a reduction in the depletion region. As soon as the formaldehyde was exposed, the trapped electrons returned back to the ZnO shell, decreasing the potential barrier. Ultimately, the holes of Co_3O_4 were ingested, with an increase in the potential barrier [65].

SnO_2/gold (Au)-doped In_2O_3 C−S NFs were prepared through a facile coaxial ES technique. The C−S SnO_2/Au-doped In_2O_3 NF sensor showed excellent sensing properties towards acetone (100 ppm), with a response and recovery time of 2 and 9 s, respectively, at 300°C. Meanwhile, Au-doped In_2O_3 NF sensors exhibited a low response. The presence of Au catalyst facilitated the additional absorption of O_2 molecules, followed by dissociation and spillover on the surface of In_2O_3 by the capture of additional free electrons, which eventually increased the gas response [66]. Furthermore, the sensor with NF morphology showed a surface area of $69.4 \text{ m}^2 \text{ g}^{-1}$, which provided many adsorption sites for the target gases. The excellent electron transport from In_2O_3 to SnO_2 resulted in a fast recovery time [67].

In another study, ZnO-SnO_2 NFs with a hollow structure were prepared by the ES method followed by annealing [68]. Fig. 6.9 shows their TEM micrographs along with their EDX pattern, thus confirming the formation of hollow NFs with the desired composition. The specific surface of ZnO-SnO_2 hollow NFs was found to be $36.8 \text{ m}^2 \text{ g}^{-1}$, whereas it showed a high response to NO_2 (1 ppm) of 46 at 90°C. Upon intimate contact, electrons flowed from SnO_2 to ZnO until a Fermi level equilibrium was achieved, resulting in band bending and then a depletion layer at the interface of ZnO and SnO_2. In an NO_2 atmosphere the amount of band bending and the thickness of the depletion layer increased, leading to an increase in the sensing resistance and appearance of a sensing signal.

6.2.4 Loaded or doped metal oxide nanofibers

Metal-doped NFs are widely used for sensing studies [69]. Palladium (Pd)-loaded SnO_2 NFs were produced for gas-sensing studies. SnO_2 NFs were produced by using ES, whereas Pd NPs were deposited by using magnetron sputtering, thus enabling the uniform dispersion of Pd after heat treatment, which enhanced the H_2 response of SnO_2 NFs. The optimum operating temperature was reduced from 310°C to 160°C, owing to the Pd decoration on the SnO_2 NFs as well as the increase in the hydrogen response (28) by a factor of 2.5 over that of pristine SnO_2 (11.5). The high sensing result was mainly attributed to the dispersion of Pd catalyst, which is capable of forming a high depletion layer because of the capture of free electron by increasing resistance. The depletion layer was then reduced with the reduction in the resistance after the H_2 gas reacted with the oxygen ionic species, which gave rise to a sensing signal. Furthermore, Pd can be converted to PdH_x, which affected the sensing properties (Fig. 6.10A). Owing to the difference between the work functions of Pd and SnO_2, a Schottky junction was formed with a space charge region (Fig. 6.10B). The work function of PdH_x is lower than that of Pd, which eventually lowers the Schottky barrier height (Fig. 6.10B). Fig. 6.10C shows the current-versus-voltage graph of Pd-coated SnO_2 NFs when exposed to hydrogen. The curve of the current response deflects, which confirms the decrease in the Schottky barrier in the presence of H_2.

Metal oxide-based nanofibers and their gas-sensing applications

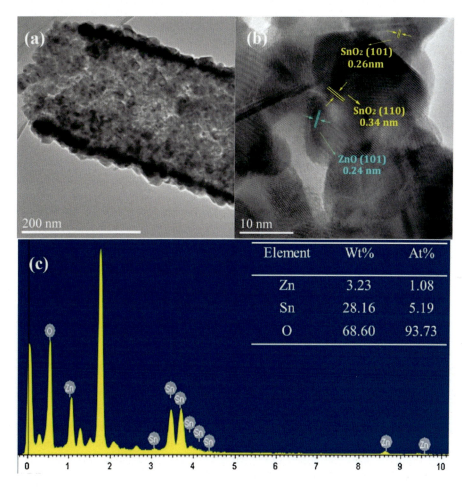

Figure 6.9 High-resolution transmission electron microscopy images of the ZnO-SnO$_2$ hollow nanofibers calcined at 500°C. (A) Low-magnification image. (B) High-magnification image showing lattice fringes. (c) EDX pattern.
Source: From S. Bai, H. Fu, Y. Zhao, K. Tian, R. Luo, D. Li, et al., On the construction of hollow nanofibers of ZnO-SnO$_2$ heterojunctions to enhance the NO$_2$ sensing properties. Sens. Actuators B: Chem., 266 (2018) 692−702.

In another study, the pristine and Pd/SnO$_2$ NFs were synthesized by the ES method. The 0.4 wt.% Pd-loaded SnO$_2$ NF sensor showed a high response to C$_2$H$_5$OH at 330°C. The high amount of oxygen ion species and the ethanol oxidation activities of the Pd catalyst are the main reasons for the high response [71].

Pt-doped In$_2$O$_3$ porous NFs were synthesized through the ES method and heat treatment at 500°C [72]. The response value of the Pt-In$_2$O$_3$ porous NF sensor was 15.1−1 ppm of acetone. The Brunauer−Emmett−Teller (BET) surface area of the sensor was found to be 212.3 m^2 g^{-1}, whereas it was 104.7 m^2 g^{-1} for the pristine

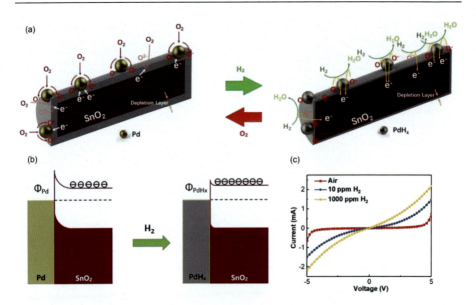

Figure 6.10 (A) Detection of H_2 gas by pristine and Pd-decorated SnO_2 nanofibers. (B) Schematic of the resistance change in the contact area between Pd and SnO_2, with the Schottky barrier shown. (C) Current response under various voltages and concentrations of H_2 [70].
Source: From F. Wang, K. Hu, H. Liu, Q. Zhao, K. Wang, Y. Zhang, Low temperature and fast response hydrogen gas sensor with Pd coated SnO_2 nanofiber rods. Int. J. Hydrog. Energy, 45 (2020) 7234–7242.

sensor. Accordingly, more adsorption sites were available on the surface of the Pt-doped In_2O_3 NF sensor, thus contributing to the higher response to the acetone vapor. Initially, in the Pt-doped In_2O_3 NF sensor, O_2 molecules are adsorbed on it and chemisorbed to the oxygen ionic species by capturing electrons [73]. Pt catalyst helps in the dissociation of molecular oxygen to oxygen ionic species due to the spillover effect. The superior selectivity to acetone was attributed to the difference in the bond dissociation energy (BDE) of the gas molecules. For example, the BDE of CH_3COCH_3 (352 kJ mol^{-1}) was smaller than that of HCHO (368 kJ mol^{-1}), NH_3 (452 kJ mol^{-1}), H_2S (376 kJ mol^{-1}), H_2 (436 kJ mol^{-1}), C_7H_8 (371 kJ mol^{-1}), CH_3OH (462 kJ mol^{-1}), and CH_3CH_2OH (462 kJ mol^{-1}), showing that acetone reacted more easily with chemisorbed oxygen ionic species than with other gas molecules.

Pristine and Rh-doped SnO_2 NFs were synthesized by ES followed by calcination treatment. It was revealed that after Rh doping, the grain size of the SnO_2 NFs decreased significantly. The 0.5 moL% Rh-doped SnO_2 NF sensor exhibited a response of 60.6–50 ppm of acetone. The higher acetone-sensing properties of the Rh-doped sensor was related to the high baseline resistance, lower grain size, and promising catalytic effects of Rh and increase in oxygen vacancies. In fact, the grains are almost completely depleted from the electrons, and upon exposure to acetone, the released electrons subsequently vary the resistance in the sensor, thus leading to an enhanced gas response. In addition, the amount of oxygen vacancies was increased upon Rh doping, because it can facilitate a larger number of active sites for the gas adsorption, diffusion, and sensing reaction on the sensor surface [74].

Silver (Ag)-doped $LaFeO_3$ NFs were produced by the ES method for formaldehyde sensing studies. It was reported that the Ag-doped sensor showed an enhanced response to formaldehyde compared with that of the pristine $LaFeO_3$ NF sensor, with fast response and recovery times of 2 and 4 s, respectively. The noble metal Ag helps in the rapid adsorption and desorption of O_2, forming oxygen ions (O^-) by capturing free electrons. Furthermore, some of the Ag can replace lathanium (La^{3+}) and generate more holes in the p-type $LaFeO_3$ NFs, which increases the main charge carriers [75].

In another study, Au/La_2O_3-doped SnO_2 NFs were synthesized by an ES process, followed by calcination at $600°C$ and the subsequent sputtering of Au. The sensor exhibited a response of $10.1-100$ ppm CO_2 at $300°C$. The high-response properties were attributed to the high surface area of NFs, the formation of $Au-SnO_2$ heterojunction, and the promotional effect of Au [76].

6.3 Conclusions and outlook

Metal oxide NFs are currently among the most promising materials for sensitive detection of volatile organic compounds and toxic gases. They are synthesized mostly through ES, which is an easy, reproducible, low cost, and flexible method that can produce not only NFs but also hollow NFs and C−S NFs with appropriate designs. Different semiconducting metal oxides, particularly SnO_2, ZnO, and TiO_2, are used for sensing studies. To enhance the sensing performance of metal oxide NFs, they can be synthesized in composite form, or a noble metal can be doped or decorated on their surface. Furthermore, with appropriate annealing and proper selection of the ES solution, a desired mesoporous NF morphology can be obtained, thus facilitating the diffusion of target gases in the inner parts of the sensing layer and resulting in high sensing reactions. C-S composite NFs are among the most promising materials for gas-sensing studies, as they contribute the maximum number of heterojunctions, which induces the resistance in the gas sensors. Hollow NFs can also offer many adsorption sites for target gas molecules. Noble metals in combination with metal oxides can significantly increase the sensing response and decrease the sensing temperature, owing to their electronic and chemical sensitizations. Overall, with further study of metal oxide NFs, more reliable gas-sensing detection devices can be fabricated.

References

[1] S. Thakur, M. Kaur, W.F. Lim, M. Lal, Fabrication and characterization of electrospun ZnO nanofibers: Antimicrobial assessment, Mater. Lett. 264 (2019) 127279.

[2] C. Ma, M. Wei, $BiVO_4$-nanorod-decorated rutile/anatase TiO_2 nanofibers with enhanced photoelectrochemical performance, Mater. Lett. 259 (2020) 126849.

[3] A.P. Shah, S. Jain, V.J. Mokale, N.G. Shimpi, High performance visible light photocatalysis of electrospun PAN/ZnO hybrid nanofibers, J. Ind. Eng. Chem. 77 (2019) 154−163.

[4] Y. Chen, X. Xu, X. Li, G. Zhang, Vacancy induced room temperature ferromagnetism in Cu-doped ZnO nanofibers, Appl. Surf. Sci. 506 (2020) 144905.

[5] W. Gu, J. Lv, B. Quan, X. Liang, B. Zhang, G. Ji, Achieving MOF-derived one-dimensional porous ZnO/C nanofiber with lightweight and enhanced microwave response by an electrospinning method, J. Alloy. Compd. 806 (2019) 983−991.

[6] V.P. Dinesh, R.S. Kumar, A. Sukhananazerin, J.M. Sneha, P.M. Kumar, P. Biji, Novel stainless steel based, eco-friendly dye-sensitized solar cells using electrospun porous ZnO nanofibers, Nano-Struct. Nano-Obj. 19 (2019) 100311.

[7] Z.S. Hosseini, H.A. Bafrani, A. Naseri, A.Z. Moshfegh, High-performance UV-Vis-NIR photodetectors based on plasmonic effect in Au nanoparticles/ZnO nanofibers, Appl. Surf. Sci. 483 (2019) 1110−1117.

[8] C.N. Pangul, S.W. Anwane, S.B. Kondawar, Samarium doped ZnO nanofibers for designing orange - red light emitting fabrics, Mater. Today: Proc. 15 (2019) 464−470.

[9] S. Mansouri, T.F. Sheikholeslami, A. Behzadmehr, Investigation on the electrospun PVDF/NP-ZnO nanofibers for application in environmental energy harvesting, J. Mater. Res. Technol. 8 (2) (2019) 1608−1615.

[10] M.Q. Khan, D. Kharaghani, N. Nishat, A. Shahzad, T. Hussain, Z. Khatri, et al., Preparation and characterizations of multifunctional PVA/ZnO nanofibers composite membranes for surgical gown application, J. Mater. Res. Technol. 8 (1) (2019) 1328−1334.

[11] W. Huang, Y. Xiao, X. Shi, Construction of electrospun organic/inorganic hybrid nanofibers for drug delivery and tissue engineering applications, Adv. Fiber Mater. 1 (1) (2019) 32−45.

[12] J. Su, G. Yang, C. Cheng, C. Huang, H. Xu, Q. Ke, Hierarchically structured TiO_2/PAN nanofibrous membranes for high-efficiency air filtration and toluene degradation, J. Colloid Interface Sci. 507 (2017) 386−396.

[13] A.V. Radhamani, K.M. Shareef, M.S.R. Rao, $ZnO@MnO_2$ core−shell nanofiber cathodes for high performance asymmetric supercapacitors, ACS Appl. Mater. Interfaces 8 (44) (2016) 30531−30542.

[14] X. He, Y. Hu, R. Chen, Z. Shen, K. Wu, Z. Cheng, et al., Foldable uniform GeOx/ZnO/C composite nanofibers as a high-capacity anode material for flexible lithium ion batteries, Chem. Eng. J. 360 (2019) 1020−1029.

[15] Y. Li, Q. Li, Z. Tan, A review of electrospun nanofiber-based separators for rechargeable lithium-ion batteries, J. Power Sources 443 (2019) 227262.

[16] T. Senthil Muthu Kumar, S. Kumar, K. Rajini, N. Siengchin, S. Ayrilmis, N. Varada Rajulu, A comprehensive review of electrospun nanofibers: Food and packaging perspective, Compos. Part. B: Eng. 175 (2019) 107074.

[17] M. Zhi, S. Lee, N. Miller, N.H. Menzler, N.J.E. Wu, An intermediate-temperature solid oxide fuel cell with electrospun nanofiber cathode, Energy Environ. Sci. 5 (5) (2012) 7066−7071.

[18] M. Mirjalili, S. Zohoori, Review for application of electrospinning and electrospun nanofibers technology in textile industry, J. Nanostructure Chem. 6 (3) (2016) 207−213.

[19] E. Stojanovska, E. Canbay, E.S. Pampal, M.D. Calisir, O. Agma, Y. Polat, et al., A review on non-electro nanofibre spinning techniques, RSC Adv. 6 (87) (2016) 83783−83801.

[20] Z.-M. Huang, Y.Z. Zhang, M. Kotaki, S. Ramakrishna, A review on polymer nanofibers by electrospinning and their applications in nanocomposites, Compos. Sci. Technol. 63 (15) (2003) 2223−2253.

[21] P. Pascariu, M. Homocianu, ZnO-based ceramic nanofibers: Preparation, properties and applications, Ceram. Int. 45 (9) (2019) 11158−11173.

[22] S. Jiang, Y. Chen, G. Duan, C. Mei, A. Greiner, S. Agarwal, Electrospun nanofiber reinforced composites: A review, Polym. Chem. 9 (20) (2018) 2685−2720.

[23] Y. Lu, X. Ou, W. Wang, J. Fan, K. Lv, Fabrication of TiO_2 nanofiber assembly from nanosheets (TiO_2-NFs-NSs) by electrospinning-hydrothermal method for improved photoreactivity, Chin. J. Catal. 41 (1) (2020) 209−218.

[24] P. Pascariu, A. Airinei, N. Olaru, L. Olaru, V. Nica, Photocatalytic degradation of rhodamine B dye using $ZnO-SnO_2$ electrospun ceramic nanofibers, Ceram. Int. 42 (6) (2016) 6775−6781.

[25] W. Zhang, G. Li, C. She, A. Liu, J. Cheng, H. Li, et al., High performance tube sensor based on PANI/Eu^{3+} nanofiber for low-volume NH_3 detection, Analytica Chim. Acta 1093 (2020) 115−122.

[26] A. Robertsam, N. Victor Jaya, Fabrication of a Low-coercivity, large-magnetoresistance PVA/Fe/Co/Ni nanofiber composite using an electrospinning technique and its characterization, J.Nanoscience.Nanotech. 20 (6) (2020) 3504−3511.

[27] C. Shan, X. Feng, J. Yang, X. Yang, H.-Y. Guan, M. Argueta, et al., Hierarchical porous carbon pellicles: Electrospinning synthesis and applications as anodes for sodium-ion batteries with an outstanding performance, Carbon 157 (2020) 308−315.

[28] L.-S. Feng, Z. Liu, N. Zhang, B. Xue, J.-X. Wang, J.-M. Li, Effect of nanorod diameters on optical properties of GaN-based dual-color nanorod arrays, Chinese Phy. Lett. 36 (2) (2019) 027802.

[29] W.E. Teo, S. Ramakrishna, A review on electrospinning design and nanofibre assemblies, Nanotechnology 17 (14) (2006) R89.

[30] K. Ghosal, A. Chandra, G. Praveen, S. Snigdha, S. Roy, C. Agatemor, et al., Electrospinning over solvent casting: Tuning of mechanical properties of membranes, Sci. Rep. 8 (1) (2018) 1−9.

[31] Q.P. Pham, U. Sharma, A.G. Mikos, Electrospinning of polymeric nanofibers for tissue engineering applications: A review, Tissue Eng. 12 (5) (2006) 1197−1211.

[32] D. Li, J.T. Mccann, Y. Xia, M. Marquez, Electrospinning: A simple and versatile technique for producing ceramic nanofibers and nanotubes, J. Am. Ceram. Soc. 89 (6) (2006) 1861−1869.

[33] L. Li, S. Peng, J.K.Y. Lee, D. Ji, M. Srinivasan, S. Ramakrishna, Electrospun hollow nanofibers for advanced secondary batteries, Nano Energy 39 (2017) 111−139.

[34] S. Khorshidi, A. Solouk, H. Mirzadeh, S. Mazinani, J.M. Lagaron, S. Sharifi, et al., A review of key challenges of electrospun scaffolds for tissue-engineering applications, J. Tissue Eng. Regen. Med. 10 (9) (2016) 715−738.

[35] V. Pillay, C. Dott, Y.E. Choonara, C. Tyagi, L. Tomar, P. Kumar, et al., A review of the effect of processing variables on the fabrication of electrospun nanofibers for drug delivery applications, J. Nanomaterials (2013) (2013) 22.

[36] Y. Dai, W. Liu, E. Formo, Y. Sun, Y. Xia, Ceramic nanofibers fabricated by electrospinning and their applications in catalysis, environmental science, and energy technology, Polym. Adv. Technol. 22 (3) (2011) 326−338.

[37] A. Haider, S. Haider, I.-K. Kang, A comprehensive review summarizing the effect of electrospinning parameters and potential applications of nanofibers in biomedical and biotechnology, Arab. J. Chem. 11 (8) (2018) 1165−1188.

[38] S.-I. Oh, J.-C. Kim, M.A. Dar, D.-W. Kim, Synthesis and characterization of uniform hollow TiO_2 nanofibers using electrospun fibrous cellulosic templates for lithium-ion battery electrodes, J. Alloy. Compd. 800 (2019) 483−489.

[39] L.A. Mercante, R.S. Andre, L.H.C. Mattoso, D.S. Corrêa, Electrospun ceramic nanofibers and hybrid nanofiber-composites for gas sensing, ACS Appl. Nano Mater. 2 (7) (2019) 4026−4042.

[40] Y. Lu, J. Huang, G. Yu, R. Cardenas, S. Wei, E.K. Wujcik, et al., Coaxial electrospun fibers: Applications in drug delivery and tissue engineering, WIREs Nanomed. Nanobiotechnol 8 (2016) 654−677.

[41] M.F. Elahi, W. Lu, G. Guoping, F. Khan, Core-shell fibers for biomedical applications —A review, J. Bioeng. & Biomed. Sci. 3 (1) (2013) 1−14.

[42] H. Qu, S. Wei, Z. Guo, Coaxial electrospun nanostructures and their applications, J. Mater. Chem. A 1 (38) (2013) 11513−11528.

[43] A.C. Rai, P. Kumar, F. Pilla, A.N. Skouloudis, S.D. Sabatino, C. Ratti, et al., End-user perspective of low-cost sensors for outdoor air pollution monitoring, Sci. Total. Environment, 607−608 (2017) 691−705.

[44] A. Dey, Semiconductor metal oxide gas sensors: A review, Mater. Sci. Eng. B 229 (2018) 206−217.

[45] H. Li, M. Li, H. Kan, C. Li, A. Quan, J. Luo, et al., Surface acoustic wave based nitrogen dioxide gas sensors using colloidal quantum dot sensitive films, Surf. Coat. Tech. 362 (2018) 78−83.

[46] H. Wan, H. Yin, L. Lin, X. Zeng, A.J. Mason, Miniaturized planar room temperature ionic liquid electrochemical gas sensor for rapid multiple gas pollutants monitoring, Sens. Actuators B: Chem. 255 (2018) 638−646.

[47] M.D. Fernández-Ramos, F. Moreno-Puche, P. Escobedo, P. García-López, L.F. Capitán-Vallvey, A. Martínez-Olmos, Optical portable instrument for the determination of CO_2 in indoor environments, Talanta 208 (2020) 120387.

[48] T. Goto, T. Itoh, T. Akamatsu, N. Izu, W. Shin, CO sensing properties of Au/ SnO_2–Co_3O_4 catalysts on a micro thermoelectric gas sensor, Sens. Actuators B: Chem. 223 (2016) 774−783.

[49] A. Mirzaei, J.-H. Kim, H.W. Kim, S.S. Kim, Gasochromic WO_3 nanostructures for the detection of hydrogen gas: An overview, Appl. Sci. 9 (9) (2019) 1775.

[50] A. Mirzaei, S. Park, G.-J. Sun, H. Kheel, C. Lee, S. Lee, Fe_2O_3/Co_3O_4 composite nanoparticle ethanol sensor, J. Korean Phys. Soc. 69 (3) (2016) 373−380.

[51] A. Mirzaei, S. Park, G.-J. Sun, H. Kheel, C. Lee, CO gas sensing properties of $In_4Sn_3O_{12}$ and TeO_2 composite nanoparticle sensors, J. Hazard. Mater. 305 (2016) 130−138.

[52] E. Akbari, K. Jahanbin, A. Afroozeh, P. Yupapin, Z. Buntat, Brief review of monolayer molybdenum disulfide application in gas sensor, Phys. B: Condens. Matter 545 (2018) 510−518.

[53] C. Feng, Z. Jiang, J. Wu, B. Chen, G. Lu, C. Huang, Pt-Cr_2O_3-WO_3 composite nanofibers as gas sensors for ultra-high sensitive and selective xylene detection, Sens. Actuators B: Chem. 300 (2019) 127008.

[54] A. Mirzaei, J.-H. Kim, H.W. Kim, S.S. Kim, How shell thickness can affect the gas sensing properties of nanostructured materials: Survey of literature, Sens. Actuators B: Chem. 258 (2018) 270−294.

[55] Y.J. Kwon, H.Y. Cho, H.G. Na, B.C. Lee, S.S. Kim, H.W. Kim, Improvement of gas sensing behavior in reduced graphene oxides by electron-beam irradiation, Sens. Actuators B: Chem. 203 (2014) 143−149.

[56] J.-H. Kim, A. Mirzaei, H.W. Kim, P. Wu, S.S. Kim, Design of supersensitive and selective ZnO-nanofiber-based sensors for H_2 gas sensing by electron-beam irradiation, Sens. Actuators B: Chem. 293 (2019) 210−223.

[57] S.M. Bukhari, J.B. Giorgi, Ni doped $Sm_{0.95}Ce_{0.05}FeO_{3-\delta}$ perovskite based sensors for hydrogen detection. Sens, Actuators B: Chem. 181 (2013) 153−158.

[58] L. Ma, S.Y. Ma, X.F. Shen, T.T. Wang, X.H. Jiang, Q. Chen, et al., $PrFeO_3$ hollow nanofibers as a highly efficient gas sensor for acetone detection, Sens. Actuators B: Chem. 255 (2018) 2546−2554.

[59] Z. Pang, Z. Yang, Y. Chen, J. Zhang, Q. Wang, F. Huang, et al., A room temperature ammonia gas sensor based on cellulose/TiO_2/PANI composite nanofibers, Colloids Surf. A: Physicochemical Eng. Asp. 494 (2016) 248−255.

[60] Y.R. Choi, Y.-G. Yoon, K.S. Choi, J.H. Kang, Y.-S. Shim, Y.H. Kim, et al., Role of oxygen functional groups in graphene oxide for reversible room-temperature NO_2 sensing, Carbon 91 (2015) 178−187.

[61] E. Espid, F. Taghipour, UV-LED photo-activated chemical gas sensors: A review, Crit. Rev. Solid. State Mater. Sci. 42 (5) (2017) 416−432.

[62] W. Li, J. Guo, L. Cai, W. Qi, Y. Sun, J.-L. Xu, et al., UV light irradiation enhanced gas sensor selectivity of NO_2 and SO_2 using rGO functionalized with hollow SnO_2 nanofibers, Sens. Actuators B: Chem. 290 (2019) 443−452.

[63] F. Qu, X. Zhou, B. Zhang, S. Zhang, C. Jiang, S. Ruan, et al., Fe_2O_3 nanoparticles-decorated MoO_3 nanobelts for enhanced chemiresistive gas sensing, J. Alloy. Compd. 782 (2019) 672−678.

[64] F. Li, X. Gao, R. Wang, T. Zhang, G. Lu, Study on TiO_2-SnO_2 core-shell heterostructure nanofibers with different work function and its application in gas sensor, Sens. Actuators B: Chem. 248 (2017) 812−819.

[65] X. Gao, F. Li, R. Wang, T. Zhang, A formaldehyde sensor: Significant role of p-n heterojunction in gas-sensitive core-shell nanofibers, Sens. Actuators B: Chem. 258 (2018) 1230−1241.

[66] F. Li, T. Zhang, X. Gao, R. Wang, B. Li, Coaxial electrospinning heterojunction SnO_2/Au-doped In_2O_3 core-shell nanofibers for acetone gas sensor, Sens. Actuators B: Chem. 252 (2017) 822−830.

[67] X. Yang, V. Salles, Y.V. Kaneti, M. Liu, M. Maillard, C. Journet, et al., Fabrication of highly sensitive gas sensor based on Au functionalized WO_3 composite nanofibers by electrospinning, Sens. Actuators B: Chem. 220 (2015) 1112−1119.

[68] S. Bai, H. Fu, Y. Zhao, K. Tian, R. Luo, D. Li, et al., On the construction of hollow nanofibers of ZnO-SnO_2 heterojunctions to enhance the NO_2 sensing properties, Sens. Actuators B: Chem. 266 (2018) 692−702.

[69] R.-A. Wu, C.W. Lin, W.J. Tseng, Preparation of electrospun Cu-doped α-Fe_2O_3 semiconductor nanofibers for NO_2 gas sensor, Ceram. Int. 42 (2017) S535−S540.

[70] F. Wang, K. Hu, H. Liu, Q. Zhao, K. Wang, Y. Zhang, Low temperature and fast response hydrogen gas sensor with Pd coated SnO_2 nanofiber rods, Int. J. Hydrog. Energy 45 (2020) 7234−7242.

[71] J.-K. Choi, I.-S. Hwang, S.-J. Kim, J.-S. Park, S.-S. Park, U. Jeong, et al., Design of selective gas sensors using electrospun Pd-doped SnO_2 hollow nanofibers, Sens. Actuators B: Chem. 150 (1) (2010) 191−199.

[72] W. Liu, Y. Xie, T. Chen, Q. Lu, S.U. Rehman, L. Zhu, Rationally designed mesoporous In_2O_3 nanofibers functionalized Pt catalysts for high-performance acetone gas sensors, Sens. Actuators B: Chem. 298 (2019) 126871.

[73] J.-H. Kim, A. Mirzaei, J.-Y. Kim, J.-H. Lee, H.W. Kim, S. Hishita, et al., Enhancement of gas sensing by implantation of Sb-ions in SnO_2 nanowires, Sens. Actuators B: Chem. 304 (2020) 127307.

[74] X. Kou, N. Xie, F. Chen, T. Wang, L. Guo, C. Wang, et al., Superior acetone gas sensor based on electrospun SnO_2 nanofibers by Rh doping, Sens. Actuators B: Chem. 256 (2018) 861−869.

[75] W. Wei, S. Guo, C. Chen, L. Sun, Y. Chen, W. Guo, et al., High sensitive and fast formaldehyde gas sensor based on Ag-doped $LaFeO_3$ nanofibers, J. Alloy. Compd. 695 (2017) 1122−1127.

[76] K.-C. Hsu, T.-H. Fang, Y.-J. Hsiao, C.-A. Chan, Highly response CO_2 gas sensor based on Au-La_2O_3 doped SnO_2 nanofibers, Mater. Lett. 261 (2020) 127144.

Metal oxide nanofibers for flexible organic electronics and sensors

7

Roohollah Bagherzadeh[1] and Nikoo Saveh Shemshaki[2]
[1]Advanced Fibrous Materials Lab, Institute for Advanced Textile Materials and Technologies, School of Advanced Materials and Processes, Amirkabir University of Technology (Tehran Polytechnic), Tehran, Iran, [2]Department of Biomedical Engineering, University of Connecticut, Storrs, CT, United States

7.1 Incorporation of nanofibers into electronic devices

The inherent properties of nanofibers (NFs), including high surface-to-volume ratio, scalability, cost-effectiveness, flexibility, various orientations, and a wide range of materials, make them novel nanostructures in different fields of study [1,2]. Electrospinning (ES) is the most popular technology for fabrication of nanofibrous structures because of the high adjustability to generate suitable morphology, density, and orientation [3]. The efficacy of electrospun NFs has been investigated in different applications, including sensors [4], energy harvesting [5], solar cells [6], photoelectrochemical water splitting [7], drug delivery [8], and tissue regeneration [9].

One of the most important aspects of ES is the wide range of materials that can be used for the fabrication of NFs. These materials include polymeric or polymeric/nonpolymeric composite solutions that can be ejected from a needle into an electric field applied between the needle and the collector to stretch and fabricate the NF. The electrospun NFs can be fabricated into an aligned or random orientation with a range of sizes from nanometer to micrometer by changing the ES parameters (Fig. 7.1).

The mechanism of ES is rather complex, despite the having a setup that can be assembled simply [10,11]. The required electric field for ES will be setup between the needle and the collector by applying a high voltage to the prepared solution inside the feed pump and spinneret. The electric field induces the electrical conductivity to the solution and distorts the surface tension because of the migration of charges to the surface. This accumulation of charge leads to the deformation of liquid droplets into a conical shape called Taylor cone [12]. Increasing the strength of the electric field to the threshold value causes the repulsive electrostatic force to overcome the surface tension and a charged fine jet of the solution to erupt from the tip of the Taylor cone. The evaporation of the solvent during the flight leads to the migration of charges to the fiber surface. This process makes the jet undergo a whipping and stretching process, which elongates the fiber and causes the formation

---

Metal Oxide-Based Nanofibers and Their Applications. DOI: https://doi.org/10.1016/B978-0-12-820629-4.00006-0
© 2022 Elsevier Inc. All rights reserved.

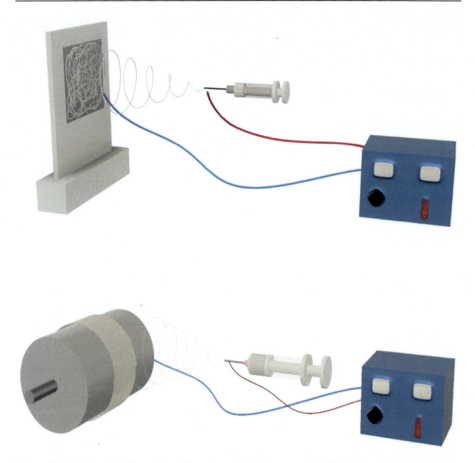

Figure 7.1 Schematic illustration of an electrospinning setup for fabrication of random and aligned nanofibers.

of fiber in the scale of nanometers to micrometers. The fibers rest on the static or rotating collector with the desired orientation. The morphology of the fibrous mat and the fiber diameter can be optimized through the principal ES parameters, including flow rate, solution viscosity, the working distance, the applied voltage, needle or nozzle size, and the environment (temperature and humidity). To fabricate novel NFs, several techniques can be used to modify the ES setup. The coaxial fibers, hollow structures, and combined NFs can be fabricated through modification of needle and spinneret. The required orientation can also be created by changing the collector.

Metal oxide NFs are currently being fabricated through direct ES of composite solution or coating modification of the surface of NFs. The ES of metal oxide NFs consist of three main procedures: (1) the preparation of a composite solution containing metal salt/alkoxide metal and polymer, (2) ES of the composite solution to fabricate NFs, (3) calcination or annealing of fabricated NFs at optimized temperatures to enhance the crystallinity, transmittance, and purity of NFs. The calcination of the as-spun NFs removes the polymeric components and subsequently reduces the average diameter of the fibers. Moreover, studies have reported the formation of nanograins during the calcination process, which have significantly change the performance of NFs by affecting their optical, electrical transport, magnetic, gas-sensing, and photocatalytic properties [13,14]. By optimizing the heating conditions, including the temperature, heating/cooling rate, and time, the size of the nanograins can be adjusted depending on their applications.

Metal oxide NFs, including titanium dioxide (Ti_2O), zinc oxide (ZnO), iron oxide (Fe_2O_3), copper oxide (CuO), and indium tin oxide (ITO), are common NFs for electronic devices. To provide appropriate viscosity for the solution and enable the ES process, polymers, such as polyvinyl pyrrolidone, polyvinyl acetate, polyvinyl alcohol (PVA), polyacrylonitrile, polyethylene oxide, and polystyrene, need to be added into the solution [15,16]. Although the calcination process mostly removes the polymers used for the fabrication of metal oxide NFs, it cannot remove them completely, and the residual polymers are a problem that needs to be addressed.

The diameter of metal oxide NFs should be around the diffusion length of the exciton, which is determined according to the application of NFs; for example, in solar cells, the diffusion length of excitons is around 10 nm. To improve the excitation dissociation and power conversion efficiency, the average diameter of metal oxide NFs should be in the range of tens of nanometers. Furthermore, the orientation of electrospun NFs had effects on the recombination of dissociated electrons and holes. The vertical orientation of aligned metal oxide NFs indicated the suppressing effects on the recombination.

7.2 Conductive and transparent nanofibrous networks as a futuristic approach toward the flexible displays

One of the principal components used in solar cells, touch screens, photoelectrochemical, transparent heaters, organic light-emitting diodes, and other electrical devices are transparent conductive electrodes (TCEs). The TCEs provide electrical conductivity and high transparency [17−21]. The surface conductivity/Sheet resistance and optical transparency are two main clues that regulate the applications of TCEs. The resistivity and transparency of TCEs depend on their applications; for example, the appropriate resistivity for touch screens is in the range of 400−1000 Ω sq^{-1} with high transparency ($>95\%$) [18] while a low resistivity and transparency in the range of visible light wavelengths are required for transparent heaters [22]. To prevent undesired voltage drops and Joule heating during device operation in solar cell displays, the required resistivity is less than 20 Ω sq^{-1} [18].

The common TCEs, such as ITO or fluorine-doped tin oxide, have high transparency and low resistivity. Because of the rising demands and low abundance of indium, the industrial ITOs are expensive and have complicated fabrication process that motivates the use of alternatives and new materials [18,20,23,24]. Furthermore, the current ITOs displays are brittle that limit their application as flexible electrodes. Different types of materials and structures have been studied to progress the performance of TCEs alternatives, including silver nanowires [25−28], Graphene composition [25,26,29], Cellulose NFs [30], and metal oxide coatings [28] alone or in combination. Among the nanostructures, NFs are one of the most interesting materials owing to their continuous nanoscale long fibers. The electrospun NFs can address the limitations of random distribution of nanoparticles/nanowires on the substrate, short length of nanowires, and the electrical contact between nanoparticles/nanowires [20,31].

Among different types of conductive materials investigating for electronic devices, including carbon nanotubes, metal/metal oxide, conductive polymers, graphene [17,20,24,31−40]. Metal oxide NFs with both align and random orientations were fabricated by ES. The combination of metal oxide NFs into electronic devices provides excellent properties, with higher transparency and sheet conductivity than the common electrodes [1]. Energy storage devices such as supercapacitors have a longer lifetime and higher charge/discharge rate than batteries and are being used in different applications. The metal oxide NFs, such as nickel dioxide, manganese dioxide (MnO_2), TiO_2, and tin dioxide (SnO_2), are commonly used as pseudocapacitors, owing to their fast and reversible faradaic redox reaction, which provide higher capacitance [41].

These conductive NFs are promising materials to address the current challenges in electronic devices, owing to their long continuous structure, high surface-to-volume ratio, and nanostructural properties [19,42,43]. The intrinsic conductivity of metal-based NFs and the contact points provides percolation paths to facilitate high conductivity. Furthermore, the nanoscale diameter of the NFs can lead to high transparency of the metallic nanofibrous electrode [24,31].

According to the literature, ES is the most common method to fabricate composite metal oxide NFs for sensors. Several metal oxide NFs have been reported for gas-sensing, including $ZnO-CuO$, $CuO-SnO_2$, TiO_2-ZnO, indium oxide−tungsten oxide ($In_2O_3-WO_3$), $ZnO-In_2O_3$, tin dioxide−cerium dioxide (SnO_2-CeO_2), and $SnO_2-In_2O_3$ [44−49].

A large surface area is one of the principal parameters for gas sensors. Therefore nanostructures with high surface areas are promising structures for improving the efficacy of sensors compared with bulk structures. The NFs have more advantages than other nanostructures, including their higher surface area, a weblike formation, and integration with a field-effect transistor configuration. These advantages increase the sensitivity and allow rapid response [50,51].

The efficacy of conductive nanostructures has been extensively investigated to fabricate cost-effective and flexible displays and electronic devices [52−54]. Flexible TCEs are promising structures for solar cells, light emissions, touch screens, and energy storage devices. Studies improved the electrochemical properties and extended TCEs intelligent displays and smart windows [53,55,56].

Some concerns limit the development of flexible TCEs. The materials and structures of the electrodes need to provide not only appropriate conductivity, capacitive behavior, and optical transmittance but also the required flexibility. Besides, in several applications, the performance of the electrode is significantly being affected by its electrochemical behavior [57]. The most important consideration for succeeding efficient, flexible TCEs is improving conductivity and flexibility while retaining high transparency.

Among all investigated materials, metal-based nanowires and NFs have attracted considerable attention, owing to their scalability and flexibility [58,59]. Besides metal oxide structures, carbon-based materials and conductive polymers have also been investigated as TCEs in electronic devices [60−62]. Besides the shape and morphology of the structure, some features such as the thickness of the layer have significant effects on the outcomes. For example, Cheng et al. investigated the efficacy of PEDOT:PSS for flexible transparent supercapacitors. It has been shown that by decreasing the thickness of the PEDOT:PSS, the capacitance of the electrode decreased while the transparency increased [63]. Different strategies have been reported to fabricate flexible metal oxide electrodes with appropriate conductivity, capacitance, and transparency by doping different elements, combining with other conducting materials, or developing hierarchical nanostructures [52,53,64,65]. Moreover, several studies evaluated the potential of using composite materials such as metal and metal oxide materials [52], metal-based fabrics and carbon-based materials [66], and metal-based nanostructures and conductive polymer [67,68] to improve the electrical performance of the electrode while keeping the high transparency. Among all the strategies, ES of metal-based materials indicated promising results for fabrication of transparent electrodes, owing to their high electrooptical performance and desirable mechanical properties [24,52,69]. Singh et al. reported the fabrication of core−shell MnO_2@AuNF as a flexible transparent supercapacitor electrode. Following the ES of gold (Au)/PVA NFs, the nanofibrous mats were coated with MnO_2 nanosheets. The Au NFs had good flexibility, and the sheet resistance and transparency of the fibrous mats were 9.58 Ω sq^{-1} and 93.13%, respectively. The core−shell MnO_2@AuNF electrode exhibited a high capacitance of 8.26 mF cm^{-1} at 5 mV s^{-1}, along with high transparency (86%), long-term cycling stability, high rate capability, and good flexibility [52].

7.3 Recent progress in electrospun metal oxide nanofibers

ES is considered a facile method to cost-effectively fabricate metal oxide NFs [1,7,20,31]. Several studies investigated the efficacy of metal oxide NFs with high transparent and low to high conductivity [13,20,31,52]. For example, the fabrication of CuO NFs through the ES of composite structures of copper acetate and PVA precursors was studied for a transparent conductive electrode (Fig. 7.2) [31]. The study reported the morphological, conductivity, and transmittance properties of both aligned and random CuO NFs. On the basis of the results, the optimized CuO NFs exhibited high transparency ($\sim$70% to $\sim$90%) and a sheet resistance in the range of $\sim$0.38−5.41 MΩ sq^{-1}. The NFs cross junctions in the random electrospun layer

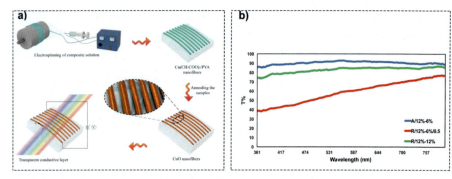

Figure 7.2 (A) The schematic illustration of CuO nanofiber fabrication, including the electrospinning setup and annealing process. (B) Diffuse transmittance spectroscopy results of random and aligned samples (R and A refer to random and aligned nanofibers, respectively). The percentages show polyvinyl alcohol and Cu(CH$_3$COO)$_2$ concentrations.
Source: N. Saveh-Shemshaki, R. Bagherzadeh, M. Latifi, Electrospun metal oxide nanofibrous mat as a transparent conductive layer. Org. Electron. 2019;70:131–139.

increased the thickness and packing density of NFs and subsequently decreased the transparency. It has been shown that the transparency of the electrode can be adjusted by changing the NFs' orientation and thickness (Fig. 7.2) [31].

Table 7.1 lists some recent studies that reported the potential of metal oxide fibrous electrodes for electronic devices.

Incorporation of metal oxide NFs presents promising results as transparent conductive layers and for the fabrication of photoanode. Several studies have investigated the efficacy of metal oxide nanostructures as a photoanode. Iron oxide or hematite (α-Fe$_2$O$_3$) has been studied as a potential material for photoanode fabrication and catalytic and gas sensors [7,84,85]. The fabrication of hematite nanostructures can overcome its current limitations, such as poor electronic conductivity, short hole diffusion length, and low absorptivity of photons near its band edge [85,86]. The potential of hematite electrospun NFs was investigated as a photoanode for photoelectrochemical water splitting [7]. Fig. 7.3 shows the photocurrent density-time curve at a constant potential (0.7 V) for hematite NFs. On the basis of the results, the calcinated hematite NFs with the average diameter of 110 nm indicated a photocurrent density of 0.53 mA cm^{-2} under dark and visible illumination conditions at voltage 1.23 V and constant intensity (900 mW cm^{-2}) [7]. The photovoltaic performance of hematite NFs provides a promising approach for using this structure with a flexible conductive substrate for solar fuel generators.

7.4 Summary and future trends

Metal oxide NFs have shown promising features for making electronic devices. The intrinsic properties of NFs and the potential of ES to control the fiber features have

Table 7.1 Some conducted studies on the metal oxide fibrous electrode for electronic devices.

Applications	Composition	Specific properties	References
Flexible supercapacitors	Manganese oxide carbon fiber−shell nanocables	Capacitance: 295.24 F/gSpecific energy: 22.2 Wh/kgStability performance (%): 96.4 (3000 cycles)	[70]
Supercapacitors	$NiCo_2O_4$−carbon fiber−$Ni(OH)_2$ core−shell NFs	Capacitance: 1925 F/gStability performance (%): 87 (5000 cycles)	[71]
Flexible supercapacitors	Carbon−MnO_2−PEDOT:PSS	Capacitance: 351 F/gStability performance (%): 84.2 (1000 cycles)	[72]
Supercapacitors	PVA−GO−MnO_2/PEDOT	Capacitance: 144.6 F/gSpecific energy: 9.60 Wh/kgStability performance (%): 91.2 (1000 cycles)	[73]
Flexible supercapacitors	Fe_3O_4−carbon fiber	Capacitance: 306 F/gSpecific energy: 13 Wh/kgStability performance (%): 85 (2000 cycles)	[74]
Supercapacitors	NiO−NFs/Ni	Capacitance: 737 F/gSpecific energy: 22.7 Wh/kg	[75]
Transparent conductive electrode	AgO	Transmittance (%): 93.5	[76]
Transparent conductive electrode	CuO	Transmittance (%): 70−90Sheet resistance: 0.38−5.41 MΩ/sq	[31]
Transparent conductive electrode/ supercapacitors	MnO_2−Au	Capacitance: 8.26 mF/cm at 5 mV/sTransmittance (%): 86Sheet resistance: 9.58 Ω/sq	[52]
Transparent conductive electrode	ITO	Transmittance (%): 81Sheet resistance: 304 Ω/sq	[77]
Transparent conductive electrode	Aluminum−ZnO	Transmittance (%): 84Sheet resistance: 190 Ω/sq	[78]

(Continued)

Table 7.1 (Continued)

Applications	Composition	Specific properties	References
Gas sensor	Aluminum−ZnO	Gas: NO_2Conc. (ppm): 0.5T (°C): 250Response (R_g/R_a): 11	[79]
Gas sensor	ZnO−Co_3O_4	Gas: H2Conc. (ppm): 10T (°C): 300Response (R_g/R_a): 133.65	[80]
Gas sensor	NiO−ZnO	Gas: H_2Conc. (ppm): 10T (°C): 200Response (R_g/R_a): 0.6	[81]
Gas sensor	ZnO	Gas: H_2Conc. (ppm): 0.1T (°C): 350Response (R_g/R_a): 10	[82]
Gas sensor	SnO_2−CuO	Gas: H2SConc. (ppm): 5T (°C): 200Response (R_g/R_a): 1395	[83]

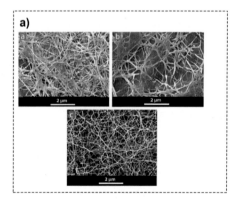
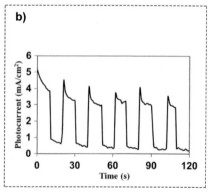

Figure 7.3 (A) scanning electronic microscope images of the calcinated hematite nanofibers with different concentration of polyvinyl alcohol/Fe(NO$_3$)$_3 \cdot$ 9H$_2$O at different calcination temperatures: (A) 450°C, (B) 550°C, and (C) 600°C. (B) Photocurrent density−time curve at 0.7 V for α-Fe$_2$O$_3$ nanofibers at a calcination temperature of 550°C under visible photoillumination.
Source: N. Saveh-Shemshaki, et al., Synthesis of mesoporous functional hematite nanofibrous photoanodes by electrospinning. Polym. Adv. Technol. 2016;27(3):358−365.

significant effects on the performance of electronic devices. Recent progress in the field of NF-based electronic devices provides a promising approach toward the application of metal oxide NFs.

This chapter described the ES principles for synthesizing metal oxide NFs and showed their great potential for various electronic applications. The performance of metal oxide NFs can be attributed to their high surface area, electrical conductivity, scalability, flexibility, and various orientations. However, the design and optimum conditions for the fabrication of NFs with appropriate properties remain a challenge that needs to be addressed. Some key factors, such as their morphology, surface area, and pore size, should be considered. By changing the morphological properties of NFs, different structures, such as core—shell NFs, hollow NFs, and porous NFs, can be fabricated. Owing to the rapidly growth of NF applications, it is expected that the design of new ES setups could synthesize more developed composite NFs with improved properties. It is expected that the development of composite nanostructures with different materials, including metal oxide, metal, carbon-based materials, and conductive polymers, can further enhance the performance of current electronic devices.

References

[1] R. Bagherzadeh, et al., Electrospun conductive nanofibers for electronics A2—Afshari, Mehdi, in Electrospun Nanofibers, Woodhead Publishing, 2017, pp. 467—519.

[2] P. Gibson, H. Schreuder-Gibson, D. Rivin, Transport properties of porous membranes based on electrospun nanofibers, Colloids Surf. A: Physicochem. Eng. Asp. 187 (2001) 469—481.

[3] J.M. Deitzel, et al., The effect of processing variables on the morphology of electrospun nanofibers and textiles, Polymer 42 (1) (2001) 261—272.

[4] A. Chinnappan, et al., An overview of electrospun nanofibers and their application in energy storage, sensors and wearable/flexible electronics, J. Mater. Chem. C. 5 (48) (2017) 12657—12673.

[5] H.-Y. Mi, et al., High-performance flexible triboelectric nanogenerator based on porous aerogels and electrospun nanofibers for energy harvesting and sensitive self-powered sensing, Nano Energy 48 (2018) 327—336.

[6] X. Ma, et al., Nanocomposite structures related to electrospun nanofibers for highly efficient and cost-effective dye-sensitized solar cells, Multifunct. Nanocomp. Energy Environ. Appl. 1 (2018) 113—133.

[7] N. Saveh-Shemshaki, et al., Synthesis of mesoporous functional hematite nanofibrous photoanodes by electrospinning, Polym. Adv. Technol. 27 (3) (2016) 358—365.

[8] V. Mouriño, Nanoelectrospun matrices for localized drug delivery, in: A.M. Asiri Inamuddin, A. Mohammad (Eds.), Applications of Nanocomposite Materials in Drug Delivery, Woodhead Publishing, 2018, pp. 491—508.

[9] N. Saveh-Shemshaki, L.S. Nair, C.T. Laurencin, Nanofiber-based matrices for rotator cuff regenerative engineering, Acta Biomaterialia 94 (2019) 64—81.

[10] S. Ramakrishna, et al., Electrospun nanofibers: solving global issues, Mater. today 9 (3) (2006) 40—50.

[11] D. Li, Y. Xia, Electrospinning of nanofibers: reinventing the wheel? Adv. Mater. 16 (14) (2004) 1151–1170.

[12] G.I. Taylor, Electrically driven jets., Proc. Roy. Soc. London. A. Math. Phys. Sci. 313 (1515) (1969) 453–475.

[13] A. Katoch, S.-W. Choi, S.S. Kim, Nanograins in electrospun oxide nanofibers, Met. Mater. Int. 21 (2) (2015) 213–221.

[14] A. Katoch, et al., Competitive influence of grain size and crystallinity on gas sensing performances of ZnO nanofibers, Sens. Actuators B: Chem. 185 (2013) 411–416.

[15] X. Lu, C. Wang, Y. Wei, One-dimensional composite nanomaterials: synthesis by electrospinning and their applications, Small 5 (21) (2009) 2349–2370.

[16] Z.U. Abideen, et al., Electrospun metal oxide composite nanofibers gas sensors: a review, J. Korean Ceram. Soc. 54 (5) (2017) 366–379.

[17] Q. Zheng, J.-K. Kim, Vol Graphene for Transparent Conductors: Synthesis, Properties and Applications, 23, Springer, 2015.

[18] M. Layani, A. Kamyshny, S. Magdassi, Transparent conductors composed of nanomaterials, Nanoscale 6 (11) (2014) 5581–5591.

[19] S. Kim, et al., Transparent conductive films of copper nanofiber network fabricated by electrospinning, J. Nanomater. 16 (1) (2015) 371.

[20] H. Wu, et al., Electrospun metal nanofiber webs as high-performance transparent electrode, Nano Lett. 10 (10) (2010) 4242–4248.

[21] S. Ye, et al., Metal nanowire networks: the next generation of transparent conductors, Adv. Mater. 26 (39) (2014) 6670–6687.

[22] J. Jang, et al., Rapid production of large-area, transparent and stretchable electrodes using metal nanofibers as wirelessly operated wearable heaters, NPG Asia Mater. 9 (9) (2017) e432. e432.

[23] Y. Leterrier, et al., Mechanical integrity of transparent conductive oxide films for flexible polymer-based displays, Thin Solid. Films 460 (1–2) (2004) 156–166.

[24] H. Wu, et al., A transparent electrode based on a metal nanotrough network, Nat. Nanotechnol. 8 (6) (2013) 421.

[25] X. Zhang, et al., Large-size graphene microsheets a protective layer. Transparent conductive silver nanowire film, Heat 69 (2014) 437–443.

[26] D. Lee, et al., High-performance flexible transparent conductive film based graphene/AgNW/graphene sandwich structure, J. Carbon 81 (2015) 439–446.

[27] T. Wang, et al., Highly transparent, conductive, and bendable Ag nanowire electrodes with enhanced mechanical stability based on polyelectrolyte adhesive layer, Langmuir 33 (19) (2017) 4702–4708.

[28] K. Zilberberg, et al., Highly robust indium-free transparent conductive electrodes based on composites of silver nanowires and conductive metal oxides, Adv. Funct. Mater. 24 (12) (2014) 1671–1678.

[29] L. Yu, C. Shearer, J.J.Cr Shapter, Recent development of carbon nanotube transparent conductive films, Chem. Rev. 116 (22) (2016) 13413–13453.

[30] M. Nogi, et al., Transparent conductive nanofiber paper for foldable solar cells, Sci. Rep. 5 (2015) 17254.

[31] N. Saveh-Shemshaki, R. Bagherzadeh, M. Latifi, Electrospun metal oxide nanofibrous mat as a transparent conductive layer, Org. Electron. 70 (2019) 131–139.

[32] J. Lee, et al., Very long Ag nanowire synthesis and its application in a highly transparent, conductive and flexible metal electrode touch panel, Nanoscale 4 (20) (2012) 6408–6414.

[33] J.L. Blackburn, et al., Transparent conductive single-walled carbon nanotube networks with precisely tunable ratios of semiconducting and metallic nanotubes, ACS Nano 2 (6) (2008) 1266–1274.

[34] H. Jung, et al., A scalable fabrication of highly transparent and conductive thin films using fluorosurfactant-assisted single-walled carbon nanotube dispersions, Carbon 52 (2013) 259–266.

[35] H.A. Becerril, et al., Evaluation of solution-processed reduced graphene oxide films as transparent conductors, ACS Nano 2 (3) (2008) 463–470.

[36] J.-S. Yu, et al., Transparent conductive film with printable embedded patterns for organic solar cells, Sol. Energy Mater. Sol. Cell 109 (2013) 142–147.

[37] W. Gaynor, et al., Smooth nanowire/polymer composite transparent electrodes, Adv. Mater. 23 (26) (2011) 2905–2910.

[38] K.M. Sawicka, A.K. Prasad, P.I. Gouma, Metal oxide nanowires for use in chemical sensing applications, Sens. Lett. 3 (1–2) (2005) 31–35.

[39] W. Wang, et al., Three-dimensional network films of electrospun copper oxide nanofibers for glucose determination, Biosens. Bioelectron. 25 (4) (2009) 708–714.

[40] G. Zhang, et al., Strongly coupled carbon nanofiber–metal oxide coaxial nanocables with enhanced lithium storage properties, Energy Environ. Sci. 7 (1) (2014) 302–305.

[41] J. Fang, et al., Applications of electrospun nanofibers for electronic devices, Handb. Smart Text. (2015) 617–652.

[42] P.-C. Hsu, et al., Performance enhancement of metal nanowire transparent conducting electrodes by mesoscale metal wires, Nat. Commun. 4 (2013) 2522.

[43] A.R. Rathmell, B.J. Wiley, The synthesis and coating of long, thin copper nanowires to make flexible, transparent conducting films on plastic substrates, Adv. Mater. 23 (41) (2011) 4798–4803.

[44] J.Y. Park, et al., Synthesis and gas sensing properties of TiO_2–ZnO core-shell nanofibers, J. Am. Ceram. Soc. 92 (11) (2009) 2551–2554.

[45] A. Katoch, et al., Mechanism and prominent enhancement of sensing ability to reducing gases in p/n core–shell nanofiber, Nanotechnology 25 (17) (2014) 175501.

[46] W. Qin, et al., Highly enhanced gas sensing properties of porous SnO_2–CeO_2 composite nanofibers prepared by electrospinning, Sens. Actuators B: Chem. 185 (2013) 231–237.

[47] A. Katoch, J.-H. Kim, S.S. Kim, TiO_2/ZnO inner/outer double-layer hollow fibers for improved detection of reducing gases, ACS Appl. Mater. Interfaces 6 (23) (2014) 21494–21499.

[48] C.-S. Lee, I.-D. Kim, J.-H. Lee, Selective and sensitive detection of trimethylamine using ZnO–In_2O_3 composite nanofibers, Sens. Actuators B: Chem. 181 (2013) 463–470.

[49] X.-J. Zhang, G.-J. Qiao, High performance ethanol sensing films fabricated from ZnO and In_2O_3 nanofibers with a double-layer structure, Appl. Surf. Sci. 258 (17) (2012) 6643–6647.

[50] E. Comini, et al., Metal oxide nanoscience and nanotechnology for chemical sensors, Sens. Actuators B: Chem. 179 (2013) 3–20.

[51] Z.U. Abideen, et al., Excellent gas detection of ZnO nanofibers by loading with reduced graphene oxide nanosheets, Sens. Actuators B: Chem. 221 (2015) 1499–1507.

[52] S.B. Singh, et al., A core–shell MnO_2@Au nanofiber network as a high-performance flexible transparent supercapacitor electrode, J. Mater. Chem. A 7 (17) (2019) 10672–10683.

[53] R.T. Ginting, M.M. Ovhal, J.-W. Kang, A novel design of hybrid transparent electrodes for high performance and ultra-flexible bifunctional electrochromic-supercapacitors, Nano Energy 53 (2018) 650–657.

[54] R.J. Tseng, et al., Polyaniline nanofiber/gold nanoparticle nonvolatile memory, Nano Lett. 5 (6) (2005) 1077−1080.

[55] D. Wei, et al., A nanostructured electrochromic supercapacitor, Nano Lett. 12 (4) (2012) 1857−1862.

[56] G. Cai, et al., Highly stable transparent conductive silver grid/PEDOT:PSS electrodes for integrated bifunctional flexible electrochromic supercapacitors, Adv. Energy Mater. 6 (4) (2016) 1501882.

[57] H. Lee, et al., Highly stretchable and transparent supercapacitor by Ag−Au core−shell nanowire network with high electrochemical stability, ACS Appl. Mater. Interfaces 8 (24) (2016) 15449−15458.

[58] S. Nam, et al., Ultrasmooth, extremely deformable and shape recoverable Ag nanowire embedded transparent electrode, Sci. Rep. 4 (1) (2014) 1−7.

[59] M. Song, et al., Highly efficient and bendable organic solar cells with solution-processed silver nanowire electrodes, Adv. Funct. Mater. 23 (34) (2013) 4177−4184.

[60] F. Chen, et al., Flexible transparent supercapacitors based on hierarchical nanocomposite films, ACS Appl. Mater. Interfaces 9 (21) (2017) 17865−17871.

[61] T.M. Higgins, J.N. Coleman, Avoiding resistance limitations in high-performance transparent supercapacitor electrodes based on large-area, high-conductivity PEDOT: PSS films, ACS Appl. Mater. Interfaces 7 (30) (2015) 16495−16506.

[62] T. Cheng, et al., High-performance free-standing PEDOT: PSS electrodes for flexible and transparent all-solid-state supercapacitors, J. Mater. Chem. A 4 (27) (2016) 10493−10499.

[63] T. Cheng, et al., Inkjet-printed flexible, transparent and aesthetic energy storage devices based on PEDOT:PSS/Ag grid electrodes, J. Mater. Chem. A 4 (36) (2016) 13754−13763.

[64] S. Zhao, et al., Rational synthesis of Cu-doped porous δ-MnO_2 microsphere for high performance supercapacitor applications, Electrochim. Acta 191 (2016) 716−723.

[65] Z. Qiao, et al., 3D hierarchical MnO_2 nanorod/welded Ag-nanowire-network composites for high-performance supercapacitor electrodes, Chem. Commun. 52 (51) (2016) 7998−8001.

[66] J. Yu, et al., Metallic fabrics as the current collector for high-performance graphene-based flexible solid-state supercapacitor, ACS Appl. Mater. Interfaces 8 (7) (2016) 4724−4729.

[67] X. Liu, et al., Highly transparent and flexible all-solid-state supercapacitors based on ultralong silver nanowire conductive networks, ACS Appl. Mater. Interfaces 10 (38) (2018) 32536−32542.

[68] J.-L. Xu, et al., Embedded Ag grid electrodes as current collector for ultraflexible transparent solid-state supercapacitor, ACS Appl. Mater. Interfaces 9 (33) (2017) 27649−27656.

[69] K. Azuma, et al., Facile fabrication of transparent and conductive nanowire networks by wet chemical etching with an electrospun nanofiber mask template, Mater. Lett. 115 (2014) 187−189.

[70] D. Zhang, et al., Highly porous honeycomb manganese oxide@carbon fibers core−−shell nanocables for flexible supercapacitors, Nano Energy 13 (2015) 47−57.

[71] L. Xu, et al., Rationally designed hierarchical $NiCo_2O_4$−C@$Ni(OH)_2$ core-shell nanofibers for high performance supercapacitors, Carbon 152 (2019) 652−660.

[72] J. Garcia-Torres, C. Crean, Ternary composite solid-state flexible supercapacitor based on nanocarbons/manganese dioxide/PEDOT:PSS fibres, Mater. Des. 155 (2018) 194−202.

[73] M.A.A.M. Abdah, N.A. Rahman, Y. Sulaiman, Enhancement of electrochemical performance based on symmetrical poly-(3, 4-ethylenedioxythiophene) coated polyvinyl alcohol/graphene oxide/manganese oxide microfiber for supercapacitor, Electrochim. Acta 259 (2018) 466–473.

[74] N. Iqbal, et al., Flexible Fe_3O_4@ carbon nanofibers hierarchically assembled with MnO_2 particles for high-performance supercapacitor electrodes, Sci. Rep. 7 (1) (2017) 1–10.

[75] M. Kundu, L. Liu, Binder-free electrodes consisting of porous NiO nanofibers directly electrospun on nickel foam for high-rate supercapacitors, Mater. Lett. 144 (2015) 114–118.

[76] H. Jo, et al., Highly transparent and conductive oxide-metal-oxide electrodes optimized at the percolation thickness of AgO_x for transparent silicon thin-film solar cells, Sol. Energy Mater. Sol. Cell 202 (2019) 110131.

[77] H. Wang, et al., Highly flexible indium tin oxide nanofiber transparent electrodes by blow spinning, ACS Appl. Mater. Interfaces 8 (48) (2016) 32661–32666.

[78] Y.-Y. Cho, C. Kuo, Optical and electrical characterization of electrospun Al-doped zinc oxide nanofibers as transparent electrodes, J. Mater. Chem. C. 4 (32) (2016) 7649–7657.

[79] A. Sanger, et al., All-transparent NO_2 gas sensors based on freestanding al-doped ZnO nanofibers, ACS Appl. Electron. Mater. 1 (7) (2019) 1261–1268.

[80] J.-H. Lee, et al., Co_3O_4-loaded ZnO nanofibers for excellent hydrogen sensing, Int. J. Hydrog. Energy 44 (50) (2019) 27499–27510.

[81] J.-H. Lee, et al., Significant enhancement of hydrogen-sensing properties of ZnO nanofibers through NiO loading, Nanomaterials 8 (11) (2018) 902.

[82] J.-H. Kim, et al., Design of supersensitive and selective ZnO-nanofiber-based sensors for H2 gas sensing by electron-beam irradiation, Sens. Actuators B: Chem. 293 (2019) 210–223.

[83] K.-R. Park, et al., Design of highly porous SnO_2–CuO nanotubes for enhancing H_2S gas sensor performance, Sens. Actuators B: Chem. 302 (2020) 127179.

[84] C.Y. Lee, et al., Anodic nanotubular/porous hematite photoanode for solar water splitting: substantial effect of iron substrate purity, ChemSusChem 7 (3) (2014) 934–940.

[85] W.J. Lee, et al., Cathodic shift and improved photocurrent performance of cost-effective Fe2O3 photoanodes, Int. J. Hydrogen Energy 39 (11) (2014) 5575–5579.

[86] Y. Qiu, et al., Efficient photoelectrochemical water splitting with ultrathin films of hematite on three-dimensional nanophotonic structures, Nano Lett. 14 (4) (2014) 2123–2129.

Role of metal oxide nanofibers in water purification

8

Ali A. El-Samak[1,2], Hammadur Rahman[2], Deepalekshmi Ponnamma[2], Mohammad K. Hassan[2], Syed Javaid Zaidi[2] and Mariam Al Ali Al-Maadeed[1,2]

[1]Materials Science and Technology Program (MATS), College of Arts & Sciences, Qatar University, Doha, Qatar, [2]Center for Advanced Materials, Qatar University, Doha, Qatar

8.1 Introduction

Energy and the environment are likely to be the main global concerns of human civilization in the forthcoming years. Hence fossil fuels present the primary source of meeting the energy demands of humanity. However, this source is finite, with a direct influence on the deterioration of the environment in the form of global warming, harmful gas emissions, and water pollution [1,2]. Scientific advancements are attempting to meet the world's energy demands through use of renewable energy sources, such as wind, solar, vibration, and heat energies [3]. Yet this does not address the harm done to the environment through water pollution, and water pollution and depletion of potable water supplies, which represent direct global crises. Scientific reports suggest that 50% of nations will face shortages of fresh water by the year 2025, and the number of struggling nations is estimated to rise to 75% by the year 2075 [4]. The focus of global research must be directed toward the challenges of water pollution, given its persistent threat to the environment, the world's economies, and humankind.

Water pollution comes in multiple forms, including chemical, groundwater, microbiological, nutrient, and surface water pollution. These pollution forms can be directly linked to human activities—such as industrial, transport, spillages, waste deposition, and agriculture [5]. Although the nature of pollution problems may vary, they are typically due to inadequate treatment, as fertilizers, pesticides, chemicals, and heavy metals are some of the main components of polluted ground and surface water [6]. Numerous methods have been adopted for removing the harmful species from polluted water such as chlorination, ion exchange, adsorption, electrochemical treatments, and ozonation. Fig. 8.1 summarizes those methods and the major shortcomings associated with each method.

Because of the importance of potable water, the viability of current practices to meet the increasing demands of water consumers must be addressed. There is a need for the development of innovative technologies and materials presents a key solution for attaining reusable and safe potable water. Although new approaches are frequently being studied, they need to be durable, lower in cost, and more effective

Metal Oxide-Based Nanofibers and Their Applications. DOI: https://doi.org/10.1016/B978-0-12-820629-4.00001-1
© 2022 Elsevier Inc. All rights reserved.

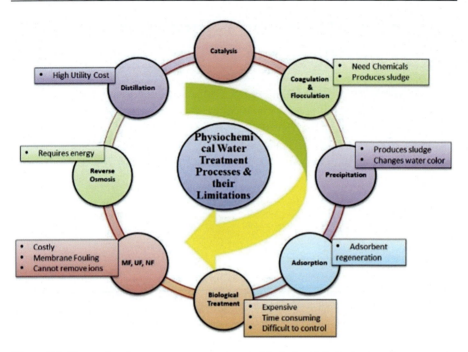

Figure 8.1 Conventional water treatment processes (circles) and their shortcomings (rectangles).
Source: Reprinted from M. Selvaraj, A. Hai, F. Banat, M.A. Haija, Application and prospects of carbon nanostructured materials in water treatment: a review. J. Water Process. Eng. 2020;33:100996 with permission.

than the currently employed methods of water treatment. Nanotechnology is classified as a technology that can be effective in resolving many of the problems faced by current water purification systems [7,8]. Hence nanotechnology resembles the fabrication and utilization of materials and system operating at the atomic and molecular scales. Nanomaterials have nanoscale dimensions ranging from 1 to 100 nm and display significantly different properties as a result of their newly obtained structure, high surface-to-volume ratio, and quantum effects, leading to the development of unprecedented opportunities to develop cost-effective and environmentally friendly water purification systems [9]. The potential impact areas of nanotechnology include sensing and detection, pollution prevention, and treatment and remediation, while the potential nanomaterials that can be used include zeolites, carbon nanotubes, dendrimers, and zeolites [10,11].

Since the agricultural and industrial wastewater contains organic pollutants, coloring agents, microorganisms, and so on, advanced treatment methods are necessary for ultimate water purification. In general, combinations of several techniques or materials with multiple properties are used for water treatment. Herein lies the major significance of metal oxide particles, since they possess multiple properties such as photodegradation, mechanical and thermal stability, and antibacterial activity. Moreover, the electronic properties of metal oxide nanoparticles are notable when the synthesis and fabrication methods are considered. Metal oxides have different mechanisms in water treatment, such as adsorption of pollutant heavy metal ions, filtering the particles on the basis of size, disinfecting and photocatalysis activities. Major achievements are made in the area of metal oxide nanofibers, as their very high specific surface area provides extensive reaction sites. This chapter aims to highlight the use of metal oxide nanofibers in the treatment and purification of water. We will discuss in depth the different types of metal oxides, their utilization in nanofiber composites, and the challenges faced in deploying this method in water purification.

8.2 Metal oxide as water purifiers

Metal oxide nanoparticles have been used for multiple reasons in different sectors, such as electronics, cosmetics, health, and water treatment. The decrease in the solid particle size causes the number of atoms constructing the particle to diminish exponentially. This leads to changes in the fundamental physical properties of the material, such as melting point, structure, and morphology [12,13]. Metal oxide nanoparticles exhibit good mechanical properties, high antibacterial activities, and efficient photocatalytic behavior that can be utilized for water treatment applications [12−14]. The combination of metal and oxygen is highly appealing, owing to the ability to shape its properties; hence it is possible to control the melting point, insulating and conducting properties, and so on. This combination allows for the unique structure of metal oxide nanoparticles to aggressively dismantle toxic biological systems [15]. For example, calcium oxide and magnesium oxide nanoparticles are productive in combating bacteria due to their special surface and alkalinity [16]. Alternatively, other metal oxides such as titanium dioxide (TiO_2) and zinc oxide (ZnO) are capable of treating wastewater through their photocatalytic activity, which allows for the formation of hydroxyl radicals that denaturalize bacteria and organic contaminants to nonharmful by-products [17]. Metal oxides present a great modification in the production of efficient membranes used for water treatment [18]. All of these properties of metal oxides will be discussed in depth in this section of the chapter.

The discovery of photocatalytic water splitting via TiO_2 electrode by Fuiishima and Honda in 1972, elicited great interest in its application [19]. Among the prioritized applications is the photocatalyst effect for the treatment of groundwater and wastewater due to its potential to dismantle a wide range of inorganic compounds [20]. The photocatalysis process of TiO_2 is initiated through the photoexcitation of the metal oxide using light with energy higher than its bandgap, thus leading to the formation of electrons in the conduction band and positive holes in the valence band. This leads to hydroxyl ions and water molecules to trap the positive holes, which proceeds to form a strong oxidizing agent in the form of hydroxyl radicals (OH^-). Additionally, the produced electrons are adsorbed by the oxygen species, resulting in the formation of superoxide species that are unstable. The resultant radicals formed through the photocatalysis process then start to react with the adsorbed species on the surface of the TiO_2, leading to the oxidation and mineralization of the contaminants to convert them into safe carbon dioxide and water by-products [21,22]. The whole mechanism is explained by Eqs. 8.1–8.9 and in Fig. 8.2A. TiO_2 can also serve as both an oxidative catalyst and a reductive catalyst for inorganic and organic pollutants. Chen et al. proved that the addition of TiO_2 coupled with ultraviolet (UV) light was largely successful in removing the total organic carbon present in wastewater contaminated with organic compounds [23]. Murgolo et al. have recently reviewed the ability of TiO_2 nanostructured mesh to degrade emerging organic pollutants and remove toxic metal ions from wastewater. This leads to the successful documentation of TiO_2 metal oxide nanoparticles in its ability to degrade organic compounds (benzenes and dioxins) and reduce toxic metal ions [Pt(II), Cr(VI)] in the presence of UV light. There are multiple synthesis route for the production of nano TiO_2. Polo et al. developed the most common hydrothermal method, which was then modified with ruthenium and platinum to ensure the reduction of TiO_2. The method was effective in managing cost, in the degradation of organic contaminates, and also in being malleable through its ability to be modified [24]. Other forms of metal oxides have also proved successful in the reduction of contaminants within wastewater. For example, Fe_3O_4 was highly successful in removing common chromium contaminants through reduction to its less toxic and mobile species, from Cr(VI) to Cr(III) [25].

$$\text{Metal oxide nanofibers} + h\upsilon \rightarrow \text{electron}(e^-) + \text{hole}(h^+) \tag{8.1}$$

$$h^+ + H_2O \rightarrow H^+ + OH\bullet \tag{8.2}$$

$$h^+ + OH^- \rightarrow OH\bullet \tag{8.3}$$

$$e^- + O_2 \rightarrow O_2{}^- \tag{8.4}$$

$$2e^- + O_2 + 2H^+ \rightarrow H_2O_2 \tag{8.5}$$

$$e^- + H_2O_2 \rightarrow OH\bullet + OH^- \tag{8.6}$$

$$\text{Pollutant in wastewater} + OH\bullet \rightarrow \text{clean water and degradation products} \tag{8.7}$$

$$\text{Pollutant in wastewater} + \text{metal oxide nanofibers}(h+) \rightarrow \text{oxidation products} \tag{8.8}$$

$$\text{Pollutant in wastewater} + \text{metal oxide nanofibers}(e^-) \rightarrow \text{reduction products} \tag{8.9}$$

However, the photoelectrochemical properties of metal oxide nanofibers are modified and potentially improved by combining with other metals or metal oxides

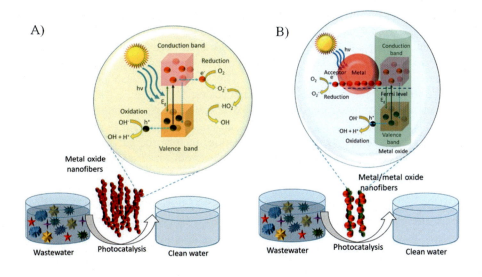

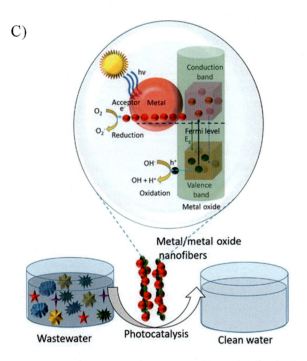

Figure 8.2 Schematic representation of electron transfer mechanism in (A) metal nanofibers, (B) metal−metal oxide nanofiber, and (C) metal oxide−metal oxide nanofiber.
Source: Reprinted from K. Mondal, Recent advances in the synthesis of metal oxide nanofibers and their environmental remediation applications. Inventions. 2017;2(2):9, with permission.

so that metal–metal oxide or metal oxide–metal oxide nanocombinations are formed. The photochemical reaction mechanisms for these systems are respectively represented in Figs. 8.2B and 8.2C. The photocarriers can ideally be quickly transferred from one metal to the other through modified band arrangements, thus allowing chemical reactions for degradation. This combination also provides enough surface adsorption sites for the pollutant molecules.

It has been estimated that water-related infectious diseases cause about 10–20 million deaths and up to 200 million nonfatal infection per year [26]. The common disinfection methods for drinking water supplies include the addition of prominent oxidants such as chlorine; this treatment is successful via deactivating viruses and bacteria, as it disables their ability to grow. However, the use of chlorine may lead to undesirable water qualities because of resulting strange tastes and odors. Furthermore, chlorine disinfectants are known to generate toxic by-products, such as trihalomethanes and aldehydes [27,28]. A successful alternative to the common chlorine disinfectant is nanobiocides, which are antimicrobial nanoparticles that are divided into three categories: engineered nanomaterials (fullerenes), natural antibacterial substances (chitosan), and metal oxides (ZnO, copper oxide (CuO), and TiO_2) [29]. ZnO and CuO are two of the strongest antimicrobial agents, as they have a terminating effect on broad-spectrum bacteria. Although the mechanism involved remains uncertain, Wang et al. suggested that the ZnO nanoparticles tend to pierce the bacterial cell membrane and disintegrate its components [30]. Additionally, ZnO has been used in pharmaceutical products such as ointments and lotions, which supports its safety as a viable water treatment method for human consumption. This led to further studies that incorporated the usage of ZnO fabricated by the Pechini method as an additive to blend films, leading to an increase in thermal stability and porosity of the film and effectively killing *Staphylococcus aureus* bacteria [31]. CuO is also being studied because of its relative inexpensiveness, physical stability, and ability to be mixed with polymers. Although its antimicrobial activity has not been deeply studied, it carries desirable traits, such as a large surface area and unique crystal morphologies, which allow it to be a potential antimicrobial agent [32]. Das et al. represented one of the studies that synthesized CuO nanoparticles via the thermal decomposition method and displayed its efficient antioxidant activity against *Escherichia coli* and *Pseudomonas aeruginosa* [33]. Another candidate for wastewater treatment is magnesium oxide (MgO), which clearly showed antibacterial activity, which is dependent on the particle size of the metal oxide. A nanoparticle smaller than 8 nm proved to be the most effective in destabilizing *E. coli* and *S. aureus* bacteria; however, the gradual increase in size (11–23 nm) led to reduction in bacterial activity [34]. Although its inclusion as a water treatment material is still unknown, MgO still presents itself as a favorable candidate because of its ease of fabrication and successful bacterial reduction abilities.

Fig. 8.3 provides a schematic illustration of the bactericidal activity of metal oxide nanofibers in the presence of visible light. During light illumination the metal oxides undergo electronic transfer reactions, and the metal ions will be generated with reactive oxygen species. Thereafter, the free radicals attack the bacterial walls and destroy the bacteria.

Role of metal oxide nanofibers in water purification

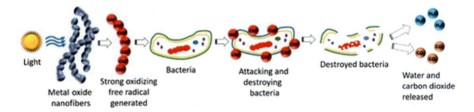

Figure 8.3 Schematic representation of antibacterial activity of metal oxide nanofibers. *Source*: Reprinted from K. Mondal, Recent advances in the synthesis of metal oxide nanofibers and their environmental remediation applications. Inventions 2017;2(2):9, with permission.

8.3 Polymer–metal oxide composite fibers for water treatment

Engineered metal oxide nanoparticles provide a large incentive for their inclusion in water treatment strategies. However, they face significant issues related to the safety and manufacture aspects of water treatment. As for the safety concerns, recent studies have raised concerns about human exposure to the residue of the metal oxides used for water treatment [35]. The nanoparticles are expected to spread at an extremely fast pace because of their small size, leading to dispersion over great distances. Therefore the safest method for using metal oxide nanoparticles is in a matrix that prevents their distribution within the water medium, such as a polymeric matrix. This section addresses the polymer–metal oxide composite fibers that are useful in water treatment.

Composites form when two or more dissimilar materials that differ in chemical and physical properties mix together [36–38]. Since the output properties depend on the nature of the materials, mixing ratios, and preparation strategies, there are different possibilities for tuning the materials choices based on the final application. The inclusion of nanoparticle in the design and synthesis of new materials expanded rapidly with the advancement in nanotechnology. Addition of these nanoparticles in polymers received a wide approval from the scientific community, owing to the remarkable change in physical and chemical properties in comparison to pure organic polymers [39]. Any polymer nanocomposite system has two phases: a matrix polymer phase and a dispersed nanoparticle phase. The matrix phase is a continuous phase that acts as support to hold the dispersed discontinuous phase within it [40]. In other words, the dispersed phase is infused within the matrix phase, leading to the formation of a composite with desired and novel properties. Generally, the matrix phase is softer and more flexible than the dispersed phase, whereas major property enhancement comes from the novel properties of the dispersed nanoparticle phase [41]. This highlights significant applications of polymer nanocomposites in different fields, and this chapter highlights the membrane-forming ability. The polymer composite membranes utilize the presence of the nanoparticles within its matrix through the enhancement of the permeability, selectivity, and strength of the polymer [42]. However, one of the limiting factors behind the implementation of nanoparticles in polymeric membranes is the aggregation behavior that is present in materials synthesized at the nano scale. It is recorded in the literature that nanoparticles with diameter less than 100 nm are difficult to disperse within a polymeric matrix, owing to the surface interactions [43]. These factors affecting the surface interactions remain unclear; however, there are reports suggesting that the nanoparticles at certain pH and ionic strength may reduce the problems associated with particle agglomerations [44].

Generally, the metal oxide nanocomposites with polymers are fabricated by many techniques, including melt mixing, solution mixing, and in situ polymerization methods. Melt mixing, including extruder mixing, is industrially more feasible, as no additional chemicals are involved. However, solution mixing offers a high rate of mixing, as both metal oxide nanoparticles and polymers are mixed as dispersed in specific solvents. This is done by mechanical stirring, magnetic stirring, and soon, but the fibrous form for the polymers is always achieved through spinning techniques. This section particularly addresses polymeric metal oxide nanocomposite fibers prepared by electrospinning.

In the case of polymer−metal oxide nanocomposites, the metal oxide nanoparticles have multifunctional properties and processability in aqueous medium and provide good mechanical stability. In a recent research, the addition of SiO_2 to poly(vinylidene fluoride) largely increased its high temperature endurance, selectivity and diffusivity [45]. Additionally, ZnO nanoparticle inclusion in thermoplastic polyurethane led to an increase in its antibacterial activity [46]. Another important metal oxide that has been incorporated within a polymer membrane for the sake of water treatment is aluminum oxide (Al_2O_3). The reason behind its incorporation is its known enhancements, including an increase in hydrophilicity and suppressing fouling reactions in the polymeric membrane [47,48]. The preparation of Al_2O_3 can be done through in situ polymerization or through surface deposition. A study compared the different filler particles on polyethersulfone membranes and found that Al_2O_3 of higher concentration was found at the surface of the membrane in comparison to other fillers (ZrO_2 and TiO_2), owing to its dense nature. This allows membranes injected with Al_2O_3 to have enhanced properties in comparison with other membranes, which has been proven by the maximum influx value of 208.9 L m^{-2} h [49,50]. Stronger membranes utilize the catalytic, adsorption, and mechanical properties of the nanomaterials for water treatment. The nanocomposite membranes are involved in water purification by extending their applicability in photodegradation, membrane separation, adsorption, and sensing the pollutants with quality monitoring.

Though the polymer nanocomposites are generated by different methods, such as solution casting and melt mixing, the polymer−metal oxide nanocomposite fibers that are useful for water treatment are achieved mainly by the well-known method of electrospinning. With recent advances, the process has infinite control over the parameters that influence the fiber formation, which in turn allowed an extension of electrospun membrane applications to industrial scale. Currently, various commercial fields utilize electrospinning, owing to its less complex fabrication strategy and the possibility of modifying the conventional substrate surfaces. Electrospinning is a crucial technique in the fiber formation industry, as it allows the conversion of polymeric solution into macroscale and nanoscale fibers that can be easily applied in multiple fields, including medicine, filtration, transportation, and structure-based products. Electrospun fibers possess desirable properties such as porosity, tunable pore size and surface properties, and the ability to retain the electrostatic charge.

However, the electrospinning process depends on different parameters, such as molecular weight distribution, structure, viscosity, conductivity, concentration, and surface tension of the polymer. The polymer concentration is directly proportional to the fiber thickness. The efficiency of the process also varies according to the size of the needle, change in voltage, and evaporation rate of the solution [51−54].

Electrospun fibers of polyaniline composite containing well-dispersed Ti and Ag nanoparticles were fabricated by Xu et al. and achieved outstanding adsorption performance and antimicrobial properties. Efficiency of methylene blue adsorption from polluted water was tested by varying the temperature, solution pH, contact time, dosage, and dye concentration. With pH variation from 3 to 8, the dye adsorption increased to 95%, which was attributed to the strong interactions between the electronegative polymeric hydroxyl groups and the electropositive dye molecules. While 25°C was identified as the best temperature for maximum dye adsorption, 97.8% removal efficiency was obtained at 5 mg L^{-1} dye concentration for an initial contact time of 20 min. The availability of surface adsorption sites and thus the dye adsorption efficiency was the greatest at 10 mg of the nanocomposite fiber dosage. In addition, the fiber offered better reusability, as the adsorbent can easily be separated from the dye solution [55]. Complete dye removal from the adsorbed fibers took only 20 min at 25°C, with a maximum retention capacity of 155.4 mg g^{-1}. Moreover, stable results were shown for up to six cycles, suggesting the potential applicability of the nanofibers in dye adsorption. Very recent studies by Song et al. [56] also illustrate the photocatalytic activity of electrospun fibers of flexible Ag@ZnO/TiO$_2$ membranes with the polymer polyvinyl pyrrolidone (PVP). Degradation efficiency was 91.6% towards the tetracycline hydrochloride within 1 h, and good antibacterial activity was also noticed against E. coli within the same time.scanning electron microscope (SEM) images given on the Fig. 8.4 demonstrates the three-dimensional network structure of the membranes. ZnO nanorods of 200−400 nm length were anchored on the fiber surface, and this firm attachment of secondary ZnO on the TiO$_2$ membranes ensures structural integrity. Owing to smaller particle size, the Ag particles are not seen in the SEM images. Ag@ZnO/TiO$_2$ exhibited the maximum photocatalytic activity with respective decomposition efficiencies of 82%, 77%, and 68% toward DC-H, OTC-H, and CIP after 60 min of simulated sunlight irradiation. Consistent degradation performance (91.6%−88.5%) was noticed for the sample for the five consecutive cycles as well.

Solar light−driven disinfection ability toward the *E. coli* bacteria for the various fibers are represented in Fig. 8.5. While no photolysis of the bacteria was observed for the light control (after exposure), respective values of 1.2 and 3.0 Log of colony forming unit (CFU) were observed for TiO$_2$ and ZnO/TiO$_2$ membranes after 60 min of light exposure. For the Ag@ZnO/TiO$_2$ membrane, 6.5 Log of CFU reduction was achieved due to the generation of numerous reactive oxygen species.

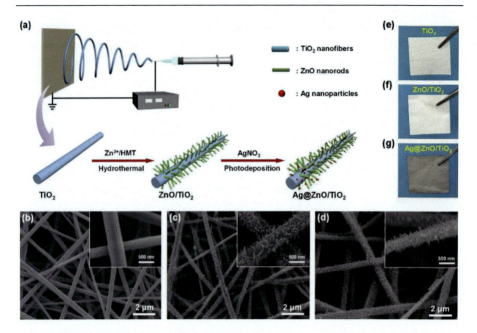

Figure 8.4 (A) schematic illustration of the preparation of soft Ag@ZnO/TiO$_2$ nanofibrous membranes. scanning electron microscope images of (B) TiO$_2$, (C) ZnO/TiO$_2$, and (D) Ag@ZnO/TiO$_2$. Optical photographs of (E) TiO$_2$, (F) ZnO/TiO$_2$, and (G) Ag@ZnO/TiO$_2$ membranes.
Source: Reprinted from J. Song, G. Sun, J. Yu, Y. Si, G. Ding, Construction of ternary Ag@ZnO/TiO2 fibrous membranes with hierarchical nanostructures and mechanical flexibility for water purification. Ceram. Int. 2020;46(1):468–475, with permission.

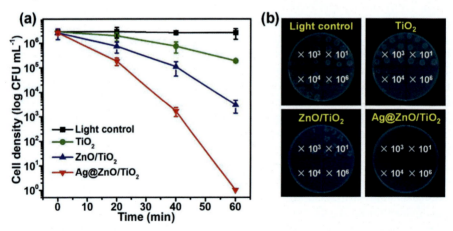

Figure 8.5 (A) Photocatalytic antibacterial efficiency against *E. coli* over TiO$_2$, ZnO/TiO$_2$, and Ag@ZnO/TiO$_2$ under simulated solar light illumination. (B) Photographs of *E. coli* colonies treated by various samples.
Source: Reprinted from J. Song, G. Sun, J. Yu, Y. Si, B. Ding, Construction of ternary Ag@ZnO/TiO2 fibrous membranes with hierarchical nanostructures and mechanical flexibility for water purification. Ceram. Int. 2020;46(1):468–475, with permission.

Modification of ZnO nanofibers by coupling with various metal oxides, sulfides, and metals improves the photocatalytic performance and thus helps in water purification. Both the temperature of catalyst processing and the dopant zinc molar ratio are crucial in determining the high photocatalytic activity of the nanofibers [57]. The authors have also compared different dyes, such as solvent dyes, vat dyes, mordants, and acid/direct dyes, and found that the azo dyes are the most commonly present pollutant dyes in wastewater streams, and nanofibers should be targeted in such a way as to degrade their structure. In fact, ZnO doping with other elements reduces the possible electron hole recombination process, thus causing the lifetime of charge carriers to increase. Huang's research group also generated nanofibers of α-Fe_2O_3 by combining electrospinning using PVP as a complexing reagent. This simple, safe and template-free method obtained material exhibited high photo catalytic efficiency to the rhodamine dye. The high stability and non-secondary pollution to the environment achieved for this nanofiber can be a considerable option for its practical application in water filtration and purification [58]. When irradiated in visible light for 90 min, the degradation efficiency reached 97.1%, 84.7%, and 55.1% for porous hollow spheres, nanofibers, and commercial α-Fe_2O_3, respectively, attributed to the variation in the surface area of the structures and thus the surface active sites.

Inorganic nanoparticles such as ZnO and CuO have good antibacterial activity, and these ions are necessary for human health [59]. CuO microparticle- and nanoparticle-modified polyurethane nanofibers were prepared by the nanospider technique, which is a modified version of electrospinning for antibacterial and water filtration [60]. The study found similar behavior for antibacterial filtration, whereas microparticles presented superior properties in comparison to nanoparticles in water filtration tests. The study concluded that microparticles (700 nm to 1 μm) of CuO were better additives for antibacterial modification of polyurethane nanofibers for filtration application than nanoparticles ($\approx$ 50 nm). The key limitation for this study was the least difference obtained between the nanoparticle and microparticle activity. This can be because of the high agglomeration of the nanoparticles, which caused deterioration in the functional properties. In another study, Shalaby et al. demonstrated the antibacterial activity of CuO, ZnO, and Ag hybrid nanoparticles and their electrospun composite membrane (170−250 nm average fiber diameter) based on polyacrylonitrile. The inhibition of bacterial growth by nanofibers loaded with ZnO, CuO, and Ag nanoparticles suggests that it can be used as effective antibacterial membranes for water filtration and disinfection along with various biomedical applications [59]. The bacterial destruction can happen by various mechanisms: (1) the transition metal oxides of ZnO and CuO on the electrospun fiber attach to the bacterial cell and increases the membrane permeability, finally leading to the decomposition of cell wall, deoxyribonucleic acid (DNA),ribonucleic acid (RNA), and the other internal cell parts; (2) generation of H_2O_2 due to the catalytic activity of the metal oxides; (3) nerutralization of the surface potential and generation of eletcron hole pairs by the metal oxides. Depending on the different types of bacterial cell, the degradation mechanism also varies due to the thinner cell walls, complex permeability barriers and the penetration rate of charged reaction intermediates.

Zhu et al. reviewed a series of polymer based antibacterial membranes by emphasizng the limitations of the current lab-based membranes and the effective functionalization using carbon and metal particles to impart antibacterial activity [61]. Fig. 8.6 deals with the synthesis of silver nanoparticles with incorporated patent ductus arteriosus (PDA) coating on the surface of polyamide reverse osmosis membrane, where PDA act as the reducing agent for the silver ions. The particles were uniformly dispersed on the membrane surface, and the ridge-and-valley structure is clear from the figure. Salt rejection rate was higher for the modified membranes, as the silver nanoparticles preferentially occupied the defect sites. The PDA coatings had a stronger influence for the flux decline, whereas the silver nanoparticles influenced the reverse salt flux. On the basis of these studies, the membrane properties are observed to be tunable by changing the thickness of PDA layer and the silver nanoparticles' growth. Zhu et al. were also successful in summarizing the significant challenges existing in the development of antibacterial membranes such

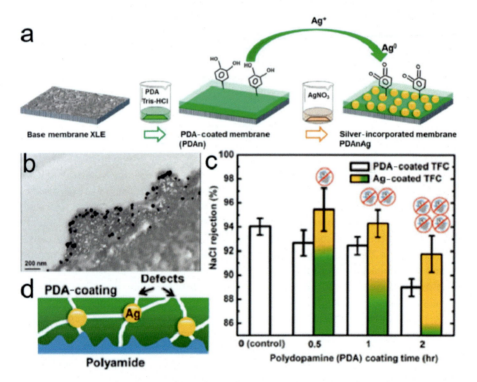

Figure 8.6 (A) Schematic diagram of in situ formation of Ag nanoparticles on a thin-film composite membrane. (B) Silver incorporated membrane patent ductus arteriosus @Ag. Scale bar is 200 nm. (C) Salt rejection of Ag-modified membranes. (D) Illustration of the formation of Ag nanoparticles at defective sites within the patent ductus arterious matrices.
Source: Reprinted from Z. Yang, Y. Wu, J. Wang, B. Cao, C.Y. Tang, In situ reduction of silver by polydopamine: a novel antimicrobial modification of a thin-film composite polyamide membrane, Environ. Sci. Technol. 2016;50:9543–9550, with permission.

as lab-scale manufacturing, use of static antibacterial assay, membrane filtration performance, and antifouling for long-term application. However, smoother, thinner, and uniform fiber coatings with modifications on the surface are being developed on a commercial scale using many modified electrospinning techniques.

Other than the electrospinning, centrifugal jet spinning is a developing area for the generation of nanofibers. This method has been developed for the fabrication of microfibers or nanofibers in a high-efficiency, high-throughput, economically feasible fashion [62,63,64]. In principle, the thinning of the solution filament into nanofibers using centrifugal jet spinning is achieved through controlled manipulation of centrifugal force, viscoelasticity, and mass transfer characteristics of the spinning solutions. Elasticity and evaporation rate of spinning solution play a critical role in determining the eventual diameter of as-prepared nanofibers. Various polymer—solvent combinations are reported in the literature, such as PVP—water, PVP—ethanol, PVP—dichloromethane (PVP—DCM), and PLLA—DCM, as spinning solutions for preparing nanofibers through centrifugal jet spinning. Among the various benefits of the centrifugal jet spinning technique, its exceptional throughput has an edge over conventional electrospinning, as its productivity is predicted to be 500 times higher, which was illustrated in a recent research reporting the use of centrifugal jet spinning for the rapid preparation of customizable multilevel-structured silica microfibers and nanofibers [63].

It was demonstrated that by utilizing thermal annealing and tuning the amount of nonsolvent in the PVP—ethanol spinning solution, continuous silica nanofiber assembly with different internal structures and cross sections of porous, hollow and solid could be easily prepared [63]. However, depending on the nature of solution and spinning parameters, various structural modifications are observed for the fibers. Fig. 8.7 shows the different fiber morphologies obtained from using various combinations of solvents and nonsolvents. As the solvent evaporates from the boundary, the solute concentration fluctuations develop from outside to inside. This generates a skin-core morphology with a TEOS—ethanol composition. For a less viscous solution, faster solvent evaporation results in a hollow core without solvent (PVP and silica) before the phase separation. But at high viscosity the thicker filaments cause slower solvent diffusion and result in nonsolvent induced spinoidal decomposition. This causes bicontinuous morphology in the core. With only ethanol as the solvent, uniform distribution of the PVP—silica fibrous structure is formed, wheras thermal annealing causes the formation of silica fiber nanotubes, silica fibers encapsulated with pores, and solid silica fibers. This points to the fact that the fiber diameter and the developed solvent gradient within the fiber during solvent evaporation influence the formation of multilayered structured fibers. Therefore pressure, humidity, nature of solvent, spin rate, and concentration of the polymer solution have to be effectively optimized for particular fiber morphologies.

The same group of researchers has identified the influence of spinning stages, fluid viscoelasticity, solvent mass transfer, and centrifugal forces on the fiber morphology [64]. A capillary number of the solution below 1 caused the formation of beads, whereas a value above two generated uniform fibers. Fiber thinning happened when the centrifugal forces were enhanced, and a series of relaxation times

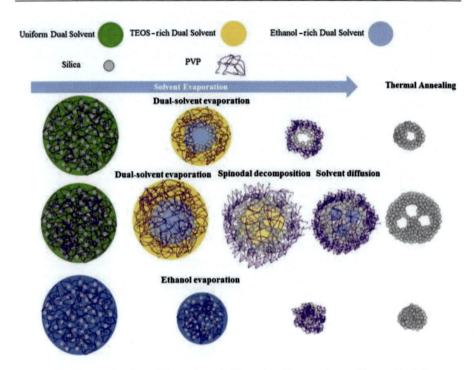

Figure 8.7 The mechanism of formation of silica microfibers and nanofibers with different morphologies within their cross sections.
Source: Reprinted from L. Ren, R. Ozisik, S.P. Kotha, Rapid and efficient fabrication of multilevel structured silica micro-/nanofibers by centrifugal jet spinning. J. Colloid Interface Sci. 2014;425:136−142, with permission.

and solvent evaporation rates were investigated. It is anticipated that the fiber structures can regulate the water purification application by acting as reverse osmosis membranes or photocatalytic antibacterial membranes.

8.4 Conclusions

Many metal oxides are used in fabricating fibers as such or by embedding with specific polymers so that environmental applications are targeted. This chapter dealt with metal oxide fibers based on ZnO, TiO$_2$, Ag$_2$O, and soon and their nanocomposite in combination with different polymers. Fibers developed by spinning methods were mainly discussed, and the close association of fiber morphology with the significant properties of the composites was explored. It is concluded that optimization of spinning parameters and other processing conditions are important in determining water flux, separation efficiency, and antibacterial activity. Depending upon the nanoparticles that are embedded, photocatalytic, and antibacterial properties are emerged as additional characteristics for the fibers. Finally, the study helps to determine the metal oxide and polymer metal oxide nanofibers that can be used for environmental applications.

Acknowledgment

This publication was made possible by NPRP grant 10-0127-170269 from the Qatar National Research Fund (a member of the Qatar Foundation). The statements made herein are solely the responsibility of the authors.

References

[1] R.P. Schwarzenbach, et al., Global water pollution and human health, Annu. Rev. Environ. Resour. 35 (2010;) 109−136.

[2] C.A. Stanford, M. Khraisheh, F. Al Momani, A.B. Albadarin, G.M. Walker, M.A. Al Ghouti, Use of nanoadvanced activated carbon, alumina and ferric adsorbents for humics removal from water: Isotherm study, Emergent Mater. (2020) 1−6.

[3] N. Panwar, S. Kaushik, S. Kothari, Role of renewable energy sources in environmental protection: a review, Renew. Sustain. Energy Rev. 15 (3) (2011) 1513−1524.

[4] V. Thavasi, G. Singh, S. Ramakrishna, Electrospun nanofibers in energy and environmental applications, Energy Environ. Sci. 1 (2) (2008) 205−221.

[5] B. Moss, Water pollution by agriculture, Philos. Trans. R. Soc. B: Biol. Sci. 363 (1491) (2008) 659−666.

[6] J. Theron, J. Walker, T. Cloete, Nanotechnology and water treatment: applications and emerging opportunities, Crit. Rev. microbiology 34 (1) (2008) 43−69.

[7] M. Selvaraj, A. Hai, F. Banat, M.A. Haija, Application and prospects of carbon nanostructured materials in water treatment: a review, J. Water Process. Eng. 33 (2020) 100996.

[8] N. Savage, M.S. Diallo, Nanomaterials and water purification: opportunities and challenges, J. Nanopart. Res. 7 (4−5) (2005) 331−342.

[9] J. Gubicza, Defect Structure and Properties of Nanomaterials, Woodhead Publishing, 2017.

[10] Y. Wu, Y. Zhang, J. Zhou, D. Gu, Recent progress on functional mesoporous materials as catalysts in organic synthesis, Emergent Mater (2020) 1−20.

[11] C. Ursino, et al., Progress of nanocomposite membranes for water treatment, Membranes 8 (2) (2018) 18.

[12] A. Hezam, K. Namratha, D. Ponnamma, Q.A. Drmosh, A.M. Saeed, K.K. Sadasivuni, et al., Sunlight-driven combustion synthesis of defective metal oxide nanostructures with enhanced photocatalytic activity, ACS Omega 4 (24) (2019) 20595−20605.

[13] H. Parangusan, D. Ponnamma, M.A. Al-Maadeed, A. Marimuthu, Nanoflower-like yttrium-doped ZnO photocatalyst for the degradation of methylene blue dye, Photochem. Photobiol. 94 (2) (2018) 237−246.

[14] H. Parangusan, D. Ponnamma, M.A. Al-Maadeed, Effect of cerium doping on the optical and photocatalytic properties of ZnO nanoflowers, Bull. Mater. Sci. 42 (4) (2019) 179.

[15] A. Heidari, Investigation of medical, medicinal, clinical and pharmaceutical applications of estradiol, mestranol (Norlutin), norethindrone (NET), norethisterone Acetate

(NETA), norethisterone Enanthate (NETE) and testosterone nanoparticles as biological imaging, cell labeling, anti−microbial agents and anti−cancer nano drugs in nanomedicines based drug delivery systems for anti−cancer targeting and treatment, Parana. J. Sci. Educ. (PJSE) 3 (4) (2017) 10−19.

[16] J. Sawai, K. Himizu, O. Yamamoto, Kinetics of bacterial death by heated dolomite powder slurry, Soil. Biol. Biochem. 37 (8) (2005) 1484−1489.

[17] I. Ani, et al., Photocatalytic degradation of pollutants in petroleum refinery wastewater by TiO_2-and ZnO-based photocatalysts: recent development, J. Clean. Prod. 205 (2018) 930−954.

[18] M.S.S.A. Saraswathi, A. Nagendran, D. Rana, Tailored polymer nanocomposite membranes based on carbon, metal oxide and silicon nanomaterials: a review, J. Mater. Chem. A 7 (15) (2019) 8723−8745.

[19] A. Fujishima, K. Honda, TiO_2 photoelectrochemistry and photocatalysis, Nature 238 (5358) (1972) 37−38.

[20] K. Hashimoto, H. Irie, A. Fujishima, TiO_2 photocatalysis: a historical overview and future prospects, Japanese J. Appl. Phys. 44 (12R) (2005) 8269.

[21] A. Mills, S. Le Hunte, An overview of semiconductor photocatalysis, J. Photochem Photobiol A: Chem. 108 (1) (1997) 1−35.

[22] K. Mondal, Recent advances in the synthesis of metal oxide nanofibers and their environmental remediation applications, Inventions 2 (2) (2017) 9.

[23] D. Chen, et al., Photocatalytic degradation of organic pollutants using TiO_2-based photocatalysts: a review, J. Clean. Prod. (2020) 121725.

[24] A.S. Polo, et al., Pt−Ru−TiO_2 photoelectrocatalysts for methanol oxidation, J. Power Sources 196 (2) (2011) 872−876.

[25] A. Khan, et al., Visible-light induced simultaneous oxidation of methyl orange and reduction of Cr (VI) with Fe (III)-Grafted $K_2Ti_6O_{13}$ photocatalyst, ChemistrySelect 3 (27) (2018) 7906−7912.

[26] M.B. Tahir, et al., Semiconductor nanomaterials for the detoxification of dyes in real wastewater under visible-light photocatalysis, Int. J. Environ. Anal. Chem. (2019) 1−15.

[27] Y. Du, et al., Formation and control of disinfection byproducts and toxicity during reclaimed water chlorination: a review, J. Environ. Sci. 58 (2017) 51−63.

[28] X. Zhou, et al., An ignored and potential source of taste and odor (T&O) issues—biofilms in drinking water distribution system (DWDS), Appl. Microbiol Biotechnol. 101 (9) (2017) 3537−3550.

[29] M. Samiei, et al., Nanoparticles for antimicrobial purposes in endodontics: a systematic review of in vitro studies, Mater. Sci. Eng: C. 58 (2016) 1269−1278.

[30] S. Wang, et al., Antibacterial activity and mechanism of Ag/ZnO nanocomposite against anaerobic oral pathogen *Streptococcus mutans*, J. Mater. Sci: Mater. Med. 28 (1) (2017) 23.

[31] C. Sánchez, et al., Nanocystalline ZnO films prepared via polymeric precursor method (Pechini), Phys. B: Condens. Matter 405 (17) (2010) 3679−3684.

[32] J. Xiaoyuan, et al., Studies of pore structure, temperature-programmed reduction performance, and micro-structure of CuO/CeO_2 catalysts, Appl. Surf. Sci. 173 (3−4) (2001) 208−220.

[33] D. Das, et al., Synthesis and evaluation of antioxidant and antibacterial behavior of CuO nanoparticles, Colloids Surf. B: Biointerfaces 101 (2013) 430−433.

[34] S. Makhluf, et al., Microwave-assisted synthesis of nanocrystalline MgO and its use as a bacteriocide, Adv. Funct. Mater. 15 (10) (2005) 1708−1715.

[35] J. Unrine, P. Bertsch, S. Hunyadi, Bioavailability, trophic transfer, and toxicity of manufactured metal and metal oxide nanoparticles in terrestrial environments, Nanosci. Nanotechnol: Environ. Health Impacts (2008) 345−366.

[36] D. Ponnamma, K.K. Sadasivuni, C. Wan, S. Thomas, M.A. AlMa'adeed (Eds.), Flexible and Stretchable Electronic Composites, Springer, 2015.

[37] D. Ponnamma, S. Thomas, Non-linear viscoelasticity of rubber composites and nanocomposites, InInfluence of Filler Geometry and Size in Different Length Scales, Springer, Switzerland, 2014.

[38] D. Ponnamma, M.A. Al-Maadeed, 3D architectures of titania nanotubes and graphene with efficient nanosynergy for supercapacitors, Mater. Des. 117 (2017) 203−212.

[39] S. Kango, et al., Surface modification of inorganic nanoparticles for development of organic−inorganic nanocomposites—a review, Prog. Polym. Sci. 38 (8) (2013) 1232−1261.

[40] S.N. Monteiro, et al., Natural-fiber polymer-matrix composites: cheaper, tougher, and environmentally friendly, JOM 61 (1) (2009) 17−22.

[41] S. Gubin, G.Y. Yurkov, I. Kosobudsky, Nanomaterials based on metal-containing nanoparticles in polyethylene and other carbon-chain polymers, Int. J. Mater. Product. Technol. 23 (1−2) (2005) 2−25.

[42] J. Jordan, et al., Experimental trends in polymer nanocomposites—a review, Mater. Sci. engineering: A 393 (1−2) (2005) 1−11.

[43] J. Jiang, G. Oberdörster, P. Biswas, Characterization of size, surface charge, and agglomeration state of nanoparticle dispersions for toxicological studies, J. Nanopart. Res. 11 (1) (2009) 77−89.

[44] R.F. Domingos, et al., Agglomeration and dissolution of zinc oxide nanoparticles: role of pH, ionic strength and fulvic acid, Environ. Chem. 10 (4) (2013) 306−312.

[45] S. Wang, et al., Preparation of a durable superhydrophobic membrane by electrospinning poly (vinylidene fluoride)(PVDF) mixed with epoxy−siloxane modified SiO_2 nanoparticles: a possible route to superhydrophobic surfaces with low water sliding angle and high water contact angle, J. colloid interface Sci. 359 (2) (2011) 380−388.

[46] P. Wagener, et al., Photoluminescent zinc oxide polymer nanocomposites fabricated using picosecond laser ablation in an organic solvent, Appl. Surf. Sci. 257 (16) (2011) 7231−7237.

[47] X.-M. Wang, X.-Y. Li, K. Shih, In situ embedment and growth of anhydrous and hydrated aluminum oxide particles on polyvinylidene fluoride (PVDF) membranes, J. Membr. Sci. 368 (1−2) (2011) 134−143.

[48] F. Hizal, et al., Nanoengineered superhydrophobic surfaces of aluminum with extremely low bacterial adhesivity, ACS Appl. Mater. Interfaces 9 (13) (2017) 12118−12129.

[49] T.A. Saleh, V.K. Gupta, Synthesis and characterization of alumina nano-particles polyamide membrane with enhanced flux rejection performance, Sep. Purif. Technol. 89 (2012) 245−251.

[50] Y.M. Mojtahedi, M.R. Mehrnia, M. Homayoonfal, Fabrication of Al_2O_3/PSf nanocomposite membranes: efficiency comparison of coating and blending methods in modification of filtration performance, Desalination Water Treat. 51 (34−36) (2013) 6736−6742.

[51] D. Ponnamma, K.K. Sadasivuni, M.A. Al-Maadeed, S. Thomas, Developing polyaniline filled isoprene composite fibers by electrospinning: Effect of filler concentration on the morphology and glass transition, Polym. Sci, Ser. A 61 (2) (2019) 194−202.

[52] A.A. Issa, M.A. Al-Maadeed, A.S. Luyt, D. Ponnamma, M.K. Hassan, Physicomechanical, dielectric, and piezoelectric properties of PVDF electrospun mats containing silver nanoparticles. C—Journal of Carbon, Research. 3 (4) (2017) 30.

[53] D. Ponnamma, H. Parangusan, A. Tanvir, M.A. AlMa'adeed, Smart and robust electrospun fabrics of piezoelectric polymer nanocomposite for self-powering electronic textiles, Mater. Des. 184 (2019) 108176.

[54] Y. Elgawady, D. Ponnamma, S. Adham, M. Al-Maas, A. Ammar, K. Alamgir, et al., Mesoporous silica filled smart super oleophilic fibers of triblock copolymer nanocomposites for oil absorption applications, Emergent Mater. 3 (3) (2020) 279−290.

[55] Z. Xu, et al., Development of a novel mixed titanium, silver oxide polyacrylonitrile nanofiber as a superior adsorbent and its application for MB removal in wastewater treatment, J. Braz. Chem. Soc. (2017).

[56] J. Song, G. Sun, J. Yu, Y. Si, B. Ding, Construction of ternary Ag@ZnO/TiO$_2$ fibrous membranes with hierarchical nanostructures and mechanical flexibility for water purification, Ceram. Int. 46 (1) (2020) 468−475.

[57] G. Panthi, et al., Electrospun ZnO hybrid nanofibers for photodegradation of wastewater containing organic dyes: a review, J. Ind. Eng. Chem. 21 (2015) 26−35.

[58] J. Huang, et al., Selective fabrication of porous iron oxides hollow spheres and nanofibers by electrospinning for photocatalytic water purification, Solid. State Sci. 82 (2018) 24−28.

[59] T. Shalaby, et al., Electrospun nanofibers hybrid composites membranes for highly efficient antibacterial activity, Ecotoxicol. Environ. Saf. 162 (2018) 354−364.

[60] G. Ungur, J. Hrůza, Modified polyurethane nanofibers as antibacterial filters for air and water purification, RSC Adv. 7 (78) (2017) 49177−49187.

[61] J. Zhu, J. Hou, Y. Zhang, M. Tian, T. He, J. Liu, et al., Polymeric antimicrobial membranes enabled by nanomaterials for water treatment, J. Membr. Sci. 550 (2018) 173−197.

[62] Z. Yang, Y. Wu, J. Wang, B. Cao, C.Y. Tang, In situ reduction of silver by polydopamine: a novel antimicrobial modification of a thin-film composite polyamide membrane, Environ. Sci. Technol. 50 (2016) 9543−9550.

[63] L. Ren, R. Ozisik, S.P. Kotha, Rapid and efficient fabrication of multilevel structured silica micro-/nanofibers by centrifugal jet spinning, J. Colloid Interface Sci. 425 (2014) 136−142.

[64] L. Ren, et al., Highly efficient fabrication of polymer nanofiber assembly by centrifugal jet spinning: process and characterization, Macromolecules 48 (8) (2015) 2593−2602.

Metal oxide nanofiber for air remediation via filtration, catalysis, and photocatalysis

9

Chin-Shuo Kang, Edward A. Evans and George G. Chase
Department of Chemical, Biomolecular and Corrosion Engineering, The University of Akron, Akron, OH, United States

9.1 Introduction

Air pollution is of increasing concern, owing to its severe effects on human health. The World Health Organization (WHO) has released a report stating that around seven million people died from air pollution exposure in 2012 [1] which corresponds to more than 10% of total global deaths that year. Worse still, research released by the WHO [2] in 2016, indicated that both indoor and outdoor air pollution are highly associated with respiratory and cardiovascular disease including lung cancer deaths, chronic obstructive pulmonary disease (COPD) deaths, ischemic heart disease and stroke and respiratory infection deaths.

Air pollution refers to the contamination of indoor and/or outdoor air by solid particles and/or harmful gas. The contaminants derive from various sources such as fossil fuel production, emission of combustion engines, and/or manufacturer and agricultural incineration [3]. Air pollutants can be categorized into two groups [4]: particulate matter (PM) and gaseous pollutants. Following the guideline from the WHO [5], PM is the mixture of fine organic and inorganic materials that float in air, such as carbon dust [6]. It is classified into three levels: PM 10 (coarse, aerodynamic diameter $2.5-10 \mu m$), PM 2.5 (fine, aerodynamic diameter $0.1-2.5 \mu m$), and PM 0.1 (ultrafine, aerodynamic diameter $\leq 0.1 \mu m$) [7]. PM can enter the respiratory tract easily, causing an increase in carcinogenic potency. Gaseous pollutants are mostly associated with industrial activities and the growing number of automobiles. Since the efficiency of combustion is not 100%, some of the gasoline that powers the automobile engine is converted into undesired by-products. The exhaust of combustion includes noncombusted hydrocarbons (HCs), incompletely combusted carbon monoxide (CO), nitrogen oxides (NO_x), sulfur dioxide (SO_2) [8], methane (CH_4) [9], and ozone (O_3). The average concentration of CO is no more than 0.001% of the atmosphere, but it is one of the most common types of toxic pollutants around the world [10]. It is generated because of incomplete combustion of fuel. The level of CO is higher in rural areas in comparison to urban area since the gasoline engine idles more often in a rural environment [11]. It may bring up the concentration of CO up to $100 \, mg \, m^{-3}$, which is 10 times higher than the

---

Metal Oxide-Based Nanofibers and Their Applications. DOI: https://doi.org/10.1016/B978-0-12-820629-4.00010-2
© 2022 Elsevier Inc. All rights reserved.

standard suggested from the WHO [12]. It may cause fatal consequence to human if exposed to 1% of CO in atmosphere over 1 min [13]. (NO_x with $1 \leq x \leq 2$) are composed mainly of nitrogen dioxide (NO_2) and nitrogen monoxide (NO). NO_x are produced by the bacterial reaction as well as manufacturing and combustion [14]. Nitrogen oxide derivatives infiltrate into the respiratory system efficiently, exacerbating heart disease and emphysema [15]. In terms of the greenhouse effect, nitrogen oxide derivatives affect the environment 3000 times more than carbon dioxide does on a molecular basis [16]. SO_2 is another pollutant that is produced by the combustion of fossil fuel. When inhaled into the lungs, SO_2 is hydrated by body fluid in the respiratory tract, producing sulfurous acid. The sulfurous derivate compounds are acid and are easily absorbed by blood and body fluid causing respiratory tract disease and lung cancer [17,18]. CH_4 is generated by gas turbines, the livestock industry, and human activities. It is responsible for roughly 20% of the heating effect generated by greenhouse gases. The concentration of CH_4 has risen 150% since the 18th century. The increasing concentration of CH_4 contributes to climate change [19] which may ultimately affect agriculture and ribonucleic acid (RNA) metabolism [20,21]. O_3 is also included as an air pollutant. It is formed by various chemical reactions involving air and volatile organic compounds in the presence of ultraviolet (UV) light [22,23]. O_3 has already been shown to have a high impact on asthma and other respiratory symptoms, such as airway obstruction and air hyperresponsiveness [24].

All in all, air pollutants have already imposed adverse health effect on both human beings and the environment [25]. To remediate the issue of air pollution, scientists have developed techniques from different perspectives. For example, one approach is to control the potential generation of air pollutants, such as by reducing the usage of fossil-fuel, seeking alternative, and renewable energy resources [26–29], and improving energy efficiency. In this chapter, the focus will be on reducing the concentration of air pollutants by fiber-based filters made of metal oxide nanofibers (NFs).

9.2 Filtration for air pollutants

9.2.1 Filtration mechanism

Before introducing filters, media made of metal oxide NFs, and their application in air remediation, a brief background of the filtration mechanism and its characterization is needed. The whole filtration process [30] removes a dispersed phase (e.g., PM) from a continuous phase (e.g., air) by making both of them through a porous filter medium. The driving force for flow is a pressure gradient between the inlet and the outlet of the filter medium. In the case of air remediation the dispersed phase is either PM particles or toxic gases, and the continuous phase is air. The filtration can be identified as either surface filtration or depth filtration (Fig. 9.1) [31]. When the size of the dispersed phase is larger than the pores of the filter medium, the pollutants are blocked by the inlet surface of the filter and cannot penetrate the filter medium. Since the dispersed phase is

Metal oxide nanofiber for air remediation via filtration, catalysis, and photocatalysis

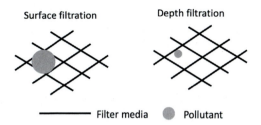

Figure 9.1 Schematic of surface filtration and depth filtration.

removed at the surface of the filter, this mechanism is called surface filtration. In depth filtration, by contrast, the pollutants are smaller than the pores of the filter and can flow into the depth of the filter. The continuous phase flows through the filter medium, while the dispersed phase is captured internally.

When contaminated air flows through the filter medium, the PM particles or toxic gases are immobilized. Immobilization allows for the removal of the contaminant by one or more mechanisms, as described later in the chapter. For capturing PM pollutants, there are four main conventional capture mechanisms of depth filtration: inertial impaction, direct interception, diffusion, and gravitational deposition [32] (Fig. 9.2). These four mechanisms are called single-fiber mechanisms. It is assumed that the filter medium is uniform such that the properties of a single fiber, including fiber diameter, surface properties, and strength, are representative of the rest of the fibers in the filter medium. The capture mechanism on a single fiber is dependent on the conditions, the dispersed phase, and the continuous phase.

The four capture mechanisms are as follows:

1. Gravitational force, which normally happens when the size of contaminant is between 20 and 50 μm. These contaminants have enough weight that they will naturally drop out from gas streamline and settle down on the fiber medium.

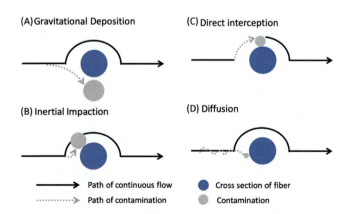

Figure 9.2 Single-fiber capture mechanisms. (A) Gravitational deposition. (B) Inertial impaction. (C) Direct interception. (D) Diffusion.

2. Inertial impaction, which occurs when the contaminant carried by gas flow encounters the fiber medium and the path of the contamination deviates from the gas streamline because it has sufficient inertia. The deviation reduces the distance between the contaminant and the fiber medium and increases the chance of the contaminant being captured. This is the main capture mechanism when the particle size is larger than 1 μm.
3. Direct interception most likely happens when the particle size is between 0.3 and 1 μm. The contaminant follows the gas streamline bending around the fiber medium. The contaminants within or equal to the distance of the particle radius will be captured.
4. Diffusion occurs when the size of the contaminant is smaller than 0.3 μm. It tends to follow Brownian motion because of collision with air molecules. The Brownian motion of contaminants may increase the chance of being captured by a fiber.

Conventionally, particles with sizes in range of 0.05−0.5 μm have weak inertia and interception, resulting in low filtration efficiency [33]. Along with the conventional capture mechanism, researches also investigate the electrostatic force for submicron particles removal. [34]. The electrostatic force enhances the attraction of the filter medium to submicron particles without modification of filter structure. Electrostatic enhancement can be obtained by particles precharging, filter charging and adapting electric field [35] in the filter medium [36].

9.2.2 Characterization of filter

The performance of the nanofibrous filter, regardless of the filter medium that is used, depends on the porosity, pore size, permeability, pressure drop, filtration efficiency, and quality factor. All of the factors are correlated with each other and can be determined by different measurements.

9.2.2.1 Porosity

The porosity of the filter medium, ε, is a ratio of volume of voids that can be occupied by fluid to the total volume of the filter medium and is determined by the weight and volume of a specimen as follows:

$$\varepsilon = \frac{V_v}{V_v + V_f} = 1 - \frac{V_s}{V_v + V_f} = 1 - \frac{\rho_B}{\rho_f} = 1 - \frac{\left(\frac{w}{Al}\right)}{\rho_f} \tag{9.1}$$

where V_v and V_f indicate the volume of void and volume of fiber in the filter medium, respectively; ρ_B and ρ_f are the bulk density of filter medium and the density of fiber; and w, A, and l represent the weight, area, and the thickness of filter, respectively. The higher the porosity, the more voids are available in the filter medium. The porosity of a nanofiber (NF)-based filter can reach higher than 90% [37]. Conventionally, the dimension of the filter medium is measured in micrometers. However, the fibrous filter is flexible and hard to tell the boundary. Furthermore, the thickness of nanofibrous filter is usually smaller than the accuracy of a micrometer. These issues lead to a significant error in computing the volume and the bulk density of the fibrous medium. An accurate porosity measurement of the filter medium relies on pycnometer measurements [38]. This is a nondestructive and chemically inert test. A pycnometer calculates the volume of fiber by the pressure difference in comparison to a known volume. The testing procedure of pycnometer is shown schematically in Fig. 9.3. First, two connected chambers, chamber 1 and chamber 2, with known volume, V_1 and V_2 are pressurized with atmosphere, P_1 and known pressure, P_2, and a valve is closed in between the two chambers. Pressure should show in relative pressure, that is, $P_1 = 0$. Then the valve is opened, and the pressure, P_3, is recorded. The equilibrium condition is expressed as follows:

$$P_2 V_2 = P_3 (V_1 + V_2) \tag{9.2}$$

Next, the filter sample is placed in chamber 1, and the procedure is repeated to obtain pressure P_4 in chamber 2 and equilibrium pressure P_5 (Fig. 9.4).

The equilibrium condition is

$$P_4 V_2 = P_5 ((V_1 - V_f) + V_2) \tag{9.3}$$

By solving Eqs. (9.2) and (9.3), the fiber sample volume V_f is obtained:

$$V_f = V_1 - P_3 V_1 (P_5 - P_4) / P_5 (P_3 - V_2)$$

9.2.2.2 Pore size and size distribution

The pore size has an impact on the direct interception capture mechanism. The pore size and size distribution are determined by the bubble point method. The pore diameter is measured by the required pressure difference to overcome the surface tension of liquid that saturates the pores [39]. The pore diameter can be determined by the following equation:

$$d_p = \frac{4\sigma \cos\theta}{\Delta P} \tag{9.4}$$

where d_p is the diameter of the pore, σ is the surface tension of the wetting liquid, θ is the liquid-filter contact angle, and ΔP is the pressure difference across the

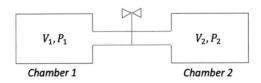

Figure 9.3 Schematic of pycnometer apparatus.

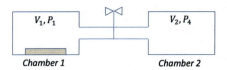

Figure 9.4 Schematic of pycnometer apparatus with testing sample in chamber.

filter. The pore size for nanofibrous filters ranges from tens of nanometers to several microns [40,41].

9.2.2.3 Surface area

Surface area affects the diffusion mechanism because the higher the surface area, the higher the chance of particles being captured [42]. The specific surface of an alumina NF can be as high as 600 m^2 g^{-1} [43,44]. The Brunauer, Emmett, and Teller (BET) method is a common method for measuring the internal surface area of a nanofibrous filter. It is based on the physical adsorption of gas molecules on the surface of the fibers and is described by the following equation:

$$\frac{1}{v\left[\left(\frac{P_0}{P}\right) - 1\right]} = \frac{c - 1}{v_m}\left(\frac{P}{P_0}\right) + \frac{1}{v_m C} \tag{9.5}$$

where v, v_m, P_0, P, and C represent the adsorbed gas quantity, monolayer adsorbed gas quantity, equilibrium pressure of adsorbate at testing temperature, saturation pressure of adsorbate at testing temperature, and BET constant, respectively.

9.2.2.4 Permeability

The permeability determines the flow rate of the carrier fluid through the filter medium. Theoretically, the permeability of a filter medium can be related to the pressure drop, ΔP, per unit thickness of filter, L, and the properties of the fluid as follows [45,46]:

$$\frac{\Delta P}{L} = \alpha\eta\mu + \beta\rho\mu^2 \tag{9.6}$$

where α, η, μ, β, and ρ are a viscous term coefficient, viscosity of the fluid, face velocity of the fluid, an inertia term, and density of the fluid, respectively. The Darcy's permeability is equal to the inverse of α. When the velocity of the incompressible fluid is slow, the value of β is close to zero. The equation is then reduced to the classic Darcy's law:

$$v = \frac{\kappa \Delta P}{\mu L} \tag{9.7}$$

The permeability is inversely proportional to the pressure drop with constant thickness and face velocity of carrier fluid.

9.2.2.5 Single-fiber efficiency

There are two important constants that correlate with single-fiber efficiency: the Kuwabara constant [47], Ku, and the Stokes number, St. Ku [48] describes the relationship between the volume fraction of the filter medium and the dimensionless drag force. St describes the motion of a particle suspension in fluid flow [49]:

$$Ku = \frac{-\ln(\alpha)}{2} - \frac{3}{4} + \alpha - \frac{\alpha^2}{4}$$

where α represent the volume fraction of the fiber in the fiber medium and

$$St = \frac{d_P^2 \rho_P v}{18 \mu d_f}$$

where $d_P, \rho_P, v, \mu,$ and d_f denote the particle diameter, particle density, fluid velocity away from the obstacle, gas dynamic viscosity, and fiber diameter, respectively. The single-fiber efficiency [32] depends upon inertial impaction (E_I), direct interception (E_R), and diffusion (E_D) as follows:

$$E_I \propto \frac{St}{2Ku^2}$$

$$E_R = \frac{(d_P/d_f)^2}{Ku}$$

$$E_D = \frac{2.7}{\left(\frac{d_f v}{D}\right)^{\frac{2}{3}}}$$

where D stands for the diffusion coefficient. As shown, the pressure drop and single-fiber efficiency for inertial impaction, direct interception, and diffusion are all inversely proportional to the fiber diameter. While pressure drop is proportional to filter efficiency in tuning the fiber diameter, the increment of filter efficiency is faster than the compensation of the pressure drop.

9.2.2.6 Quality factor

The performance of a filter is expressed by the penetration of dispersed phase (η) and the pressure drop (P) across the filter medium. It is a measurement of filtration efficiency per unit energy required. This is a general evaluation for a PM filter. A quality factor is then given as follows:

$$QF = -\frac{\ln(1-\eta)}{\Delta P}$$

where ΔP is the pressure drop and η is the penetration rate.

Penetration of dispersed phase (η) is defined as the concentration of particles that have been captured by the filter divided by inlet particles concentration:

$$\eta = \frac{(N_{in} - N_{out})}{N_{in}}\frac{Q(C_{in} - C_{out})}{QC_{in}} = 1 - \frac{C_{out}}{C_{in}}$$

where the subscripts in and out denote the inlet and outlet of the filter medium, respectively, and N, Q, and C represent the quantity of dispersed phase, the volumetric flow rate, and the concentration of the dispersed phase, respectively. The pressure drop is dependent on the structure of the filter medium. For the fiber-based filter, the fiber diameter, surface properties of fiber, packing density, filter thickness, and pore size matter.

9.2.3 Nanofibrous particulate matter filter

The design of a nanofibrous PM filter depends significantly on the PM particle size and corresponding mechanisms described earlier. Nanofibrous filters have several advantages over microfiber filters, including small pore size, low filter resistance, and high filtration efficiency [50]. The fiber in nano scale further enhances the performance of the filter. Comparing to filters made of fiber at macro scale, NF-based filters dramatically increase the internal surface area. The increased surface area not only improves the probability of capture and reaction but also enhances the surface-loading capacity if used as a catalyst carrier in high-temperature filtration. In addition, the pore size of a NF-based filter will be 100 times smaller than melt-blown (normally at micro scale) fiber filter, leading to reduction of airflow resistance up to 156 times that of macrofiber filter [51]. Xiong et al. [52] reinforced a cotton fiber PM filter by adding ultralong hydroxyapatite NFs. The hydroxyapatite NFs were 20 nm in diameter with over 200 µm in length. By modifying the ratio between the cotton and the hydroxyapatite NF, the filtration efficiency was increased to 95% for both PM 2.5 and PM 10. Furthermore, the pressure drop through the hybrid filter medium was lower than that of the commercial breathing mask. Lee fabricated perovskite-type oxide NFs in a web structure and compared them to a particle structure for soot oxidation (Fig. 9.5). Lee et al. [53] found that the active surface area for a web structure is twice as many particles per gram, since the particles are agglomerated and compacted to maintain the porous structure. A higher surface area leads not only to a greater chance of capturing soot particles but also to more reaction sites to oxidize the soot.

Another factor that makes NF-based filter media preferable is that the drag force is negligible when the fiber size is small enough. The movement of the air molecules is

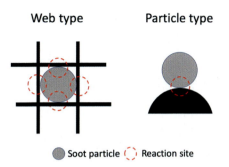

Figure 9.5 Schematic of web sturcture and particle sturcture soot oxidation.

strongly related to the size of the fiber and the flow field [54]. Conventionally, it is assumed that the continuous airflow around the fiber is in a nonslip condition in filtration theory. However, when the size of fiber is reduced to a certain degree, the nonslip assumption is no longer valid, and the slip flow model is adapted in filtration theory. The Knudsen number [55], Kn, describes the rarefaction of a flow. It relates the molecular mean free length, λ, to the overall flow field, d_f, and is defined as follows:

$$\mathrm{Kn} = \frac{2\lambda}{d_f}$$

When $0.1 > \mathrm{Kn} > 0.01$, slip flow is dominant. The mean free path for air molecules, λ, at standard temperature and pressure is 34 nm [56]. According to the definition of the Knudsen number, when the fiber diameter is smaller than 6.8 μm, slip flow is dominant. There is no doubt that the NF-based filter and partial micron fiber filter are applicable for the slip flow condition. Owing to the slip of airflow at the surface of fiber, the drag force decreases, resulting in lower pressure drop across the filter medium. Moreover, in the slip flow condition, the streamline of airflow is closer to the fiber surface compared to the nonslip surface. The slip flow condition increases the chance that particles will collide with the fiber, increasing the direct interception of particles. Li et al. [44] assessed a commercial filter for viral aerosol removal. The filter is made of alumina NFs with a diameter of around 2−4 nm, which leads to a surface area up to 600 m^2 g^{-1} and 92% porosity. In addition, the electropositive surface helped to retain electronegative microorganisms. The alumina NF filter demonstrates around 95% removal efficiency with the size of feed aerosol controlled at 30 nm and 2.5 inches H$_2$O pressure drop. Compared to a commercial high efficiency particulate air (HEPA) (Millipore, AP1504700) filter, while the alumina NF filter has slightly lower separation efficiency, the pressure drop is significantly reduced, owing to the slip surface condition. As a result, the quality factor for alumina NF filters is three times higher than that of HEPA filters.

Metal oxide NFs have the same advantages for filtration as other NFs but also have outstanding thermal stability. Since most PM and gaseous pollutants are generated by combustion, the gas flows can be hot. Generally, the operation temperature for a diesel exhaust filter is around 250°C−400°C. The glass transition temperature for most polymers is in this temperature range [57]. A filter made of polymer may not be able to sustain this kind of working condition. If a polymer-based filter is used for these conditions, the gas stream must be cooled before the filtration step. By contrast, filters made of metal oxide materials are thermally stable in this range. Wang et al. [58] fabricated highly flexible, high-tensile-strength, and self-standing alumina fibrous membrane as a PM filter. The filter is composed of alumina NF 230 nm in diameter. The filter shows above 99% of filtration efficiency for 300 nm particles with a 0.95-inch H_2O pressure drop. In addition, the filter remained thermally stable up to 900°C and is able to sustain a 2.98-MPa tensile strength.

9.2.4 Gas filter

While the filters for removing gases share some basics features with filters for removing PM, there are differences across the types of filter, such as target pollutants, characterization, and materials. A filter design for PM capturing may or may not be suitable for gaseous pollutant, since the nature of the pollutant is different. For gaseous filters, the gaseous pollutants are captured by physical adsorption and chemical adsorption. Physical adsorption [59] is responsible for capturing the gaseous pollutants on the internal surface of the filter medium by Van der Waal force and/or electrostatic force. Physical adsorption may adapt the conventional capture mechanisms by assuming that the PM particle size is extremely small [60]. By contrast, in chemical adsorption [61] the gaseous pollutants are bound to the surface of the filter by a chemical bond. For example, γ-aluminum oxide is able to adsorb NO_2:

$$NO_2 + Al - O(\text{lattice}) \rightarrow Al - NO_3$$

The evaluation of gaseous filters is also different from that of PM filters, especially for chemical adsorption-type filters. The efficiency characterization of chemical adsorption-type filters is determined by the reduction of concentrations of pollutant molecules downstream or the mass change of the filter medium. The pollutant is injected with a stream of carrier air and flows through the filter continuously. A measurement is made when the downstream concentration of the pollutant or the mass of the filter reaches steady state. In contrast to PM filters that physically hinder PM particles from penetrating the filter medium, gaseous filters chemically degrade or convert the toxic gaseous molecules into nontoxic molecules. Metal oxide NFs can serve three major functions as filter media in gaseous filtration: catalyst, supporting matrix for noble metal catalyst, and photocatalyst. One should be aware that catalysis and photocatalysis of metal oxide NFs happens only after gaseous pollutants are adsorbed by the surface via the mechanisms described previously.

9.2.4.1 Metal oxide nanofibers as catalyst support matrix and catalyst

Exhaust gas pollutants are composed mostly of NO_x, HC, and CO. To deal with exhaust gas pollutants, a catalytic reaction via oxidation and reduction of gaseous pollutants has been investigated. Metal oxide NFs can serve as a catalyst, owing to their high surface area, meaning more reaction sites and quick reversible redox properties. Examples include Cu^{2+}/Cu^+, Ce^{+4}/Ce^{+3}, and Mn^{+4}/Mn^{+3} [62,63]. Xu et al. [64] fabricated $Cu_{0.1}Ce_{0.9}O_{2-x}$ NFs. The ceramic oxide NF was fabricated by electrospinning [65] with sol−gel chemistry [66]; this synthesis approach is a common method of fabricating ceramic fibers at submicron scale. With 12,000 $g_{cat}^{-1}h^{-1}$ space velocity, the concentration of CO is reduced to less than 10% of inlet CO concentration with over 90% of O_2 to CO selectivity at 120°C. Du et al. [67] followed the same approach but made $In_2O_3/Pd/Co_3O_4$ NFs in core-shell morphology with a shell thickness of about 20−30 nm. They successfully reduced the reaction temperature to 56°C with 90% of O_2 reduction under 10,000 $g_{cat}^{-1}h^{-1}$ space velocity. For NO_x reduction, Marani et al. [68] demonstrated NO_x-selective catalytic reduction on $V_2O_5/WO_3/TiO_2$ NFs with around 90% of NO_x conversion at 350°C. Marani also tried various recipes of $V_2O_5/WO_3/TiO_2$ NFs. The results indicated that above 350°C the activation energy is close to 0. The reaction rate may be driven by transport phenomena of NO. Huang et al. [69] electrospun $La_{0.75}Sr_{0.25}MnO_3$ NF for both CO and CH_4 oxidation. $La_{0.75}Sr_{0.25}$ MnO_3 NF was proved to have superior catalytic activity in comparison to nanoparticles; this was attributed to more Mn^{4+} sites, more oxygen vacancies, and better thermal stability.

Metal oxides can also be used as catalyst support, but not all of metal oxides are suitable as catalyst support. At first, researchers focused on the oxidation of HC and CO using copper and nickel catalysts [70]. However, copper and nickel are easily poisoned by the lead additive and sulfuric compounds in fuel. Other metals, such as ruthenium, iridium, and osmium, were also considered, but these metals become volatile in forming metal oxides during catalytic reaction with exhaust gas. The mass loss of the catalyst is greater than was expected. Researchers then moved on to platinum (pt), palladium (Pd), and rhodium (Rh). These were found to be suitable exhaust catalysts with several advantages [71]. First, Pt, Pd, and Rh do not have volatile oxides that will cost the loss of catalyst. Second, they are highly reactive and significantly reduce the residence time required to convert exhaust pollutants. Third, they remain chemically inert to sulfur compounds and supporting materials such as alumina oxide even with elevated working temperature. Fourth, they are able to reduce NO_x by CO, leading to the elimination of both gaseous pollutants simultaneously. A three-way catalyst (TWC) is a catalytic device made of combinations of Pt, Pd, and rhodium Rh for exhaust pollutants conversion and is believed to be the most efficient and cost-effective approach [72]. The TWC earns its name by eliminating three pollutants simultaneously via oxidation of CO and HC and reduction of NO_x. Typically usage of catalyst are 1−2 g of Pt, 0.5−1 g of Pd and 0.1−0.2 of Rh per TWC unit [73]. To make the TWC less sensitive to air/fuel (A/F) ratio fluctuation and improve the performance of catalyst, oxygen storage components (OSC) were incorporated into TWC in 1981 [74]. An OSC is a component that can be either oxidized or reduced reversibly. An OSC absorbs oxygen when the A/F ratio is in a lean condition and then releases oxygen when it is in a rich condition. It helps to control the exhaust gas around the best stoichiometric ratio (A/F ratio values equal to 14.6) [75]. The performance of a TWC is enhanced by introducing an OSC as a buffer into the catalytic system. Cerium (Ce) is the most popular choice as OSC, owing to the manipulative Ce^{+3} and Ce^{+4} oxidation states depending on the accessibility of oxygen [76]. Ce dominates the oxygen storage capacity at high temperature compared to catalysts used in TWC, such as Pt and Rh [77]. The reactions involved in OSC are expressed as follows:

$$CeO_2 + CO \rightarrow Ce_2O_3$$

$$Ce_2O_3 + 0.5O_2 \rightarrow 2CeO_2$$

The catalytic conversion and the stability of the catalyst is highly affected by the support and the concentration of catalyst. The catalytic activity of Rh on supporting materials was reported to be $Rh/TiO_2 > Rh/Al_2O_3 > Rh/SiO_2$ at 400°C [78]. With the choice of supporting materials including ZrO_2, Al_2O_3, TiO_2, CeO_2, and SiO_2, Souza et al. [79] reported similar results in which a higher total CO oxidation conversion rate was achieved for zirconia-supported Pt catalyst in comparison to Al_2O_3, SiO_2, ZrO_2, and CeO_2 at 150°C. Zirconia-supported Pt catalyst presented 100% total oxidation of CO, while others supported are all below 20%. This may be attributed to the capacity of CO adsorption. The ratio of active site to CO adsorption (CO/Pt) for Pt/ZrO_2, Pt/Al_2O_3, and Pt/CeO_2 are 1.2, 0.7, and 0.2, respectively. Wang et al. [80] tried to improve the CO oxidation performance of alumina-supported Pd catalyst. It was found that the CO oxidation activity performance is better when an Al_2O_3 NF substrate decorated with a trace amount of SnO_2 is used rather than either pure Al_2O_3 or SnO_2 substrate. With the decoration of SnO_2, both surface area and active oxygen species are enhanced, leading to better Pd dispersion and stronger metal-support interaction [81].

Fibrous membrane has been proved to be an efficient catalytic support. It allows the reagents to flow continuously through the reactor, resulting in a reduction in reaction time and separation of the product [82]. Metal oxide NFs are highly recommended as a catalyst and support for high-temperature application for the same reasons as mentioned previously, such as high surface area and thermal stability. Swaminathan et al. [83] reported fabrication and catalytic efficiency evolution of TWC composed of noble metal oxide NFs by electrospinning [65] with sol−gel chemistry. In the report, Pt and Rh at a fixed ratio of 5 to 1 was incorporated on the support of micron alumina fibers. Small amounts of Pd were also applied for cost reduction. Pt, Rh, and Pd are major components of TWC [84]. The scanning electron microscope (SEM) images revealed that the alumina fiber size and noble metal oxide particle size are around 77 and 3 nm, respectively. Catalytic filter with 0.01, 0.05, and 0.1 g of $PdO/PtO/Rh_2O_3/Al_2O_3$ NFs were analyzed for degradation of gaseous pollutants, including CO, NO, CO_2, N_2O, N_2, and O_2, with working temperatures from room temperature to 400°C. The results showed that the decomposition temperature for NO decreased from 400°C to 250°C with increased addition level of $PdO/PtO/Rh_2O_3/Al_2O_3$ NFs from 0.01 to 0.1 g. Shahreen et al. [85] produced titania NF−supported Pd catalyst using electrospinning with sol−gel chemistry. To test the NO decomposition ability, a disk-shaped nonwoven porous filter medium was made of the titania NF−supported Pd catalyst incorporating titania and alumina microfibers. Fibers at micro scale served as mechanical support structures. It was found that the reaction temperature of NO decomposition was lower with higher loading of the Pd catalyst. The decomposition temperature was reduced from 450°C to 300°C with increased loading of the Pd catalyst from 0.5% to 2.7%. Wang et al. [86] described an improvement in both CO reduction and mechanical properties for nanofibrous Pd/TiO2 catalyst with a CeO_2 promoter. The molar ratio of Ce and Ti was 5−100. By adding CeO_2, the reductions in both CeO_2 and PdO were enhanced by the stronger metal-support interaction in the CeO_2-TiO_2 composite, which resulted in better catalytic efficiency of CO oxidation at low temperature. It achieved a 100% CO conversion rate at 200°C, and catalytic stability was observed up to 30 h. For the mechanical properties the tensile strength of the CeO_2-TiO_2 composite was measured up to 1.38 MPa, which is three times higher than that of pure TiO_2. The thermal stability was also improved. It showed that the tensile strength of CeO_2-TiO_2 composite retained 1.22 MPa after heat treatment at 400°C for 20 h

9.2.4.2 Metal oxide nanofibrous photocatalysts

Photocatalytic oxidation (PCO) by metal oxide is one of the promising approaches to eliminating gaseous pollutants. Most of the PCO purification research and pollutant removal mechanisms have been developed on the foundation of surface reactions on TiO_2 [87−89]. TiO_2 in fibrous morphology has been one of the most popular metal oxide photocatalyst for air purification, owing to its nontoxicity, low cost, chemical stability, and outstanding physicochemical properties [87,90]. TiO_2 has two common crystal forms, that is, anatase and rutile. The bandgaps (E_g) for anatase and rutile are around 3.2 and 3.02 eV, respectively. The wavelength of exposure light should be lower than 380 and 410 nm to activate photocatalysis. When photocatalysts are exposed to light sources with energy equal to or higher than the bandgap, electrons are stimulated. The photoexcited electrons will transfer from the valance band to the conduction band, which leaves positive electron holes in the valance band. The electron and hole pairs (e^-/h^+) are generated. The photoexcited electron is a powerful reducing agent, while the electron hole serves as a strong oxidation agent [91]. The activation equation and oxidation and reduction reaction are expressed as follows:

$$Activation : TiO_2 + h\upsilon \rightarrow h^+ + e^-$$

$$Oxidation \ reaction : h^+ + OH^- \rightarrow OH^\bullet$$

$$Reduction \ reaction : e^- + O_{2ads} \rightarrow O_2^- ads$$

where the subscript *ads* refers to molecules that are adsorbed by the surface of TiO_2. The hydroxyl radical oxidized from the absorbed moisture or OH^- is the dominant oxidant that degrades organic compounds. The net reaction can be described as follows:

$$VOC + O_2 + OH^\bullet \rightarrow nCO_2 + mH_2O$$

The whole photocatalysis schematic is shown in Fig. 9.6.

The degradation of volatile organic compounds by photocatalysis is mainly attributed to an interfacial charge transfer mechanism [92] between surface of photocatalyst and volatile organic compounds. The degradation of volatile organic compounds may be oxidized or reduced directly by the surface electron or hole or with the reactive oxidants generated by photocatalysis [90]. Almost all of gaseous pollutant including NO_x [93], SO_x [94], CO [95], and O_3 [96] can be photocatalytically oxidized.

Several factors contribute to the performance of PCO, such as airflow rate [97], inlet contaminant concentration, and light intensity. Airflow directly affects the residence time and mass transfer rate. With an increase in the airflow rate, the residence time decreases, resulting in reduction in the chance of the organic compound being adsorbed by the photocatalyst for PCO and lower degradation efficiency. Also, the concentration gradient between the bulk and surface of the photocatalyst is smaller with a higher flow rate [98]. It was reported that the relationship between airflow rate and volatile organic compound removal efficiency has three distinct regimes [99,100]. At a low airflow rate, an increase in airflow rate increases the contaminant removal rate. It reveals that the mass transfer to the surface of the photocatalyst limits overall contaminant removal. At an intermediate airflow rate, the contaminant removal rate is significantly affected by the airflow rate, which means that the surface reaction kinetic is a rate-limiting step. At a high airflow rate, the contaminant removal rate decreases with the increase in airflow rate, indicating that the residence time is reduced for mass transfer. For the contaminant concentration the removal efficiency increases with lower initial contaminant concentration.

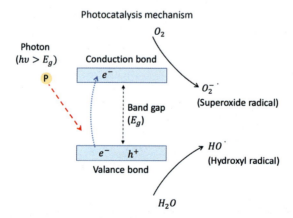

Figure 9.6 Schematic of photocatalysis mechanism.

Since the photocatalyst has constant active sites, adsorption is limited. Moreover, concentrated contaminant tends to adsorb or scatter the light, which contributes to quantum yield reduction [97]. Generally, the photocatalytic efficiency increases with increase in light intensity. However, excessive light intensity has a negative effect, owing to more electron hole recombination [101].

Typically, photocatalytic performance is carried out by degrading organic compounds, such as methylene and rhodamine B. Metal oxide NFs are widely used in various structures, such as hollow, porous, and solid. Zhao et al. [102] demonstrated that mesoporous TiO_2 hollow NFs have higher photodegradation efficiency in comparison to commercial TiO_2 spherical particles and mesoporous powders with the same surface area and particle size. To broaden the use of photocatalysts, binary and ternary compounds in nanofibrous form have been explored. For example, TiO_2 is blended with various content of zinc oxide (ZnO) and made in NF form [103]. The best UV light absorption efficiency is found with 15.76 wt.% of ZnO. With 365 nm UV light irradiation, almost 100% of rhodamine B and 85% of phenol are decomposed. $SrTiO_3/TiO_2$ NFs perform better photocatalytic activity in rhodamine B photodegradation. This may be because of the reduction in electron recombination [104]. Other metal oxides have also been investigated as photocatalysts. ZnO is a great candidate photocatalyst with a bandgap energy of 3.37 eV However, ZnO suffers from the fast electron hole recombination, which limits the photocatalytic performance. Pascariu et al. [105] reported that with doping of SnO_2 with a molar ratio of Sn/Zn of 0.03, a huge improvement of photodegradation efficiency is achieved, owing to the lower electron−hole recombination hindered by SnO_2. Li et al. [106] electrospun CeO_2−ZnO NFs with an average diameter of around 50 nm. CeO_2−ZnO NFs showed almost 100% rhodamine B photodegradation within 3 h of UV treatment, while pure CeO_2 and ZnO NFs achieved only 17% and 82.3% photodegradation, respectively, under the same test conditions. ZnO−SnO_2 NF has also been used as a photocatalyst.

9.3 Conclusion

In this chapter, formation and environmental concerns of air pollutions were introduced. Metal oxide NFs have already been proved to be among the most effective and promising solutions for air remediation. Fibrous metal oxide has several advantages, such as a large surface area, stronger mechanic strength, chemical stability, thermal stability, and photocatalytic ability. Metal oxide NFs have been adapted for air remediation in form of filter, catalyst, catalyst support, and photocatalyst. The capture mechanism and the characterization of filter media have been addressed to help in design of filters. Depending on the nature of the air pollutants, various remediation strategies are employed. For PM, properties of the filter medium, such as porosity, pore size, permeability, and pressure, are carefully designed for different target size of particles. For gaseous pollutants, metal oxide NFs play an important role as catalyst support and catalyst. Metal oxide NFs such alumina have been reported as great catalyst substrate to enhance the stability and photocatalytic performance of noble metal catalysts. The optical properties of metal oxides have also been investigated. Light-induced electron−hole pairs on the surface of metal oxide NFs are strong oxidants and reductants that contribute to the PCO of air pollutants.

References

[1] WHO, 7 Million Premature Deaths Annually Linked to Air Pollution, World Health Organization, Geneva, Switzerland, 2014.

[2] WHO, Global Health Observatory (GHO) Data. 2016, Child Mortal Causes Death WHO, Geneva, 2016.

[3] DO THEY WHY. Healthy Environments for Healthier Populations.

[4] J.A. Bernstein, N. Alexis, C. Barnes, I.L. Bernstein, A. Nel, D. Peden, et al., Health effects of air pollution, J. Allergy Clin. Immunol. 114 (5) (2004) 1116−1123.

[5] WHO, WHO Air Quality Guidelines for Particulate Matter, Ozone, Nitrogen Dioxide and Sulfur Dioxide: Global Update 2005: Summary of Risk Assessment, World Health Organization, Geneva, 2006.

[6] A. Don Porto Carero, P.H.M. Hoet, L. Verschaeve, G. Schoeters, B. Nemery, Genotoxic effects of carbon black particles, diesel exhaust particles, and urban air particulates and their extracts on a human alveolar epithelial cell line (A549) and a human monocytic cell line (THP-1), Env. Mol. Mutagen. 37 (2) (2001) 155−163.

[7] B. Brunekreef, S.T. Holgate, Air pollution and health, Lancet 360 (9341) (2002) 1233−1242.

[8] D.A. Vallero, Fundamentals of Air Pollution, Academic Press, 2014.

[9] T.V. Choudhary, S. Banerjee, V.R. Choudhary, Catalysts for combustion of methane and lower alkanes, Appl. Catal. A Gen. 234 (1−2) (2002) 1−23.

[10] A. Ernst, J.D. Zibrak, Carbon monoxide poisoning, N. Engl. J. Med. 339 (22) (1998) 1603−1608.

[11] L.Y. Chan, Y.M. Liu, Carbon monoxide levels in popular passenger commuting modes traversing major commuting routes in Hong Kong, Atmos. Environ. 35 (15) (2001) 2637−2646.

[12] J.A. Raub, M. Mathieu-Nolf, N.B. Hampson, S.R. Thom, Carbon monoxide poisoning —a public health perspective, Toxicology 145 (1) (2000) 1−14.

[13] M.K. Khair, W.A. Majewski, Diesel Emissions and Their Control. SAE Technical Paper, 2006.

[14] E. Robinson, R.C. Robbins, Gaseous nitrogen compound pollutants from urban and natural sources, J. Air Pollut. Control. Assoc. 20 (5) (1970) 303−306.

[15] K. Bedard, K.-H. Krause, The NOX family of ROS-generating NADPH oxidases: physiology and pathophysiology, Physiol. Rev. 87 (1) (2007) 245−313.

[16] T. Boningari, P.G. Smirniotis, Impact of nitrogen oxides on the environment and human health: Mn-based materials for the NO_x abatement, Curr. Opin. Chem. Eng. 13 (2016) 133−141.

[17] I.B. Andersen, G.R. Lundqvist, P.L. Jensen, D.F. Proctor, Human response to controlled levels of sulfur dioxide, Arch. Env. Heal. An. Int. J. 28 (1) (1974) 31−39.

[18] Z. Meng, G. Qin, B. Zhang, DNA damage in mice treated with sulfur dioxide by inhalation, Env. Mol. Mutagen. 46 (3) (2005) 150−155.

[19] Y. Fang, V. Naik, L.W. Horowitz, D.L. Mauzerall, Air pollution and associated human mortality: the role of air pollutant emissions, climate change and methane concentration increases from the preindustrial period to present, Atmos. Chem. Phys. (2013).

[20] K.A. Garrett, S.P. Dendy, E.E. Frank, M.N. Rouse, S.E. Travers, Climate change effects on plant disease: genomes to ecosystems, Annu. Rev. Phytopathol. 44 (2006) 489–509.

[21] G.C. Nelson, H. Valin, R.D. Sands, P. Havlík, H. Ahammad, D. Deryng, et al., Climate change effects on agriculture: economic responses to biophysical shocks, Proc. Natl Acad. Sci. 111 (9) (2014) 3274–3279.

[22] D. McKee, Tropospheric Ozone: Human Health and Agricultural Impacts, CRC Press, 1993.

[23] A.M. Hough, R.G. Derwent, Changes in the global concentration of tropospheric ozone due to human activities, Nature. 344 (6267) (1990) 645–648.

[24] D.B. Peden, The role of oxidative stress and innate immunity in O_3 and endotoxin-induced human allergic airway disease, Immunol. Rev. 242 (1) (2011) 91–105.

[25] J.J. Kim, Ambient air pollution: health hazards to children, Pediatrics. 114 (6) (2004) 1699–1707.

[26] D. Barlev, R. Vidu, P. Stroeve, Innovation in concentrated solar power, Sol. Energy Mater. Sol Cell. 95 (10) (2011) 2703–2725.

[27] I. Yüksel, Hydropower for sustainable water and energy development, Renew. Sustain. Energy Rev. 14 (1) (2010) 462–469.

[28] T.A.A. Adcock, S. Draper, T. Nishino, Tidal power generation–a review of hydrodynamic modelling, Proc. Inst. Mech. Eng. Part. A J. Power Energy. 229 (7) (2015) 755–771.

[29] M.R. Patel, Wind and Solar Power Systems: Design, Analysis, and Operation, CRC Press, 2005.

[30] R.C. Brown, Air Filtration: An Integrated Approach to the Theory and Applications of Fibrous Filters, Pergamon, 1993.

[31] C.N. Davies, Filtration of aerosols, J. Aerosol Sci. 14 (2) (1983) 147–161.

[32] T. Grafe, M. Gogins, M. Barris, J. Schaefer, R. Canepa, Nanofibers in filtration applications in transportation, in: Filtration 2001 International Conference and Exposition, Chicago, Illinois, 2001.

[33] C.-S. Wang, Electrostatic forces in fibrous filters—a review, Powder Technol. 118 (1–2) (2001) 166–170.

[34] A. Jaworek, A. Krupa, T. Czech, Modern electrostatic devices and methods for exhaust gas cleaning: a brief review, J. Electrostat 65 (3) (2007) 133–155.

[35] B. Tan, L. Wang, X. Zhang, The effect of an external DC electric field on bipolar charged aerosol agglomeration, J. Electrostat 65 (2) (2007) 82–86.

[36] Q. Yao, S.-Q. Li, H.-W. Xu, J.-K. Zhuo, Q. Song, Reprint of: studies on formation and control of combustion particulate matter in China: a review, Energy 35 (11) (2010) 4480–4493.

[37] A. Podgorski, A. Bałazy, L. Gradoń, Application of nanofibers to improve the filtration efficiency of the most penetrating aerosol particles in fibrous filters, Chem. Eng. Sci. 61 (20) (2006) 6804–6815.

[38] S.S. Sreedhara, N.R. Tata, A novel method for measurement of porosity in nanofiber mat using pycnometer in filtration, J. Eng. Fiber Fabr. 8 (4) (2013). 155892501300800420.

[39] A. Hernández, J.I. Calvo, P. Prádanos, F. Tejerina, Pore size distributions in microporous membranes. A critical analysis of the bubble point extended method, J. Memb. Sci. 112 (1) (1996) 1–12.

[40] K.M. Yun, C.J. Hogan Jr, Y. Matsubayashi, M. Kawabe, F. Iskandar, K. Okuyama, Nanoparticle filtration by electrospun polymer fibers, Chem. Eng. Sci. 62 (17) (2007) 4751–4759.

[41] H. Ma, K. Yoon, L. Rong, Y. Mao, Z. Mo, D. Fang, et al., High-flux thin-film nanofibrous composite ultrafiltration membranes containing cellulose barrier layer, J. Mater. Chem. 20 (22) (2010) 4692–4704.

[42] Z.-M. Huang, Y.-Z. Zhang, M. Kotaki, S. Ramakrishna, A review on polymer nanofibers by electrospinning and their applications in nanocomposites, Compos. Sci. Technol. 63 (15) (2003) 2223–2253.

[43] H.U. Shin, A.B. Stefaniak, N. Stojilovic, G.G. Chase, Comparative dissolution of electrospun Al_2O_3 nanofibres in artificial human lung fluids, Env. Sci. Nano 2 (3) (2015) 251–261.

[44] H.-W. Li, C.-Y. Wu, F. Tepper, J.-H. Lee, C.N. Lee, Removal and retention of viral aerosols by a novel alumina nanofiber filter, J. Aerosol Sci. 40 (1) (2009) 65–71.

[45] L. Green, Fluid flow through porous metals, J. Appl. Mech. 18 (1951) 39–45.

[46] R.S. Barhate, C.K. Loong, S. Ramakrishna, Preparation and characterization of nanofibrous filtering media, J. Memb. Sci. 283 (1–2) (2006) 209–218.

[47] Z.G. Liu, P.K. Wang, Pressure drop and interception efficiency of multifiber filters, Aerosol Sci. Technol. 26 (4) (1997) 313–325.

[48] S. Kuwabara, The forces experienced by randomly distributed parallel circular cylinders or spheres in a viscous flow at small Reynolds numbers, J. Phys. Soc. Jpn. 14 (4) (1959) 527–532.

[49] B.V. Ramarao, C. Tien, S. Mohan, Calculation of single fiber efficiencies for interception and impaction with superposed Brownian motion, J. Aerosol Sci. 25 (2) (1994) 295–313.

[50] V.V. Kadam, L. Wang, R. Padhye, Electrospun nanofibre materials to filter air pollutants—a review, J. Ind. Text. 47 (8) (2018) 2253–2280.

[51] H.L. Schreuder-Gibson, P. Gibson, Y.-L. Hsieh, Transport properties of electrospun nonwoven membranes, Int. Nonwovens J. (2)(2002). 1558925002OS-01100206.

[52] Z.-C. Xiong, R.-L. Yang, Y.-J. Zhu, F.-F. Chen, L.-Y. Dong, Flexible hydroxyapatite ultralong nanowire-based paper for highly efficient and multifunctional air filtration, J. Mater. Chem. A 5 (33) (2017) 17482–17491.

[53] C. Lee, Y. Jeon, S. Hata, J.-I. Park, R. Akiyoshi, H. Saito, et al., Three-dimensional arrangements of perovskite-type oxide nano-fiber webs for effective soot oxidation, Appl. Catal. B Env. 191 (2016) 157–164.

[54] H.-C. Yeh, B.Y.H. Liu, Aerosol filtration by fibrous filters—I. Theoretical, J. Aerosol Sci. 5 (2) (1974) 191–204.

[55] G. Karniadakis, A. Beskok, N. Aluru, Vol Microflows and Nanoflows: Fundamentals and Simulation, 29, Springer Science & Business Media, 2006.

[56] S.G. Jennings, The mean free path in air, J. Aerosol Sci. 19 (2) (1988) 159–166.

[57] C.E. Wilkes, J.W. Summers, C.A. Daniels, M.T. Berard, PVC Handbook, vol. 184, Hanser Munich, 2005.

[58] Y. Bai, X. Mao, J. Song, X. Yin, J. Yu, B. Ding, Self-standing Ag_2O@YSZ-TiO_2 pn nanoheterojunction composite nanofibrous membranes with superior photocatalytic activity, Compos. Commun. 5 (2017) 13–18.

[59] K.E. Noll, Adsorption Technology for Air and Water Pollution Control, CRC Press, 1991.

[60] R.M. Flores, in: R.M. Flores (Ed.), Chapter 4—Coalification, Gasification, and Gas Storage. Coal Coalbed Gas, Elsevier Boston, MA, 2014, pp. 167–233.

[61] V. Inglezakis, S. Poulopoulos, Vol Adsorption, Ion Exchange and Catalysis, 3, Elsevier, 2006.

[62] J. Zhu, Q. Gao, Z. Chen, Preparation of mesoporous copper cerium bimetal oxides with high performance for catalytic oxidation of carbon monoxide, Appl. Catal. B Env. 81 (3−4) (2008) 236−243.

[63] A. Alvarez, S. Ivanova, M.A. Centeno, J.A. Odriozola, Sub-ambient CO oxidation over mesoporous Co_3O_4: effect of morphology on its reduction behavior and catalytic performance, Appl. Catal. A Gen. 431 (2012) 9−17.

[64] S. Xu, D. Sun, H. Liu, X. Wang, X. Yan, Fabrication of Cu-doped cerium oxide nanofibers via electrospinning for preferential CO oxidation, Catal. Commun. 12 (6) (2011) 514−518.

[65] C. Eid, A. Brioude, V. Salles, J.-C. Plenet, R. Asmar, Y. Monteil, et al., Iron-based 1D nanostructures by electrospinning process, Nanotechnology 21 (12) (2010) 125701.

[66] J. Livage, M. Henry, C. Sanchez, Sol−gel chemistry of transition metal oxides, Prog. solid. state Chem. 18 (4) (1988) 259−341.

[67] X. Du, F. Dong, Z. Tang, J. Zhang, The synthesis of hollow In_2O_3@Pd-Co_3O_4 core/shell nanofibers with ultra-thin shell for the low-temperature CO oxidation reaction, Appl. Surf. Sci. 505 (2020) 144471.

[68] D. Marani, R.H. Silva, A. Dankeaw, K. Norrman, R.M.L. Werchmeister, D. Ippolito, et al., NOx selective catalytic reduction (SCR) on self-supported V−W-doped TiO_2 nanofibers, N. J. Chem. 41 (9) (2017) 3466−3472.

[69] K. Huang, X. Chu, W. Feng, C. Zhou, W. Si, X. Wu, et al., Catalytic behavior of electrospinning synthesized $La_{0.75}Sr_{0.25}MnO_3$ nanofibers in the oxidation of CO and CH_4, Chem. Eng. J. 244 (2014) 27−32.

[70] S. Bhattacharyya, R.K. Das, Catalytic control of automotive NO_x: a review, Int. J. Energy Res. 23 (4) (1999) 351−369.

[71] K.C. Taylor, Automobile Catalytic Converters, Catalysis, Springer, 1984, pp. 119−170.

[72] X. Yao, F. Gao, Y. Cao, C. Tang, Y. Deng, L. Dong, et al., Tailoring copper valence states in CuO δ/γ-Al_2O_3 catalysts by an in situ technique induced superior catalytic performance for simultaneous elimination of NO and CO, Phys. Chem. Chem Phys 15 (36) (2013) 14945−14950.

[73] M.V. Twigg, Catalyst Handbook, Routledge, 2018.

[74] M.V. Twigg, Controlling automotive exhaust emissions: successes and underlying science, Philos. Trans. R. Soc. A Math. Phys. Eng. Sci. 363 (1829) (2005) 1013−1033.

[75] M. Iwamoto, H. Hamada, Removal of nitrogen monoxide from exhaust gases through novel catalytic processes, Catal. Today 10 (1) (1991) 57−71.

[76] E. Aneggi, C. de Leitenburg, M. Boaro, P. Fornasiero, A. Trovarelli, Catalytic applications of cerium dioxide, Cerium Oxide (CeO_2): Synthesis, Properties and Applications, Elsevier, 2020, pp. 45−108.

[77] P. Lööf, B. Kasemo, K.-E. Keck, Oxygen storage capacity of noble metal car exhaust catalysts containing nickel and cerium, J. Catal. 118 (2) (1989) 339−348.

[78] H. Fujitsu, H. Ishibashi, N. Ikeyama, I. Mochida, Remarkable activity enhancement of Rh/Al_2O_3 prepared from RhCl3 for CO−H_2 reaction by the pretreatment of high temperature evacuation, Chem. Lett. 17 (4) (1988) 581−584.

[79] M.M.V.M. Souza, N.F.P. Ribeiro, M. Schmal, Influence of the support in selective CO oxidation on Pt catalysts for fuel cell applications, Int. J. Hydrogen Energy 32 (3) (2007) 425−429.

[80] X. Wang, J.S. Tian, Y.H. Zheng, X.L. Xu, W.M. Liu, X.Z. Fang, Tuning Al_2O_3 surface with SnO_2 to prepare improved supports for Pd for CO oxidation, ChemCatChem 6 (6) (2014) 1604–1611.

[81] M.G. Sanchez, J.L. Gazquez, Oxygen vacancy model in strong metal-support interaction, J. Catal. 104 (1) (1987) 120–135.

[82] Y. Dai, W. Liu, E. Formo, Y. Sun, Y. Xia, Ceramic nanofibers fabricated by electrospinning and their applications in catalysis, environmental science, and energy technology, Polym. Adv. Technol. 22 (3) (2011) 326–338.

[83] S. Swaminathan, Metal oxide nanofibers as filters, catalyst and catalyst support structures, University of Akron, 2010.

[84] C.E. Hori, H. Permana, K.Y.S. Ng, A. Brenner, K. More, K.M. Rahmoeller, et al., Thermal stability of oxygen storage properties in a mixed CeO_2–ZrO_2 system, Appl. Catal. B Env. 16 (2) (1998) 105–117.

[85] L. Shahreen, G.G. Chase, A.J. Turinske, S.A. Nelson, N. Stojilovic, NO decomposition by CO over Pd catalyst supported on TiO_2 nanofibers, Chem. Eng. J. 225 (2013) 340–349.

[86] W. Li, Y. Wang, B. Ji, X. Jiao, D. Chen, Flexible Pd/CeO_2–TiO_2 nanofibrous membrane with high efficiency ultrafine particulate filtration and improved CO catalytic oxidation performance, RSC Adv. 5 (72) (2015) 58120–58127.

[87] A. Fujishima, K. Honda, Electrochemical photolysis of water at a semiconductor electrode, Nature 238 (5358) (1972) 37–38.

[88] Y. Boyjoo, H. Sun, J. Liu, V.K. Pareek, S. Wang, A review on photocatalysis for air treatment: from catalyst development to reactor design, Chem. Eng. J. 310 (2017) 537–559.

[89] R.I. Bickley, G. Munuera, F.S. Stone, Photoadsorption and photocatalysis at rutile surfaces: II. Photocatalytic oxidation of isopropanol, J. Catal. 31 (3) (1973) 398–407.

[90] Z. Shayegan, C.-S. Lee, F. Haghighat, TiO_2 photocatalyst for removal of volatile organic compounds in gas phase—a review, Chem. Eng. J. 334 (2018) 2408–2439.

[91] D. Vildozo, C. Ferronato, M. Sleiman, J.-M. Chovelon, Photocatalytic treatment of indoor air: optimization of 2-propanol removal using a response surface methodology (RSM), Appl. Catal. B Env. 94 (3–4) (2010) 303–310.

[92] S.G. Kumar, L.G. Devi, Review on modified TiO_2 photocatalysis under UV/visible light: selected results and related mechanisms on interfacial charge carrier transfer dynamics, J. Phys. Chem. A 115 (46) (2011) 13211–13241.

[93] E. Luévano-Hipólito, A. Martínez-de la Cruz, Q.L. Yu, H.J.H. Brouwers, Precipitation synthesis of WO_3 for NO_x removal using PEG as template, Ceram. Int. 40 (8) (2014) 12123–12128.

[94] Z. Topalian, G.A. Niklasson, C.-G. Granqvist, L. Osterlund, Spectroscopic study of the photofixation of SO_2 on anatase TiO_2 thin films and their oleophobic properties, ACS Appl. Mater. Interfaces 4 (2) (2012) 672–679.

[95] M.A. Bollinger, M.A. Vannice, A kinetic and DRIFTS study of low-temperature carbon monoxide oxidation over Au—TiO_2 catalysts, Appl. Catal. B Env. 8 (4) (1996) 417–443.

[96] P. Kopf, E. Gilbert, S.H. Eberle, TiO_2 photocatalytic oxidation of monochloroacetic acid and pyridine: influence of ozone, J. Photochem. Photobiol. A Chem. 136 (3) (2000) 163–168.

[97] M. Hussain, N. Russo, G. Saracco, Photocatalytic abatement of VOCs by novel optimized TiO_2 nanoparticles, Chem. Eng. J. 166 (1) (2011) 138–149.

[98] L. Yang, Z. Liu, J. Shi, Y. Zhang, H. Hu, W. Shangguan, Degradation of indoor gaseous formaldehyde by hybrid VUV and TiO_2/UV processes, Sep. Purif. Technol. 54 (2) (2007) 204−211.

[99] C.A. Korologos, C.J. Philippopoulos, S.G. Poulopoulos, The effect of water presence on the photocatalytic oxidation of benzene, toluene, ethylbenzene and m-xylene in the gas-phase, Atmos. Env. 45 (39) (2011) 7089−7095.

[100] K. Wang, Y. Hsieh, Heterogeneous photocatalytic degradation of trichloroethylene in vapor phase by titanium dioxide, Env. Int. 24 (3) (1998) 267−274.

[101] J.-S. Kim, K. Itoh, M. Murabayashi, Photocatalytic degradation of trichloroethylene in the gas phase over TiO_2 sol−gel films: analysis of products, Chemosphere 36 (3) (1998) 483−495.

[102] S. Zhan, D. Chen, X. Jiao, C. Tao, Long TiO_2 hollow fibers with mesoporous walls: sol − gel combined electrospun fabrication and photocatalytic properties, J. Phys. Chem. B 110 (23) (2006) 11199−11204.

[103] R. Liu, H. Ye, X. Xiong, H. Liu, Fabrication of TiO_2/ZnO composite nanofibers by electrospinning and their photocatalytic property, Mater. Chem. Phys. 121 (3) (2010) 432−439.

[104] T. Cao, Y. Li, C. Wang, C. Shao, Y. Liu, A facile in situ hydrothermal method to $SrTiO_3$/TiO_2 nanofiber heterostructures with high photocatalytic activity, Langmuir 27 (6) (2011) 2946−2952.

[105] P. Pascariu, A. Airinei, N. Olaru, L. Olaru, V. Nica, Photocatalytic degradation of Rhodamine B dye using ZnO−SnO_2 electrospun ceramic nanofibers, Ceram. Int. 42 (6) (2016) 6775−6781.

[106] C. Li, R. Chen, X. Zhang, S. Shu, J. Xiong, Y. Zheng, et al., Electrospinning of CeO_2−ZnO composite nanofibers and their photocatalytic property, Mater. Lett. 65 (9) (2011) 1327−1330.

Section 3

Piezoelectric application of metal oxide nanofibers 10

Tutu Sebastian and Frank Clemens
Laboratory for High Performance Ceramics, Empa, Swiss Federal Laboratories for
Materials Science and Technology, Dübendorf, Switzerland

10.1 Introduction

Piezoelectric and ferroelectric materials have found their way into a plethora of applications, owing to their unique ability to transverse electromechanical signals. Ferroelectric materials are a subclass of piezoelectric materials by which existing applications can be extended to dielectric capacitors, sonar and ultrasonic transducers, thermal imaging, surveillance devices, medical diagnostic transducers, positive temperature coefficient sensors, ultrasonic motors, thin-film capacitors, ferroelectric memories, piezoelectric transformers, and so on [1,2]. Although they are very widely used, their applicability can be further enhanced by increasing the material properties of these materials along with adaptability into complex and thin structures with improved flexibility. One of the most common strategies to achieve this is to explore the viability of one-dimensional (1D) structures, such as nanofibers (NFs), nanowires, nanotubes, and nanobelts in various applications. These materials have become the focus of intensive research, owing to their unique application in the fabrication of electronic, optoelectronic, and sensor devices when compared to bulk structures. It has been reported that the reduction in size to nano-scale (<100 nm in one dimension) improves the performance of piezoelectric materials [3]. At nano-scale, crystal defects such as dislocation and vacancies can be eliminated, and perfect single crystallinity can be easily achieved. In 2004, Zhao et al. reported that during piezoelectric measurement a zinc oxide (ZnO) nanobelt with a dimension of tens of nanometers in thickness, hundreds of nanometers in width, and tens of micrometers in length displayed a piezoelectric coefficient, d_{33} of ZnO nanobelts (27 pm V^{-1}), a much higher value compared to the bulk ZnO (9.7 pm V^{-1}) [4]. This enhancement in properties is generally attributed to the high specific surface area of the nanostructures along with the possibility of forming single-domain structures. Nanostructures have relatively high specific surface areas and hence have more surface atoms and higher levels of surface energy compared to their bulk counterparts [5]. Therefore these materials are being widely examined these days and found to have great potential to be used in energy harvesters, nano-generators, sensors, nonvolatile memory devices, microelectromechanical systems, and so on. This chapter will discuss the recent advancements in the synthesis and properties of some of the most important metal oxide NFs, such as lead zirconate

Metal Oxide-Based Nanofibers and Their Applications. DOI: https://doi.org/10.1016/B978-0-12-820629-4.00002-3
© 2022 Elsevier Inc. All rights reserved.

titanate (PZT), barium titanate (BT), potassium sodium niobate (KNN), and ZnO, with particular emphasis on their piezoelectric applications.

10.2 Inorganic piezoelectric materials and their properties

All solid materials, when subjected to stress, produce a proportional strain. However, in a piezoelectric material there is an additional creation of an electrical charge. Piezoelectricity is the property of certain materials characterized by the development of an electrical charge proportional to an applied mechanical stress or the development of a displacement under an applied electric field. Of the 32 existing crystal classes, 21 are noncentrosymmetric [6]. A noncentrosymmetric crystal class is mandatory to achieve polarity and therefore a charge displacement within the lattice. Among these 21 noncentrosymmetric crystal classes, 20 exhibit piezoelectricity.

Of these 20 piezoelectric classes, 10 of them are characterized by having a permanent electric dipole moment or a spontaneous polarization even in the absence of a stress (see Fig. 10.1). This spontaneous polarization exists when there is a separation between the centers of the positive and negative charges in a unit cell.

For piezoelectricity the magnitude of polarization is linearly proportional to the magnitude of applied stress. The sign depends on the type of stress (tensile or compressive). For commercial devices, either the direct or converse effect of piezoelectric material is used. For the direct effect, mechanical energy is converted into electrical energy (generator). If a force is applied to the piezoelectric material, a surface charge is developed by dielectric displacement. This surface charge can

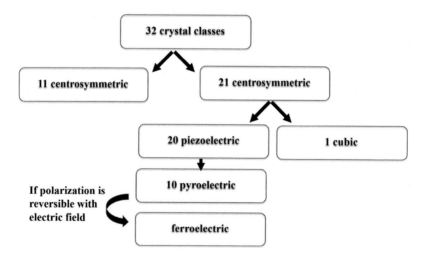

Figure 10.1 Classification of crystal class.

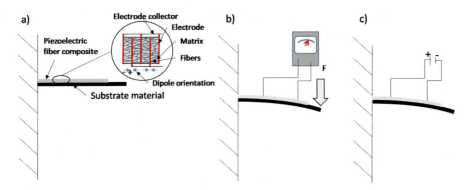

Figure 10.2 Schematic illustration of the (A) piezoelectric ceramic fiber composite, (B) direct piezoelectric effect, and (C) converse piezoelectric effect.

flow through the circuit, generating a voltage. For the converse or indirect effect, electrical energy is converted to a mechanical motion (motor). When an electric field is applied to the piezoelectric material, a displacement occurs in the material. A schematic of the direct and converse piezoelectric effects is given in Fig. 10.2. If a force is applied to prevent this distortion, an elastic tension builds up.

In a ferroelectric crystal (unit cell), this displacement occurs along a unique polar axis, but in general, the dipole moment cannot be detected on the material surfaces, owing to the redistribution of charges within the crystal itself and into the surrounding medium. However, the spontaneous polarization is temperature dependent and forms at a certain temperature (Curie temperature, T_c) and is caused by phase transition. This results in a change from paraelectric to ferroelectric material behavior (Fig. 10.3A). Ferroelectricity means spontaneous random polarization of the dipoles at a certain temperature and the reorientation of dipoles in the material by an applied electric field. It is worthwhile to mention that all ferroelectric crystals must be piezoelectric, while the reverse does not hold true.

The reorientation of the polarization under a strong electrical field is not linear; a hysteretic behavior is associated with significant changes to the domain structure. Domains are a band of uniformly polarized regions (dipoles of unit cells) that are aligned in a single direction. A single grain may consist of several such domains. Two domains are separated by an interface called a domain wall. Certain activation energy is necessary to reorientate the dipoles of the unit cells under an applied external field. Thus under a small introduced field (electric or mechanical), the energy is not high enough to align the domains in the direction of the applied field. Hence it behaves linearly. However, with an increase in the applied field, more and more domains tend to align in the direction of applied field, thereby increasing the polarization (poling). Coercivity in ferroelectrics is the intensity of the applied electrical field (E_c) required to reduce the overall polarization of the domains. Complete switching of domains is associated with saturated polarization, P_s. If the field is removed, most but not all domains do not switch back, giving remanent polarization, P_r, which is typically slightly less than P_s. The effect of the

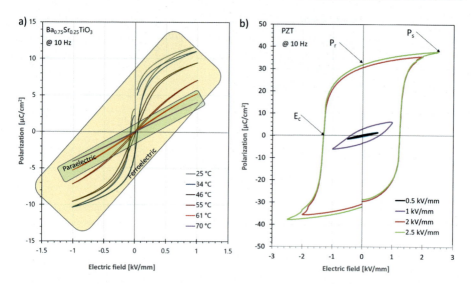

Figure 10.3 Ferroelectric behavior of (A) Ba$_{0.75}$Sr$_{0.25}$TiO$_3$ above and below the Curie temperature and (B) lead zirconate titanate in a small and a large electrical field.

temperature and the electrical field on the polarization behavior of two ferroelectric materials is illustrated in Fig. 10.3.

The orientation of the dipoles (poling) is the process of applying a large direct current (DC) electric field, several kilovolts per millimeter, to a sample immersed in silicone oil, usually at elevated temperatures. Its purpose is to align domains in individual grains in the direction of the applied field. Owing to higher mobility of the domain walls at higher temperatures, the decrease in coercive field facilitates poling at lower electric fields to avoid an electrical breakdown during the poling process.

Despite not showing any ferroelectric behavior, quartz (SiO$_2$) and ZnO are some of the interesting piezoelectric materials that have been studied over the last decades. Being a naturally n-type semiconductor, ZnO has attracted much attention from researchers, owing to its unique properties, especially in the nanostructure form. ZnO nanostructure provides the possibility of integration with other p-type materials. Moreover, the highly stable noncentrosymmetric hexagonal wurtzite structure leads to a relatively large piezoelectric coefficient, a high modulus of elasticity and a high piezoelectric tensor [7]. ZnO exhibits a d_{33} of about 12.4 pC/N for the bulk and between 8 and 12 pC/N for thin films [8]. Although they have a lower value compared to other classical piezoelectric materials, aligned ZnO nanowires are particularly interesting in fabrication of nanogenerators that convert nano-scale mechanical energy into electric energy [8]. ZnO possesses the richest family of different morphologies, which are possible to obtain by using a variety of physical and chemical synthesis techniques [9]. Hence nanostructure ZnO is reserved as a prospective material in many applications such as sensing, energy-harvesting, and photonics. ZnO also has biocompatibility and inherent piezoelectricity and is cheap to synthesize, all of which make it useful for energy-harvesting applications [9,10].

Ceramics are polycrystalline materials, and the net piezoelectric effect is zero due to the random orientation of grains; even though each crystallite will develop an electrical charge under mechanical stress [11]. However, by a so-called poling process, a high DC voltage is applied across the ferroelectric ceramic that aligns the individual dipoles in the direction of the applied electric field and yields a nonzero net polarization.

Since 1950, PZT ($Pb(Zr_{1-x} Ti_x)O_3$) has been the most dominant piezoelectric ceramic [12]. It is particularly used with a composition around the morphotropic phase boundary, (MPB), a phase boundary where rhombohedral and tetragonal phases coexist, at $x = 0.48$. At MPB the properties such as piezoelectric coefficients, dielectric permittivity, and coupling factors show a peak maximum [1,12,13]. Large signal piezoelectric coefficients as high as $800 \, pm \, V^{-1}$ have been reported for doped PZT (PZT 5H) [14]. Further modifications of doping PZT with acceptors and donors have been performed quite extensively for the manufacture of hard and soft PZTs to cater to the need for required applications. Excellent and comprehensive reviews on piezoelectric ceramics on both modified and undoped PZT ceramics are given by Jaffe et al. [15] and Herbert et al. [13].

Since the European Union (EU) in 2003, included PZT as a hazardous substance that should be replaced by safe materials, a search for potential lead-free piezoelectric ceramics started in the early 2000s [16]. As one of the most studied electroceramic materials and one of the first developed piezoelectric ceramic materials, $BaTiO_3$ was the first choice. However, owing to the limited operation temperature, the low remanent polarization, and several polymorphic transformations it is not widely used as a piezoelectric material. $BaTiO_3$ finds its application mainly as a ferroelectric material for multilayered ceramic capacitors, owing to its high dielectric constant and low coercive field. It has a Curie temperature of 120°C, above which the charge separation disappears and the material becomes paraelectric. Hence researchers started to focus more on materials such as KNN and sodium bismuth titanate (NBT) as a possible PZT replacement. An overview of room temperature values of the dielectric, piezoelectric, and electromechanical properties have been compiled for these family of materials [14]. Analogous to PZT, an MPB separating ferroelectric rhombohedral and tetragonal phases is observed in the NBT-based system, leading to enhanced properties. Apart from that, the MPB is strongly curved when compared to the PZT system [14]. Though an orthorhombic-orthorhombic MPB boundary does occur in KNN at $x \sim 0.5$, the enhancement in piezoelectric properties is not comparable to PZT with little or no enhancement in dielectric properties [14]. Although the comprehensive properties of KNN-based piezoelectric ceramics are still inferior to those of their PZT counterparts, KNN is believed to be one of the most promising lead-free piezoelectric materials that can replace PZT. A significant increase in some or all of the typical figure of merit coefficients (piezoelectric coefficients, mechanical quality factor) of the materials is considered essential for the advancement of the next-generation piezoelectric applications. Apart from MPB compositions, development of single-crystal materials near the MPB started to gain attention in the late 1990s. A huge increase in d_{33} coefficients between 1500 and 2500 pC/N and electromechanical coupling coefficient, $k_{33} > 0.9$ were reported for PZN-PT and PMN-PT single crystals, respectively, by Park and Shrout. These high values were obtained for crystals primarily grown in $<001>$ orientation [17,18]. However, owing to the slow growth rate, high cost, and difficulty in controlling crystal stoichiometry the trend shifted more toward the processing of textured piezoelectric ceramics. This resulted in ceramics with properties that lie between single crystals and randomly oriented ceramics [19].

Great interest in the miniaturization of piezoelectric generators has been observed since 2006, when Wang and Song reported the development of nanogenerators that can generate several millivolts of voltage by bending a ZnO nanowire with the tip of an atomic force microscope [20]. 1D piezoelectric materials have gained particular attention because of their enhanced piezoelectric properties, excellent mechanical properties, and tunable electric properties and possible applicability in nanogenerators, sensors, actuators, electronic devices, and so on. The piezoelectric coefficients, their performance as nanogenerators, and the structure-dependent electromechanical properties of these 1D materials can be assessed through piezoresponse force microscopy, atomic force microscopy (AFM), and in situ scanning/ transmission electron microscopy, respectively [21]. The morphology and crystal structure could be decisive parameters in selecting these 1D materials for a particular application. A detailed study of these factors should reveal information about the underlying mechanism and ways to improve the performance of these materials. However, such studies are seldom found in the literature. One such study investigated the electromechanical properties of individual indium arsenide (InAs) nanowires grown along different crystallographic directions using in situ SEM, and the crystal structures of these nanowires were assessed through transmission electron microscopy (TEM). It was shown that InAs nanowires showed maximum piezoelectric and piezoresistive effects when the nanowires were grown along $<0001>$ crystallographic directions [22].

10.3 Synthesis of one-dimensional nanostructures

The fabrication of nanostructures with well-aligned morphology and high piezoelectric responses is still a challenge [23]. In this section, several fabrication techniques that have been employed to synthesize nanostructured piezoelectric materials are considered.

10.4 Hydrothermal synthesis

Hydrothermal synthesis is one of the most popular techniques for synthesizing nanostructured materials. This involves a heterogeneous chemical reaction in aqueous media above room temperature (over 100°C) and higher pressure levels (over 1 bar), generally achieved by using an autoclave oven [24,25]. The product must undergo posttreatment sometimes to complete a crystal structure transition [26]. Generally, a temperature gradient is maintained between the opposite ends of the growth chamber to facilitate the formation of certain metastable phases. The solute dissolves at the hotter end and is deposited on a seed crystal at the cooler end, growing the desired crystal.

Hydrothermal synthesis of PZT was a widely researched topic with an early focus on both polycrystalline powders and thin films [27]. The optimal hydrothermal conditions for preparing PZT NFs were proposed by Chen et al. with a pH value of 13, reaction temperature 190°C, hydrothermal reactor filling 60%−70%, reaction time 10 h, and the activated carbon as the template [28]. During the synthesis of PZT 1D nanostructures, the effect of the addition of the two polymeric surfactants, polyvinyl alcohol (PVA) and polyacrylic acid (PAA), was studied in detail. It was observed that PZT nanorods were produced if only PVA was used as the surfactant, while isotropic nanoparticles were produced if only PAA was used [29]. A mixture of PVA and PAA with mainly PAA is therefore the optimal composition for obtaining high-aspect-ratio PZT nanostructures. This interesting behavior is due to the fact that PVA molecules are considered to adsorb on the particle surface by hydrogen bonds, whereas PAA adsorbs on the surface by chemical bonds [29]. An increase in the aspect ratio of PZT nanorods was also observed with increasing PVA content while using it as a surfactant [29,30]. The use of surfactants helps to suppress the radial growth of PZT, leading to the formation of 1D structures. Xu et al. successfully synthesized the PMN-PT, $0.72Pb(Mg_{1/3}Nb_{2/3})O_3 - 0.28PbTiO_3$, nanowires using hydrothermal synthesis, and its piezoelectric coefficient has reached an average value of 373 ± 5 pm V^{-1}, which is three times larger than the highest reported value of 1D PZT nanostructures [31].

Hydrothermal reactions were used for producing high-aspect-ratio and highly aligned ferroelectric $BaTiO_3$ nanowire arrays [32]. A textured TiO_2 film, was synthesized through a two-step hydrothermal reaction and deposited on titanium foil upon which highly aligned nanowires are grown via homoepitaxy and converted to $BaTiO_3$ [32]. Nanowires of BT were prepared by hydrothermal reaction and the fabricated composite with poly(vinylidene fluoride-trifluoroethylene) matrix showed a higher dielectric constant than a similar composite prepared by using $BaTiO_3$ nanoparticles [33]. Tu et al. prepared $BaTiO_3$ nanowires using a two-step hydrothermal conversion and further incorporating those nanowires into polyarylene ether nitrile (PEN) matrix to give a composite with high-energy storage density application in film capacitors [34]. Tetragonal $BaTiO_3$ nanotube arrays were synthesized by using the template-assisted hydrothermal method combined with an annealing process. A relative dielectric permittivity of 1000 and a dielectric loss as low as 0.02 at 1 kHz at room temperature could be achieved on the nanotube arrays with this technique [35]. Typically, $BaTiO_3$ bulk ceramics display a relative permittivity of about 1600 and a dielectric loss less than 0.05 at 1 kHz, depending on the grain size [12].

During the past decade, lead-free KNN started to gain much attention for the piezoelectric application as a possible replaceable candidate for PZT. In 2010, Wang et al. synthesized KNN nanostructures by using the hydrothermal method and

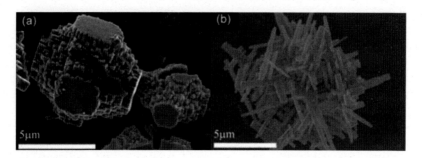

Figure 10.4 Field emission scanning electron microscope images of $K_{0.50}Na_{0.50}NbO_3$ (A) steplike microstructure and (B) nanorods.
Source: Reprinted from Z. Wang, H. Gu, Y. Hu, K. Yang, M. Hu, D. Zhou, et al., Synthesis, growth mechanism and optical properties of (K,Na)NbO3 nanostructures, CrystEngComm. 2010;12:3157−3162. https://doi.org/10.1039/C000169D. Copyright (2010) with permission from The Royal Society of Chemistry.

analyzed their growth mechanisms and optical properties [36]. The morphology of such KNN nanomaterials is shown in Fig. 10.4. The morphology and structure were found to vary a lot according to the temperatures and reaction times even under the same reaction conditions [36,37]. 1D KNN nanostructures were also synthesized as aligned arrays on a substrate. Xu et al. reported the fabrication of KNN nanorods grown on different substrates, namely, STO/KNN (100) ($SrTiO_3$/KNN) and STO/KNN (111) substrates [38]. An overview of some hydrothermal synthesized 1D nanostructured ferroelectric materials is given in Table 10.1.

Table 10.1 Overview of hydrothermal synthesized ferroelectric one-dimensional nanostructures.

Raw material	Solvent	Composition	Diameter (nm)	References
Zirconium propoxide, titanium isopropoxide, lead nitrate	KOH (aqueous)	PZT nanowires	50−200	[39]
Tetrabutyl titanate, zirconium oxychloride, lead nitrate	KOH (aqueous)	PZT nanowires	Nano-scale but nonuniform	[40]
Tetrabutyl titanate, barium hydroxide octahydrate	KOH (aqueous)	$BaTiO_3$ nanowires	40	[33]

(*Continued*)

Table 10.1 (Continued)

Raw material	Solvent	Composition	Diameter (nm)	References
Titanium oxide, sodium hydroxide, barium hydroxide octahydrate	NaOH (aqueous)	$BaTiO_3$ nanowires	250	[34]
Barium hydroxide octahydrate, tetrabutyl titanate	NaOH (aqueous)	$BaTiO_3$ nanowires	60–300	[41]
Sodium hydroxide, niobium oxide	NaOH (aqueous)	$NaNbO_3$ nanowires	180	[42]
Sodium hydroxide, niobium oxide	NaOH	$NaNbO_3$ nanowires	10	[43]
Sodium hydroxide, potassium hydroxide, niobium pentoxide	KOH + NaOH (aqueous)	KNN nanorods	220–646	[44]

10.5 Electrospinning

Electrospinning is a technique that has been very widely exploited to produce NFs of ceramic, polymeric, and composite materials. This technique is a simple, versatile, low-cost, and effective method in which a high voltage is applied to a precursor solution as it is ejected from a metallic needle. High-voltage potential forces the droplet to deform into a conical shape known as a Taylor cone at the tip of the needle. When the solution properties are properly tuned, repulsion between the surface charges produces a stable jet that undergoes whipping and deformation and produces thin fibers with nano-scale diameters [45]. With this technique it is possible to produce ferroelectric NFs with diameters ranging from tens of nanometers to hundreds of micrometers, using a mixture of polymers, inorganic salts, and/or metal alkoxides. Electrospun fibers have a very high surface-to-volume ratio and a relatively defect-free structure at the molecular level, which aids in making high-mechanical-performance composite materials. Organic NFs can be obtained directly from this process, while inorganic NFs must undergo a postheat treatment. Often sol−gel-based precursor is used for the electrospinning process. Therefore a metallo-organic reactant and a polymer solution are used to tune the viscosity properties for fabricating ceramic NFs with different sizes, compositions, and morphologies, such as tubular, ribbons, and porosity, in a single step [46].

Electrospinning of PZT based NFs has been reported since the early 2000s. Wang et al. synthesized PZT by electrospinning using metallo-organic decomposition techniques and calcined at 850°C to get PZT NFs [47]. The formation of a perovskite PZT phase was observed as low as 550°C after calcination of the electrospun PZT/polyvinyl acetate composite fibers [48]. However, considerable interest has been observed in electrospinning PZT by using polyvinylpyrrolidone (PVP) as the polymer matrix. The PVP/PZT composite NFs were prepared by electrospinning and then annealed at 650°C to obtain PZT NFs with a perovskite phase and diameter of around 60 nm. Finally, PZT NFs were treated with an electric field of $4 \, kV \, mm^{-1}$ at a temperature over 140°C for about 24 h to attain the required piezoelectric properties. The poled NFs had orientated electric dipoles along the electric field with an output voltage of 1.63 V on applying a periodic finger pressure [49]. The effect of the poling process on the piezo-electric properties of PZT was studied by Chen et al.[50]. An increase in voltage output from 25 to 70 mV was observed on displacing the sample about 2 cm after 90 min of polarization under a $3 \, kV \, mm^{-1}$ electric field. The fabricated device generated an output voltage of about 100 mV in response to the falling ball (8.34 g) impact [50]. PZT nanowires were also fabricated by using precursor salts containing tetrabutyl titanate, zirconium acetylacetonate, lead subacetate, and PVP at a specific ratio. After electrospinning, PZT nanowires were prepared by calcination of the composite fiber at 650°C for about 3 h [51]. Electrospinning was also used to prepare vertically aligned Pb $(Zr_{0.52}Ti_{0.48})O_3$ nanowire arrays [52].

The first stand-alone electrospun complex oxide, lead-free ferroelectric NFs was claimed to be done by Yuh et al. in 2005 [53]. Perovskite $BaTiO_3$ NFs were achieved by electrospinning after heat treatment at 750°C for 1 h. The obtained NFs displayed diameters in the range of 80−190 nm, and the length exceeded 0.1 mm. $BaTiO_3$ fibrils less than 50 nm in diameter and fibers with ribbon-like morphology can be obtained by adjusting the concentration of alkoxide precursor in the electrospinning solution [54]. It was realized that a coarse surface morphology could be obtained when the diameter of the $BaTiO_3$ NFs was smaller than 100 nm, which refers to poor mechanical quality. Hence it was advised for $BaTiO_3$ NFs to sinter in nitrogen atmosphere for better surface morphology, microstructure, and elasticity [55]. The $BaTiO_3$ NFs, synthesized by electrospinning and incorporated into $BaTiO_3$ macrofibers, resulted in enhancing the piezoelectric properties, as reported by Sebastian et al. [56]. A novel sol−gel electrospinning method for ultra-flexible crystalline $BaTiO_3$ NF films was developed recently followed by calcination. It facilitates the formation of perovskite $BaTiO_3$ crystals with intricate grain boundaries at low temperatures by growing them within polymer NF templates [57]. The developed $BaTiO_3$ film has a polymer-like softness of 50 mN, a large Young's modulus of 61 MPa, and an elastic strain of 0.9%. For the electrospinning of $BaTiO_3$, researchers often use PVP as the polymer carrier. During the thermal decomposition of $BaTiO_3−PVP$ composite NFs, it is indicated that most of the acetate and organic groups were removed approximately at 700°C. The electrospinning process was also extended into coaxial mode to synthesize $BaTiO_3$ nanotubes [58]. A schematic representation of the setup is given in Fig. 10.5. $BaTiO_3$ precursor/ PVP−ethanol and heavy mineral oil were used as the shell solution and core liquid, respectively. XRD analysis revealed that the crystallization of $BaTiO_3$ nanotubes occurred between 550°C and 630°C [58]. Using Raman spectroscopy, Hedayati et al. demonstrated the presence of a small amount of tetragonal phase at early stages of crystallization of electrospun $BaTiO_3$ nanotubes [58].

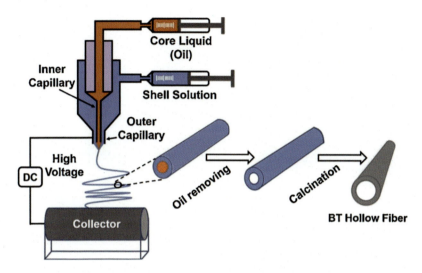

Figure 10.5 Schematic illustration of coaxial electrospinning and subsequent processes to achieve BaTiO₃ nanotubes.
Source: Reprinted from M. Hedayati, E. Taheri-Nassaj, A. Yourdkhani, M. Borlaf, J. Zhang, M. Calame, et al., BaTiO3 nanotubes by co-axial electrospinning: Rheological and microstructural investigations, J. Eur. Ceram. Soc. 2020;40:1269−1279. https://doi.org/ https://doi.org/10.1016/j.jeurceramsoc.2019.11.078. Copyright (2020) with permission from Elsevier.

Fig. 10.6 shows the TEM analysis of BaTiO₃ nanotubes produced by the coaxial electrospinning process along with the selected area electron diffraction (SAED) patterns and high-resolution TEM images of lattice fringes of BaTiO₃ nanotubes calcined at 850°C. BaTiO₃ nanotubes of uniform wall thickness, about 41 nm were obtained after calcination at 850°C, and SAED patterns reveal the polycrystalline structure of individual grains [58].

Lead-free KNN fibers were fabricated by using the electrospinning technique, and the effect of calcination temperature on the final phase composition was analyzed by Lusiola et al. [59]. A solution of potassium acetate, sodium methoxide, and niobium ethoxide dissolved in methanol, acetylacetone, and acetic acid was mixed with PVP dissolved in methanol, producing a viscous solution for electrospinning. Cylindrical and homogeneous KNN NFs (diameter range: 350−470 nm) were successfully produced by using the electrospinning process after calcining from 700°C to 1050°C [59]. KNN NF webs with random and aligned configurations were prepared by the electrospinning process from a modified PVP solution [60]. It was found that the perovskite phase with the preferred {100} orientation along the radial direction has a lower crystallization temperature of 472°C in the KNN NFs compared to 570°C for the KNN bulk gel and thus effectively suppressed the loss of alkali ions, leading to a well-controlled chemical

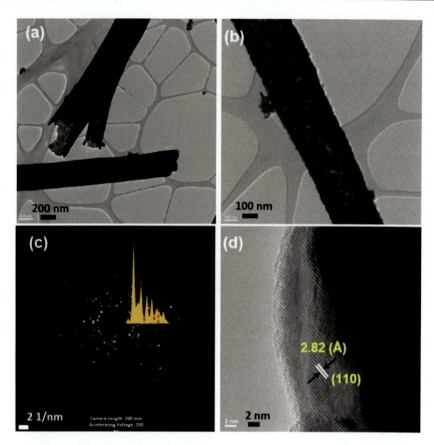

Figure 10.6 (A, B) transmission electron microscope morphological images of BaTiO$_3$ nanotube calcined at 850°C. (C) selected area electron diffraction patterns showing some planes of tetragonal phase of barium titanate. (D) High-resolution transmission electron microscope image showing lattice fringes.
Source: Reprinted from M. Hedayati, E. Taheri-Nassaj, A. Yourdkhani, M. Borlaf, J. Zhang, M. Calame, et al., BaTiO3 nanotubes by co-axial electrospinning: rheological and microstructural investigations, J. Eur. Ceram. Soc. 2020;40:1269–1279. https://doi.org/https://doi.org/10.1016/j.jeurceramsoc.2019.11.078. Copyright (2020) with permission from Elsevier.

stoichiometry. The high surface-to-volume ratio of NFs reduced the energy barrier due to surface-induced heterogeneous nucleation and resulted in the lower crystallization temperature [60]. Several interesting 1D lead-free ferroelectric nanostructures have been prepared over the years via the electrospinning method, such as (Bi$_{1/2}$Na$_{1/2}$)TiO$_3$-based NFs [61,62], Ba$_{0.6}$Sr$_{0.4}$TiO$_3$ NFs [63], BiFeO$_3$-based NFs [64,65], (Bi$_{3.15}$Nd$_{0.85}$Ti$_3$O$_{12}$) NFs [66], and (Bi$_{3.25}$La$_{0.75}$Ti$_3$O$_{12}$) NFs [67]. An overview of some ferroelectric and piezoelectric 1D nanostructures produced by electrospinning is given in Table 10.2.

Table 10.2 Overview of one dimenstional nanostructured ferroelectric and piezoelectric materials produced by electrospinning.

Raw material	Polymer	Composition	Diameter (nm)	References
Lead acetate trihydrate, zirconium nitrate pentahydrate, tetrabutyl titanate	PVP	$PbZr_{0.52}T_{0.48}O_3$	50−600	[68]
Lead acetate trihydrate, zirconium acetonate, tetrabutyl titanate	PVP	$PbZr_{0.52}T_{0.48}O_3$	100−300	[69]
Lead acetate trihydrate, zirconium n-propoxide, titanium isopropoxide	PVP	$PbZr_{0.52}T_{0.48}O_3$	100	[70]
Lead acetate trihydrate, lanthanum acetate trihydrate, zirconium acetate, titanium butoxide	PVP	$(Pb_{0.93}La_{0.07})(Zr_{0.6}Ti_{0.4})_{0.9825}O_3$	300	[71]
Barium acetate, calcium acetate hydrate, titanium isobutaoxide, zirconium acetylacetonate	PVP	$(Ba_{0.85}Ca_{0.15})(Ti_{0.9}Zr_{0.1})O_3$	120−180	[72]
Barium acetate, calcium acetate, titanium isopropoxide, zirconium propoxide	PVP	$(Ba_{0.85}Ca_{0.15})(Ti_{0.9}Zr_{0.1})O_3$	200	[73]
Barium acetate, calcium acetate monohydrate, tetrabutyl titanate, zirconium acetylacetonate, yttrium acetate tetrahydrate	PVP	$(Ba_{0.85}Ca_{0.15})(Ti_{0.9}Zr_{0.1})O_3-0.2 \, mol\%Y$	350	[74]
Barium acetate, titanium isopropoxide	PVP	$BaTiO_3$	60−200	[56]
Barium acetate, titanium isopropoxide	PVP	$BaTiO_3$	100−300	[75]
Barium acetate, titanium isopropoxide	PVP	$BaTiO_3$	200	[76]
Potassium acetate, sodium acetate, niobium ethoxide	PVP	$K_{0.5}Na_{0.5}NbO_3$	350	[73]
Sodium methoxide, potassium acetate, niobium ethoxide	PVP	$K_{0.5}Na_{0.5}NbO_3$	472	[59]
Zinc acetate dihydrate	PVA	ZnO	200	[77]
Zinc acetate dihydrate	PVP	ZnO	300−550	[78]
Bismuth nitrate, potassium nitrate, sodium acetate, titanium butoxide	PVP	$(Na_{0.82}K_{0.18})_{0.5}Bi_{0.5}TiO_3$	300−900	[79]
Bismuth nitrate, sodium nitrate, potassium nitrate, tetrabutyl titanate	PVP	$Bi_{0.5}Na_{0.5}TiO_3-Bi_{0.5}K_{0.5}TiO_3$	150	[80]

10.6 Molten salt synthesis

In molten salt synthesis an inorganic molten salt serves as a medium that enhances the reaction rate and reduces the reaction temperature. Owing to the fast mobility of the reactant species at medium and short diffusion distances, the reactions are usually carried out at moderate temperatures ($600°C-800°C$) within $1-2$ h [76,81]. Moreover, the morphology of the complex materials can be controlled by altering the process parameters, and 1D nanostructures can thus be obtained [82]. Single-crystalline $BaTiO_3$ nanowires and $SrTiO_3$ nanocubes were prepared with a simple one-step solid-state chemical reaction in the presence of NaCl and a nonionic surfactant. The mixture was heat-treated at $820°C$ for 3.5 h and suggested as a technique for large-scale synthesis of single-crystalline perovskite nanostructures [83].A surfactant-free synthesis of $MTiO_3$ (M = Pb, Ba, Sr) perovskite nanostrips was proposed in a nonaqueous molten salt medium. The fabricated nanostrips had a width of $50-200$ nm, a thickness of $20-50$ nm, and a length up to tens of micrometers [84]. Lead-free KNN single-crystalline nanorods with a high piezoelectric coefficient were also synthesized by using molten salt synthesis. Nanorods with a diameter in the range of $200-700$ nm and a length of $10-20$ μm were obtained [85]. Long single-crystal $BaTiO_3$ nanowires that have a uniform cylindrical structure with length from 20 to 80 μm and diameter from 100 nm to 1 μm with a tetragonal structure were also prepared in the molten salt medium [86]. Xue et al. have given a detailed review of recent progress in the molten salt synthesis of low-dimensional perovskite oxide nanostructures, structural characterization, properties, and functional applications [87].

10.7 Sol−gel template synthesis

Sol−gel template synthesis has been considered a simple and cheap method of fabricating 1D ferroelectric nanostructures. The sol−gel template method was used to fabricate PZT nanowires with diameters of about 45 nm and lengths of about 6 μm utilizing nanochannel alumina templates and heating at $700°C$ for 30 min. The same method was employed for fabricating PZT nanotubes. However, after thermal treatment, anodic aluminum oxide is dissolved out to obtain PZT nanotubes with an excellent aspect ratio and wall thickness [88]. The $BaTiO_3$ and $SrTiO_3$ nanorods were produced by solution-phase decomposition of bimetallic alkoxide precursors in the presence of coordinating ligands [89]. Well-isolated nanorods with diameters ranging from 5 to 60 nm and lengths reaching more than 10 μm could be achieved by this synthesis [90]. BT nanorods were also successfully synthesized by a combined sol−gel and surfactant-template method at low temperatures. Single-crystalline cubic perovskite $BaTiO_3$ nanorods with diameters ranging from 20 to 80 nm and lengths reaching more than 10 μm were fabricated by this route. Sol−gel processing with anodic aluminum oxide templates was extended to synthesize KNN nanotube arrays. The prepared samples had a monoclinic phase with an outer diameter of about 200 nm and a wall thickness of about 20 nm [91].

10.8 Material and structural characterizations

The use of piezoelectric NFs in applications strongly relies on the piezoelectric properties. Hence comprehensive materials characterization techniques are required to evaluate the molecular/crystal structure and piezoelectric properties of NFs. After the synthesis of piezoelectric NFs, typically their compositions, structures, and physical properties have to be evaluated. A series of characterization techniques are typically used to evaluate the chemistry, structures, and electromechanical properties of piezoelectric and ferroelectric NFs.

10.9 X-ray diffraction

X-ray diffraction (XRD) is an analytical technique that is commonly used for the qualitative and quantitative analysis of the crystalline structure of materials along with the unit cell and lattice strain. XRD patterns are experimentally or computationally based on crystal structure and Bragg's law. A major difficulty during the fabrication of PZT materials is the high partial pressure and the accompanying evaporation of lead monoxide during sintering, which results in a change in the crystalline phase composition of the fiber during thermal treatment above 1000°C. It is obvious that a high surface-to-volume ratio will promote the lead loss during sintering, which results in an overall change in the chemical composition [92]. To overcome this issue, lead-rich powder atmospheres are used during sintering, which results in a compositional and microstructure gradient across the fiber radius [93,94]. Compared to the bulk materials, removal of compositional gradient by mechanical grinding is not a viable option. Hence the performance of the fibers could be influenced, depending on the fiber diameter (larger influence as diameter decreases) [93,95]. Wang et al. have used XRD to observe the perovskite phase evolution with increase in temperature. At 600°C a coexistence of pyrochlore and perovskite phase appears, and above 850°C the pyrochlore phase was completely replaced by the perovskite phase [96]. In the case of $BaTiO_3$ nanotubes, increasing the calcination temperature to 975°C leads to a peak splitting of (200) planes to (002) and (200) (around $2\theta = 45$ degrees). This can be considered as the identifier for the tetragonal phase of $BaTiO_3$. However, the lack of an observable peak splitting in the $BaTiO_3$ samples heat-treated up to 950°C cannot be considered as evidence of the cubic $BaTiO_3$. Owing to the low c/a ratio in the nanograins of tetragonal BT, it is usually difficult to distinguish the tetragonal phase from the cubic phase by using the XRD measurements [97,98]. The Rietveld refinements on synchrotron radiation data revealed small tetragonality at lower temperatures [58].

10.10 Raman spectroscopy

Raman spectroscopy is a nondestructive spectroscopic technique that is used to observe vibrational, rotational, and various low-frequency modes in a system and provides information on the phase composition, crystallographic state, and orientation of materials. This technique can be used as a tool to identify substances from their characteristic spectral patterns (fingerprinting) [99]. Since the spatial resolution of most Raman spectrometers is close to micrometer scale, the results can be comparable to synchrotron XRD and neutron diffraction. In Raman scattering, inelastic scattering of light occurs when the light interacts with a molecule or crystal and forms a spectral line. Infrared spectroscopy and Raman scattering are often complementary and may be used together, to give a better view of the vibrational structure of a molecule, as they give quite different intensity patterns [100]. Raman spectroscopy has been used to characterize material properties such as crystallographic orientation, including the detection of the formation of PZT structures [96]. In the case of $BaTiO_3$ nanotubes, Raman spectroscopy was used to gain more insight into the effect of calcination temperature on the crystalline structure. Raman spectra revealed the existence of small tetragonal crystalline phases in $BaTiO_3$ already at an early stage of crystallization, while XRD or synchrotron diffraction detects only the cubic $BaTiO_3$ crystalline phase [58].

10.11 Atomic force microscopy

AFM is a powerful technique for the characterization of a variety of properties at nanoscales to atomic scales. It is a particularly suitable tool to investigate mechanical properties of nanomaterials, including NFs. A cantilever with a sharp-ended tip typically scans the surface of a sample. Applying a small electrical voltage between the support, on which the sample is placed, AFM can be used to evaluate electromechanical properties and the piezoelectric coefficient [101]. Owing to a very sharp-ended probe tip, a very high electric field can be generated by the application of a relatively low-voltage potential. Hence AFM works as piezoforce microscopy (PFM), and a local polarization of the sample can be investigated. By using electric potential across the ferroelectric nano-sized material, a controlled local switching of polarization in the ferroelectric crystal will occur, and the dynamics of domain nucleation, growth, and interaction can be investigated at a nano-scale resolution [102]. Structural and ferroelectric characterization of continuous well-aligned NFs of $BaTiO_3$ produced by the electrospinning technique in which the ferroelectric properties and domain structure was reported [75].

PFM has been used for the characterization of PZT fibers. Wang et al. have characterized the polarization domains of electrospun PZT fibers and have found that the domains of original fibers were randomly oriented to the substrate on which the fiber is deposited [103]. A postpoling process was necessary for nanogenerator applications. The direct piezoelectric coefficient of the polycrystalline $BiFeO_3$ NFs measured by using PFM showed a d_{33} of 11 pC/N, which is a smaller value compared to epitaxial films: 22 pC/N for 60-nm films and 43 pC/N for 400-nm films. It is believed that the NFs impose a clamping factor that decreases the $BiFeO_3$ piezoelectric property and restricts its application as NFs [65].

10.12 Potential applications

10.12.1 Nanogenerators

A sustainable energy source has been in extremely increasing demand to provide a solution for meeting current energy demands. The piezoelectric nanogenerator is considered to be a possible source and was first demonstrated in ZnO nanowire arrays [20]. Subsequently, both DC [104,105] and AC [106] nanogenerators with ZnO nanowire arrays were developed. A p-type nanowire was proved experimentally to generate an output signal 10 times stronger than that of a n-type ZnO nanowire [107]. To enhance the output power of the nanogenerator, the nanowires are stacked together in a systematic order. Schematics of two effective (vertical and lateral) integrations, developed by Wang et al., are shown in Fig. 10.7 [108–110]. In comparison to the energy conversion efficiency, it was shown that the laterally integrated nanowire could generate higher voltage than the vertical one [111,112]. Although the output voltage of nanogenerators has been raised to 1.2 V by using an integration of millions of ZnO nanowires [113], the output power that was generated was still too small to power any conventional electronic components; this can overcome by using materials with a higher piezoelectric coefficient. Recently, ZnO nanogenerators were fabricated by integrating 2D thin sheets, 1D nanorods, and 0D nanoparticles that are produced by hydrothermal synthesis, and the properties are compared. The maximum piezoelectric power output of 2.4 μW m^{-2} is obtained for

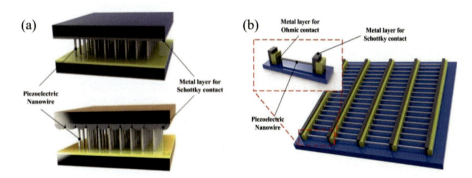

Figure 10.7 Geometrical configuration of a piezoelectric nanogenerator. (A) Vertically integrated. (B) Laterally integrated.
Source: Reprinted from M. Ani Melfa Roji, G. Jiji, T. Ajith Bosco Raj, A retrospect on the role of piezoelectric nanogenerators in the development of the green world, RSC Adv. 2017;7:33642–33670. https://doi.org/10.1039/C7RA05256A; S. Sripadmanabhan Indira, C. Aravind Vaithilingam, K.S. Oruganti, F. Mohd, S. Rahman, Nanogenerators as a sustainable power source: state of art, applications, and challenges, Nanomater. 2019;9. https://doi.org/10.3390/nano9050773. Copyright (2017) and Copyright (2019), respectively. This is an open-access article distributed under the terms of the Creative Commons Attribution License, which permits unrestricted use, distribution, and reproduction in any medium, provided the original author and source are credited.

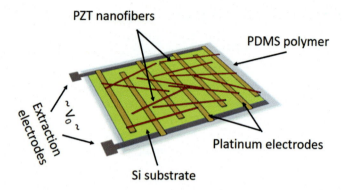

Figure 10.8 Schematic view of the lead zirconate titanate nanofiber generator.
Source: Reprinted (adapted) from X. Chen, S. Xu, N. Yao, Y. Shi, 1.6 V nanogenerator for mechanical energy harvesting using PZT nanofibers, Nano Lett. 2010;10:2133−2137. https://doi.org/10.1021/nl100812k. Copyright (2010) with permission from American Chemical Society.

2D nanosheets integrated nanogenerator, owing to the networking of sheets and the presence of in-between vacant spaces [114].

Long ferroelectric PZT NFs were produced by electrospinning and deposited laterally on comb-shaped platinum electrodes, and the whole device was covered in polydimethylsiloxane (PDMS) for flexibility by Chen et al. [49]. The peak output voltage from this nanogenerator was 1.63 V, and the output power was 0.03 μW with a load resistance of 6 MΩ (Fig. 10.8). A nanogenerator based on parallel PZT nanowires was fabricated by Wu et al. and generated a high output voltage and short-circuit current of 6 V and 45 nA, respectively [51]. The high piezoelectric constant (d_{33}) and the unique hierarchical structure of the PMN-PT nanowires were exploited to create a nanogenerator that generates an output voltage up to 7.8 V and an output current up to 2.29 μA (current density of 4.58 μA cm^{-2}). A nanogenerator with vertically aligned electrospun PZT NFs was fabricated that can generate a voltage output as high as 209 V and a current density of 23.5 μA cm^{-2}, which was directly used to stimulate a frog's sciatic nerve and induce contraction of its gastrocnemius [52].

Flexible lead-free piezoelectric nanogenerators have been investigated using BaTiO$_3$ NFs. The NFs were aligned vertically, horizontally, or randomly in PDMS elastomer as a matrix. A schematic of the fabrication of nanogenerators based on BaTiO$_3$ NFs in three alignment directions is shown in Fig. 10.9. The vertically aligned BaTiO$_3$ NFs showed the best piezoelectric performance. By applying a mechanical stress of 0.002 MPa, a maximum voltage output of 2.67 V and a current of 261.40 nA could be achieved [115].

Another study indicated that electrospun BaTiO$_3$ NFs with an average diameter of 45 nm embedded within polydimethylsiloxane produced a maximum voltage and power output of 7.94 V_{p-p} and 1.95 μW cm^{-2}, respectively, by applying a strain of 0.16% [116].

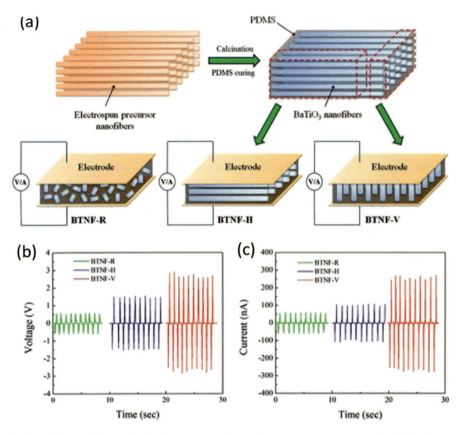

Figure 10.9 (A) Schematic fabrication procedure of the nanogenerators based on BaTiO$_3$ nanofibers in three alignment modes. (B) Voltage and (C) current outputs of the BaTiO$_3$ nanogenerators under a periodic mechanical compression.
Source: Reprinted from J. Yan, Y.G. Jeong, High performance flexible piezoelectric nanogenerators based on BaTiO3 nanofibers in different alignment modes, ACS Appl. Mater. Interfaces. 2016;8:15700−15709. https://doi.org/10.1021/acsami.6b02177. Copyright (2016) with permission from American Chemical Society.

A lead-free nanogenerator based on KNN electrospun flexible fibrous webs infiltrated with a PVDF polymer matrix could generate around 1.9 V [117]. However, using a melt-spun KNN nanorod-based nanogenerator, the output voltage could be increased up to 3.7 V [118]. Recently, a hybrid piezoelectric nanogenerator based on a combination of ferroelectric (KNN) and piezoelectric (ZnO) nanorods embedded in a PVDF matrix has been exploited by Bairagi et al. [119]. A voltage of 25 V and a current of 1.81 μA could be observed in compression mode (when pressure is applied by finger tapping). By using a standard sewing machine, a voltage of 8.31 V and a current of 5 μA could be achieved in compression mode [119]. A summary of inorganic NF nanogenerators is given in Table 10.3.

Table 10.3 Summary of inorganic nanofiber nanogenerators.

Inorganic nanofiber	Fiber diameter (nm)	Polymer	Force/strain	Electric outputs	Reference
ZnO nanorod arrays	~100	—	Finger pressure	current density 1 μAcm^{-2}	[104]
ZnO nanowire arrays	~40	—	Ultrasonic wave excitation	~-0.7 mV	[105]
ZnO nanowires	—	Polymethyl-methacrylate (PMMA)	Strain of 0.19%	1.26 V, 28.8 nA	[113]
ZnO nanorods	57	—	20 N	~0.25 V	[120]
Co−ZnO NFs	—	Poly(vinylidene fluoride-hexafluoropropylene) (PVDF-HFP)	Finger pressure	2.8 V	[121]
ZnO nanowires + PDMS	30	Polyimide	Finger pressure	0.7 V, 7 nA	[122]
Zinc Ferrite nanorods	~50	PVDF-HFP	Finger pressure	8.5 V	[123]
ZnO nanorods	1000	—	Finger pressure	185 mV, 5 nA	[114]
PZT nanowires	60	PDMS	Finger pressure	1.63 V	[49]
PZT NFs	50−120	Epoxy resin	Steel ball with a mass of 8.34 g	~100 mV	[50]
PZT nanowire array	16	PDMS	0.53 MPa	209 V, 53 μA	[52]
$(Pb_{0.93}La_{0.07})(Zr_{0.6}Ti_{0.4})_{0.9825}O_3$ nanowires	300	PDMS	Bending/releasing	5 V, 42 nA	[71]
$BaTiO_3$ NFs	354.1	PDMS	0.002 MPa	2.67 V, 261.4 nA	[115]
$BaTiO_3$ NFs	45	PDMS	0.16% bending strain	7.94 V_{p-p}	[116]

BaTiO$_3$ nanotube arrays	O.D ~ 130I. D ~ 82	PDMS	0.283% bending strain	0.7 to 1.0 V, 10 to 20 nA	[124]
BaTiO$_3$ nanowires	~ 156	PDMS	5-mm maximum horizontal displacement	7.0 V, 360 nA	[125]
0.5Ba(Zr$_{0.2}$Ti$_{0.8}$)O$_3$−0.5(Ba$_{0.7}$Ca$_{0.3}$) TiO$_3$ nanowires	175	PDMS	Stretch/release using linear motor	3.25 V, 55 nA	[126]
Ba$_{0.85}$Ca$_{0.15}$Zr$_{0.1}$Ti$_{0.9}$O$_3$ NFs	80−250	room temperature− vulcanizing silicone elastomer	Finger pressure	~ 2.68 V, ~ 1.1 μA	[127]
Dopamine-modified Ba$_{0.85}$Ca$_{0.15}$Zr$_{0.1}$Ti$_{0.9}$O$_3$ NFs	120−180	PVDF-HFP	Finger pressure	~ 10 V, ~ 0.78 μA	[72]
(Ba$_{0.85}$Ca$_{0.15}$)(Ti$_{0.9}$Zr$_{0.1}$)O$_3$, (Ba$_{0.85}$Ca$_{0.15}$)(Ti$_{0.9}$Zr$_{0.1}$)O$_3$−Y NFs	350	PDMS	Finger pressure	3 V, 85 nA	[74]
(K,Na)NbO$_3$ NFs	100	PDMS	Finger pressure	1.6 V	[128]
KNN-based nanofibrous web	25	PVDF	Finger pressure	1.9 V	[117]
0.96 (K$_{0.48}$Na$_{0.52}$)(Nb$_{0.95}$Sb$_{0.05}$) O$_3$−0.04Bi$_{0.5}$(Na$_{0.82}$K$_{0.18}$)$_{0.5}$ZrO$_3$ NFs	250	PDMS	Dynamic pressure at 1 Hz	10 V	[129]
KNN + ZnO + PVDF	400	PVDF	Finger pressure, standard sewing machine	25 V, 1.81 μA 8.31 V, 5 μA	[119]

10.12.2 High-energy-density storage devices

The interest in high-performance electrical storage devices has increased in the last years because of the increased energy demand resulting from hybrid electric vehicles, transportable electrical and electronic devices, and stationary power systems. To increase the energy density of a capacitor, the breakdown strength and the dielectric displacement of the material must be increased. One interesting option to simultaneously increase both properties is the development of a ferroelectric ceramic-polymer composite. Polymer offers high flexibility, high breakdown strength, and low dielectric loss tangent, whereas ceramic offers high dielectric permittivity. $BaTiO_3$ NFs have become an interesting filler material for high-energy-density capacitors based on ceramic-polymer composites. Electrospun $BaTiO_3$ NFs were successfully introduced into polyimide matrix after calcination and obtained high dielectric permittivity. A dielectric permittivity of 27 at 100 Hz and very low tan δ (0.015) was obtained at 30 vol% loading of NFs [130]. It was realized that improving the interface and compatibility between the nanofiller and polymer matrix can improve the dielectric properties. Liu et al. functionalized $BaTiO_3$ NFs with 3-aminopropyltriethoxysilane (APS) before incorporating them into the polymer matrix polyvinylidene fluoride (PVDF). The dielectric permittivity of the composite improved up to 24 at 1 kHz along with a tan δ of 0.018 at 7.5 vol.% NF loading. The energy storage density of this composite was calculated to be about $5.6 \, J \, cm^{-3}$ with 2.5 vol.% of $BaTiO_3$ NFs with APS at $330 \, kV \, mm^{-1}$ [131]. Dopamine-modified $BaTiO_3$ nanotubes were added to PVDF which increases the dielectric permittivity to 47.05 with 10.8 vol.% $BaTiO_3$ NFs at 1 kHz and a higher energy storage density of $7.03 \, J \, cm^{-3}$ could be achieved at 2.1 vol % $BaTiO_3$ nanotubes loading at an electric field of $350 \, kV \, mm^{-1}$ [42]. A novel strategy of functionalizing $BaTiO_3$ nanowires with bioinspired fluoro-polydopamine and a fluoropolymer polymer matrix resulted in a high energy density of about $12.9 \, J \, cm^{-3}$, at a reasonably small electric field ($540 \, kV \, mm^{-1}$) with 5 vol.% $BaTiO_3$ nanowires loading [132]. A comprehensive review covering the latest advances in polymer matrix nanocomposites incorporating different 1D nanofillers for high energy storage applications was published recently [133].

10.12.3 Structural health monitoring

Structural health monitoring is analogous to mimicking the human nervous system to detect flaws in a structure or material. It is used primarily for crack detection in structures and materials in a nondestructive way. Piezoelectric acoustic emission is one of the most widely exploited techniques where an ultrasonic wave is allowed to pass through the structure and is analyzed for structural defects. Active fiber composites (AFCs) containing PZT microfibers embedded in a polymer matrix gained high attention in the early part of this century [134,135]. However, the high stiffness of AFCs makes it difficult to mount them on curved surfaces. Embedding 1D NFs seems to be an efficient way to tackle this problem. An electrospun PZT NF composite with high piezoelectric voltage constant (g_{33}, $0.079 \, Vm \, N^{-1}$) imparts high flexibility, sensitivity, and mechanical strength and makes it possible to mount electrospun NF-based AFCs on a curved surface [136]. In another paper, electrospun PZT NFs with diameters varying from 50 to 120 nm were embedded into an approximately 5-μm-thick polymeric film with interdigitated electrodes. Significant improvement in electromechanical coupling (3.7 times) was observed after 90 min of polarization under an external electric field of ~ 3 V μm^{-1} [50]. The influence of poling lead-free piezoelectric NFs, $78Bi_{0.5}Na_{0.5}TiO_3-22SrTiO_3$, on sensor characteristics was studied by simulating several situations for structural health-monitoring sensor applications [137].

10.13 Summary and outlook

In recent years, huge interest in piezoelectric and ferroelectric nanostructured 1D materials has been observed, particularly for energy-related applications. Therefore in this chapter we discussed the recent developments in 1D piezoelectric and ferroelectric materials and their synthesis, characterizations, and applications. Since early 2000, attempts have been made to replace lead, which is toxic, with environmental friendly materials as a result of worldwide initiatives such as EU directives on waste electrical and electronic equipment and recycling of hazardous waste substances [138,139]. Therefore the development of lead-free materials is becoming more and more popular for the development of nanostructured 1D materials. KNN is considered to be next in line, although its sinterability issues need to be overcome.

For 1D nanostructure fabrication, electrospinning is probably the most preferred technique, owing to its versatility and low production cost. The morphology change from fiber to tube (hollow fiber) can be done with a mere change in the setup, making this technique attractive. However, an industrial scale-up is still a concern because of low yield and parameter optimization for reproducibility. Considerable interest in energy harvesters, nanogenerators, and high-energy-density devices can be seen as application fields for 1D piezoelectric and ferroelectric nanostructures.

References

[1] G.H. Haertling, Ferroelectric ceramics: history and technology, J. Am. Ceram. Soc. 82 (1999) 797−818. Available from: https://doi.org/10.1111/j.1151-2916.1999.tb01840.x.

[2] C.A. Rosen, Ceramic transformers and filters, in Proceedings of the Electronic Components Symposium, 1957, pp. 205−211. http://ci.nii.ac.jp/naid/10004082548/en/ (accessed 17.06.2020).

[3] G. Zhang, S. Xu, Y. Shi, Electromechanical coupling of lead zirconate titanate nanofibres, Micro Nano Lett 6 (2011) 59−61. Available from: https://doi.org/10.1049/mnl.2010.0127.

[4] M.H. Zhao, Z.L. Wang, S.X. Mao, Piezoelectric characterization individual zinc oxide nanobelt probed by piezoresponse force microscope, Nano Lett 4 (2004). Available from: https://doi.org/10.1021/nl035198a.

[5] L. Liang, X. Kang, Y. Sang, H. Liu, One-dimensional ferroelectric nanostructures: synthesis, properties, and applications, Adv. Sci. 3 (2016). Available from: https://doi.org/10.1002/advs.201500358.

[6] A.M. Glazer, Crystallography: A Very Short Introduction, Oxford University Press, 2016.

[7] L. Kou, W. Guo, C. Li, Piezoelectricity of ZnO and its nanostructures, in: 2008 Symposium of Piezoelectricity, Acoustics Waves, Device Applications, 2008, pp. 354−359. https://doi.org/10.1109/SPAWDA.2008.4775808.

[8] Z.L. Wang, The new field of nanopiezotronics, Mater. Today. 10 (2007) 20−28. Available from: https://doi.org/10.1016/S1369-7021(07)70076-7.

[9] Z.L. Wang, ZnO nanowire and nanobelt platform for nanotechnology, Mater. Sci. Eng. R. Rep 64 (2009) 33−71. Available from: https://doi.org/10.1016/j.mser.2009.02.001.

[10] M. Willander, O. Nur, Q.X. Zhao, L.L. Yang, M. Lorenz, B.Q. Cao, et al., Zinc oxide nanorod based photonic devices: recent progress in growth, light emitting diodes and lasers, Nanotechnology 20 (2009) 332001. Available from: https://doi.org/10.1088/0957-4484/20/33/332001.

[11] D. Damjanovic, Ferroelectric, dielectric and piezoelectric properties of ferroelectric thin films and ceramics, Rep. Prog. Phys 61 (1998) 1267−1324. Available from: https://doi.org/10.1088/0034-4885/61/9/002.

[12] H. Jaffe, Piezoelectric ceramics, J. Am. Ceram. Soc. 41 (1958) 494−498. Available from: https://doi.org/10.1111/j.1151-2916.1958.tb12903.x.

[13] A.J. Moulson, J.M. Herbert, Piezoelectric ceramics, Electroceramics (2003) 339−410. Available from: https://doi.org/10.1002/0470867965.ch6.

[14] T.R. Shrout, S.J. Zhang, Lead-free piezoelectric ceramics: alternatives for PZT? J. Electroceram 19 (2007) 113−126. Available from: https://doi.org/10.1007/s10832-007-9047-0.

[15] B. Jaffe, W.R. Cook, H. Jaffe, Chapter 6—Properties of $PbTiO_3$, $PbZrO_3$, $PbSnO_3$, and $PbHfO_3$ plain and modified, Piezoelectric Ceramics, Academic Press, 1971, pp. 115−134. Available from: https://doi.org/10.1016/B978-0-12-379550-2.50010-7.

[16] EU Directive 2002/95/EC: Restriction of the use of certain hazardous substances in electrical and electronic equipment (RoHS), Off. J. Eur. Union 2003;46(L37):19−23.

[17] S.E. Park, T.R. Shrout, Characteristics of relaxor-based piezoelectric single crystals for ultrasonic transducers, IEEE Trans. Ultrason. Ferroelectr. Freq. Control. 44 (1997) 1140−1147. Available from: https://doi.org/10.1109/58.655639.

[18] S.E. Park, T.R. Shrout, Ultrahigh strain and piezoelectric behavior in relaxor based ferroelectric single crystals, J. Appl. Phys. 82 (1997) 1804−1811. Available from: https://doi.org/10.1063/1.365983.

[19] G.L. Messing, S. Trolier-McKinstry, E.M. Sabolsky, C. Duran, S. Kwon, B. Brahmaroutu, et al., Templated grain growth of textured piezoelectric ceramics, Crit. Rev. Solid. State Mater. Sci. 29 (2004). Available from: https://doi.org/10.1080/10408430490490905.

[20] Z.L. Wang, J. Song, Piezoelectric nanogenerators based on zinc oxide nanowire arrays, Sci. 80 (2006). Available from: https://doi.org/10.1126/science.1124005.

[21] X. Li, M. Sun, X. Wei, C. Shan, Q. Chen, 1D piezoelectric material based nanogenerators: methods, materials and property optimization, Nanomaterials (2018). Available from: https://doi.org/10.3390/nano8040188.

[22] K. Zheng, Z. Zhang, Y. Hu, P. Chen, W. Lu, J. Drennan, et al., Orientation dependence of electromechanical characteristics of defect-free InAs nanowires, Nano Lett 16 (2016) 1787−1793. Available from: https://doi.org/10.1021/acs.nanolett.5b04842.

[23] P. Sá, J. Barbosa, I. Bdikin, B. Almeida, A.G. Rolo, E. de, et al., Ferroelectric characterization of aligned barium titanate nanofibres, J. Phys. D. Appl. Phys 46 (2013) 105304. Available from: https://doi.org/10.1088/0022-3727/46/10/105304.

[24] Y. Ozeren, E. Mensur-Alkoy, S. Alkoy, Sodium niobate particles with controlled morphology synthesized by hydrothermal method and their use as templates in KNN fibers, Adv. Powder Technol. 25 (2014) 1825−1833. Available from: https://doi.org/10.1016/j.apt.2014.07.012.

[25] A. Rabenau, The role of hydrothermal synthesis in preparative chemistry, Angew. Chemie Int. (Ed.) English. 1985;24:1026−1040. https://doi.org/10.1002/anie.198510261.

[26] S. Ji, H. Liu, Y. Sang, W. Liu, G. Yu, Y. Leng, Synthesis, structure, and piezoelectric properties of ferroelectric and antiferroelectric NaNbO$_3$ nanostructures, CrystEngComm 16 (2014) 7598−7604. Available from: https://doi.org/10.1039/C4CE01116C.

[27] K. Shimomura, T. Tsurumi, Y. Ohba, M. Daimon, Preparation of lead zirconate titanate thin film by hydrothermal method, Jpn. J. Appl. Phys. 30 (1991) 2174−2177. Available from: https://doi.org/10.1143/jjap.30.2174.

[28] C. Chen, X. Han, J. Liu, Z. Ding, Preparation and piezoelectric properties of textured PZT ceramics using PZT nano fibers, in: 2011 Symposium on Piezoelectricity, Acoustic Waves Device Applications, 2011, pp. 126−129. https://doi.org/10.1109/SPAWDA.2011.6167208.

[29] G. Xu, Z. Ren, P. Du, W. Weng, G. Shen, G. Han, Polymer-assisted hydrothermal synthesis of single-crystalline tetragonal perovskite PbZr$_{0.52}$Ti$_{0.48}$O$_3$ nanowires, Adv. Mater. 17 (2005) 907−910. Available from: https://doi.org/10.1002/adma.200400998.

[30] Y. Deng, J.X. Zhou, Y.L. Du, K.R. Zhu, Y. Hu, D. Wu, et al., Polymer-assisted hydrothermal synthesis of single crystal Pb$_{1-x}$La$_x$TiO$_3$ nanorods, Mater. Lett. 63 (2009) 937−939. Available from: https://doi.org/10.1016/j.matlet.2009.01.043.

[31] S. Xu, G. Poirier, N. Yao, PMN-PT nanowires with a very high piezoelectric constant, Nano Lett 12 (2012) 2238−2242. Available from: https://doi.org/10.1021/nl204334x.

[32] C.C. Bowland, M.H. Malakooti, Z. Zhou, H.A. Sodano, Highly aligned arrays of high aspect ratio barium titanate nanowires via hydrothermal synthesis, Appl. Phys. Lett. 106 (2015) 222903. Available from: https://doi.org/10.1063/1.4922277.

[33] Y. Feng, W.L. Li, Y.F. Hou, Y. Yu, W.P. Cao, T.D. Zhang, et al., Enhanced dielectric properties of PVDF-HFP/BaTiO$_3$-nanowire composites induced by interfacial polarization and wire-shape, J. Mater. Chem. C. 3 (2015) 1250−1260. Available from: https://doi.org/10.1039/C4TC02183E.

[34] L. Tu, Y. You, C. Liu, C. Zhan, Y. Wang, M. Cheng, et al., Enhanced dielectric and energy storage properties of polyarylene ether nitrile composites incorporated with barium titanate nanowires, Ceram. Int. 45 (2019) 22841−22848. Available from: https://doi.org/10.1016/j.ceramint.2019.07.326.

[35] L. Wang, X. Deng, J. Li, X. Liao, G. Zhang, C. Wang, et al., Hydrothermal synthesis of tetragonal BaTiO$_3$ nanotube arrays with high dielectric performance, J. Nanosci. Nanotechnol. 14 (2014) 4224−4228. Available from: https://doi.org/10.1166/jnn.2014.7782.

[36] Z. Wang, H. Gu, Y. Hu, K. Yang, M. Hu, D. Zhou, et al., Synthesis, growth mechanism and optical properties of (K,Na)NbO$_3$ nanostructures, CrystEngComm 12 (2010) 3157−3162. Available from: https://doi.org/10.1039/C000169D.

[37] C. Sun, X. Xing, J. Chen, J. Deng, L. Li, R. Yu, et al., Hydrothermal synthesis of single crystalline (K,Na)NbO$_3$ powders, Eur. J. Inorg. Chem. 2007 (2007) 1884−1888. Available from: https://doi.org/10.1002/ejic.200601131.

[38] Y. Xu, Q. Yu, J.-F. Li, A facile method to fabricate vertically aligned (K,Na)NbO$_3$ lead-free piezoelectric nanorods, J. Mater. Chem. 22 (2012) 23221−23226. Available from: https://doi.org/10.1039/C2JM35090D.

[39] J. Wang, A. Durussel, C.S. Sandu, M.G. Sahini, Z. He, N. Setter, Mechanism of hydrothermal growth of ferroelectric PZT nanowires, J. Cryst. Growth 347 (2012) 1−6. Available from: https://doi.org/10.1016/j.jcrysgro.2012.03.022.

[40] Y. Lin, Y. Liu, H.A. Sodano, Hydrothermal synthesis of vertically aligned lead zirconate titanate nanowire arrays, Appl. Phys. Lett. 95 (2009) 122901. Available from: https://doi.org/10.1063/1.3237170.

[41] B. Xie, H. Zhang, H. Kan, S. Liu, M.-Y. Li, Z. Li, et al., Mechanical force-driven growth of elongated $BaTiO_3$ lead-free ferroelectric nanowires, Ceram. Int. 43 (2017) 2969−2973. Available from: https://doi.org/10.1016/j.ceramint.2016.11.049.

[42] Z. Pan, L. Yao, J. Zhai, B. Shen, H. Wang, Significantly improved dielectric properties and energy density of polymer nanocomposites via small loaded of $BaTiO_3$ nanotubes, Compos. Sci. Technol. 147 (2017) 30−38. Available from: https://doi.org/10.1016/j.compscitech.2017.05.004.

[43] Q. Gu, K. Zhu, N. Zhang, Q. Sun, P. Liu, J. Liu, et al., Modified solvothermal strategy for straightforward synthesis of cubic $NaNbO_3$ nanowires with enhanced photocatalytic H2 evolution, J. Phys. Chem. C. 119 (2015) 25956−25964. Available from: https://doi.org/10.1021/acs.jpcc.5b08018.

[44] S. Bairagi, S.W. Ali, Flexible lead-free PVDF/SM-KNN electrospun nanocomposite based piezoelectric materials: significant enhancement of energy harvesting efficiency of the nanogenerator, Energy 198 (2020) 117385. Available from: https://doi.org/10.1016/j.energy.2020.117385.

[45] D. Li, Y. Xia, Electrospinning of nanofibers: reinventing the wheel? Adv. Mater. 16 (2004) 1151−1170. Available from: https://doi.org/10.1002/adma.200400719.

[46] R. Ramaseshan, S. Sundarrajan, R. Jose, S. Ramakrishna, Nanostructured ceramics by electrospinning, J. Appl. Phys. 102 (2007) 111101. Available from: https://doi.org/10.1063/1.2815499.

[47] Y. Wang, S. Serrano, J.J. Santiago-Aviles, Electrostatic synthesis and characterization of $Pb(ZrxTi_{1-x})O_3$ micro/nano-fibers, MRS Proc 702 (2001). U10.2.1. https://doi.org/DOI: 10.1557/PROC-702-U10.2.1.

[48] N. Dharmaraj, C.H. Kim, H.Y. Kim, $Pb(Zr_{0.5},Ti_{0.5})O_3$ nanofibres by electrospinning, Mater. Lett. 59 (2005) 3085−3089. Available from: https://doi.org/10.1016/j.matlet.2005.05.040.

[49] X. Chen, S. Xu, N. Yao, Y. Shi, 1.6 V Nanogenerator for mechanical energy harvesting using PZT nanofibers, Nano Lett 10 (2010) 2133−2137. Available from: https://doi.org/10.1021/nl100812k.

[50] X. Chen, S. Guo, J. Li, G. Zhang, M. Lu, Y. Shi, Flexible piezoelectric nanofiber composite membranes as high performance acoustic emission sensors, Sens. Actuators A Phys 199 (2013) 372−378. Available from: https://doi.org/10.1016/j.sna.2013.06.011.

[51] W. Wu, S. Bai, M. Yuan, Y. Qin, Z.L. Wang, T. Jing, Lead zirconate titanate nanowire textile nanogenerator for wearable energy-harvesting and self-powered devices, ACS Nano 6 (2012) 6231−6235. Available from: https://doi.org/10.1021/nn3016585.

[52] L. Gu, N. Cui, L. Cheng, Q. Xu, S. Bai, M. Yuan, et al., Flexible fiber nanogenerator with 209 V output voltage directly powers a light-emitting diode, Nano Lett 13 (2013) 91−94. Available from: https://doi.org/10.1021/nl303539c.

[53] J. Yuh, J.C. Nino, W.M. Sigmund, Synthesis of barium titanate ($BaTiO_3$) nanofibers via electrospinning, Mater. Lett. 59 (2005) 3645−3647. Available from: https://doi.org/10.1016/j.matlet.2005.07.008.

[54] J.T. McCann, J.I.L. Chen, D. Li, Z.-G. Ye, Y. Xia, Electrospinning of polycrystalline barium titanate nanofibers with controllable morphology and alignment, Chem. Phys. Lett. 424 (2006) 162−166. Available from: https://doi.org/10.1016/j.cplett.2006.04.082.

[55] Y. Zhuang, X. Wei, Y. Zhao, J. Li, X. Fu, Q. Hu, et al., Microstructure and elastic properties of $BaTiO_3$ nanofibers sintered in various atmospheres, Ceram. Int. 44 (2018) 2426−2431. Available from: https://doi.org/10.1016/j.ceramint.2017.10.213.

[56] T. Sebastian, A. Michalek, M. Hedayati, T. Lusiola, F. Clemens, Enhancing dielectric properties of barium titanate macrofibers, J. Eur. Ceram. Soc. 39 (2019) 3716−3721. Available from: https://doi.org/10.1016/j.jeurceramsoc.2019.05.040.

[57] J. Yan, Y. Han, S. Xia, X. Wang, Y. Zhang, J. Yu, et al., Polymer template synthesis of flexible $BaTiO_3$ crystal nanofibers, Adv. Funct. Mater. 29 (2019) 1907919. Available from: https://doi.org/10.1002/adfm.201907919.

[58] M. Hedayati, E. Taheri-Nassaj, A. Yourdkhani, M. Borlaf, J. Zhang, M. Calame, et al., $BaTiO_3$ nanotubes by co-axial electrospinning: rheological and microstructural investigations, J. Eur. Ceram. Soc. 40 (2020) 1269−1279. Available from: https://doi.org/10.1016/j.jeurceramsoc.2019.11.078.

[59] T. Lusiola, L. Gorjan, F. Clemens, Preparation and characterization of potassium sodium niobate nanofibers by electrospinning, Int. J. Appl. Ceram. Technol. 15 (2018) 1292−1300. Available from: https://doi.org/10.1111/ijac.12883.

[60] Y.M. Yousry, K. Yao, X. Tan, A.M. Mohamed, Y. Wang, S. Chen, et al., Structure and high performance of lead-free $(K_{0.5}Na_{0.5})NbO_3$ piezoelectric nanofibers with surface-induced crystallization at lowered temperature, ACS Appl. Mater. Interfaces 11 (2019) 23503−23511. Available from: https://doi.org/10.1021/acsami.9b05898.

[61] D. Liu, Y. Song, Z. Xin, G. Liu, C. Jin, F. Shan, High-piezocatalytic performance of eco-friendly $(Bi_{1/2}Na_{1/2})TiO_3$-based nanofibers by electrospinning, Nano Energy 65 (2019) 104024. Available from: https://doi.org/10.1016/j.nanoen.2019.104024.

[62] S. Ji, J. Yun, Fabrication and characterization of aligned flexible lead-free piezoelectric nanofibers for wearable device applications, Nanomaterials 8 (2018) 206. Available from: https://doi.org/10.3390/nano8040206.

[63] S. Maensiri, W. Nuansing, J. Klinkaewnarong, P. Laokul, J. Khemprasit, Nanofibers of barium strontium titanate (BST) by sol−gel processing and electrospinning, J. Colloid Interface Sci 297 (2006) 578−583. Available from: https://doi.org/10.1016/j.jcis.2005.11.005.

[64] S. You, B. Zhang, Enhanced magnetic properties of cobalt-doped bismuth ferrite nanofibers, Mater. Res. Exp. 7 (2020) 46102. Available from: https://doi.org/10.1088/2053-1591/ab83a4.

[65] A. Queraltó, R. Frohnhoven, S. Mathur, A. Gómez, Intrinsic piezoelectric characterization of $BiFeO_3$ nanofibers and its implications for energy harvesting, Appl. Surf. Sci. 509 (2020) 144760. Available from: https://doi.org/10.1016/j.apsusc.2019.144760.

[66] M. Liao, X.L. Zhong, J.B. Wang, H.L. Yan, J.P. He, Y. Qiao, et al., Nd-substituted bismuth titanate ferroelectric nanofibers by electrospinning, J. Cryst. Growth. 304 (2007) 69−72. Available from: https://doi.org/10.1016/j.jcrysgro.2007.01.038.

[67] M. Tang, W. Shu, F. Yang, J. Zhang, G. Dong, J. Hou, The fabrication of La-substituted bismuth titanate nanofibers by electrospinning, Nanotechnology 20 (2009) 385602. Available from: https://doi.org/10.1088/0957-4484/20/38/385602.

[68] Y. Zhang, X. Liu, J. Yu, M. Fan, X. Ji, B. Sun, et al., Optimizing the dielectric energy storage performance in P(VDF-HFP) nanocomposite by modulating the diameter of PZT nanofibers prepared via electrospinning, Compos. Sci. Technol. 184 (2019) 107838. Available from: https://doi.org/10.1016/j.compscitech.2019.107838.

[69] X. Hou, S. Zhang, J. Yu, M. Cui, J. He, L. Li, et al., Flexible piezoelectric nanofibers/polydimethylsiloxane-based pressure sensor for self-powered human motion monitoring, Energy Technol 8 (2020) 1901242. Available from: https://doi.org/10.1002/ente.201901242.

[70] J. He, X. Guo, J. Yu, S. Qian, X. Hou, M. Cui, et al., A high-resolution flexible sensor array based on PZT nanofibers, Nanotechnology 31 (2020) 155503. Available from: https://doi.org/10.1088/1361-6528/ab667a.

[71] C.C. Jin, X.C. Liu, C.H. Liu, H.L. Hwang, Q. Wang, Preparation and structure of aligned PLZT nanowires and their application in energy harvesting, Appl. Surf. Sci. 447 (2018) 430−436. Available from: https://doi.org/10.1016/j.apsusc.2018.03.251.

[72] K.S. Chary, V. Kumar, C.D. Prasad, H.S. Panda, Dopamine-modified $Ba_{0.85}Ca_{0.15}Zr_{0.1}Ti_{0.9}O_3$ ultra-fine fibers/PVDF-HFP composite−based nanogenerator: synergistic effect on output electric signal, J. Aust. Ceram. Soc. (2020). Available from: https://doi.org/10.1007/s41779-020-00458-0.

[73] A. Jalalian, A.M. Grishin, S.X. Dou, Ferroelectric and ferromagnetic nanofibers: synthesis, properties and applications, J. Phys. Conf. Ser. 352 (2012) 12006. Available from: https://doi.org/10.1088/1742-6596/352/1/012006.

[74] Y. Wu, F. Ma, J. Qu, Y. Luo, C. Lv, Q. Guo, et al., Vertically-aligned lead-free BCTZY nanofibers with enhanced electrical properties for flexible piezoelectric nanogenerators, Appl. Surf. Sci. 469 (2019) 283−291. Available from: https://doi.org/10.1016/j.apsusc.2018.10.229.

[75] P. Sá, I. Bdikin, B. Almeida, A.G. Rolo, D. Isakov, Production and PFM characterization of barium titanate nanofibers, Ferroelectrics 429 (2012) 48−55. Available from: https://doi.org/10.1080/00150193.2012.676950.

[76] X. Xu, Z. Wu, L. Xiao, Y. Jia, J. Ma, F. Wang, et al., Strong piezo-electro-chemical effect of piezoelectric $BaTiO_3$ nanofibers for vibration-catalysis, J. Alloy. Compd 762 (2018) 915−921. Available from: https://doi.org/10.1016/j.jallcom.2018.05.279.

[77] A. Baranowska-Korczyc, A. Reszka, K. Sobczak, B. Sikora, P. Dziawa, M. Aleszkiewicz, et al., Magnetic Fe doped ZnO nanofibers obtained by electrospinning, J. Sol−Gel Sci. Technol. 61 (2012) 494−500. Available from: https://doi.org/10.1007/s10971-011-2650-1.

[78] J. Guo, Y. Song, D. Chen, X. Jiao, Fabrication of ZnO nanofibers by electrospinning and electrical properties of a single nanofiber, J. Dispers. Sci. Technol. 31 (2010) 684−689. Available from: https://doi.org/10.1080/01932690903212222.

[79] Y.Q. Chen, X.J. Zheng, X. Feng, S.H. Dai, D.Z. Zhang, Fabrication of lead-free $(Na_{0.82}K_{0.18})_{0.5}Bi_{0.5}TiO_3$ piezoelectric nanofiber by electrospinning, Mater. Res. Bull 45 (2010) 717−721. Available from: https://doi.org/10.1016/j.materresbull.2010.02.013.

[80] Q. Yang, D. Wang, M. Zhang, T. Gao, H. Xue, Z. Wang, et al., Lead-free $(Na_{0.83}K_{0.17})_{0.5}Bi_{0.5}TiO_3$ nanofibers for wearable piezoelectric nanogenerators, J. Alloy. Compd 688 (2016) 1066−1071. Available from: https://doi.org/10.1016/j.jallcom.2016.07.131.

[81] T. Phatungthane, G. Rujijanagul, Observation of very high dielectric constant in $Sr(Fe_{1-x}Al_x)_{0.5}Nb_{0.5}O_3$ ceramics prepared by molten salt technique, Ceram. Int. 41 (2015) S841−S845. Available from: https://doi.org/10.1016/j.ceramint.2015.03.225.

[82] P.-H. Xiang, Y. Kinemuchi, K. Watari, Synthesis of layer-structured ferroelectric Bi_3NbTiO_9 plate-like seed crystals, Mater. Lett. 59 (2005) 1876−1879. Available from: https://doi.org/10.1016/j.matlet.2005.02.003.

[83] Y. Mao, S. Banerjee, S.S. Wong, Large-scale synthesis of single-crystalline perovskite nanostructures, J. Am. Chem. Soc. 125 (2003) 15718−15719. Available from: https://doi.org/10.1021/ja038192w.

[84] H. Deng, Y. Qiu, S. Yang, General surfactant-free synthesis of $MTiO_3$ (M = Ba, Sr, Pb) perovskite nanostrips, J. Mater. Chem. 19 (2009) 976−982. Available from: https://doi.org/10.1039/B815698K.

[85] L.-Q. Cheng, K. Wang, Q. Yu, J.-F. Li, Structure and composition characterization of lead-free $(K, Na)NbO_3$ piezoelectric nanorods synthesized by the molten-salt reaction, J. Mater. Chem. C. 2 (2014) 1519−1524. Available from: https://doi.org/10.1039/C3TC32148G.

[86] H. Zhao, G. Yang, Z. Wang, X. Cao, L. Gu, N. Zhao, Molten salt route of single crystal barium titanate nanowires, J. Exp. Nanosci. 10 (2015) 1126−1136. Available from: https://doi.org/10.1080/17458080.2014.980445.

[87] P. Xue, H. Wu, Y. Lu, X. Zhu, Recent progress in molten salt synthesis of low-dimensional perovskite oxide nanostructures, structural characterization, properties, and functional applications: a review, J. Mater. Sci. Technol. 34 (2018) 914−930. Available from: https://doi.org/10.1016/j.jmst.2017.10.005.

[88] W.-S. Jung, Y.-H. Do, M.-G. Kang, C.-Y. Kang, Energy harvester using PZT nanotubes fabricated by template-assisted method, Curr. Appl. Phys. 13 (2013) S131−S134. Available from: https://doi.org/10.1016/j.cap.2013.01.009.

[89] L. Sun, Y. Zhang, J. Li, T. Yi, X. Yang, Graphene-oxide-directed hydrothermal synthesis of ultralong $M(VO_3)n$ composite nanoribbons, Chem. Mater. 28 (2016) 4815−4820. Available from: https://doi.org/10.1021/acs.chemmater.6b02058.

[90] J.J. Urban, W.S. Yun, Q. Gu, H. Park, Synthesis of single-crystalline perovskite nanorods composed of barium titanate and strontium titanate, J. Am. Chem. Soc. 124 (2002) 1186−1187. Available from: https://doi.org/10.1021/ja017694b.

[91] G.U.H.-S. Qian Zhe-Li Hu Yong-Ming, Zhou Di, Wang Zhao, Xia Hua-Ting, Fabrication and characterization of $K_{0.5}Na_{0.5}NbO_3$ nanotube arrays by sol−gel AAO template method, J. Inorg. Mater. n.d.;25:687−690. http://www.jim.org.cn.

[92] J. Heiber, F. Clemens, T. Graule, D. Hülsenberg, Thermoplastic extrusion to highly-loaded thin green fibres containing $Pb(Zr,Ti)O_3$, Adv. Eng. Mater. 7 (2005) 404−408. Available from: https://doi.org/10.1002/adem.200500052.

[93] J. Heiber, F. Clemens, U. Helbig, A. Meuron, C. Soltmann, T. Graule, et al., Properties of $Pb(Zr,Ti)O_3$ fibres with a radial gradient structure, Acta Mater 55 (2007) 6499−6506. Available from: https://doi.org/10.1016/j.actamat.2007.08.004.

[94] F.J. Clemens, J. Heiber, T. Graule, M. Piechowiak, L. Kozielski, D. Czekaj, Microstructural and electromechanical comparison of different piezoelectric PZT based single fibers and their 1−3 composites, in: 2010 IEEE International Symposium on Applied Ferroelectrics, 2010, pp. 1−4. https://doi.org/10.1109/ISAF.2010.5712231.

[95] R. Dittmer, F. Clemens, A. Schoenecker, U. Scheithauer, M.R. Ismael, T. Graule, Microstructural analysis and mechanical properties of $Pb(Zr,Ti)O_3$ fibers derived by different processing routes, J. Am. Ceram. Soc. 93 (2010) 2403−2410. Available from: https://doi.org/10.1111/j.1551-2916.2010.03742.x.

[96] Y. Wang, J.J. Santiago-Avilés, A review on synthesis and characterization of lead zirconate titanate nanofibers through electrospinning, Integr. Ferroelect. 126 (2011) 60−76. Available from: https://doi.org/10.1080/10584587.2011.574988.

[97] M.B. Smith, K. Page, T. Siegrist, P.L. Redmond, E.C. Walter, R. Seshadri, et al., Crystal structure and the paraelectric-to-ferroelectric phase transition of nanoscale $BaTiO_3$, J. Am. Chem. Soc. 130 (2008) 6955−6963. Available from: https://doi.org/10.1021/ja0758436.

[98] M. Yashima, T. Hoshina, D. Ishimura, S. Kobayashi, W. Nakamura, T. Tsurumi, et al., Size effect on the crystal structure of barium titanate nanoparticles, J. Appl. Phys. 98 (2005) 14313. Available from: https://doi.org/10.1063/1.1935132.

[99] T. Lusiola, F. Clemens, Fabrication of one-dimensional ferroelectric nano- and micro-structures by different spinning techniques and their characterization, Nanoscale Ferroelectr. Multiferroics. (2016) 232−268. Available from: https://doi.org/10.1002/9781118935743.ch9.

[100] D. Swain, V.S. Bhadram, P. Chowdhury, C. Narayana, Raman and X-ray investigations of ferroelectric phase transition in NH_4HSO_4, J. Phys. Chem. A. 116 (2012) 223−230. Available from: https://doi.org/10.1021/jp2075868.

[101] A.G. Agronin, Y. Rosenwaks, G.I. Rosenman, Piezoelectric coefficient measurements in ferroelectric single crystals using high voltage atomic force microscopy, Nano Lett 3 (2003) 169−171. Available from: https://doi.org/10.1021/nl0258933.

[102] A. Gruverman, O. Kolosov, J. Hatano, K. Takahashi, H. Tokumoto, Domain structure and polarization reversal in ferroelectrics studied by atomic force microscopy, J. Vac. Sci. Technol. B Microelectron. Nanom. Struct. Process. Meas. Phenom. 13 (1995) 1095−1099. Available from: https://doi.org/10.1116/1.587909.

[103] Y. Wang, R. Furlan, I. Ramos, J.J. Santiago-Aviles, Synthesis and characterization of micro/nanoscopic $Pb(Zr_{0.52}Ti_{0.48})O_3$ fibers by electrospinning, Appl. Phys. A. 78 (2004) 1043−1047. Available from: https://doi.org/10.1007/s00339-003-2152-2.

[104] M.-Y. Choi, D. Choi, M.-J. Jin, I. Kim, S.-H. Kim, J.-Y. Choi, et al., Mechanically powered transparent flexible charge-generating nanodevices with piezoelectric ZnO nanorods, Adv. Mater. 21 (2009) 2185−2189. Available from: https://doi.org/10.1002/adma.200803605.

[105] X. Wang, J. Song, J. Liu, Z.L. Wang, Direct-current nanogenerator driven by ultrasonic waves, Sci. (80) 316 (2007). Available from: https://doi.org/10.1126/science.1139366. 102 LP − 105.

[106] R. Yang, Y. Qin, L. Dai, Z.L. Wang, Power generation with laterally packaged piezoelectric fine wires, Nat. Nanotechnol. 4 (2009) 34−39. Available from: https://doi.org/10.1038/nnano.2008.314.

[107] M.-P. Lu, J. Song, M.-Y. Lu, M.-T. Chen, Y. Gao, L.-J. Chen, et al., Piezoelectric nanogenerator using p-type ZnO nanowire arrays, Nano Lett 9 (2009) 1223−1227. Available from: https://doi.org/10.1021/nl900115y.

[108] Ani Melfa Roji M., Jiji G., Ajith Bosco Raj T., A retrospect on the role of piezoelectric nanogenerators in the development of the green world, RSC Adv 7 (2017) 33642−33670. Available from: https://doi.org/10.1039/C7RA05256A.

[109] S. Sripadmanabhan Indira, C. Aravind Vaithilingam, K.S. Oruganti, F. Mohd, S. Rahman, Nanogenerators as a sustainable power source: state of art, applications, and challenges, Nanomater 9 (2019). Available from: https://doi.org/10.3390/nano9050773.

[110] Y. Zi, Z.L. Wang, Nanogenerators: an emerging technology towards nanoenergy, APL. Mater 5 (2017) 74103. Available from: https://doi.org/10.1063/1.4977208.

[111] A. Yu, H. Li, H. Tang, T. Liu, P. Jiang, Z.L. Wang, Vertically integrated nanogenerator based on ZnO nanowire arrays, Phys. Status Solidi − Rapid Res. Lett 5 (2011) 162−164. Available from: https://doi.org/10.1002/pssr.201105120.

[112] G. Zhu, R. Yang, S. Wang, Z.L. Wang, Flexible high-output nanogenerator based on lateral ZnO nanowire array, Nano Lett 10 (2010) 3151−3155. Available from: https://doi.org/10.1021/nl101973h.

[113] S. Xu, Y. Qin, C. Xu, Y. Wei, R. Yang, Z.L. Wang, Self-powered nanowire devices, Nat. Nanotechnol. 5 (2010) 366−373. Available from: https://doi.org/10.1038/nnano.2010.46.

[114] J. Kaur, H. Singh, Fabrication and analysis of piezoelectricity in 0D, 1D and 2D zinc oxide nanostructures, Ceram. Int. 46 (2020) 19401−19407. Available from: https://doi.org/10.1016/j.ceramint.2020.04.283.

[115] J. Yan, Y.G. Jeong, High performance flexible piezoelectric nanogenerators based on $BaTiO_3$ nanofibers in different alignment modes, ACS Appl. Mater. Interfaces. 8 (2016) 15700−15709. Available from: https://doi.org/10.1021/acsami.6b02177.

[116] P. Shirazi, G. Ico, C. Anderson, M. Ma, B.S. Kim, J. Nam, et al., Size-dependent piezoelectric properties of electrospun $BaTiO_3$ for enhanced energy harvesting, Adv. Sustain. Syst. 1 (2017) 1700091. Available from: https://doi.org/10.1002/adsu.201700091.

[117] A. Teka, S. Bairagi, M. Shahadat, M. Joshi, S. Ziauddin Ahammad, S. Wazed Ali, Poly(vinylidene fluoride) (PVDF)/potassium sodium niobate (KNN)−based nanofibrous web: A unique nanogenerator for renewable energy harvesting and investigating the role of KNN nanostructures, Polym. Adv. Technol. 29 (2018) 2537−2544. Available from: https://doi.org/10.1002/pat.4365.

[118] S. Bairagi, S.W. Ali, A unique piezoelectric nanogenerator composed of melt-spun PVDF/KNN nanorod-based nanocomposite fibre, Eur. Polym. J. 116 (2019) 554−561. Available from: https://doi.org/10.1016/j.eurpolymj.2019.04.043.

[119] S. Bairagi, S.W. Ali, A hybrid piezoelectric nanogenerator comprising of KNN/ZnO nanorods incorporated PVDF electrospun nanocomposite webs, Int. J. Energy Res. 44 (2020) 5545−5563. Available from: https://doi.org/10.1002/er.5306.

[120] M.S. Al-Ruqeishi, T. Mohiuddin, B. Al-Habsi, F. Al-Ruqeishi, A. Al-Fahdi, A. Al-Khusaibi, Piezoelectric nanogenerator based on ZnO nanorods, Arab. J. Chem 12 (2019) 5173−5179. Available from: https://doi.org/10.1016/j.arabjc.2016.12.010.

[121] H. Parangusan, D. Ponnamma, M.A.A. Al-Maadeed, Stretchable electrospun PVDF-HFP/Co-ZnO nanofibers as piezoelectric nanogenerators, Sci. Rep. 8 (2018) 754. Available from: https://doi.org/10.1038/s41598-017-19082-3.

[122] Y.F. Wang, W.L. Yang, Z.Y. Hou, Y. Wang, A flexible piezoelectric nanogenerator based on free-standing polydimethylsiloxane/ZnO nanowire hybrid film, Mater. Sci. Forum. 977 (2020) 244−249. Available from: https://doi.org/10.4028/http://www.scientific.net/MSF.977.244.

[123] I. Chinya, A. Pal, S. Sen, Flexible, hybrid nanogenerator based on zinc ferrite nanorods incorporated poly(vinylidene fluoride-co-hexafluoropropylene) nanocomposite for versatile mechanical energy harvesting, Mater. Res. Bull. 118 (2019) 110515. Available from: https://doi.org/10.1016/j.materresbull.2019.110515.

[124] C.K. Jeong, J.H. Lee, D.Y. Hyeon, Y. Kim, S. Kim, C. Baek, et al., Piezoelectric energy conversion by lead-free perovskite $BaTiO_3$ nanotube arrays fabricated using electrochemical anodization, Appl. Surf. Sci. 512 (2020) 144784. Available from: https://doi.org/10.1016/j.apsusc.2019.144784.

[125] K.-I. Park, S. Bin Bae, S.H. Yang, H.I. Lee, K. Lee, S.J. Lee, Lead-free $BaTiO_3$ nanowires-based flexible nanocomposite generator, Nanoscale 6 (2014) 8962−8968. Available from: https://doi.org/10.1039/C4NR02246G.

[126] W. Wu, L. Cheng, S. Bai, W. Dou, Q. Xu, Z. Wei, et al., Electrospinning lead-free $0.5Ba(Zr_{0.2}Ti_{0.8})O_3−0.5(Ba_{0.7}Ca_{0.3})TiO_3$ nanowires and their application in energy harvesting, J. Mater. Chem. A. 1 (2013) 7332−7338. Available from: https://doi.org/10.1039/C3TA10792B.

[127] K.S. Chary, H.S. Panda, C.D. Prasad, Fabrication of large aspect ratio $Ba_{0.85}Ca_{0.15}Zr_{0.1}Ti_{0.9}O_3$ superfine fibers-based flexible nanogenerator device: synergistic effect on curie temperature, harvested voltage, and power, Ind. Eng. Chem. Res 56 (2017) 10335−10342. Available from: https://doi.org/10.1021/acs.iecr.7b02182.

[128] Z. Wang, Y. Zhang, S. Yang, Y. Hu, S. Wang, H. Gu, et al., $(K,Na)NbO_3$ nanofiber-based self-powered sensors for accurate detection of dynamic strain, ACS Appl. Mater. Interfaces 7 (2015) 4921−4927. Available from: https://doi.org/10.1021/am5090012.

[129] R. Zhu, J. Jiang, Z. Wang, Z. Cheng, H. Kimura, High output power density nanogenerator based on lead-free $0.96(K_{0.48}Na_{0.52})(Nb_{0.95}Sb_{0.05})O_3-0.04Bi_{0.5}(Na_{0.82}K_{0.18})$ $0.5ZrO_3$ piezoelectric nanofibers, RSC Adv 6 (2016) 66451−66456. Available from: https://doi.org/10.1039/C6RA12123C.

[130] Y.-H. Wu, J.-W. Zha, Z.-Q. Yao, F. Sun, R.K.Y. Li, Z.-M. Dang, Thermally stable polyimide nanocomposite films from electrospun $BaTiO_3$ fibers for high-density energy storage capacitors, RSC Adv 5 (2015) 44749−44755. Available from: https://doi.org/10.1039/C5RA06684K.

[131] S. Liu, S. Xue, W. Zhang, J. Zhai, Enhanced dielectric and energy storage density induced by surface-modified $BaTiO_3$ nanofibers in poly(vinylidene fluoride) nanocomposites, Ceram. Int. 40 (2014) 15633−15640. Available from: https://doi.org/10.1016/j.ceramint.2014.07.083.

[132] G. Wang, X. Huang, P. Jiang, Bio-inspired fluoro-polydopamine meets barium titanate nanowires: a perfect combination to enhance energy storage capability of polymer nanocomposites, ACS Appl. Mater. Interfaces. 9 (2017) 7547−7555. Available from: https://doi.org/10.1021/acsami.6b14454.

[133] H. Zhang, M.A. Marwat, B. Xie, M. Ashtar, K. Liu, Y. Zhu, et al., Polymer matrix nanocomposites with 1D ceramic nanofillers for energy storage capacitor applications, ACS Appl. Mater. Interfaces. 12 (2020) 1−37. Available from: https://doi.org/10.1021/acsami.9b15005.

[134] A.A. Bent, N.W. Hagood, J.P. Rodgers, Anisotropic actuation with piezoelectric fiber composites, J. Intell. Mater. Syst. Struct. 6 (1995) 338−349. Available from: https://doi.org/10.1177/1045389X9500600305.

[135] A.A. Bent, N.W. Hagood, Piezoelectric fiber composites with interdigitated electrodes, J. Intell. Mater. Syst. Struct. 8 (1997) 903−919. Available from: https://doi.org/10.1177/1045389X9700801101.

[136] X. Chen, S. Xu, N. Yao, W. Xu, Y. Shi, Potential measurement from a single lead zirconate titanate nanofiber using a nanomanipulator, Appl. Phys. Lett. 94 (2009) 253113. Available from: https://doi.org/10.1063/1.3157837.

[137] S.H. Ji, J.H. Cho, J.-H. Paik, J. Yun, J.S. Yun, Poling effects on the performance of a lead-free piezoelectric nanofiber in a structural health monitoring sensor, Sens. Actuators A Phys 263 (2017) 633−638. Available from: https://doi.org/10.1016/j.sna.2017.07.016.

[138] S.C.L. Koh, T. Ibn-Mohammed, A. Acquaye, K. Feng, I.M. Reaney, K. Hubacek, et al., Drivers of United States toxicological footprints trajectory 1998−2013, Sci. Rep. 6 (2016) 39514. Available from: https://doi.org/10.1038/srep39514.

[139] F. Cucchiella, I. D'Adamo, S.C. Lenny Koh, P. Rosa, Recycling of WEEEs: an economic assessment of present and future e-waste streams, Renew. Sustain. Energy Rev. 51 (2015) 263−272. Available from: https://doi.org/10.1016/j.rser.2015.06.010.

Memristive applications of metal oxide nanofibers

Shangradhanva E. Vasisth, Parker L. Kotlarz, Elizabeth J. Gager and Juan C. Nino
University of Florida, Gainesville, FL, United States

11.1 Introduction

Metal oxide nanofibers (MONFs) and metal oxide nanowires (MONWs) offer an array of benefits, including a high surface-to-volume ratio and a high specific area. Therefore as summarized in Fig. 11.1, they have been used for several applications, including gas sensors, photocatalysts, supercapacitors, and biomedical devices. While other applications have been reported (e.g., piezotronic harvesting [1], methanation of CO_2 [2], and removal of radioactive ions [3]), the four broad groups briefly discussed here represent well-documented research, specifically on NFs.

For example, in electrochemical applications, gas sensors are used for a variety of functions, such as food processing and air quality monitoring [4]. Electrospun MONFs have been used most often, owing to their high response, high stability, and low cost. While traditional MONF gas sensors have poor selectivity at

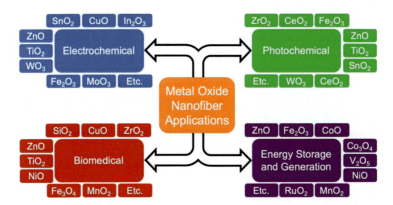

Figure 11.1 Common materials used in well-documented applications of metal oxide nanofibers and nanowires used in electrochemical applications involve gas sensors and memristors used in photochemical applications involve water splitting, dye-sensitive solar cells, wastewater treatment, and photocatalysis. Nanofibers and nanowires can also be used in cheap and durable batteries and as pseudocapacitors and are used as physiological sensors, scaffolds, or antibacterials for biomedical applications.

Metal Oxide-Based Nanofibers and Their Applications. DOI: https://doi.org/10.1016/B978-0-12-820629-4.00018-7
© 2022 Elsevier Inc. All rights reserved.

extremely low levels, using a heterojunction material in which two dissimilar metals are used or doping the NFs with dopants such as iron (Fe), cobalt (Co), or copper (Cu) can resolve this problem [5]. By using a heterojunction material, MONFs have been shown to successfully detect gas levels at low concentrations (5–1000 ppm) that their previous counterparts were unable to sense. The heterojunction uses n-type and p-type metal oxides to detect gas through monitoring a change in resistance [5]. For an excellent review on gas sensors, the reader can consult reviews done by Abideen et al. [6] on composite NFs and Dey et al. [7] on semiconductor metal oxides.

In photocatalysis applications, in which substances are degraded or reacted under light, MONFs can be engineered for use in applications such as hydrogen production, solar cells, and water remediation. The use of MONFs in water splitting has been shown to be a successful way to produce H_2 as a result of their increase in surface-to-volume ratio and specific area as the material decreases in size. For example, an electrospun TiO_2 NF can be used for water splitting via photon adsorption, which generates a charge that then migrates within the metal oxide followed by a redox reaction [8]. Additionally, MONFs, most commonly TiO_2, can be applied to augment dye-sensitive solar cells, which produce electrical light through the conversion of visible light. Using hollow mesoporous TiO_2 NFs prepared by coaxial electrospinning has been shown to increase the efficiency of solar cells [9]. Moreover, ZnO can be used as a MONF to eliminate organic pollutants from wastewater. ZnO is used in this regard because of its wide bandgap, high photocatalytic activity, chemical stability, environmental friendliness, and low cost. These NFs can be doped to create a heterojunction material, generally using other metal oxides such as TiO_2 and SnO_2 to increase the photocatalytic properties. Fabrication includes dopants that are incorporated into the precursors of the spinning solution for electrospinning [10]. An overview of the main advances in this field can be seen in the reviews by Hoang et al. [11] and Reddy et al. [12] on NW arrays and heterostructures for water splitting, respectively.

In terms of energy storage, MONFs can also be used as the electrodes of supercapacitors, which offer the ability to store and release energy quickly in a manner that is relatively cheap and safe over long lifetimes. For example, RuO_2 NFs, fabricated by electrospinning, display the highest capacity among the metal oxides and have performed very well in long-cycle testing [13]. MnO_2, Fe_2O_3, and Co_3O_4 also offer potential. Subsequently, polycationic MONFs demonstrating pseudocapacitive behavior have also been observed [14].

Finally, synthesis of MONFs for biomedical applications has become a recent focus because of their versatility and usefulness as a biocompatible material. Electrospun NFs have been applied to electrochemical biosensing and tissue engineering. An important consideration for biomedical applications of MONFs is toxicity. While expansive clinical testing is still needed on MONFs for biomedical uses, some promising low-toxicity MONFs include ZnO and TiO_2. With increased sensitivity due to a NF's large surface-to-volume ratio, a plethora of MONFs have been utilized for sensing important physiological components, including glucose, adenine, and breast cancer biomarkers [15]. TiO_2 nanocomposite fibers have also been employed as a scaffold for tissue engineering and wound dressing [16]. Antibacterial applications of MONFs have also been explored, notably in combination with nanocellulose [17].

11.2 Recent trends

One of the vastly expanding fields of MONFs includes various electrical advancements in complex electronics. MONFs in piezoelectric applications include a ZnO NW array [18] and ZnO in a field effect transistor [19]. Further advancements include augmentation of secondary batteries with MONFs being applied in various batteries such as lithium ion batteries as cathodes and anodes [20]. One notable application of MONFs in electronics includes neuromorphic architectures. These are analog/digital hybrid circuits in integrated systems that are fabricated to emulate computational abilities of biological systems artificially [21]. Neuromorphic systems were proposed because of two limiting factors in current computing systems: computing devices leading up to a technological and physical limit to accommodate more transistors [22,23] and the von Neumann bottleneck [24], in which, owing to the physical separation between processing and memory units, the data transfer rate is limited. Therefore utilizing biological style computing, neuromorphic systems can surpass these limitations. Nonvolatile memory devices, especially devices exhibiting resistive switching, have been proposed to act as synapses in neuromorphic systems. These are two-terminal electrical components in which the resistance is a function of the amount and direction of the current [20]. The fabricated devices described by Shimeng Yu [21] have high on-chip density ($4-12\,F^2$ per bit), increased energy efficiency (consumption of $<10\,fJ$ per programming pulse), and low leakage power consumption while performing fast-parallel computing [22]. NFs and NWs have been the functional blocks of memory devices in which the wire may be the switching layer or an access to the electrode or both. For example, Zhu et al. [23] have demonstrated excitatory postsynaptic current, paired-pulse facilitation, and spoke-timing-dependent plasticity in ZnSnO NFs. However, reducing power consumption and enhancing scalability beyond what is described by Yu [21] has been difficult. Present architectures offer a synaptic density of 10^6-10^8 connections cm^{-2} [24,25] but Nino and coworkers have proposed a novel, patented architecture (memristive NW neural network, or MN^3) that utilizes conductive core and metal oxide shell NWs to achieve a theoretical synaptic density of 10^9 connections cm^{-2} [26,27].

In this chapter we present an introduction to memristors and resistive switching, fabrication, and resistive switching characteristics, and some applications of metal oxide memristors. We present a promising approach for future devices, such as core—shell wires, in which attempts have been made to scale the switching voltages down as a result of a thin metal oxide layer. Finally, we provide a brief review of fabrication and preliminary memristive performance of MN^3 core—shell wires.

11.3 Memristors and resistive switching

Memristive or resistive switching behavior has been observed in numerous binary and complex oxides. Some examples of state-of-the-art crossbar and thin-film architectures of binary oxides and complex oxides are listed in Table 11.1. These devices are capacitor-like and are experimentally fabricated with a metal–insulator–metal (MIM) structure. Here, the component "M" can be any compound or metal with high conductivity, and "I" is the intermediate layer or layers of oxides or other switching material present between two electrodes [20]. These structures have been widely fabricated as crossbar arrays, which were originally proposed to establish electrical connection for switching systems (e.g., telecommunication switching systems) [28]. The systems have straight electrode bars arranged perpendicularly in a gridlike pattern with the switching material sandwiched between the electrodes at the intersection. The goal of these structures was to not only connect a single pair of circuits but also connect pairs from various circuits (in-plane or stacked). Highly dense crossbar memories must have a particular set of characteristics: offer repeatable and reliable switching between at least two states, require less power, and survive billions of switching cycles [20]. Also, during device operation, exhibiting excellent thermal stability is imperative. Stacked crossbar structures fabricated by Kim et al. [29] with NbO_2 switching material displayed reproducible IV for over 100 cycles. The device also showed exceptional thermal stability up to 160°C when compared to VO_2 devices.

To understand memristive performance, devices are initially characterized in IV sweep mode, as this enables determination of the voltage and current thresholds and the operation mode of switching between a high resistance state (HRS) and a low resistance state (LRS). Most devices have to undergo the process of electroforming before the first write-read operation. Usually, a higher voltage and stronger current are employed to activate the device. To change the resistance states in these devices, the devices have to be single electron transistor (SET) and remember every situation

Table 11.1 Active layer thickness, single electron transistor/remember every situation encourages transformation potential, cyclic endurance, and resistance retention of binary and complex oxides in crossbar and thin-film architectures.

Memristive oxide	Active layer thickness (nm)	V_{SET} (V)	V_{RESET} (V)	Endurance (cycles)	Retention
HfO_2 [30]	5	1.3	− 3.05	1.2×10^{11}	$> 2.6 \times 10^6$ s
Ta_2O_{5-x}/TaO_{2-x} [31]	40	6	− 4.5	10^{12}	10 years
ZnO [32]	60	− 6.2	6.3	250	10^6 s
$Nb:SrTiO_3$ [33]	5×10^5	2.5	− 3.5	10^6	10^5 s
$Pr_{0.7}Ca_{0.3}MnO_3$ [34]	100	− 5	5	100	10^4 s

encourages transmission (RESET). During the SET process, the device is relatively turned ON, implying that the device transitions into LRS from HRS. Subsequently, during the RESET process, the device is relatively turned OFF, implying that the device transitions into HRS from LRS. There are three main operation modes in resistive switching with differing IV sweep: unipolar resistive switching (URS), bipolar resistive switching (BRS), and complementary resistive switching (CRS). These modes are presented schematically in Fig. 11.2 and discussed in brief detail next.

URS behavior is known mainly for its SET and RESET operation occurring with one voltage polarity. Initially, from HRS the devices are SET, and the process occurs when a threshold potential is reached while the current is being limited. A compliance current is usually applied to prevent excessive current flow, leading to dielectric breakdown of the device. To RESET the devices, the compliance current must be removed for complete transition back into HRS.

A majority of systems in literature operate in BRS mode. Initially, devices in HRS are triggered to SET at a threshold voltage, often in the presence of a compliance current. To read the state of the system, potentials much smaller than the threshold are applied. The device is RESET back to HRS at a certain potential threshold in the opposite polarity.

CRS (explained in detail in Section 11.5), however, is obtained by connecting two memristors exhibiting BRS in an antiserial manner, as suggested by Linn et al. [35]. However, in some cases, this behavior can also be observed by controlling processing and operation of random access memory (RAM) [36].

The overarching switching mechanism of memristors involves either an interface-type path or a filamentary conducting path; the latter make the majority of memristors. These mechanisms are described only briefly here. For additional details and fundamentals of memristors, the reader is directed to reference [20]. In the interface-type path, the memristive switching occurs at the boundary between the metal electrode and the oxide. In a filamentary conducting path, a conducting filament forms between the electrodes. The rupturing and reforming of the filament path result in the switching mechanism [37]. The conducting filament path can be influenced through thermochemical switching, whereby thermal effects can influence the rupturing and forming of filaments through Joule heating (Figs. 11.3a−c) [38].

The memristive switching mechanisms can be further organized into electrochemical metallization, valence change, or purely electronic. Electrochemical metallization (Figs. 11.3d−f), also known as cation migration, involves an electrochemically active electrode and an inert counterelectrode whereby the cation (e.g., Cu or Ag) becomes oxidized and migrates toward the counterelectrode, forming a filament conducting path. Valence change, or anion migration, primarily involves oxygen anions and is described through oxygen vacancies. An interface-type path (Figs. 11.3g−i) can form as a result of the accumulation of oxygen vacancies at the boundary, or a filament conducting path can form via migration of the oxygen vacancies, leading to a compositional change in the memristor and ultimately a conducting filament [38]. It is important to note that these two types of pathways are not mutually exclusive, as was demonstrated by Yang et al. [39]. Finally, purely electronic switching mechanisms occur as a result of atomic displacements caused by polarization reversal.

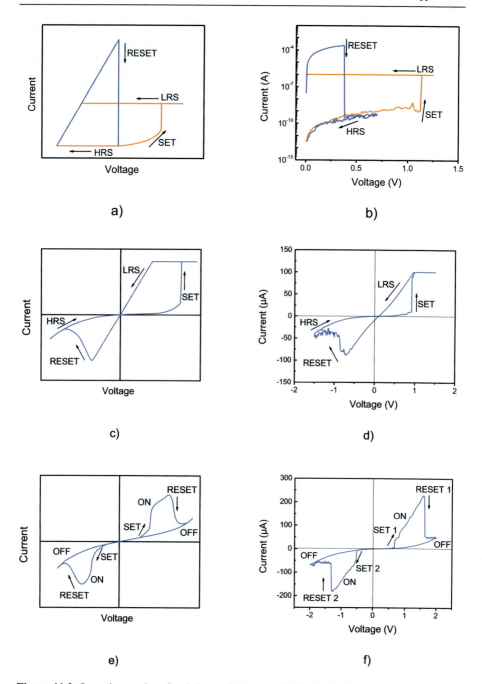

Figure 11.2 Operation modes of resistive switching. (a, b) Ideal unipolar resistive switching and unipolar resistive switching observed in alternating ZrO$_2$/HfO$_2$ nanolayers, where the device is single electron transistor at a threshold voltage and compliance into a low-resistance

(*Continued*)

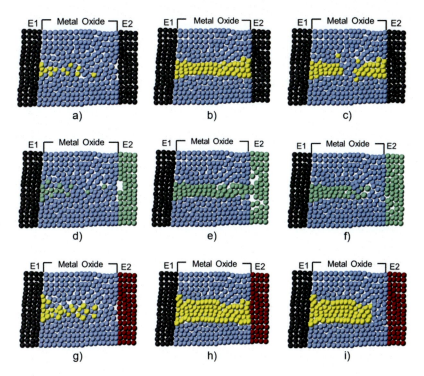

Figure 11.3 Schematic representations for mechanisms of resistive switching. (a–c) Electroforming, single electron transistor, and remember every situation encourages transformation in thermochemical switching mechanism, where the metal oxide sandwiched between similar electrodes (E1 and E2), and oxygen vacancies growing from E1 in part a), forming a filament in part b), and breaking of filament in part c) in the metal oxide. (d–f) Electroforming, single electron transistor, and remember every situation encourages transformation in electrochemical switching mechanism, where the metal oxide sandwiched between an active electrode (E2) and an inert electrode (E1) with atoms showing the dissolution of active electrode into the metal oxide to result in deposition at the inert electrode in part d, forming a filament in part e, and breaking of filament in part f. (g–i) Electroforming, single electron transistor, and remember every situation encourages transformation in valence change or interface-type switching mechanism, where the metal oxide sandwiched between dissimilar electrodes (E1 and E2) and oxygen vacancies growing in part g, forming a filament in part e, and breaking of filament in part f in the metal oxide.

◀ state and the device is remember every situation encourages transformation with voltage of same polarity at a higher compliance into a high-resistance state. (c, d) Ideal bipolar resistive switching and bipolar resistive switching observed in Pt/HfO$_2$ core–shell nanowire, where the device is single electron transistor at a threshold voltage and compliance into a low-resistance state and the device is remember every situation encourages transformation with voltage of opposite polarity at a higher compliance into a high-resistance state. (e, f) Ideal complementary resistive switching and complementary resistive switching observed in Pt/HfO$_2$ core–shell nanowire, where as a result of the antiserial arrangement, the two bipolar resistive switching memristors undergo two single electron transistors and remember every situation encourages transformations. One of the memristors undergoes single electron transistor and remember every situation encourages transformation at single electron transistor 1 and remember every situation encourages transformation 2, and the other undergoes single electron transistor and remember every situation encourages transformation at single electron transistor 2 and remember every situation encourages transformation 1. ON and OFF represent when both memristors are in low resistance state and high resistance state, respectively.

11.4 Resistive switching in metal oxide nanofibers

Although resistive switching has been observed in a variety of materials (detailed in Fig. 11.4), the mechanisms of switching have yet to be explored thoroughly. However, the fabrication of one-dimensional (1D) structures has provided a novel way to explore these mechanisms at the nanoscale where the switching events are localized. These structures also enable the fabrication of highly dense and easily scalable devices for various applications. A brief overview of metal oxide NW systems that have been shown to exhibit memristive switching behavior is presented next, starting with NWs or nanorods of NiO that were fabricated by using anodized aluminum oxide (AAO) membranes. A summary of all metal oxide NW memristors is given in Table 11.2.

11.4.1 NiO

Using the rationale that memristive switching behavior is reproducible at low voltages, Kim et al. [40] designed a crossbar structure of NiO NWs that exhibited unipolar switching. It was found that for a 1-µm NW the forming voltage is 2.5 V, which is much lower than what occurs in NiO thin films. They also showed a minimum obtainable RESET current of 0.23 mA for a 1-µm NW. In addition, they showed that the SET and RESET potentials increase with increasing length of the NW.

Herderick et al. [41] used a template-based method to produce a large number of nearly identical MIM Au−NiO−Au NWs. In doing so, they were able to show reproducible results in multiple fabricated devices. The devices exhibited bipolar behavior with an HRS/LRS ratio of 10,500 and exhibited resistive switching between −6 and 5.2 V. The HRS to LRS switching is due to the migration of Ni vacancies away from the interface.

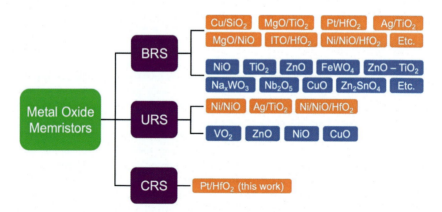

Figure 11.4 Oxides used in the fabrication metal oxide nanofibers. The materials in blue represent individual one-dimensional metal oxide nanofibers, and the materials in orange are core−shell structures, written as core/shell (e.g., Pt/HfO$_2$).

Table 11.2 Summary of metal oxide nanowires, nanowire diameter, single electron transistor/remember every situation encourages transformation thresholds, cyclic endurance, resistance retention, and type of switching available in literature.

Memristive oxide	Nanowire diameter (nm)	V_{SET} (V)	V_{RESET} (V)	Endurance (cycles)	Retention	Type of switching
NiO [40]	70	1.2	0.52	N.A.	N.A.	URS
NiO [41]	250	5.2	-6	N.A.	N.A.	BRS
NiO [42]	N.A.	10	5	N.A.	10^4 s	BRS
TiO$_2$ [43]	100	1.75	-2	N.A.	N.A.	BRS
TiO$_2$ [44]	100	10	-5	10^6	10^3 s	BRS
ZnO$-$TiO$_2$ [45]	100	10	-10	N.A.	N.A.	BRS
TiO$_2$ [46]	50$-$100	7.5	-7.5	N.A.	N.A.	BRS
TiO$_2$ [47]	100$-$200	8	-8	10	N.A.	BRS
CuO [48]	30/50	1	-1	N.A.	N.A.	BRS
CuO [49]	$\sim$50	$\sim$1.5	0.78	N.A.	N.A.	URS
CuO [49,58]	$\sim$50	$\sim$1.1	-1	50	N.A.	BRS
ZnO [50]	500	10	-10	20	N.A.	BRS
ZnO [51]	$\sim$150	3	-1.6	7	2×10^6 s	BRS
ZnO [52]	$\sim$150	0.5	0.2	25	N.A.	URS
ZnO [53]	50$-$80	-1	1	N.A.	N.A.	BRS
ZnO [54]	104	<4	-1	300	$>10^3$	BRS
ZnO [55]	N.A.	2.5	-2.5	N.A.	N.A.	BRS
ZnO [56]	200	-6	6	80	N.A.	BRS
Na-doped ZnO [57]	$\sim$300	40	-40	10^3	10^5 s	BRS
ZnO[58]	$\sim$50	22	-30	300	N.A.	BRS
Ga-doped ZnOS [58]	$\sim$100	-25	40	500	N.A.	BRS
Sb-doped ZnO [58]	$\sim$100	40	-40	>2000	N.A.	BRS
ZnO [59]	N.A.	7	12	100	N.A.	URS
ZnO [60]	$\sim$20000	0.19$-$0.4	0.6$-$0.8	100	N.A.	URS

(Continued)

Table 11.2 (Continued)

Memristive oxide	Nanowire diameter (nm)	V_{SET} (V)	V_{RESET} (V)	Endurance (cycles)	Retention	Type of switching
ZnO [61]	100	<40	40	N.A.	N.A.	URS
ZnO [62]	~ 50	<10	>-10	100	4×10^3 s	BRS
ZnO [63]	50–150	10	-5	N.A.	N.A.	BRS
ZnO [64]	50/100	40	-40	N.A.	2×10^3 s	BRS
ZnO [65]	100	40	-30	N.A.	2×10^3 s	BRS
ZnO nanowire bundle [66]	30	5	-5	N.A.	N.A.	BRS
Nb_2O_5 [67]	80–200	2.5	-6	N.A.	N.A.	BRS
VO_2 [68]	~ 160	~ 0.34	0.1	N.A.	N.A.	URS
WO_3	100–200	<9	<-9	N.A.	N.A.	BRS
Zn_2SnO_4 [69]	150	1.6	-1	14	5 months	BRS
$FeWO_4$ [70]	80	1	-1	100	N.A.	BRS
Na_xWO_3 [71]	300	<10	>-10	N.A.	N.A.	BRS

Note: Some have multiple single electro transistorSET/remember every situation encourages transformationRESET thresholds due to doping in the study or due to additional types of switching.

Klamchuen et al. [42] used vapor−liquid−solid (VLS) growth to produce a single crystalline NiO NW with controlled diameter and spatial position. The use of a single crystalline NW when compared to polycrystalline NiO NWs showed an improvement in resistive switching characteristics. For example, the resistance remained stable for longer retention times. Also, a narrow distribution of SET voltage was observed in single crystalline NWs when compared to polycrystalline devices. By using a material flux window, it is shown that various other metal oxides can be grown through VLS that were not able to be grown previously.

11.4.2 TiO$_2$

Titanium dioxide is one of the most studied and promising materials to be used as an active material in resistive random access memory (RRAM) devices [72]. Depending on either oxygen vacancy diffusion or the fuse-antifuse mechanism (another term indicating a thermochemical switching mechanism), these devices can exhibit BRS or URS type of switching. Therefore, Xia et al. [43] integrated crossbars of wire memristors fabricated via nanoimprint lithography into hybrid reconfigurable logic circuits on a complementary metal oxide semiconductor (CMOS) platform. The configuration bits and switches of the network were composed of the wired memristors and were successfully connected to a CMOS component to obtain a device akin to a field programmable gate array. The CMOS connections programmed the memristors, leading to an architecture where the transistors and memristors are complementary similar to a 1 transistor−1 resistor (1T-1R) configuration. This successful integration enables fabrication of nonvolatile random access memory (NV-RAM), neuromorphic architectures where the transistors act as neurons and the memristors act as synapses.

To better understand the underlying switching mechanisms and multilevel switching processes, individual wires have been fabricated via pulsed laser deposition (PLD) and electron beam lithography (EBL) [44], cationic exchange [45], spraying a dilute solution of NFs [46], and hydrothermal synthesis [47]. For example, Nagashima et al. [44] studied the material dependence of resistive switching in planar-type devices. They discovered that TiO$_{2-x}$ memristive NWs (100 nm in diameter) did not exhibit resistive switching at room temperature and atmospheric pressure. Resistive switching occurred only when the ambient pressure was lowered below 5×10^{-3} Pa. ON and OFF currents both increased with decreasing pressure as the oxidizing effect of the surroundings was reduced. Additionally, 300 nm of SiO$_2$ passivation was used to isolate the NW successfully. With successful isolation these devices were stable for 10^6 cycles. Despite some state-of-the-art crossbar systems displaying stability of over 10^{11} cycles [30] this is the most encouraging result so far in NW systems. To obtain the best crystalline TiO$_2$, Huang et al. [45] transformed single crystalline ZnO NWs into ZnO−TiO$_2$ hybrid NWs by solid-solid cationic exchange reaction. The wires were about 100 nm wide and exhibited resistive switching between 10 and −10 V.

Multilevel switching in TiO$_2$ was initially explored by O'Kelly et al. [46] by spraying a dilute solution of NWs (50−100 nm in diameter and 5−20 μm in length) and depositing Au electrodes. There exists a clear saturated SET state, and when the device is RESET, the conductance can be precisely controlled (with voltage pulses) to obtain multiple memory states. However, Lin et al. [47] later used femtosecond laser irradiation to localize plasmons at the metal-insulator interface to obtain crystalline defects and release oxygen. The junctions are mechanically bonded, and with an engineered NW, multilevel switching is achieved. Here, TiO_2 NWs were fabricated via hydrothermal synthesis and have a diameter of $100-200$ nm and a length of $1-6$ μm. They also discovered that with femtosecond irradiation, the current response is stabilized when excitation is programmed with different voltage cycles. Also, multilevel current states can be achieved by applying specific potentials under optimized conditions. Ultimately, eight levels of current states were obtained in the device at low irradiation before current saturation. Therefore, these devices are excellent candidates to scale memory devices with multilevel current states.

11.4.3 CuO

CuO NWs offer another approach for designing memristive systems. Fan et al. [48] synthesized CuO NWs using a copper foil with a 2-h incubation at 500°C in a box oven. The CuO NW memristive-forming process used the electron beam of transmission electron microscopy (TEM). The electron beam caused oxygen vacancies to form via the deoxygenation of CuO to CuO_{1-x}. The NW was shown to have bipolar switching through the movement of oxygen vacancies under external bias, which was confirmed by using electron energy loss spectroscopy. CuO NWs also demonstrated sensing capabilities with the switching threshold voltage changing as a function of external applied stress (bending the NW) [48]. Liang et al. [49] later demonstrated both unipolar and bipolar switching in CuO NWs. The NWs were formed via electrochemical deposition on an AAO template. The forming process involved oxidation at 400°C at 10 h for partially oxidized wires and 20 h for fully oxidized wires. Device fabrication included drop casting the wires on a SiO_2/Si substrate and using e-beam evaporation to deposit the Ni electrodes with more detailed procedures in the original literature. The fully oxidized CuO NWs did not exhibit resistive switching, but the partially oxidized CuO_x did demonstrate both bipolar and unipolar switching. Unipolar switching in CuO NWs was unstable and irreversible. Bipolar switching was reversible and resulted from the migration of oxygen vacancies from an external bias and the rupture of the conducting filament due to Joule heating. Unipolar switching resulted from oxygen vacancy migration but is believed to immediately reform the filament that broke via Joule heating.

11.4.4 ZnO

Zinc oxide NFs are by far the most studied when compared to other metal oxide NFs. These fibers have been fabricated via electrospinning [73], physical vapor deposition [50], chemical vapor deposition [51−54], hydrothermal synthesis [55,56], and, most commonly, the VLS method [57−65,74]. Studies performed on single ZnO NWs have revealed that these wires exhibit both bipolar and unipolar switching. Akin to other oxide systems, the conduction mechanism is dependent on the electrode at the end of the wires. Inert electrodes induce oxygen vacancy motion to form a conduction path, whereas the presence of active electrodes, such as Cu [51,53] or Ag [54,57,58,66] results in the metal's dissolution along the length of the NW to form a conductive path. Jasmin et al. [53] also discovered that coating a thin layer of polyacrylic acid, a polymer, onto ZnO NWs resulted in enhanced nonlinearity in IV plots due to modification of surface states on the NW. Sun et al. [56] discovered another way to improve the resistive switching behavior in hydrothermally synthesized ZnO nanorods by annealing them in hydrogen. This caused a thin layer of oxygen vacancies to develop on the surface, thereby making it a reservoir. As a result, a large hysteresis loop was obtained and raised the resistance ON-OFF ratio from approximately $10-10^4$. Apart from single NWs and nanorods, NW mesh and NW bundles of ZnO have been studied for resistive switching by Puzyrev et al. [66]. They discovered switching between two Ag electrodes due to the formation of Ag bridges between individual wires when they intersect. These devices exhibit a high ON−OFF ratio of 10^5 among NWs, making them ideal candidates for memory applications. However, high SET and RESET potentials at -6 and 6 V do not make them compatible with standard CMOS technology.

In addition to single state switching, multilevel switching has also been achieved, making these wires ideal candidates for RRAM devices. Milano et al. [54] achieved five states by varying compliance current, whereas Lai et al. [62] achieved three states by varying compliance current on argon plasma-treated wires. They also discovered that SET and RESET voltages decrease when the wires are plasma-treated. Finally, Lee et al. [63] achieved multiple states in the wire by controlling the direction and number of positive and negative pulses.

To make the devices self-complianced (in which the device prevents the abrupt increase in current as potential increases) and self-rectifying (in which the material prevents sneaking of current between nodes), Qi et al. [57] doped ZnO NWs with Na, and Wang et al. [58] doped ZnO NWs with Ga and Sb. In Na-doped wires, Ag was used for electrical contact. The Na-doping and Ag partial retraction from the Ag filament provided the self-complianced behavior, whereas the asymmetric contact due to doping between the wire and Ag electrodes made it self-rectifying. The wires were approximately 200 nm in diameter with SET and RESET potentials at 40 and -40 V. Ga- and Sb-doped wires showed similar behavior, with Ga-doped wires exhibiting n-type conduction and Sb-doped wires exhibiting p-type conduction behavior with possible application to multilevel switching.

To further understand switching mechanisms, high-resolution transmission electron microscopy (HRTEM) and in situ TEM have been performed by several groups. Only a representative sample of the body of work in literature is discussed (Fig. 11.5). When active electrodes such as Ag are present, the switching is due to the presence of Ag nanoclusters along the ZnO NW, as was observed by Milano et al. [54] and Qi et al. [57]. Huang et al. [60] noticed crystal defects, namely, stacking faults and dislocations that extended beyond 10 nm, in a ZnO NW displaying URS. Chiang et al. [59] and Huang et al. [61] observed the interfaces between the NW and the electrode when conduction was dependent on oxygen vacancy. Chiang et al. [59] noticed that when Ti

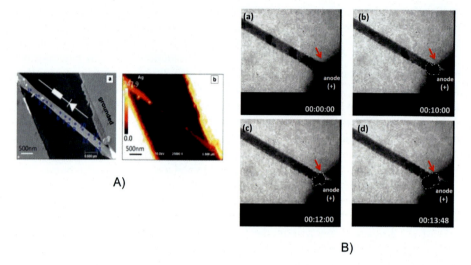

Figure 11.5 (A) Scanning electron microscope image in subpanel a) and a high spatial resolution auger electron spectroscopy (HSR-AES) map in subpanel b) of Ag migration along a Na-doped zinc oxide nanowire revealing higher Ag content on the biased side. (B) In-situ TEM images obtained from a real time video indicating on a ZnO nanowire the initial state in subpanel a), oxygen vacancy accumulation starting and growing in subpanels b), c) at Au electrode and completed accumulation at Au electrode in subpanel d).
Source: Panels adapted with permission from [57] (Copyright 2013 Royal Society of Chemistry) and [61] (Copyright 2013 American Chemical Society).

is present as an electrode, TiO_2 is formed at the interface. Huang et al. [61] observed that when inert electrodes such as Au are present for electrical contact, oxygen migrates toward the anode, thereby creating a gradient along the length of the wire and thus influencing the switching process.

As a result of extensive knowledge of resistive switching in ZnO NWs, this behavior has been used to enhance various electronic applications, including piezoelectrically modulated resistive memory (PRM), gas sensing, and artificial synapses. Wu and Wang [50] deposited ZnO NFs onto a polyethylene terephthalate substrate. They controlled resistive switching in NWs by using the strain-induced polarization at the metal-oxide interface under deformation due to the piezotronic effect (tuning of the metal-semiconductor interface by piezoelectric charges induced by strain). This occurs only because of ZnO's ability to enable ionic polarization under mechanical strain. They controlled different logic levels in the wire by varying strain, which resulted in different threshold potentials for the PRM to switch from HRS to LRS. Next, Zhang et al. [65] utilized the resistive switching behavior of ZnO to sense NO_2 (0.2–10 ppm) and NH_3 (50–500 ppm) gases. The switching between ON and OFF states enabled the selective detection of these gases. The ON current increased when the wires were exposed to NO_2 at all concentrations while there was no response in OFF current. Whereas the OFF current increased when the wires were exposed to NH_3, the ON current remained unchanged. Zhang et al. also discovered that the recovery time (desorption of gases) can be improved by increasing the pulse time and width for RESET and SET operations, which resulted in eight times reduction in recovery time with comparable sensitivity to state-of-the-art NW sensors. Finally, Milano et al. [54] realized short-term plasticity and similarity between the Ca^+ dynamics in neural synapses and Ag^+ dynamics in their devices. They noted a gradual increase in conductance with increased number of pulses and the spontaneous relaxation of the device to ground state after paired-pulse facilitation and low spiking frequency is insufficient to sustain higher conductive states.

11.4.5 Nb_2O_5

Grishin et al. [67] used an ethoxide-rout sol-gel-assisted electrospinning technique to produce continuous highly crystalline Nb_2O_5 NWs. The samples were spun and then annealed directly on a Pt-coated Si wafer. Bipolar behavior was observed with ON−OFF current ratios as high as 2×10^4. The switching behavior was determined by Poole-Frenkel emission from NbO_x complexes trapped between Nb_2O_5 and Pt.

11.4.6 VO_2

Bae et al. [68] produced a two-terminal memristor from a single VO_2 NW fabricated by hydrothermal method followed by a thermal annealing process. The NWs that were produced were between 80 and 160 nm in length. VO_2 was chosen as a promising material for memristive devices because of its fast response time, large range of resistance values, and low power consumption. They produced a device that used self-Joule heating generated by a low bias voltage of 0.34 V.

11.4.7 WO_3

Zhou et al. [75] fabricated a Au/WO_3/Au sandwiched device to explore the idea that humidity strongly influences the memristive behavior. The WO_3 NWs, which were 100−200 nm in diameter and up to 5 μm in length, were synthesized by a hydrothermal method and fabricated on an n-doped Si substrate. When the bias sweep range was less than ± 3 V and the relative humidity was less than 51%, almost no electrical hysteresis was observed on the IV curves. When the relative humidity was greater than 51%, the electrical hysteresis was enhanced even at the low bias sweep range. The memristive performance was determined to be enhanced through the Grotthuss mechanism as a result of an increase in the number of water molecules absorbed on the WO_3 surface, which changes the barrier height and increases the electrical current.

11.4.8 Complex Oxide Nanofibers

Resistive switching has been observed previously in complex oxides thin films such as Nb:$SrTiO_3$ [33] and $Pr_{0.7}Ca_{0.3}MnO_3$ [34]. However, with the need to enhance computing speed and density of memories, Dong et al. [69] fabricated the first complex oxide NWs with Zn_2SnO_4. These NWs were fabricated by modifying the vapor phase transport method in a horizontal furnace with a quartz tube [76]. Quartz boats with ZnO and SnO powders were heated to 800°C for 60 min with a Si substrate with Au films placed downstream. The vapors were guided toward the substrate by flowing pure Ar and O_2. The NWs that were obtained had diameters of upto 200 nm and lengths of several tens of micrometers. The as-grown NWs were then dispersed in ethanol and sonicated to recover individual fibers and were transferred onto a doped Si substrate with thermally grown SiO_2. Pd and Cu were deposited by using photolithography, deposition, and lift-off techniques. Tests performed on a 150-nm wire revealed forming at 5 V and subsequent BRS between 1.6 V (V_{SET}) and -1 V (V_{RESET}) with Cu as the biasing electrode. The wires displayed high ON-OFF ratios ($> 10^5$), fast switching speeds (<20 ns), and greater retention times (> 5 months).

Interest in the multifunctional nature of $FeWO_4$ led Sun and Li [70] to focus on investigating its resistive switching characteristics. They grew $FeWO_4$ as a NW array on a Ti substrate via hydrothermal synthesis. Briefly, $FeCl_3 \cdot 6H_2O$ and $Na_2WO_4 \cdot 2H_2O$ precursors were dissolved in deionized water with a cationic surfactant and placed in an autoclave with a clean Ti substrate at 140°C for 72 h. Ag was deposited on top to be used as the biasing electrode. The NWs have a diameter of around 80 nm and are on average 6 μm in length. These NWs exhibit low-threshold switching where the BRS occurs between 0.9 (V_{SET}) and -0.9 V (V_{RESET}) while having a high ON−OFF ratio of 10^4 for 100 cycles.

Building on their previous work studying resistive switching in hexagonal WO_3, Lei et al. [71] synthesized hexagonal sodium tungsten bronze (Na_xWO_3) NWs using the hydrothermal method. The precursors sodium tungstate ($Na_2WO_4 \cdot 2H_2O$), oxalic acid ($H_2C_2O_2$), sodium sulfate (Na_2SO_4), and hydrochloric acid (HCl) were dissolved and placed in an autoclave. The precipitates were washed to obtain NWs. Individual wires were placed on a doped Si substrate with thermally grown SiO_2. Next, Au was deposited for electrical contact by using photolithography, deposition, and lift-off techniques. Tests performed on a 300-nm-diameter and 4-μm-long wire revealed forming at 8 V and subsequent BRS between 4 V (V_{SET}) and -4 V (V_{RESET}). The mechanism of switching was confirmed via energy-dispersive X-ray spectroscopy to be the motion of Na^+ ions along the length of the wire. Motion of these ions changes Na^+ concentrations at the electrodes, resulting in changed Schottky barrier heights and lattice distortions.

11.5 Core−shell nanowires

Core−shell NWs, summarized in Table 11.3, were initially fabricated by Oka et al. [77] as a solution to the unreliability of thin-film devices when scaled down to nanoscale dimensions. They studied magnesium oxide (MgO) core and nickel(II) oxide (NiO) shell systems for resistive switching behavior. NiO was deposited onto MgO NWs, and the NWs were subsequently placed on a Si substrate before Pt was

Table 11.3 Summary of core−shell metal oxide nanowires (written as core/shell, e.g., Pt/HfO$_2$), core nanowire diameter, shell thickness, single electron transistor/remember every situation encourages transformation thresholds, cyclic endurance, resistance retention, and type of switching, as reported in literature.

Memristive oxide	Core diameter (nm)	Shell thickness (nm)	V_{SET} (V)	V_{RESET} (V)	Endurance (cycles)	Retention	Type of switching
MgO/NiO	10	10	− 20	20	6	10^4 s	BRS
Ni/NiO [78]	75	4.5	0.5−1	0.5−3.5	25	3.5×10^6 s	URS
Ni/NiO [79]	75	5	1.5	0.5	50	N.A.	URS
Ni/NiO [80]	∼10.6	4−8	5	3	20	N.A.	URS
MgO/Co$_3$O$_4$ [81]	10	5	25	− 10	10^8	10^4 s	BRS
MgO/Co$_3$O$_4$ [82]	10	5	∼35	− 30	10^8	10^4 s	BRS
Au/Ga$_2$O$_3$ [83]	40	40	5.4	− 8.5	100	30,000 s	BRS
Cu/SiO$_2$ [84]	∼150	15	2.3	− 1.2	10^4	10^6 s	BRS
Cu/SiO$_2$ [85]	N.A.	18	3	− 3	10^4	10 years	BRS
MgO/TiO$_2$ [86]	10	5	30	− 10	N.A.	10^4 s	BRS
Ag/TiO$_2$ [87]	63	15	0.5	− 0.5	150	10^3 s	BRS
Ag/TiO$_2$ [86,90]	63	15	0.96	0.5	15	>10^6 s	URS
Ag/TiO$_2$ [88]	75	25	0.1	− 0.1	200	10^5 s	BRS
ZnO/polyacrylic acid [89]	50−130	20	1.5	− 1.5	N.A.	N.A.	BRS
ITO/HfO$_2$ [90]	80−250	510	11.85	− 1 − 1.35	$10^3$30	10^4 sN.A.	BRS
Ni/NiO/HfO$_2$ [91]	55 (Ni) + 10 (NiO)	20	2	1.7	200	10^7 s	URS
Ni/NiO/HfO$_2$ [90]	55 (Ni) + 10 (NiO)	20	2	−1.6	200	10^7 s	BRS
Pt/HfO$_2$ [this work]	40	7	1.1	− 1.5	1000	10^4 s	BRS
Pt/HfO$_2$ [this work]	40	7	2	−2	N.A.	N.A.	CRS

Note: Some have multiple single electron transistor/remember every situation encourages transformation thresholds due to varying the thickness in the study or due to additional types of switching.

deposited for electrical contact. The wires displayed BRS between -20 and 20 V with a retention for 10^4 s but had poor endurance. Inspired by this work, He et al. [78] and Cagli et al. [79] explored the resistive switching behavior of Ni core and NiO shell NWs.

He et al. [78] deposited Ni into AAO template pores and subsequently oxidized them in air to fabricate Ni/NiO core−shell NWs. They deposited several Au electrodes before and after oxidation to have electrical contact with the core and the shell. The wires displayed URS behavior with SET and RESET at 1.1 and 0.6 V. The devices retained resistance for 40 days and had an endurance of 25 cycles. They discovered that both Joule heating and electric field drive the oxygen vacancies in the oxide to form a filament of oxygen vacancies. To truly understand the behavior of multiple wires in a crossbar architecture, Cagli et al. [79] studied the switching behavior at the intersection of two core−shell wires. The core−shell wires were fabricated in a similar fashion as the previous work, and two wires were aligned perpendicular to each other. Au was deposited at the ends of both wires for electrical contact. Since the electrodes were deposited on the shell, they had to form two individual memristors. Initial electrical probing did not indicate switching at the junction. The wires were formed individually and displayed URS behavior. Subsequently, when one electrode from each wire was probed, they were able to control the switching behavior at the intersection. This configuration provides flexible wires with high-density and scalability to fabricate crossbar memory arrays. Attempts to program NW networks have been made. Bellew et al. [80] used core−shell Ni/NiO NWs and varied the separation between the electrodes. This introduced a physical programmability in which the size of the connectivity cells varied, thereby enabling control of properties. However, the SET and RESET currents remained consistent despite varied distance, as only one junction was switching ON and OFF, and this was comparable to results in other works in the literature.

Later, in an effort to obtain reliable multistate metal oxide resistive switching devices, Nagashima et al. [81] synthesized NWs with a MgO core and a Co_3O_4 shell. An Au-assisted VLS technique was utilized to grow MgO on a MgO single-crystal substrate. These NWs were 5 μm long with a diameter of 10 and 5 nm of Co_3O_4 was sputtered on the wires to create a core−shell NW. The core−shell NWs were then placed on SiO_2/Si substrate, and Pt/Au was deposited for electrical contact. The individual wires showed endurance for 10^8 cycles and a HRS and LRS retention for 10^4 s under vacuum. They were also able to achieve five different resistance states by varying the compliance current between 10^{-10} and 10^{-8} A. In a subsequent work, they demonstrated that the SET current of these wires increases in an oxidizing ambience [82].

To reduce the operation current and therefore overall power consumption, Hsu and Chou [83] fabricated Au core−Ga_2O_3 shell NWs. They employed a VLS technique to assemble the core−shell NWs. The Au has a diameter of 40 nm, and the shell is 40 nm thick. When probed electrically, the wires exhibit BRS behavior between 8 and -8 V, and the ON/OFF ratio at 2 V was more than three. Thermionic emission and space-charge-limited mechanisms account for charge transport at HRS and LRS states in the NWs.

Owing to the success of Cu/SiO_2 thin films in exhibiting stable resistive switching, Wiley's group explored the properties of Cu core and SiO_2 shell memristors [84]. Cu was coated from solution to fabricate the NWs. The devices displayed BRS behavior between 2 and -2 V. When pulsed at a high speed (50 ns), the devices displayed endurance for more than 10^4 cycles and a retention time of four days. Subsequently, they printed a polymer-NW composite as a resistive switching layer to create a programmable crossbar array [85]. These composites with 3-μs write speeds displayed an endurance for more than 10^4 cycles and a retention time of 10 years. They were also able to program 16 states onto a 4-bit memory by writing at 5 V and reading at 0.7 V.

Since TiO_2 has been widely studied for resistive switching, Nagashima et al. [86] fabricated a MgO core with a TiO_2 shell to analyze the type of carriers that are present during charge transport. Retention studies under controlled atmosphere were performed, and it was found that the ON current decreased when an oxidizing gas was present for a wire exhibiting BRS. This response can therefore be interpreted as electron carriers compensating by oxidizing using the ambient gas resulting in an n-type conduction.

TiO_2 as a shell was also used with an Ag core for cation-driven resistive random access memory (ReRAM) [87] and wearable electronics [88]. Manning et al. [87] synthesized core−shell NWs with polycrystalline and amorphous rutile TiO_2 structures, and electrical contact was provided by Ag electrodes, since voids are formed in the core if Au electrodes are used. They discovered that by controlling the compliance current when probing, the wires can exhibit URS and BRS. BRS is observed at a lower compliance setting and demonstrated 150 stable cycles and a retention time of up to 10^3 s. URS was observed at a higher compliance setting and showed a higher ON-OFF ratio of 10^7 and a retention time greater than 10^6 s. This allows for flexibility in function between short-term and long-term memories.

Kim et al. [88] embedded Ag/TiO_2 core−shell wires in a polyvinyl alcohol matrix for potential use in wearable electronics applications. The resulting composite exhibited BRS with SET and RESET at 0.1 and -0.3 V. At a read voltage of 0.05 V, the devices were stable for 200 cycles and showed resistance retention for more than 10^7 s. The BRS was attributed to Ag ion movement in the TiO_2 shell. Also, because the field is concentrated and random filament formation is suppressed, owing to the composite, the operating voltage remained low. The devices were also mechanically resilient with a stable ON−OFF ratio for bending and relaxing cycles of approximately 5×10^4.

To enhance the switching properties of ZnO NWs, Porro et al. [89] coated ZnO NWs of diameters in the range 50−130 nm with 50 nm of polyacrylic acid to result in a metal oxide core and a polymeric shell structure. The wires displayed BRS behavior with SET and RESET occurring at 1.25 and -1.5 V. The switching occurs as a result of electrical interaction of ZnO with the polymer at the tip of the Cu (top) electrode. The multiple states that were obtained were tuned by controlling the redox reactions at the oxide−polymer interface.

HfO_2 NWs have been explored only as core−shell structures. Huang et al. [90] fabricated Sn-doped In_2O_3 (ITO) core and HfO_2 shell by depositing hafnia on ITO

NWs. The wires displayed BRS with SET and RESET occurring at 1 and -1 V for a 5-nm-thick HfO_2 shell. The device remained stable for 10^3 cycles and had an ON$-$OFF ratio of 10 with retention of resistance up to 10^4 s. However, as was predicted, for a thicker shell of 10 nm, SET and RESET voltages were larger: 1.85 and -1.35 V, respectively. With a thicker shell, the ON$-$OFF ratio increased to 10^2, but endurance was reduced to 30 cycles. This was attributed to higher voltages leading to breakdown of devices.

Huang et al. [91] from Wu's group took functioning Ni/NiO core$-$shell wires and coated them with 20 nm of HfO_2, using atomic layer deposition because of its excellent resistive switching characteristics. When probed, the devices could be operated in both directions. The SET potential was constant at 2 V with a compliance current of 10 μA. However, the device could be RESET in two different ways: by sweeping to 1.7 V at a higher compliance current or by sweeping to -1.6 V. As a result, the device can be used in both URS and BRS mode. The devices were stable for 200 cycles and retained resistance for 10^6 s. High-resolution transmission electronic microscopy (HRTEM) and scanning transmission electron microscopy (STEM) analysis reveals that switching occurs due to the migration of Hf^{4+}, Ni^{2+}, and O^{2-} ions in the NWs between the core and the top electrode. Therefore, by adding HfO_2 to the NW array, the density can be improved.

In our recent work at the Nino Research Group, resistive switching in a Pt (40 nm diameter) core and HfO_2 (7 nm thick) shell was explored. Owing to its current use in CMOS devices as a gate dielectric and compatible fabrication with semiconductor fabrication methods, HfO_2 has proven to be a viable candidate to produce resistive switching devices [92$-$94]. The current state-of-the-art crossbar device with a HfO_2 memristive layer has low programming voltage, a switching speed of less than 5 ns, and a record endurance of 1.2×10^{11} cycles with retention that can be extrapolated to 10 years. Additionally, 24 resistance levels have been programmed [30]. Several combinations of electrodes have been used for HfO_2 resistive switching devices. However, Pt (core) and Ti (top electrode) were selected, as they serve different roles to enable resistive switching in HfO_2. A thin Ti buffer layer added adjacent to HfO_2 by Lee et al. [94] suggested that the Ti enabled movement of oxygen vacancies in HfO_2 by absorbing oxygen. Another work performed by Lee et al. [95] on resistive switching in HfO_2 with Au and Pt electrodes suggested that when the device is biased at the Pt$-HfO_2$ interface, a dearth of oxygen was observed. Given that resistive switching in HfO_2 is based on movement of oxygen vacancies to form a conductive filament, Pt and Ti were the ideal candidates.

In an effort to tune the switching performance, a core$-$shell structure was employed. This has previously been suggested, as this structure provides access in the radial direction to the MIM structures rather than in the axial direction. Besides, a conductive core serves as a potential use to obtain multibit memory in crossbar architectures [90,96]. Additionally, previous work done by Sharath et al. [97] on the effect of HfO_2 thickness on resistive switching has suggested that with increasing thickness the forming voltage increases, where 4.7 V was required to form 10 nm of HfO_2 and there was an increase of 0.26 V for every 1 nm of oxide added. Also, Huang et al. [90] observed better lower operating voltage with better retention

Memristive applications of metal oxide nanofibers 267

and endurance when 5 nm of HfO_2 shell was used instead of 10 nm. As a result, the thickness of the HfO_2 shell in our work was chosen to be 7 nm. The wires were defined on a Si/SiO_2 substrate via e-beam lithography. The electrodes, Pt and Ti, and the shell, HfO_2, were sputtered on with additional lithography, deposition, and lift-off techniques as shown in Fig. 11.6a. It can be observed from the schematic in Fig. 11.6b that as a result of the symmetric arrangement of the top electrodes with the core and the shell, there is a presence of two memristors connected in an antiserial manner with an electrode between them in the wires.

Vourkas and Sirakoulis [98] detailed the modeling and simulation of current-voltage behavior in their book for back-to-back memristors connected in an antiserial manner. They suggested that because the memristors possess opposite polarities, they exhibit opposite behavior, and the dynamic IV response of both the memristors

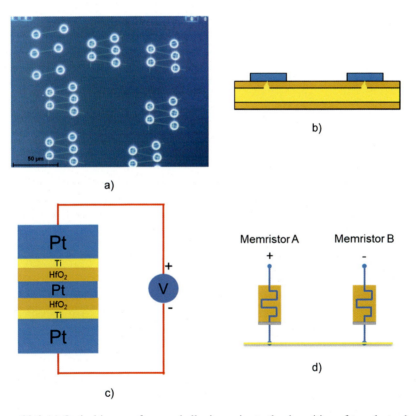

Figure 11.6 (a) Optical image of core–shell wires prior to the deposition of top electrodes (circular patterns). The pattern was fabricated to study resistive switching in individual wires and in the future study the effect of forming on resistive switching between interconnected wires. (b) Schematic of core (Pt)-shell (HfO_2) wires indicating the probable evolution of oxygen vacancies between the core and the top electrodes. (c) An equivalent crossbar structure detailing the several layers present in the core–shell wire. (d) An equivalent circuit diagram representing two memristors of similar polarity connected in an antiserial manner.

depends on their initial state. In our devices, when voltage was biased in sweep mode, the new core–shell wires displayed BRS behavior with SET and RESET occurring at 1.1 and −1.5 V. The initial cycle is shown in Fig. 11.7a, and the subsequent 10 cycles are shown in Fig. 11.7b. Next, to explore the repeatability of the wires, they were subjected to pulse mode operations. The SET and RESET potential were set at −1.5 and 1.5 V with read voltage at 0.3 V. Under this operation the device showed an endurance for at least 1000 cycles, as shown in Fig. 11.7c. Retention studies were

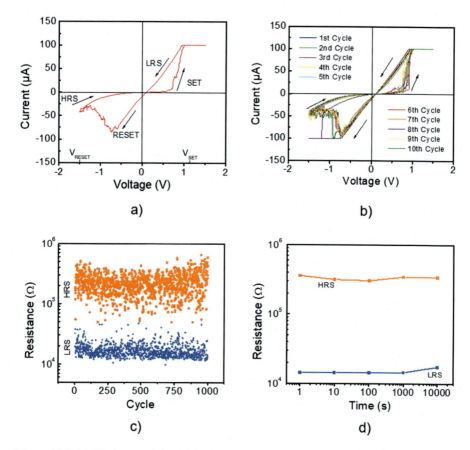

Figure 11.7 (a) IV characteristics of the first cycle representing bipolar resistive switching behavior in the core–shell wires with single electro transistor at 1.1 and remember every situation encourages transformation at −1.5 V with the compliance set at 100 μA. (b) The subsequent 10 cycles displaying stable switching in the core–shell wires at a compliance of 100 μA. (c) Cyclic endurance obtained for at least 1000 cycles at a pulse single electron transistor/remember every situation encourages transformation voltage of −1.5 and 1.5 V and a read pulse voltage of 0.3 V. (d) Resistance retention obtained for 10^4 s for both low resistance scale and high resistance scale after reading the state of the system by applying a read pulse at 0.3 V after either a single electron transistor pulse of −1.5 V or a remember every situation encourages transformation pulse of 1.5 V.

performed by applying a single SET/RESET pulse and reading the state of the system at 10^x intervals (where $x = 0, 1, 2, 3$, etc.). The retention for this system was 10^4 s, as shown in Fig. 11.7d. These results, although steps in the right direction, are nowhere near the performance of the systems described previously. A way to improve the performance would be to increase the thickness of the HfO_2 shell. In a previous work performed on a $Au/HfO_2/Pt$ thin-film system with 10 nm thick HfO_2, the researchers observed that, owing to a thin active layer, the filaments were eliminated during RESET, causing fluctuations in RESET voltage. However, the device would SET at the same threshold voltage despite lack of filaments [99]. Since the thickness is lower in our core–shell system, it is possible that, owing to repeated operation, a permanent filament is formed, causing the device to lose its resistive switching behavior.

From work performed by Linn et al. [35], we understand that systems with an electrode between two switching layers display BRS in the eight-wise and counter-eight-wise direction due to the presence of two electrodes. Therefore, when necessary threshold voltages are achieved, the devices will display CRS-type behavior. Similarly, when the threshold potential was raised to 2 V in Pt/HfO_2 core–shell wires, both the memristors were activated, and CRS-type switching behavior was observed. Stable CRS behavior is shown for 10 cycles in Fig. 11.8.

To program and store information in the device, states 0 and 1 are defined. Here, state 0 occurs when memristor A is in HRS and memristor B is in LRS, and state 1 occurs when memristor A is in LRS and memristor B is in HRS. Since one memristor is in HRS, the device is effectively in HRS in both these cases. The device is

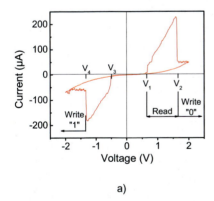
a)

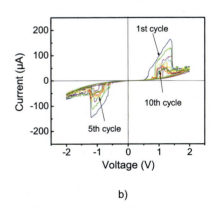
b)

Figure 11.8 (a) Complementary resistive switching behavior observed in core–shell Pt/HfO_2 system with zero and one as two defined states to store information, where state 0 is when memristor A is in high resistance scale and memristor B is in low resistance scale and state 1 is when memristor A is in low resistance scale and memristor B is in high resistance scale. As a result, the device is effectively in high resistance scale in both these cases. The device is ON when both the memristors are in low resistance scale, which results in an increase in current. The read voltage has to be greater than V_1, the first threshold voltage, and smaller than V_2. Since during every read operation the state written is erased, a pulse $V < V_4$ must be applied to restore state 1 or a pulse $V > V_2$ must be applied to restore state 0. (b) Next 10 cycles.

ON when both memristors are in LRS, which results in an increase in current. The read voltage must be greater than V_1, the first threshold voltage, and smaller than V_2. Since during every read operation the state written is erased, a pulse $V < V_4$ must be applied to restore state 1, or a pulse $V > V_2$ must be applied to restore state 0, as shown in Fig. 11.8. Therefore, owing to the ability of the wires to display both BRS and CRS types of switching, the dynamic control of the resistance states in the NWs is enabled. Also, a spike-read scheme is used, owing to the CRS nature, in which the current spikes can be observed in the ON state. Therefore the number of voltage levels required for programming the device is reduced, as the ON state is transient [35].

11.6 Perspective and outlook

As a result of continuous efforts made toward scaling of memristive devices through miniaturization, metal oxide NWs are potential candidates to replace traditional crossbar or thin-film architectures. These NW devices fabricated via the bottom-up approach [100] have been previously discussed by Ielmini et al. [96] and recently by Milano et al. [101]. In this chapter we have reviewed the resistive switching behavior in metal oxide NWs, displaying a large variety of oxides used and the different mechanisms they follow.

An important factor to consider in understanding the behavior of these NWs systems is that the properties such as forming voltage, SET voltage, RESET voltage, endurance, retention, and resistance states are all dependent on the techniques of fabrication, which determine the properties of the material and the electrodes placed for electrical contact. In general, the NW resistive switching devices operate at a wide range of potentials with |SET/RESET| potentials under 1 V in some cases and up to 40 V in others. A general trend observed in 1D NWs showed that reducing operating voltages did not enhance the endurance of the devices. The best result was observed in TiO_2 1D planar devices with SiO_2 passivation layer [44], where the resistance was stable for 10^6 cycles. Despite core−shell structures offering better tunability, the device endurance in most of them was less than 200. Among the best of them, reducing operating voltages did not increase endurance. For example, work done on MgO/Co_3O_4 core−shell NWs by Nagashima et al. [81,82] displayed an endurance of 10^8 cycles with SET and RESET voltages greater than 25 V and greater than 10 V, respectively. In work done by Wiley's group [84,85], Cu/SiO_2 core−shell NWs displayed an endurance for 10^4 cycles at SET and RESET voltages less than 3 V.

Since high variability in switching voltages and low device endurance are persistent, future work should focus on optimizing the structure and operation of wires to make them CMOS-compatible. Although an insight into the switching mechanism was provided in $Ni/NiO/HfO_2$ core−shell structures, the switching mechanism is widely predicted to be filamentary but has never been established. Therefore, these configurations can enable the study of switching events due to electrical and ionic transport but can also help us to correlate structural changes with applied bias. They also provide opportunities to study ionic dynamics for future implementation in neuromorphic hardware, as synaptic capabilities [102] and 1/f like noise [103] akin to noise in human brain have been well established in crossbar HfO_2 systems.

A major advantage is the control of thickness while fabricating core−shell wires, where either the deposition of the shell or the growth of the active layer around the core electrode can be controlled. Also, the availability of a conductive core helps in establishing an interconnection in flexible architectures beyond the conventional crossbar array. This control over thickness can also help in obtaining CMOS-compatible devices that switch at low voltages.

At present, to increase the integration density of crossbar architectures from $4F^2$ (F is minimum feature size), the crossbars are stacked on top of each other where cells are independent to obtain a density of $4F^2/n$ (n is number of crossbars). However, since the active elements are connected in a row to others on the top and bottom electrode bars, there exist sneak paths for current to flow. Additionally, owing to multiple sneak paths, scalability is limited, as the numbers of rows and columns cannot be increased [35]. As a result, some devices with self-rectification have been proposed here [57,58] where self-rectification was observed by doping the oxide NWs. However, another solution presented by Linn et al. [35] required a complementary resistive switch in crossbar devices. This comprised two BRS memristive devices to be connected in an antiserial manner. The resulting device behaved as a voltage divider and operated between predetermined thresholds. Since the switch is essentially OFF at HRS if one of the elements is in HRS, the device is able to prevent sneak paths in crossbar architectures. The Pt/HfO_2 core−shell NW system discussed here exhibits CRS-type behavior and therefore can be used in neuromorphic architectures beyond crossbar configurations.

The device performance of nonvolatile memories is affected by leakage currents, threshold potentials, and write/read voltages. These issues are further accentuated in scaling devices below 10 nm. An alternative in the bottom-up approach suggested by these devices will affect scaling and drive highly dense devices, as it allows for reduction in the active area dimension and maximizes array density [96,104−106]. As was suggested earlier, the density can also be improved by 3D stacking of multiple layers of NWs.

Overall, memristive systems based on core−shell NWs have proven to be fertile ground for the development of novel neuromorphic devices and for the fundamental investigation of the driving mechanisms in resistive switching phenomena.

References

[1] H. Lee, H. Kim, D.Y. Kim, Y. Seo, ACS Omega 4 (2019) 2610−2617.
[2] A.C. Ferreira, J.B. Branco, Int. J. Hydrogen Energy 44 (2019) 6505−6513.
[3] T. Wen, Z. Zhao, C. Shen, J. Li, X. Tan, A. Zeb, et al., Sci. Rep. 6 (2016) 20920.
[4] R.S. Andre, R.C. Sanfelice, A. Pavinatto, L.H.C. Mattoso, D.S. Correa, Mater. Des. 156 (2018) 154−166.

[5] D.R. Miller, S.A. Akbar, P.A. Morris, Sens. Actuators B: Chem. 204 (2014) 250–272.

[6] Z.U. Abideen, J.-H. Kim, J.-H. Lee, J.-Y. Kim, A. Mirzaei, H.W. Kim, et al., J. Korean Ceram. Soc. 54 (2017) 366–379.

[7] A. Dey, Mater. Sci. Eng. B 229 (2018) 206–217.

[8] R. Singh, S. Dutta, Fuel 220 (2018) 607–620.

[9] G. Massaglia, M. Quaglio, Mater. Sci. Semicond Process. 73 (2018) 13–21.

[10] S.-J. Choi, L. Persano, A. Camposeo, J.-S. Jang, W.-T. Koo, S.-J. Kim, et al., Macromol. Mater. Eng. 302 (2017) 1600569.

[11] S. Hoang, P.-X. Gao, Adv. Energy Mater. 6 (2016) 1600683.

[12] C.V. Reddy, K.R. Reddy, N.P. Shetti, J. Shim, T.M. Aminabhavi, D.D. Dionysiou, Int. J. Hydrog. Energy (2019).

[13] X. Lu, C. Wang, F. Favier, N. Pinna, Adv. Energy Mater. 7 (2017) 1601301.

[14] O. Crosnier, N. Goubard-Bretesché, G. Buvat, L. Athouël, C. Douard, P. Lannelongue, et al., Curr. Opin. Electrochem. 9 (2018).

[15] K. Mondal, A. Sharma, RSC Adv. 6 (2016) 94595–94616.

[16] K. Ghosal, C. Agatemor, Z. Špitálsky, S. Thomas, E. Kny, Chem. Eng. J. 358 (2019) 1262–1278.

[17] J. Li, R. Cha, K. Mou, X. Zhao, K. Long, H. Luo, et al., Adv. Healthcare Mater. 7 (2018) 1800334.

[18] Z.L. Wang, J. Song, Science 312 (2006) 242–246.

[19] J. Backus, Commun. ACM 21 (1978) 613–641.

[20] D. Ielmini, R. Waser, Resistive Switching: From Fundamentals of Nanoionic Redox Processes to Memristive Device Applications, John Wiley & Sons, 2015.

[21] S. Yu, in: S. Yu (Ed.), Neuro-Inspired Computing Using Resistive Synaptic Devices, Springer International Publishing, Cham, 2017, pp. 1–15.

[22] S. Yu, Proc. IEEE 106 (2018) 260–285.

[23] Y. Zhu, B. Shin, G. Liu, F. Shan, IEEE Electron. Device Lett. 40 (2019) 1776–1779.

[24] B.V. Benjamin, P. Gao, E. McQuinn, S. Choudhary, A.R. Chandrasekaran, J.-M. Bussat, et al., Proc. IEEE 2014;102:699–716.

[25] G.-Y. Jung, E. Johnston-Halperin, W. Wu, Z. Yu, S.-Y. Wang, W.M. Tong, et al., Nano Lett. 6 (2006) 351–354.

[26] J.C. Nino, J. Kendall, Memristive Nanofiber Neural Networks, US10614358B2, 2020.

[27] R.D. Pantone, J.D. Kendall, J.C. Nino, Neural Netw. 106 (2018) 144–151.

[28] W.E.P. Goodwin, Electric connecting device, US2667542A, 1954.

[29] S. Kim, X. Liu, J. Park, S. Jung, W. Lee, J. Woo, et al., in: 2012 Symposium on VLSI Technology (VLSIT), 2012, pp. 155–156.

[30] H. Jiang, L. Han, P. Lin, Z. Wang, M.H. Jang, Q. Wu, et al., Sci. Rep. 6 (2016) 1–8.

[31] M.-J. Lee, C.B. Lee, D. Lee, S.R. Lee, M. Chang, J.H. Hur, et al., Nat. Mater. 10 (2011) 625–630.

[32] A. Kumar, M. Das, V. Garg, B.S. Sengar, M.T. Htay, S. Kumar, et al., Appl. Phys. Lett. 110 (2017) 253509.

[33] D. Seong, M. Jo, D. Lee, et al., Solid-State Lett. 10 (2007) H168.

[34] D.-J. Seong, M. Hassan, H. Choi, J. Lee, J. Yoon, J.-B. Park, et al., IEEE Electron. Device Lett. 30 (2009) 919–921.

[35] E. Linn, R. Rosezin, C. Kügeler, R. Waser, Nat. Mater. 9 (2010) 403–406.

[36] F. Nardi, S. Balatti, S. Larentis, D.C. Gilmer, D. Ielmini, IEEE Trans Electron. Devices 60 (2013) 70–77.

[37] A. Sawa, Mater. Today 11 (2008) 28–36.

[38] Y. Yang, W. Lu, Nanoscale 5 (2013) 10076–10092.

[39] J.J. Yang, M.D. Pickett, X. Li, D.A.A. Ohlberg, D.R. Stewart, R.S. Williams, Nat. Nanotechnol. 3 (2008) 429−433.

[40] S.I. Kim, J.H. Lee, Y.W. Chang, S.S. Hwang, K.-H. Yoo, Appl. Phys. Lett. (n.d.) 4.

[41] E.D. Herderick, K.M. Reddy, R.N. Sample, T.I. Draskovic, N.P. Padture, Appl. Phys. Lett. 95 (2009) 203505.

[42] A. Klamchuen, M. Suzuki, K. Nagashima, H. Yoshida, M. Kanai, F. Zhuge, et al., Nano Lett. 15 (2015) 6406−6412.

[43] Q. Xia, W. Robinett, M.W. Cumbie, N. Banerjee, T.J. Cardinali, J.J. Yang, et al., Nano Lett. 9 (2009) 3640−3645.

[44] K. Nagashima, T. Yanagida, K. Oka, M. Kanai, A. Klamchuen, S. Rahong, et al., Nano Lett. 12 (2012) 5684−5690.

[45] C.-W. Huang, J.-Y. Chen, C.-H. Chiu, W.-W. Wu, Nano Lett. 14 (2014) 2759−2763.

[46] C. O'Kelly, J.A. Fairfield, J.J. Boland, ACS Nano 8 (2014) 11724−11729.

[47] L. Lin, L. Liu, K. Musselman, G. Zou, W.W. Duley, Y.N. Zhou, Adv. Funct. Mater. 26 (2016) 5979−5986.

[48] Z. Fan, X. Fan, A. Li, L. Dong, Nanoscale 5 (2013) 12310−12315.

[49] K.-D. Liang, C.-H. Huang, C.-C. Lai, J.-S. Huang, H.-W. Tsai, Y.-C. Wang, et al., ACS Appl. Mater. Interfaces 6 (2014) 16537−16544.

[50] W. Wu, Z.L. Wang, Nano Lett. 11 (2011) 2779−2785.

[51] Y. Yang, X. Zhang, M. Gao, F. Zeng, W. Zhou, S. Xie, et al., Nanoscale 3 (2011) 1917−1921.

[52] K.R.G. Karthik, R. Ramanujam Prabhakar, L. Hai, S.K. Batabyal, Y.Z. Huang, S.G. Mhaisalkar, Appl. Phys. Lett. 103 (2013) 123114.

[53] A. Jasmin, S. Porro, A. Chiolerio, C.F. Pirri, C. Ricciardi, in: 2015 IEEE 15th International Conference on Nanotechnology (IEEE-NANO), 2015, pp. 496−498.

[54] G. Milano, M. Luebben, Z. Ma, R. Dunin-Borkowski, L. Boarino, C.F. Pirri, et al., Nat. Commun. 9 (2018) 5151.

[55] Z.J. Chew, L. Li, Mater. Lett. 91 (2013) 298−300.

[56] Y. Sun, X. Yan, X. Zheng, Y. Liu, Y. Zhao, Y. Shen, et al., ACS Appl. Mater. Interfaces 7 (2015) 7382−7388.

[57] J. Qi, J. Huang, D. Paul, J. Ren, S. Chu, J. Liu, Nanoscale 5 (2013) 2651−2654.

[58] B. Wang, T. Ren, S. Chen, B. Zhang, R. Zhang, J. Qi, et al., J. Mater. Chem. C. 3 (2015) 11881−11885.

[59] Y.-D. Chiang, W.-Y. Chang, C.-Y. Ho, C.-Y. Chen, C.-H. Ho, S.-J. Lin, et al., IEEE Trans. Electron. Dev 58 (2011) 1735−1740.

[60] Y. Huang, Y. Luo, Z. Shen, G. Yuan, H. Zeng, Nanoscale Res. Lett. 9 (2014) 381.

[61] Y.-T. Huang, S.-Y. Yu, C.-L. Hsin, C.-W. Huang, C.-F. Kang, F.-H. Chu, et al., Anal. Chem. 85 (2013) 3955−3960.

[62] Y. Lai, P. Xin, S. Cheng, J. Yu, Q. Zheng, Appl. Phys. Lett. 106 (2015) 031603.

[63] S. Lee, J.-B. Park, M.-J. Lee, J.J. Boland, AIP Adv. 6 (2016) 125010.

[64] R. Zhang, W. Pang, Q. Zhang, Y. Chen, X. Chen, Z. Feng, et al., Nanotechnology 27 (2016) 315203.

[65] R. Zhang, W. Pang, Z. Feng, X. Chen, Y. Chen, Q. Zhang, et al., Sens. Actuators B: Chem. 238 (2017) 357−363.

[66] Y.S. Puzyrev, X. Shen, C.X. Zhang, J. Hachtel, K. Ni, B.K. Choi, et al., Appl. Phys. Lett. 111 (2017) 153504.

[67] A.M. Grishin, A.A. Velichko, A. Jalalian, Appl. Phys. Lett. 103 (2013) 053111.

[68] S.-H. Bae, S. Lee, H. Koo, L. Lin, B.H. Jo, C. Park, et al., Adv. Mater. 25 (2013) 5098−5103.

[69] H. Dong, X. Zhang, D. Zhao, Z. Niu, Q. Zeng, J. Li, et al., Nanoscale 4 (2012) 2571–2574.

[70] B. Sun, C.M. Li, Chem. Phys. Lett. 604 (2014) 127–130.

[71] L. Lei, Y. Yin, C. Liu, Y. Zhou, Y. Peng, F. Zhou, et al., Solid. State Ion. 308 (2017) 107–111.

[72] D.-H. Kwon, K.M. Kim, J.H. Jang, J.M. Jeon, M.H. Lee, G.H. Kim, et al., Nat. Nanotechnol. 5 (2010) 148–153.

[73] Y. Feng, W. Hou, X. Zhang, P. Lv, Y. Li, W. Feng, J. Phys. Chem. C. 115 (2011) 3956–3961.

[74] Y. Lai, W. Qiu, Z. Zeng, S. Cheng, J. Yu, Q. Zheng, Nanomaterials 6 (2016) 16.

[75] Y. Zhou, Y. Yin, Y. Peng, W. Zhou, H. Yuan, Z. Qin, et al., Mater. Res. Express 1 (2014) 025025.

[76] J. Wang, S. Xie, W. Zhou, MRS Bull. 32 (2007) 123–126.

[77] K. Oka, T. Yanagida, K. Nagashima, H. Tanaka, T. Kawai, J. Am. Chem. Soc. 131 (2009) 3434–3435.

[78] L. He, Z.-M. Liao, H.-C. Wu, X.-X. Tian, D.-S. Xu, G.L.W. Cross, et al., Nano Lett. 11 (2011) 4601–4606.

[79] C. Cagli, F. Nardi, B. Harteneck, Z. Tan, Y. Zhang, D. Ielmini, Small 7 (2011) 2899–2905.

[80] A.T. Bellew, A.P. Bell, E.K. McCarthy, J.A. Fairfield, J.J. Boland, Nanoscale 6 (2014) 9632–9639.

[81] K. Nagashima, T. Yanagida, K. Oka, M. Taniguchi, T. Kawai, J.-S. Kim, et al., Nano Lett. 10 (2010) 1359–1363.

[82] K. Nagashima, T. Yanagida, M. Kanai, T. Kawai, in: 2015 IEEE 15th International Conference on Nanotechnology (IEEE-NANO), 2015, pp. 120–123.

[83] C.-W. Hsu, L.-J. Chou, Nano Lett. 12 (2012) 4247–4253.

[84] P.F. Flowers, M.J. Catenacci, B.J. Wiley, Nanoscale Horiz. 1 (2016) 313–316.

[85] M.J. Catenacci, P.F. Flowers, C. Cao, J.B. Andrews, A.D. Franklin, B.J. Wiley, J. Elec Materi 46 (2017) 4596–4603.

[86] K. Nagashima, T. Yanagida, M. Kanai, K. Oka, A. Klamchuen, S. Rahong, et al., Jpn. J. Appl. Phys. 51 (2012) 11PE09.

[87] H.G. Manning, S. Biswas, J.D. Holmes, J.J. Boland, ACS Appl. Mater. Interfaces 9 (2017) 38959–38966.

[88] Y. Kim, W. Jeon, M. Kim, J.H. Park, C.S. Hwang, S.-S. Lee, Appl. Mater. Today 19 (2020) 100569.

[89] S. Porro, F. Risplendi, G. Cicero, K. Bejtka, G. Milano, P. Rivolo, et al., J. Mater. Chem. C. 5 (2017) 10517–10523.

[90] C.-H. Huang, W.-C. Chang, J.-S. Huang, S.-M. Lin, Y.-L. Chueh, Nanoscale 9 (2017) 6920–6928.

[91] T.-K. Huang, J.-Y. Chen, Y.-H. Ting, W.-W. Wu, Adv. Electron. Mater. 4 (2018) 1800256.

[92] D. Walczyk, Ch Walczyk, T. Schroeder, T. Bertaud, M. Sowińska, M. Lukosius, et al., Microelectronic Eng. 88 (2011) 1133–1135.

[93] C. Walczyk, D. Walczyk, T. Schroeder, T. Bertaud, M. Sowinska, M. Lukosius, et al., IEEE Trans. Electron. Devices 58 (2011) 3124–3131.

[94] H.Y. Lee, P.S. Chen, T.Y. Wu, Y.S. Chen, C.C. Wang, P.J. Tzeng, et al., in: 2008 IEEE International Electron Devices Meeting, 2008, pp. 1–4.

[95] S. Lee, W.-G. Kim, S.-W. Rhee, K. Yong, J. Electrochem. Soc. 155 (2007) H92.

[96] D. Ielmini, C. Cagli, F. Nardi, Y. Zhang, J. Phys. D: Appl. Phys. 46 (2013) 074006.

[97] S.U. Sharath, J. Kurian, P. Komissinskiy, E. Hildebrandt, T. Bertaud, C. Walczyk, et al., Appl. Phys. Lett. 105 (2014) 073505.

[98] I. Vourkas, G. Sirakoulis, Memristor-Based Nanoelectronic Computing Circuits and Architectures: Foreword by Leon Chua, Springer International Publishing, 2016.

[99] P. Gonon, M. Mougenot, C. Vallée, C. Jorel, V. Jousseaume, H. Grampeix, et al., J. Appl. Phys. 107 (2010) 074507.

[100] W. Lu, C.M. Lieber, Nat. Mater. 6 (2007) 841−850.

[101] G. Milano, S. Porro, I. Valov, C. Ricciardi, Adv. Electron. Mater. 5 (2019) 1800909.

[102] S. Yu, Y. Wu, R. Jeyasingh, D. Kuzum, H.-S.P. Wong, IEEE Trans. Electron. Devices 58 (2011) 2729−2737.

[103] S. Yu, R. Jeyasingh, Yi Wu, H.-S. Philip Wong, in: 2011 International Electron Devices Meeting, 2011, pp. 12.1.1−12.1.4.

[104] D. Ielmini, F. Nardi, C. Cagli, Nanotechnology 22 (2011) 254022.

[105] R. Meyer, L. Schloss, J. Brewer, R. Lambertson, W. Kinney, J. Sanchez, et al., in: 2008 9th Annual Non-Volatile Memory Technology Symposium (NVMTS), 2008, pp. 1−5.

[106] M.-J. Lee, Y. Park, B.-S. Kang, S.-E. Ahn, C. Lee, K. Kim, et al., in: 2007 IEEE International Electron Devices Meeting, 2007, pp. 771−774.

Metal oxide nanofibers in solar cells

12

JinKiong Ling and Rajan Jose
Nanostructured Renewable Energy Material Laboratory, Faculty of Industrial Sciences & Technology, University Malaysia Pahang, Pahang, Malaysia

12.1 Introduction: role of nanofibers in various types of solar cells

Solar cells, or photovoltaic (PV) cells, are electrical devices that are capable of converting solar energy into electrical energy by engaging valence electrons in a semiconducting material to absorb radiant energy and mobilize them by a material interface of differing Fermi energy. Their capability to convert solar radiation into electrical energy without the need for combustion as well as complex mechanical parts offers clean energy generation with a low carbon footprint. In fact, PV cells can be viewed as one of the most promising sustainable energy generators. With its low cost, durability, and light weight, it could be widely utilized for self-powered portable devices. With half a century's worth of efforts, the photoconversion efficiency (PCE) of silicon PV (Si PV) cells has improved to 27.6% (silicon heterostructure) and 44.0% (concentrator three-junction Si PV cells) [1] with a drastic price reduction from $76.67/W in 1977 to $0.36/W in 2015 [2]. Surprisingly, PV cells had a market value of $38.9 billion in 2017 alone, accounting for a whopping $188.5 billion during the 2012−2017 period [3]. With continuous effort through beneficial governmental tariff policies, the market for PV technologies are expected to escalate, with increasing numbers of people using PV cells as their daily energy source. However, Si PV cells suffer from some drawbacks as well. The production process of Si PV cells involves high-temperature ingot formation and toxic chemical by-products, which are energy consuming and costly and contribute to the carbon footprint. As a matter of fact, the production process alone releases $\sim 20-40$ g CO_2-eq/kWh [4,5]. Further lowering of cost was made difficult with the involvement of energy-consuming processes for silicon ingot formation during the fabrication. Even though the overall carbon footprint of Si PV cells was lower than that of coal-fired energy (975.2 g CO_2-eq/kWh) [4], a more environmentally friendly production process would be desirable. Recently, organic molecule−sensitized PV cells was introduced as a promising alternative to replace Si PV due to its low-cost processing that does not involved any toxic chemicals. Other than having a low cost, sensitized PV cells can be integrated into different applications, such as building integrated PV cells, automotive integrated PV cells, and flexible PV cells [6−9], further extending the market reach of PV cells.

Metal Oxide-Based Nanofibers and Their Applications. DOI: https://doi.org/10.1016/B978-0-12-820629-4.00007-2
© 2022 Elsevier Inc. All rights reserved.

With the improvements in material syntheses methods, nanostructured material (material with one of its dimensions within the nanometer scale) was widely adopted in PV cells because of its excellent properties compared to its bulk counterparts. One of the most widely adopted nanostructures has been the one-dimensional (1D) material (threadlike morphology with a diameter less than 100 nm and length upto a few microns). Some applications of 1D nanomaterials in PV cells are shown in Fig. 12.1. In Si and thin-film PV cells, 1D nanostructures find their usefulness as antireflective coating, which plays an important role in reducing the reflective loss from the protective glass or plastic substrate, improving the PCE of the PV cells [10,11]. A direct coating of SiO_2 nanofibers (NFs) on glass substrate for Si PV cells through an electrospinning process was shown to increase the light transmittance from 91% to 96%, resulting in a 3% growth in the short-circuit current density [12]. However, control of the thickness of the antireflective coating is crucial, as a higher coating thickness reduces the transmittance, which is detrimental to the performance of PV cells. Nanofiber (NF)-coated flexible substrate was also observed to scatter light, allowing the fabrication of high-performing, flexible, thin-film PV cells [13]. In fact, light scattering is related to the improvement in light absorption of the photoactive material, which boosted the photocurrent of the PV cell, improving the overall PCE [14,15]. Besides coating, NFs were also developed into various components of sensitized PV cells, such as the conductive flexible substrate [16], photoanode materials [17–19], solid-state electrolyte [20–22], and low-cost counterelectrode [23–25], with the aim of reducing material cost (conductive substrate and counterelectrode) as well as improving stability (solid-state electrolyte). Different from these components, NF photoanode materials are crucial in determining the performance of sensitized PV cells, owing to their important role in extracting and transporting the excited electron. This chapter focuses on the role of metal oxide NFs as one of the major photoactive components in sensitized PV. The basic mechanism of charge excitation, extraction, and transport is thoroughly discussed,

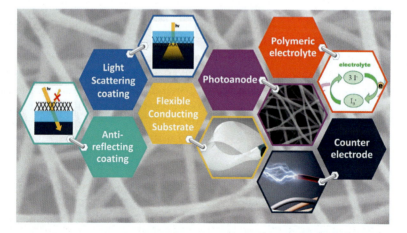

Figure 12.1 Application of one-dimensional nanomaterials in photovoltaic.

and effect of energy trap states on the carrier transportation is established. Some techniques to improve the carrier transportation, which leads to a significant improvement in the PCE of sensitized PV cells, are reviewed. Subsequently, challenges in developing highly performing sensitized PV are discussed to point out major improvements required in future research.

12.2 Photoconversion mechanism in sensitized photovoltaic cells

Sensitized PV cells convert sunlight into energy through utilizing organic molecules (referred to as dye throughout the chapter) or semiconducting clusters (quantum dots) as their primary light-absorbing material. A handful of reviews on dyes for sensitized PV cells have been reported elsewhere [26–28]. A brief schematic of a typical sensitized PV cell, namely, a dye-sensitized solar cell (DSSC), is illustrated in Fig. 12.2. The components were discussed in detail in a previous review [29]. The active components (parts that were involved in the photoconversion process) of sensitized PV cells can be generalized into those involved in light absorption and exciton generation in dye molecules and those involved in charge separation, extraction, and transportation within the photoanode materials. When the dye is illuminated, electrons at the highest occupied molecular orbital (HOMO) are excited to the lowest unoccupied molecular orbital (LUMO) of the dye. The excited electrons are then transferred from the LUMO of the dye to the conduction band minimum (CBM) of the photoanode materials within a picosecond time scale. When the electrons are transferred to the photoanode, the dye molecules are in an oxidized state with no further electron excitation allowed until they are reduced to neutral molecules. The oxidized dye molecules are then regenerated by the redox electrolyte with the help of a catalyst (a thin platinum coating in most cases). Numerous redox electrolytes, both liquid and polymeric, have been investigated for sensitized PV cells and are detailed elsewhere [30–32]. This dye oxidation-regeneration cycle repeats continuously during the PV operation of sensitized PV cells. During the

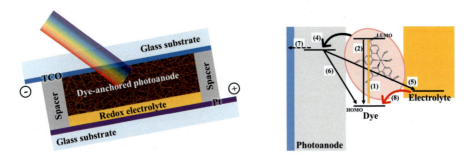

Figure 12.2 Schematic of sensitized photovoltaic device architecture and all the processes involved in solar energy conversion.

operation of sensitized PV cells, several processes play a crucial role in determining the performance of the sensitized PV cell and are listed as follows:

1. Photon absorption, which depends heavily on the intensity of the solar radiation, the energy gap, and the absorption coefficient of the dye molecule.
2. Radiative recombination, in which the excited dye relaxes back to its ground state (occurs within several nanoseconds).
3. Diffusion length of the exciton.
4. Transfer of electron from the LUMO of the dye to the conduction band of the photoanode (occurs within several picoseconds).
5. Electron back transfer, in which the electron at the CBM of photoanode is captured by the oxidized species in the redox electrolytes.
6. Interfacial charge recombination, in which the electron at the CBM of photoanode recombines with the positive charge in the oxidized dye molecules.
7. Transport of electron within the photoanode material, depending on the mobility and lifetime of the electron and nonradiative relaxation, in which the energy of the electron is lost as a result of electron−phonon interaction.
8. The rate at which the oxidized dye molecules are regenerated by the redox electrolyte.

Among these, factors 4−7 are controlled by the photoanode materials, while factors 1−3 and 8 are affected by the dye molecules as well as redox electrolyte, respectively. The higher number of processes involving the photoanode, compared to both dye molecules and redox electrolyte, indicates that the former play a substantial role in determining the performance of the sensitized PV cell.

12.2.1 Excitation of electrons

To understand the excitation process of electrons in any semiconducting materials, light should be viewed as a package of energy (known as photons) with distinct energy instead of a light wave with different frequencies. In general, when the photon falls on the semiconductor, there are four possible scenarios:

1. The photons have lower energy than the energy gap of the dye. By absorbing these photons, the electrons at the HOMO of the dye do not gain enough energy to leap across the energy gap to the LUMO. The absorbed photon will be released back and pass through the material as if there were no interaction with the material.
2. The photons have energy similar to that of the energy gap of the dye. The electrons at the HOMO gain enough energy to leap across the energy gap to the LUMO while leaving behind a positively charged hole at the HOMO. An electron hole pair is generated
3. The photons have larger energy than the energy gap of the dye. The electrons gain enough energy to be excited to a delocalized energy level above the LUMO (state with higher energy level than LUMO). The electron will then be relaxed back to the LUMO via nonradiative relaxation in which energy is released as heat that caused lattice vibration (also known as a phonon). An electron hole pair is generated as well.
4. Excited electrons meet the lifetime of the excited state. This causes the electrons to be relaxed back from LUMO to HOMO, releasing photons of energy similar to the energy gap of the dye, a process known as radiative recombination.

The three different scenarios of electron—photon interaction are illustrated in Fig. 12.3.

12.2.2 Generation of photovoltage

For all materials there are two distinctive energy bands, the conduction band and the valence band formed as a consequence of mixing up of molecular orbitals during the growth of crystals from their components. The gap between the conduction and valence band energies, known as energy gap, determines the nature of the materials (conductor, semiconductor, or insulator). Within the energy gap there is a Fermi energy level (defined as the highest energy state that can be reached by electrons at absolute zero). The Fermi energy level can be estimated according to Eq. (12.1), where N_C and N_V are the effective density of electrons in conduction and valence band, respectively; E_C and E_V are the conduction and valence band

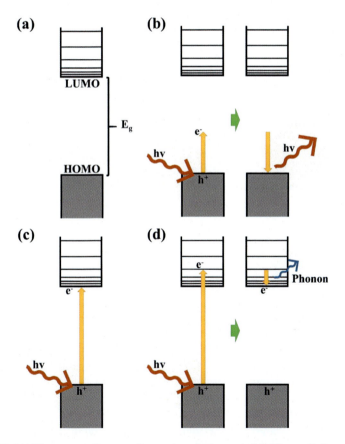

Figure 12.3 (A) The energy diagram of a semiconductor material. The excitation of electrons at the valence band when the incident photon has energy (B) less than, (C) equal to, and (D) greater than the energy gap.

energies, respectively; k_B is the Boltzmann's constant; and T is the absolute temperature.

$$N_C exp\left(\frac{E_F - E_C}{k_B T}\right) = N_V exp\left(\frac{E_V - E_F}{k_B T}\right) \tag{12.1}$$

Rearranging the equation and taking the logarithm give the following:

$$\frac{2E_F - E_C - E_V}{k_B T} = \ln\left(\frac{N_V}{N_C}\right) \tag{12.2}$$

$$E_F = \frac{E_g}{2} + \frac{k_B T}{2}\ln\left(\frac{N_V}{N_C}\right) \tag{12.3}$$

In intrinsic semiconductors the density of states in conduction and valence band are similar ($N_C = N_V$), where according to Eq. (12.3), the Fermi energy level position is at the middle of conduction and valence band energies. When the material is doped with an electron donor (where $N_C > N_V$), the Fermi energy moves closer to the conduction band or toward the valence band if doped with a hole donor (where $N_C < N_V$). However, when the material is placed in a nonequilibrium state (illuminated or biased), additional electrons and holes could be generated. The changes in the carrier concentration generates two different Fermi energy levels, with energy level inequivalent to the intrinsic Fermi levels. The newly formed Fermi levels are known as quasi-Fermi levels, $E_{F,n}$ and $E_{F,h}$ for electron and hole, respectively, with the former closer to the CBM and the latter closer to the valence band maximum (VBM) according to Eqs. (12.4) and (12.5):

$$E_F - E_i = \frac{K_B T}{q}\ln\left(\frac{n}{n_i}\right) \tag{12.4}$$

$$E_i - E_F = \frac{K_B T}{q}\ln\left(\frac{p}{n_i}\right) \tag{12.5}$$

where E_i is the intrinsic Fermi energy level, q is the electron charge, and n and p are the density of electron and hole, respectively. In a sensitized PV cell, the Fermi energy level of the photoanode realigns with the electron quasi-Fermi level of the dye, causing the conduction band to bend across the photoanode-dye interface. As the LUMO of dye is at lower energies than the CBM of the photoanode, the electrons transfer across the interface and collect at the photoanode. Subsequently, the electron transport across the external circuit, driven by the difference in potential between the CBM of the photoanode and the work function of the redox electrolyte, which also determines the working voltage or open-circuit voltage (V_{OC}) of the sensitized PV. However, the V_{OC} does not depend solely on the potential difference but is also affected by the charge recombination rate. The trap states at the donor–acceptor (dye-photoanode) interface is reported to have caused significantly reduction in V_{OC} [33], and suppressing the surface trap state was demonstrated as an effective way to enhance V_{OC} [34]. The accumulation of charge at the donor–acceptor interface leads to an increase in charge recombination and limits the V_{OC} by altering the Fermi level of the photoanode material [35]. To achieve largest possible V_{OC} as well as reducing the V_{OC} deficit (defined as E_g/eV_{OC}), the energy difference between CBM (photoanode) and LUMO (dye) as well as that between work function (electrolyte) and HOMO (dye) need to be reduced, while improving the diffusion coefficient of the carriers.

12.2.3 Charge extraction and transport

In Si PV cells, charge extraction after electron excitation is achieved by band-bending generated at the p–n junction through selective doping. When an electron hole pair is generated, the electrons diffuse from p-type (region with high hole density) to n-type (region with high electron density), following the induced electric field, with the hole moving in the opposite direction. As a result, both electron and hole are separated by a depletion region (an insulating region where charges are diffused away), which aids in reducing the recombination of the electron hole pair. To further improve the charge separation, a thin layer of selective doping on the surface is carried out, resulting in a p-type/intrinsic/n-type (p–i–n) junction [36]. The intrinsic region further increases the thickness of the depletion zone, thereby significantly reducing the charge recombination rate. A similar mechanism is adopted in thin-film PV cells. However, as was mentioned previously, dye molecules do not have the capability to transport or separate the charge on their own. Therefore a photoanode and an electrolyte are required, as n-type and p-type region, respectively, for effective charge separation process. It is noteworthy that the transferred electrons at the CBM of photoanode are subjected to recombination with the positive charges in either the redox electrolyte (process 5) or the dye molecules (process 6). To eliminate the recombination of electrons after extraction, the electrons need to be transferred to the current collector as soon as possible, a situation that requires the photoanode to be highly conductive metal oxide with an optimum energy gap. A photoanode material with a narrow energy gap would interfere with the absorption of visible light in dye molecules, which could significantly affect the photocurrent of the sensitized PV cells [37]. Therefore metal oxides with an energy gap of $\sim 3-3.5$ eV are preferred as the photoanode of sensitized PV cells. The thickness of the photoanode needs to be controlled so that light can reach the bottom layer of dye molecules to maximize the charge carrier density of the excited electrons and not exceeding the diffusion length of the electron [38,39]. The electron diffusion length is the maximum possible distance that an electron can travel prior to recombination. These conditions require careful engineering of the morphology, crystal structure, and electronic properties of the metal oxide photoanode to achieve sensitized PV cells with superior PV performance.

12.2.4 Generation of photocurrent

The photocurrent can be defined as the current generated per unit active area (unit in mA/cm^2) by the sensitized PV cells when zero bias is applied across the device. The photocurrent can be significantly affected by the active surface area of the photoanode, and the carrier transport properties of the photoanode [40–42]. Generally, the photocurrent of PV cells can be summarized as follows:

$$J = n_e \mu_e \nabla E_F \tag{12.6}$$

where J is the current density, n_e is the concentration of electron transferred from the dye, μ_e is the mobility of the electron, and ∇E_F is the potential gradient across the photoanode and redox electrolyte. Dye molecules are capable of exciting only one electron with the absorption of one photon. Therefore to increase the electron concentration, the most direct technique will be to increase the number of dyes anchored on the photoanode by enhancing the active surface area of the metal oxide photoanode. However, these electrons need to be extracted and transported, or a layer of space charge will be developed at the interfaces, inducing an electric field that would impede transfer of carriers [35]. The electron also needs to be transported to the current collector after being transferred from the dye, where a photoanode with high electron mobility is preferred. Better electron mobility can be achieved by improving the conductivity of the metal oxide photoanode, either by material engineering (doping or composite), morphology engineering (from bulk to 1D or 0D), or enhancing material crystallinity [43–47]. Functioning as the driving force, greater potential gradient would be crucial in increasing the tendency of carriers to be transported. Throughout the photocurrent generation, the photoanode plays an important role in influencing all the parameters. Therefore it is vital to reduce carrier losses by improving the diffusion coefficient in the photoanode materials to ensure a high-performing sensitized PV cell.

12.3 Metal oxide nanofibers as photoanode in dye-sensitized solar cells

Among all the nanostructures, 1D (NFs, nanotubes, nanowire, nanorods, etc.) and 0D (nanoparticles) were the most widely studied morphologies in sensitized PV cells, mostly owing to their superior surface area [48]. Compared to their 0D counterparts, 1D NFs have attracted more attention because of their anisotropic electron transport properties. Structurally, the metal oxide is arranged in such a way that the diameter is constrained within the nanometer range (> 500 nm) but the length can grow upto micrometer scale. Electronically, the electrons are not one-dimensionally confined. Given that the diameter of the NF can be as large as 100 nm, much greater than the Bohr exciton radius, the electrons move freely in all three dimensions. However, when the electrons are transferred from the dye to the surface of the metal oxide, the hole within is attracted and accumulates at the circumference of the NFs. Metal oxide is also a semiconducting material with an energy gap larger than the visible light range and can be excited by ultraviolet light. Such accumulation results in a region of high hole concentration at the circumference and lower hole concentration at the core. The Fermi energy level at the circumference moves farther away from the conduction band as a result of the increasing amount of hole and vice versa for the Fermi energy level at the core. The realignment of the Fermi energy level eventually leads to conduction band-bending as shown in Fig. 12.4. As the result, a potential gradient is formed where the injected electrons are directed and transported along the core of the NFs. This phenomenon shields the electrons from recombining with the holes, either from the dye molecules or redox electrolyte, at the surface of the NFs. The existence of the induced electric field around the circumference of the NFs results in anisotropic transport properties, directing the flow of electron along the length of the NFs, which was also attributed to the better PCE in NF-based PV cells compared to its nanoparticle-based counterpart [49]. In this circumstance the electrons are confined in a quasi-1D manner and could still diffuse perpendicular to the length of NF but only for a small distance, owing to the opposition force from the band-bending–induced electric field. A detailed energy diagram between the nanoparticles and NFs was provided by Archana et al. [50]. Such an anisotropic property is not observed in nanoparticles, owing to their extremely small diameter where charges are even distributed. With larger diameter, hole accumulation at the circumference could occur, inducing similar conduction band-bending. In this case the electrons are confined at the core of the nanoparticles, owing to its isotropic characteristic, recombining upon reaching its lifetime. Even though NFs show interesting anisotropic carrier transport

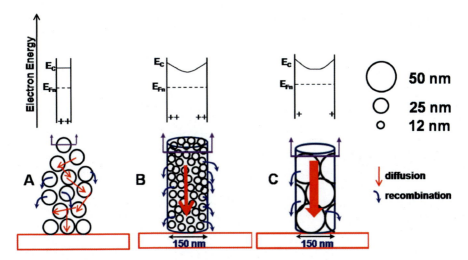

Figure 12.4 Origin of anisotropic transport property in nanofibers compared to nanoparticles.
Source: Copyright © 2009, American Chemical Society.

properties, the high surface area offer by nanostructures is mostly accompanied by a large surface energy trap state. The electrons trapped by these energy state lead to loss in energy and show a lower diffusion coefficient, leading to a high recombination rate, reducing the concentration of active electrons, which is detrimental to the performance of the PV.

12.4 Reducing energy trap states

At the surface of the materials, a different electronic state is generated owing to the difference in atomic atmosphere. Within the bulk the atoms are surrounded by other atoms, with all their charges balanced out and neutralized by the neighboring atoms. At the surface, the charges of the atom were neutralized by the bottom layer of atoms, but the charges facing toward the vacuum were not counterbalanced. These dangling bonds at the surface of the material induced different delocalized electronic states lying right below the conduction band (Fig. 12.5), trapping the electron and eventually increasing the recombination rate [51]. The trap density is at its highest below the conduction band and is exponentially reduced toward the Fermi energy level [52]. With an increase in surface area the delocalized trap states become more significant, especially in nanostructured materials, where higher trap density could severely impair electron transport [53]. Diffusion of electrons within

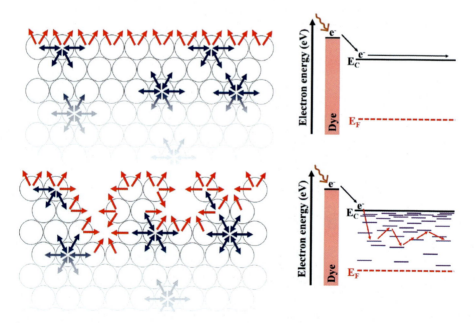

Figure 12.5 Delocalized trap states originated from the surface dangling bonds, with electron transporting within the states via a multitrapping mechanism.

the delocalized energy trap states is still possible through multiple trapping mechanisms, where the carriers are trapped and detrapped continuously. Such motion causes the carriers to move with a much lower diffuse coefficient, which eventually leads to recombination at the end of the carrier lifetime [54]. To ensure effective transportation of electrode and high performance in a sensitized PV cell, the delocalized energy trap states need to be reduced without significantly reducing the surface area of the photoanode. With all the gathered information, two techniques were discussed to reduce the energy trap states: improving the crystallinity of the metal oxide to reduce the defect density and raising the Fermi energy level to reduce the depth of delocalized energy trap state.

12.4.1 Improving crystallinity through high sintering

High electrical conductivity is tightly correlated to its crystallinity, especially the density of oxygen vacancies [55]. Metal oxide with poor crystallinity usually has a higher amount of oxygen vacancies, which could act as defect trap states, deteriorating the electrical conductivity [56,57]. By manipulating the calcination process parameters, that is, longer sintering duration, higher sintering temperature, and sintering in an oxygen-rich atmosphere, improvement in crystallinity and larger grain size were reported through reducing the oxygen vacancies [58]. However, the effect of oxygen vacancies on the electronic properties varies according to the semiconducting nature of the metal oxide. In n-type (electron-rich) semiconductors, the oxygen vacancies supply higher hole concentration, reducing the electron density and deteriorating the electron diffusion coefficient. The situation is different with p-type (electron-deficit) semiconductors, in which holes are the major carrier. In this case, higher oxygen vacancies were favorable in increasing the carrier concentration and improving the hole diffusion coefficient. Previous studies demonstrated that intentionally introducing oxygen vacancies into MoO_3 (p-type semiconductor) through sintering in vacuum showed significant improvement in its electronic properties [59]. However, in sensitized PV cells the photoanode consists of n-type metal oxide, owing to favorable energy level alignment, in which the reduction in the oxygen vacancies was preferable in reducing the density of delocalized trap states and improving the diffusion coefficient of electrons.

Long duration or high-temperature calcination is correlated with Ostwald ripening, in which the nanoparticles agglomerate and grow in size [60], reducing the surface area-to-volume ratio of the metal oxide particles. The surface area-to-volume ratio is the total surface area available in a given volume. The ratio increases with decreasing particle size, indicating that materials with smaller dimensions offer a higher surface area than their larger-sized counterparts. Golsheikh et al. previously reported that the crystallinity and particle size of zinc oxide (ZnO) nanoparticles increased from 15 (500°C) to 18 (600°C) to 22 nm (700°C), accompanied by reductions in surface area (26.7, 19.7, and 14.8 $m^2 g^{-1}$, respectively) when the sol−gel-prepared precursor was calcined at higher temperature [61]. The study demonstrated that the PCE increased from 500°C to 600°C (PCE = 3.06% and 3.44%, respectively) but started deteriorating at 700°C (PCE = 3.12%). Zhao et al.

reported similar results in which the concentration of anchored dye on TiO_2 nanoparticles decreased with increasing sintering temperature, a clear indicator of reducing surface area [62]. A trend in the reduction in charge transport resistance is observed when the sintering temperature is increased from 350°C to 500°C, with an exceptional increase in 600°C. This is attributed to the higher charge transport resistance at 600°C to the collapse of mesopores during crystal growth, generating more delocalized surface states. The transformation of TiO_2 nanoparticles from anatase to rutile phase at high-temperature (600°C−650°C) could also play some role in increasing the delocalized surface state, which directly deteriorates the charge transport [63].

Surprisingly, NFs showed different particulate growth during sintering. Considering large-scale production of NFs, electrospinning was widely adopted as the most versatile technique for synthesis of NFs. Therefore the following discussion focuses on NFs photoanode synthesized via electrospinning. For the synthesis of metal oxide NFs through electrospinning, the metal precursors are dissolved in a polymeric solution, and the resulting solution is electrospun, where the polymer will act as the template for the formation of NFs upon a high-temperature calcination and sintering process. During calcination (sintering) the Ostwald ripening of primary particle within the NFs still occurs during calcination, but the diameter of the NFs remains unchanged [50], attributed to the polymeric template, which helped to restrain any changes on the dimension of NFs during sintering. The postcalcination diameter of the NFs was reported to be affected by the ratio between the metal precursors and dissolved polymer but not by the calcination condition [64]. As a result, metal oxide NFs with higher crystallinity are obtained without severely affecting the dimension of the NFs while preserving its high surface area-to-volume ratio. Hafez et al. recently showed that the crystallinity of the $Ba_3Ti_4Nb_4O_{21}$ NFs improved substantially with increasing sintering temperature, while the NFs' diameter remain unchanged [65]. Different from nanoparticle analogs, this work demonstrated significant enhancement in surface area $(7.2-17.2 \text{ m}^2 \text{ g}^{-1})$ with increasing temperature, attributed to the increasing surface porosity. The high-temperature ramp caused the polymer template to decompose at a high rate, leaving insufficient time for the inorganic precursors to reorient, leading to a rough surface with high porosity. These reports concluded that NFs could achieve high crystallinity without severely affecting their dimension and surface area, a suitable morphology as a photoanode for sensitized PV cells.

12.4.2 Raising the Fermi energy level

The delocalized energy trap states were located below the conduction band, and its density reduced exponentially toward the Fermi energy level [52], with carrier diffused via multitrapping mechanism within the delocalized states located above the Fermi level and below the conduction band. Upon excitation to the conduction band, electrons could be trapped at these delocalized energy states by releasing part of their energy as phonons (in the form of heat). More energy is lost if the electrons are trapped in deeper delocalized energy states, impairing their diffusion coefficient, which eventually leads to recombination. Eq. (12.7) formulates the calculation for electron transport properties, where v_d is the electron drift velocity, μ_e is the mobility of electron, and E is the applied potential gradient.

$$v_d = \mu_e E \tag{12.7}$$

The diffusion coefficient (D_n) is given as follows:

$$D_n = l^2/2\tau \tag{12.8}$$

where l is the travel distance of carrier and τ is the lifetime of the electron. The expression of diffusion coefficient can be rewritten by incorporating Eq. (12.7)

$$D_n = l(\mu_e E) \tag{12.9}$$

The electron mobility is related to its kinetic energy (KE) as follows:

$$KE = \frac{m_e^* \mu_e}{2} E \tag{12.10}$$

where m_e^* is the effective mass of the electron. More kinetic energy is lost when the electron is trapped at deeper delocalized trap state. Because of the direct proportionality between the kinetic energy and electron mobility, electrons trapped at deeper delocalized trap state would hinder the transport of electrons severely and reduce the diffusion coefficient. Although the improvement in crystallinity could reduce the density of the delocalized energy trap states, the depth of these delocalized states remains undisturbed. Therefore raising the Fermi energy level closer to that of the conduction band becomes crucial in reducing the depth of the delocalized trap states, ensuring a better diffusion coefficient of the electrons and low charge recombination rate.

12.4.3 N-type doping induced diffusion coefficient improvement

Manipulating the electronic properties of metal oxide through doping has been a widely studied topic [66–69]. However, the effect of doping depends heavily on the properties of the dopants (n-type or p-type doping) [70,71]. Akubuiro et al. studied the effect of doping on TiO_2, concluding that electrical conductivity increases when the host material is doped with higher valency (M^{5+}/M^{6+}), with the lowest conductivity when doped with lower valency (M^{3+}/M^{2+}) [72]. A significant improvement in the photocurrent of sensitized PV cells was observed when the TiO_2 photoanode is doped with dopants of higher valency. Some studies on this aspect are summarized in Table 12.1. A minute amount of niobium doping (5% of Nb^{5+}) on TiO_2 showed a 12% improvement in photocurrent compared to the undoped TiO_2, attributed to an increase in the electron concentration (3.43×10^{16} compared to $1.12 \times 10^{16} \, cm^{-3}$) as well as an improved diffusion coefficient

Table 12.1 Photovoltaic performance of n-type doped nanofiber photoanode for sensitized photovoltaic cells.

Electrode	Dopant	Dye	V_{OC} (V)	J_{SC} (mA/cm²)	FF (%)	PCE (%)	References
TiO₂	Undoped	N3	0.82	8.1	67.0	4.17	[73]
	1 w.% W		0.85	12.4	61.0	5.97	
	2 wt.% W		0.88	15.4	66.3	8.99	
TiO₂	Undoped	N719	0.79	11.87	70.0	6.60	[74]
	2.5 mol% Nb		0.74	15.75	64.0	7.50	
	5 mol% Nb		0.70	17.67	63.0	7.80	
	7.5 mol% Nb		0.69	15.91	63.0	6.90	
	10 mol% Nb		0.65	11.79	57.0	4.40	
ZnO	undoped	N719	0.49	1.41	41.2	0.19	[75]
	1 wt.% Al		0.51	3.13	38.8	0.41	
	2 wt.% Al		0.55	2.49	40.8	0.37	
	3 wt.% Al		0.47	2.11	40.5	0.27	
	4 wt.% Al		0.45	2.36	36.0	0.24	
ZnO	Undoped	N719	0.68	3.57	67.5	1.63	[76]
	5 mol% Co		0.76	5.36	73.1	2.97	
	10 mol% Co		0.71	4.38	72.9	2.01	
TiO₂	Undoped	N719	0.71	11.25	70.6	5.62	[77]
	Nb-doped		0.72	11.43	73.6	6.05	
TiO₂	Undoped	N719	0.70	15.14	67.8	7.14	[78]
	Nb-doped		0.71	16.26	66.2	7.69	
TiO₂	Undoped	N719	0.74	15.30	73.2	8.4	[79]
	0.5% Nb		0.74	16.3	71.5	8.7	
	1% Nb		0.72	16.0	67.8	8.0	
	2% Nb		0.70	16.0	61.9	7.1	

$(3.84 \times 10^{-3}$ compared to 4.67×10^{-4} cm² s^{-1}) [80]. However, the undoped TiO₂ device had the highest V_{OC} and showed better PV performance than the doped analogs. The niobium doping increased the electron density, raising the Fermi energy level and reducing the depth of delocalized trap state, leading to an improvement in the carrier diffusion coefficient [52]. However, with the raising of the Fermi level energy, the conduction band decreased during the realignment of the Fermi level, resulting in decreased V_{OC}, as illustrated in Fig. 12.6. Surprisingly, doping TiO₂ with tungsten (W^{6+}) offered a better electron diffusion coefficient, and therefore better photocurrent, without deteriorating the V_{OC} of the device [73], which was attributed to the difference in ionic radii between Ti^{4+} and W^{6+}. The larger difference in ionic radii between Ti^{4+} and Nb^{5+} ($\sim$6%) compared to that of W^{6+} ($<$0.005%) caused the lattice to strain, contributing to a higher charge recombination in Nb-doped TiO₂ [74]. As was discussed previously in the chapter, a higher charge recombination eventually reduces the V_{OC} of the sensitized PV cell, which can be deduced from the low fill factor of the device.

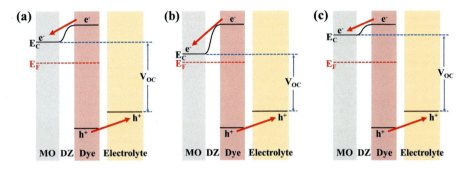

Figure 12.6 The energy diagram of the photoanode (A) without doping, (B) with n-type doping, and (C) with p-type doping, explaining the changes on V_{OC} upon doping.

12.4.4 P-type doping induced Schottky-barrier

Following the foundation developed in the previous section, doping n-type semiconducting host with dopants of lower valency could reduce the carrier concentration, leading to lower Fermi level and poorer carrier transport properties, detrimental to the performance of sensitized PV cells. However, reports have shown otherwise. High charge recombination resistance and V_{OC} were demonstrated when the photoanode materials were doped with a dopant of lower valency [81,82]. In some other cases the photocurrent substantially improved (or even doubled [75]) with negligible changes in V_{OC} [83,84]. In general, the property improvements offered by p-type doping are better than those achieved with n-type doping, as summarized in Table 12.2. To unravel the underlying mechanism responsible for such improvements in the performance of p-type doped photoanodes, understanding the effect of doping on the Fermi energy level and position of the conduction band are crucial. With lesser valency, p-type doping reduced the density of electrons and lowered the Fermi energy level. The lowering of Fermi energy level deepens the depth of delocalized trap states, which is detrimental to the carrier diffusion coefficient. However, a better diffusion coefficient was previously demonstrated in p-type doped photoanodes compared to their undoped counterparts, despite reduced electron density [81,84]. These reports clearly indicate that there are other determining factors affecting the performance besides the lowering of the Fermi level. The enhanced V_{OC} indicates that the uplift of the conduction band due to the lowering of the Fermi level could have played a role. Indeed, the shifting of the flat band potential to a more negative value with an increase in weight ratio of p-type dopant was observed, raising the conduction band energy of the photoanode [81,87]. The raising of the conduction band of the photoanode would increase the potential gradient against the work function of the redox electrolyte, as shown in Fig. 12.6. According to Eqs. (12.6), (12.7) and (12.9), a higher potential gradient increased the current density, drift velocity, and diffusion coefficient of the carrier, improving its diffusion length and reduced charge recombination. Certainly, the enhancement of potential gradient could be compensated by the reduced carrier mobility due to

Table 12.2 Photovoltaic performance of n-type doped nanofiber photoanode for sensitized photovoltaic cells.

Electrode	Dopant	Dye	V_{OC} (V)	J_{SC} (mA/cm²)	FF (%)	PCE (%)	References
ZnO	Undoped	N719	0.58	0.53	53.1	0.16	[85]
	1 wt.% Li		0.63	1.72	54.1	0.59	
	2 wt.% Li		0.75	1.00	53.6	0.41	
	3 wt.% Li		0.71	1.37	53.5	0.51	
ZnO	Undoped	N719	0.74	2.08	42.9	0.56	[86]
	1 at.% Ga		0.74	2.41	42.6	0.76	
	2 at.% Ga		0.75	2.85	41.5	0.88	
	3 at.% Ga		0.76	3.96	40.7	1.23	
	4 at.% Ga		0.76	3.12	41.4	0.98	
TiO₂	Undoped	N719	0.64	6.75	35.0	1.54	[82]
	N-doped		0.75	11.16	56.0	4.70	
TiO₂	Undoped	N719	0.80	7.57	55.0	3.30	[84]
	Ag-doped		0.78	9.51	56.0	4.13	
TiO₂	Undoped	N3	0.82	9.68	66.0	5.20	[81]
	2 at.% Ni		0.88	12.01	63.8	6.75	
	5 at.% Ni		0.83	7.57	64.5	4.07	
TiO₂	Undoped	N719	0.67	11.14	64.0	4.74	[83]
	3 wt.% Ag		0.63	12.55	61.0	4.83	
	5 wt.% Ag		0.64	13.77	59.0	5.22	
	7 wt.% Ag		0.65	12.80	60.0	5.00	
SnO₂	Undoped	N3	0.47	7.92	55.0	2.03	[87]
	0.5 at.% Al		0.49	9.75	55.0	2.62	
	1 at.% Al		0.51	11.02	51.0	2.82	
	2 at.% Al		0.56	12.83	51.0	3.56	
	3 at.% Al		0.54	10.05	46.0	2.42	

the deeper delocalized trap state. From Bhattacharjee and Hung's work on doping ZnO with lithium, the lattice parameter and crystal volume expanded with increasing amounts of Li doping [85]. Elements with lower valency had lower electronegativity, which exerted a smaller binding energy on the electron [88]. In this case the electrons have more freedom to be transported, thereby improving the diffusion coefficient even with a lower carrier density.

12.4.5 Homovalent ion substitution

Other than n-type and p-type doping in which elements with different valency are utilized to manipulate the carrier density within the host, doping of homovalent ions (ions with similar valency as the host) also demonstrated promising improvements. Substituting TiO₂ with a small amount of tin (Sn^{4+}) and zirconium (Zr^{4+}) ions at concentrations less than 10% provided increasing carrier density, leading to improvement in photocurrent when adopted as a photoanode in sensitized PV cells [89–91]. However, the similar valency between the host and the dopants indicated that the mechanism in increasing the carrier density of the host was different from that of n-type and p-type doping. It was then revealed that such an increase in carrier density can be attributed to the difference in the atomic number between the host and dopant, with the completely filled d orbital and partial filled p orbital of the dopant as the major contributors [92]. Consequently, the Fermi energy level is lifted closer to the conduction band, reducing the depth of the delocalized trap states, improving the electron transport properties, which enhances the photocurrent. The raising of the Fermi energy level is also reflected by the lowering of V_{OC}, similar to that reported when the host is doped with n-type dopants. Surprisingly, the enhancement of photocurrent is accompanied by a lower carrier lifetime and a higher charge recombination rate [92]. The mismatch in the ionic radii between Zr^{4+} and Ti^{4+} is reported to introduce disorders through lattice straining, generating midenergy gap states, which act as a trap. As a result, both V_{OC} and the fill factor were reduced, deteriorating the overall performance of sensitized PV cells. Detrimental effects of lattice strain on the carrier transport and its relationship with energy trap states were also reported elsewhere, where the trap states' density increases with higher doping concentration [93–95]. To harvest the benefits of homovalent substitution while suppressing the detrimental effect of lattice distortion, the doping concentration needs to be kept at lower values.

12.4.6 Composite fibers

Often, the properties that are required for a given material functionality are not delivered by a single material; such is the case of the photoanode requirements of sensitized PV cells. As is demonstrated in Eq. (12.6), an acceptable photoanode should offer a high surface area to anchor large amount of dyes, high electron mobility for faster charge transport, and a Fermi level such that it makes minimal band banding at the dye-photoelectrode interface to minimize the loss in potential at this interface. In addition to increasing the crystallinity and raising the Fermi level through appropriate doping, a third strategy has recently been adopted through fabrication of a composite containing materials of different electrochemical properties. For example, TiO_2 could be synthesized with a high surface area and has favorable band alignment with most of the reported dyes; however, the charge mobility is orders of magnitude lower than that of some other metal oxide semiconductors, such as ZnO or SnO_2 [96]. The ZnO has similar conduction band energy as that of TiO_2, whereas SnO_2 has a much lower conduction band energy level compared to TiO_2, resulting in a higher loss in potential at the dye-SnO_2 interface and results in a lower V_{OC} [97]. However, both SnO_2 and ZnO offer orders of magnitude higher charge mobility (150 cm^2 V^{-1} s^{-1} for nanocrystals) [98,99]. ZnO, by contrast, has inferior device stability as the divalent nature of the Zn ion makes them to easily soluble in acidic and humid conditions [100,101]. Zinab et al. prepared composite metal oxides in SnO_2–TiO_2 and SnO_2–ZnO NFs by electrospinning and used them as a photoanode in dye-sensitized solar cells [102,103]. Despite a binary composite, the composite materials showed a single bandgap and single oxidation reduction events besides high charge mobility. In terms of PCE, the SnO_2-ZnO composite fibers showed a PCE of approximately 5.65%, whereas that for the SnO_2-TiO_2 composite was approximately 8.50%. These values are much higher than those of their single-component counterparts: SnO_2 ($\sim$3.90%), ZnO ($\sim$1.38%), and TiO_2 NFs ($\sim$5.10%). Ling et al. fabricated a composite containing highly insulating Al_2O_3 with much higher conduction band energy and highly conducting SnO_2 but with a lower energy level for the CBM [18]. Interestingly, a composite fiber containing equal amounts of SnO_2 and Al_2O_3 when used as a photoanode showed orders of magnitude higher photocurrent than pure Al_2O_3 fibers even though the two materials (50/50 composite and pure Al_2O_3) showed similar electrical conductivity. These developments offer numerous opportunities to tailor advanced materials for a variety of applications, such as charge storage and catalysis.

12.5 Challenges

Although electrospinning has been widely adopted as the most versatile technique in synthesizing metal oxide NFs, it requires the metal precursors to be dispersed in a polymeric solution, which needs to be removed during the calcination process at temperatures of approximately 400°C–550°C. The removal of the polymeric solution results in huge wastage, and burning them in the atmosphere would increase the carbon load. These factors would increase the material cost of sensitized PV cells and their sustainability. Alternatively, polymerless electrospinning would be favorable in term of low wastage and low cost. Other than that, the ceramic metal oxide photoanodes studied previously were brittle and thus unsuitable for flexible devices, which were expected to be among the major applications for PV technologies. The flexible PV cell requires all the components to retain their properties and adherence to each other under multiple cycles of bending. Even though conducting polymer NFs were studied as photoanodes for flexible PV cells, their lower electrical conductivity failed to offer PCE similar to that of metal oxide analogs. Incorporating metal oxide within polymeric fibers could be a good idea in forming highly conducting yet flexible photoanodes. However, highly crystallized metal oxide requires high-temperature annealing, a temperature where polymer decomposition occur. Therefore synthesizing highly crystallized metal oxide−embedded polymer NFs would be challenging. The high annealing temperature required for photoanode coating, usually around 500°C, was also too high for a flexible substrate. Even though low coating annealing temperature was demonstrated previously, the metal oxide retained its amorphous crystal phase and showed poor PV performance [104,105]. A new coating procedure without the need for high-temperature sintering ($<$150°C) is preferred.

12.5.1 Conclusion and outlook

Undeniably, metal oxide NFs had played an important role in improving the performance of sensitized PV cells, owing to their anisotropic carrier transport properties.

However, like other nanostructured materials, the high surface area was accompanied by increasing density of delocalized energy trap states, lying between the conduction band and the Fermi energy level. Crystallization improvements via calcination as well as doping were reviewed as the most appropriate techniques to eradicate the density of delocalized energy trap state. The former decreased the density of delocalized trap states, whereas the latter reduced the depth of the states. Doping was observed to alter the carrier density and Fermi energy level of the host metal oxide, where the V_{OC} and photocurrent were greatly affected. With the mounting interest in developing flexible and portable PV cells, the focus of development in metal oxide NFs is expected to switch toward improving their flexibility and bending tolerance. Embedding metal oxide nanostructures within the conducting polymer matrix or fabricating metal oxide NFs network at fine diameter (100 nm) could be the future materials for flexible sensitized PV cells. Without a doubt, metal oxide NFs will play a crucial role in future flexible PV cells.

References

[1] Best Research-Cell Efficiency Chart. 2019, NREL.

[2] P. Diamandis, Solar energy revolution: a massive opportunity. *Forbes* 2018.

[3] GlobalData, Solar PV Module, Update 2018—Global Market Size, Competitive Landscape and Key Country Analysis to 2022, 2018, p. 239.

[4] P. Wu, et al., Review on life cycle assessment of greenhouse gas emission profit of solar photovoltaic systems, Energy Proc 105 (2017) 1289−1294.

[5] A. Louwen, et al., Re-assessment of net energy production and greenhouse gas emissions avoidance after 40 years of photovoltaics development, Nat. Commun. 7 (1) (2016) 13728.

[6] S. Yoon, et al., Application of transparent dye-sensitized solar cells to building integrated photovoltaic systems, Build. Environ. 46 (10) (2011) 1899−1904.

[7] H.M. Lee, J.H. Yoon, Power performance analysis of a transparent DSSC BIPV window based on 2 year measurement data in a full-scale mock-up, Appl. Energy 225 (2018) 1013−1021.

[8] M. Saifullah, J. Gwak, J.H. Yun, Comprehensive review on material requirements, present status, and future prospects for building-integrated semitransparent photovoltaics (BISTPV), J. Mater. Chem. A 4 (22) (2016) 8512−8540.

[9] H.C. Weerasinghe, F. Huang, Y.-B. Cheng, Fabrication of flexible dye sensitized solar cells on plastic substrates, Nano Energy 2 (2) (2013) 174−189.

[10] F. Meng, et al., Improved photovoltaic performance of monocrystalline silicon solar cell through luminescent down-converting $Gd_2O_2S:Tb_3 +$ phosphor, Prog. Photovol: Res. Appl. 27 (7) (2019) 640−651.

[11] Y. Wang, et al., Photovoltaic efficiency enhancement of polycrystalline silicon solar cells by a highly stable luminescent film, Sci. China Mater. 63 (4) (2020) 544−551.

[12] H.K. Raut, et al., Porous SiO_2 anti-reflective coatings on large-area substrates by electrospinning and their application to solar modules, Sol. Energy Mater. Sol. Cell 111 (2013) 9−15.

[13] Z.-Q. Fang, et al., Light management in flexible glass by wood cellulose coating, Sci. Rep. 4 (1) (2014) 5842.

[14] H. Liu, et al., Enhanced photovoltaic performance of dye-sensitized solar cells with TiO_2 micro/nano-structures as light scattering layer, J. Mater. Sci: Mater. Electron. 27 (5) (2016) 5452−5461.

[15] Y. Wang, et al., Enhanced light scattering and photovoltaic performance for dye-sensitized solar cells by embedding submicron SiO_2/TiO_2 core/shell particles in photoanode, Ceram. Int. 39 (5) (2013) 5407−5413.

[16] R. Wang, et al., Highly transparent, thermally stable, and mechanically robust hybrid cellulose-nanofiber/polymer substrates for the electrodes of flexible solar cells, ACS Appl. Energy Mater. 3 (1) (2020) 785−793.

[17] Q. Wali, et al., Multiporous nanofibers of SnO_2 by electrospinning for high efficiency dye-sensitized solar cells, J. Mater. Chem. A 2 (41) (2014) 17427−17434.

[18] J. Ling, et al., Photocurrents in crystal-amorphous hybrid stannous oxide/alumina binary nanofibers, J. Am. Ceram. Soc. 102 (10) (2019) 6337−6348.

[19] F. Mohtaram, et al., Electrospun ZnO nanofiber interlayers for enhanced performance of organic photovoltaic devices, Sol. Energy 197 (2020) 311−316.

[20] M.A.K.L. Dissanayake, et al., Effect of PbS quantum dot-doped polysulfide nanofiber gel polymer electrolyte on efficiency enhancement in CdS quantum dot-sensitized TiO_2 solar cells, Electrochim. Acta 347 (2020) 136311.

[21] S.S. Dissanayake, et al., Performance of dye sensitized solar cells fabricated with electrospun polymer nanofiber based electrolyte, Mater. Today: Proc. 3 (2016) S104−S111.

[22] J.J. Kaschuk, et al., Electrolyte membranes based on ultrafine fibers of acetylated cellulose for improved and long-lasting dye-sensitized solar cells, Cellulose 26 (10) (2019) 6151−6163.

[23] B. Kilic, et al., Carbon nanofiber based CuO nanorod counter electrode for enhanced solar cell performance and adsorptive photocatalytic activity, J. Nanopart. Res. 22 (2) (2020) 52.

[24] X. Zhao, et al., Bi_2S_3 nanoparticles densely grown on electrospun-carbon-nanofibers as low-cost counter electrode for liquid-state solar cells, Mater. Res. Bull. 125 (2020) 110800.

[25] G. Veerappan, W. Kwon, S.-W. Rhee, Carbon-nanofiber counter electrodes for quasi-solid state dye-sensitized solar cells, J. Power Sources 196 (24) (2011) 10798−10805.

[26] G. Richhariya, et al., Natural dyes for dye sensitized solar cell: A review, Renew. Sustain. Energy Rev. 69 (2017) 705−718.

[27] C.-P. Lee, C.-T. Li, K.-C. Ho, Use of organic materials in dye-sensitized solar cells, Mater. Today 20 (5) (2017) 267−283.

[28] N. Robertson, Optimizing dyes for dye-sensitized solar cells, Angew. Chem. Int. (Ed.) 45 (15) (2006) 2338−2345.

[29] K. Sharma, V. Sharma, S.S. Sharma, Dye-sensitized solar cells: fundamentals and current status, Nanoscale Res. Lett. 13 (1) (2018) 381.

[30] Z. Yu, et al., Liquid electrolytes for dye-sensitized solar cells, Dalton Trans. 40 (40) (2011) 10289−10303.

[31] J. Wu, et al., Electrolytes in dye-sensitized solar cells, Chem. Rev. 115 (5) (2015) 2136−2173.

[32] M.Y.A. Rahman, et al., Polymer electrolyte for photoelectrochemical cell and dye-sensitized solar cell: a brief review, Ionics 20 (9) (2014) 1201−1205.

[33] N. Shintaku, M. Hiramoto, S. Izawa, Effect of trap-assisted recombination on open-circuit voltage loss in phthalocyanine/fullerene solar cells, Org. Electron. 55 (2018) 69−74.

[34] D. Galli, et al., Suppressing the surface recombination and tuning the open-circuit voltage of polymer/fullerene solar cells by implementing an aggregative ternary compound, ACS Appl. Mater. Interfaces 10 (34) (2018) 28803–28811.

[35] D. Prochowicz, et al., Correlation of recombination and open circuit voltage in planar heterojunction perovskite solar cells, J. Mater. Chem. C. 7 (5) (2019) 1273–1279.

[36] R. Singh, B.J. Baliga, P-I-N Diode, in Cryogenic Operation of Silicon Power Devices, Springer, Boston, MA, 1998.

[37] P. Narchi, et al., Cross-sectional investigations on epitaxial silicon solar cells by kelvin and conducting probe atomic force microscopy: effect of illumination, Nanoscale Res. Lett. 11 (1) (2016) 55.

[38] M.C. Kao, et al., The effects of the thickness of TiO_2 films on the performance of dye-sensitized solar cells, Thin Solid. Films 517 (17) (2009) 5096–5099.

[39] T.H. Meen, et al., Optimization of the dye-sensitized solar cell performance by mechanical compression, Nanoscale Res. Lett. 9 (1) (2014) 523.

[40] M. Pazoki, et al., The effect of dye coverage on the performance of dye-sensitized solar cells with a cobalt-based electrolyte, Phys. Chem. Chem. Phys. 16 (18) (2014) 8503–8508.

[41] M. Sajedi Alvar, et al., Enhancing the electron lifetime and diffusion coefficient in dye-sensitized solar cells by patterning the layer of TiO_2 nanoparticles, J. Appl. Phys. 119 (11) (2016) 114302.

[42] C. He, et al., Determination of electron diffusion coefficient and lifetime in dye-sensitized solar cells by electrochemical impedance spectroscopy at high fermi level conditions, J. Phys. Chem. C. 112 (48) (2008) 18730–18733.

[43] U. Godavarti, V.D. Mote, M. Dasari, Role of cobalt doping on the electrical conductivity of ZnO nanoparticles, J. Asian Ceram. Societies 5 (4) (2017) 391–396.

[44] A.N. Afaah, et al., Electrically conductive nanostructured silver doped zinc oxide (Ag: ZnO) prepared by solution-immersion technique, in: AIP Conference Proceedings, 2016, 1733(1), p. 020055.

[45] H.R. Baek, I.S. Eo, Electrochemical properties of TiO_2-metal oxide composites for dye-sensitized solar cell, Curr. Appl. Phys. 17 (6) (2017) 854–857.

[46] A.E. Shalan, et al., Nanofibers as promising materials for new generations of solar cells, in: A. Barhoum, M. Bechelany, A. Makhlouf (Eds.), Handbook of Nanofibers, Springer International Publishing, Cham, 2018, pp. 1–33.

[47] D. Joly, et al., Electrospun materials for solar energy conversion: innovations and trends, J. Mater. Chem. C. 4 (43) (2016) 10173–10197.

[48] J.S. Shaikh, et al., Nanoarchitectures in dye-sensitized solar cells: metal oxides, oxide perovskites and carbon-based materials, Nanoscale 10 (11) (2018) 4987–5034.

[49] J.-Y. Liao, et al., Effect of TiO_2 morphology on photovoltaic performance of dye-sensitized solar cells: nanoparticles, nanofibers, hierarchical spheres and ellipsoid spheres, J. Mater. Chem. 22 (16) (2012) 7910–7918.

[50] P.S. Archana, et al., Improved electron diffusion coefficient in electrospun TiO_2 nanowires, J. Phys. Chem. C. 113 (52) (2009) 21538–21542.

[51] C. Giansante, I. Infante, Surface traps in colloidal quantum dots: a combined experimental and theoretical perspective, J. Phys. Chem. Lett. 8 (20) (2017) 5209–5215.

[52] P.S. Archana, et al., Near band-edge electron diffusion in electrospun Nb-doped anatase TiO_2 nanofibers probed by electrochemical impedance spectroscopy, Appl. Phys. Lett. 98 (15) (2011) 152106.

[53] K. Schwarzburg, F. Willig, Influence of trap filling on photocurrent transients in polycrystalline TiO_2, Appl. Phys. Lett. 58 (22) (1991) 2520–2522.

[54] K.D. Benkstein, et al., Influence of the percolation network geometry on electron transport in dye-sensitized titanium dioxide solar cells, J. Phys. Chem. B 107 (31) (2003) 7759−7767.

[55] N.K. Singh, R. Rajkumari, Effect of Annealing on Metal-Oxide Nanocluster, in Concepts of Semiconductor Photocatalysis, IntechOpen, 2019.

[56] Y. Shin, et al., Effect of oxygen vacancies on electrical conductivity of $La_{0.5}Sr_{0.5}FeO_{3-\delta}$ from first-principles calculations, J. Mater. Chem. A 8 (9) (2020) 4784−4789.

[57] F. Gunkel, et al., Oxygen vacancies: the (in)visible friend of oxide electronics, Appl. Phys. Lett. 116 (12) (2020) 120505.

[58] D.V. Christensen, et al., The role of oxide interfaces in highly confined electronic and ionic conductors, APL. Mater. 7 (1) (2018) 013101.

[59] N.K. Elumalai, et al., Enhancing the stability of polymer solar cells by improving the conductivity of the nanostructured MoO_3 hole-transport layer, Phys. Chem. Chem. Phys. 15 (18) (2013) 6831−6841.

[60] T. Tadros, Ostwald ripening, in: T. Tadros (Ed.), Encycl. Colloid Interface Sci, Springer Berlin Heidelberg, Berlin, Heidelberg, 2013, p. 820.

[61] A.M. Golsheikh, et al., Effect of calcination temperature on performance of ZnO nanoparticles for dye-sensitized solar cells, Powder Technol. 329 (2018) 282−287.

[62] D. Zhao, et al., Effect of annealing temperature on the photoelectrochemical properties of dye-sensitized solar cells made with mesoporous TiO_2 nanoparticles, J. Phys. Chem. C. 112 (22) (2008) 8486−8494.

[63] M. Kaur, N.K. Verma, Performance of dye-sensitized solar cell fabricated using titania nanoparticles calcined at different temperatures, Mater. Sci-Poland 31 (3) (2013) 378−385.

[64] M. Mazloum-Ardakani, R. Arazi, The investigation on different light harvesting layers and their sufficient effect on the photovoltaic characteristics in dye sensitized solar cell, Nanochemistry Res. 2 (1) (2017) 20−28.

[65] A.M. Hafez, et al., Highly porous $Ba_3Ti_4Nb_4O_{21}$ perovskite nanofibers as photoanodes for quasi-solid state dye-sensitized solar cells, Sol. Energy 206 (2020) 413−419.

[66] I. Pradeep, et al., Effect of Al doping concentration on the structural, optical, morphological and electrical properties of V_2O_5 nanostructures, N. J. Chem. 42 (6) (2018) 4278−4288.

[67] Ş. Baturay, et al., n-Type conductivity of CuO thin films by metal doping, Appl. Surf. Sci. 477 (2019) 91−95.

[68] K. Neuhaus, R. Dolle, H.-D. Wiemhöfer, The effect of transition metal oxide addition on the conductivity of commercially available Gd-doped ceria, J. Electrochem. Soc. 167 (4) (2020) 044507.

[69] A. Jamil, et al., Effect of titanium doping on conductivity, density of states and conduction mechanism in ZnO thin film, Appl. Phys. A 125 (4) (2019) 238.

[70] D.M. Smyth, The effects of dopants on the properties of metal oxides, Solid. State Ion. 129 (1) (2000) 5−12.

[71] Z. Zhou, et al., Doping effects on the electrical conductivity of bismuth layered Bi_3TiNbO_9-based ceramics, J. Appl. Phys. 100 (4) (2006) 044112.

[72] E.C. Akubuiro, X.E. Verykios, Effects of altervalent cation doping on electrical conductivity of platinized titania, J. Phys. Chem. Solids 50 (1) (1989) 17−26.

[73] P.S. Archana, et al., Tungsten doped titanium dioxide nanowires for high efficiency dye-sensitized solar cells, Phys. Chem. Chem. Phys. 16 (16) (2014) 7448−7454.

[74] X. Lü, et al., Improved-performance dye-sensitized solar cells using Nb-doped TiO_2 electrodes: efficient electron injection and transfer, Adv. Funct. Mater. 20 (3) (2010) 509−515.

[75] B. Sutanto, et al., Enhancement ZnO nanofiber as semiconductor for dye-sensitized solar cells by using Al doped, in: AIP Conference Proceedings, 2016, 1717(1), p. 040006.

[76] G. Kanimozhi, et al., A novel electrospun cobalt-doped zinc oxide nanofibers as photoanode for dye-sensitized solar cell, Mater. Res. Express 6 (2) (2018) 025041.

[77] Y. Horie, et al., Enhancement of carrier mobility by electrospun nanofibers of Nb-doped TiO_2 in dye sensitized solar cells, Electrochim. Acta 105 (2013) 394−402.

[78] S. Lee, et al., Nb-doped TiO_2: a new compact layer material for TiO_2 dye-sensitized solar cells, J. Phys. Chem. C. 113 (16) (2009) 6878−6882.

[79] A.K. Chandiran, et al., Doping a TiO_2 photoanode with Nb5 + to enhance transparency and charge collection efficiency in dye-sensitized solar cells, J. Phys. Chem. C. 114 (37) (2010) 15849−15856.

[80] P.S. Archana, et al., Structural and electrical properties of Nb-doped anatase TiO_2 nanowires by electrospinning, J. Am. Ceram. Soc. 93 (12) (2010) 4096−4102.

[81] P.S. Archana, et al., Random nanowires of nickel doped TiO_2 with high surface area and electron mobility for high efficiency dye-sensitized solar cells, Dalton Trans. 42 (4) (2013) 1024−1032.

[82] M. Motlak, et al., High-efficiency electrode based on nitrogen-doped TiO_2 nanofibers for dye-sensitized solar cells, Electrochim. Acta 115 (2014) 493−498.

[83] E.M. Jin, et al., Enhancement of the photoelectric performance of dye-sensitized solar cells using Ag-doped TiO_2 nanofibers in a TiO_2 film as electrode, Nanoscale Res. Lett. 7 (1) (2012) 97.

[84] J. Li, et al., Silver nanoparticle doped TiO_2 nanofiber dye sensitized solar cells, Chem. Phys. Lett. 514 (1) (2011) 141−145.

[85] R. Bhattacharjee, I.M. Hung, Effect of different concentration Li-doping on the morphology, defect and photovoltaic performance of Li−ZnO nanofibers in the dye-sensitized solar cells, Mater. Chem. Phys. 143 (2) (2014) 693−701.

[86] Y. Dou, et al., Enhanced photovoltaic performance of ZnO nanorod-based dye-sensitized solar cells by using Ga doped ZnO seed layer, J. Alloy. Compd. 633 (2015) 408−414.

[87] Y. Duan, et al., Enhancing the performance of dye-sensitized solar cells: doping SnO_2 photoanodes with Al to simultaneously improve conduction band and electron lifetime, J. Mater. Chem. A 3 (6) (2015) 3066−3073.

[88] Y. Song, et al., Calculation of theoretical strengths and bulk moduli of bcc metals, Phys. Rev. B 59 (22) (1999) 14220−14225.

[89] J. Wang, et al., Increases in solar conversion efficiencies of the ZrO_2 nanofiber-doped TiO_2 photoelectrode for dye-sensitized solar cells, Nanoscale Res. Lett. 7 (1) (2012) 98.

[90] Y. Duan, et al., Sn-doped TiO_2 photoanode for dye-sensitized solar cells, J. Phys. Chem. C. 116 (16) (2012) 8888−8893.

[91] K.-h Park, et al., 204% enhanced efficiency of ZrO_2 nanofibers doped dye-sensitized solar cells, Appl. Phys. Lett. 97 (2) (2010) 023302.

[92] P.S. Archana, et al., Charge transport in zirconium doped anatase nanowires dye-sensitized solar cells: trade-off between lattice strain and photovoltaic parameters, Appl. Phys. Lett. 105 (15) (2014) 153901.

[93] S. Sadhu, A. Patra, Lattice strain controls the carrier relaxation dynamics in CdxZn1−xS alloy quantum dots, J. Phys. Chem. C. 116 (28) (2012) 15167−15173.

[94] T.W. Jones, et al., Lattice strain causes non-radiative losses in halide perovskites, Energy & Environ. Sci. 12 (2) (2019) 596–606.

[95] C.E. Barnes, et al. The effect of hydrostatic pressure on trapping centers in strained-layer superlattice structures, in: Proceedings of the 17th International Conference on the Physics of Semiconductors, Springer New York, New York, NY, 1985.

[96] T. E, et al., Enhanced electrical conductivity of TiO_2/graphene: the role of introducing Ca2 + , J. Alloy. Compd. 827 (2020) 154280.

[97] Q. Wali, R. Jose, Chapter 7—SnO_2 dye-sensitized solar cells, in: S. Thomas, et al. (Eds.), Nanomaterials for Solar Cell Applications, Elsevier, 2019, pp. 205–285.

[98] F. Yakuphanoglu, Electrical conductivity, Seebeck coefficient and optical properties of SnO_2 film deposited on ITO by dip coating, J. Alloy. Compd. 470 (1) (2009) 55–59.

[99] M. Maruthupandy, et al., Investigation on the electrical conductivity of ZnO nanoparticles-decorated bacterial nanowires, Adv. Nat. Sci.: Nanosci. Nanotechnol. 7 (4) (2016) 045011.

[100] H. Singh, et al., Enhanced moisture sensing properties of a nanostructured ZnO coated capacitive sensor, RSC Adv. 8 (7) (2018) 3839–3845.

[101] N.-F. Hsu, M. Chang, C.-H. Lin, Synthesis of ZnO thin films and their application as humidity sensors, Microsyst. Technol. 19 (11) (2013) 1737–1743.

[102] Z.H. Bakr, et al., Synergistic combination of electronic and electrical properties of SnO_2 and TiO_2 in a single SnO_2–TiO_2 composite nanofiber for dye-sensitized solar cells, Electrochim. Acta 263 (2018) 524–532.

[103] Z.H. Bakr, et al., Characteristics of ZnO–SnO_2 composite nanofibers as a photoanode in dye-sensitized solar cells, Ind. Eng. Chem. Res. 58 (2) (2019) 643–653.

[104] Y. Galagan, et al., Roll-to-roll slot die coated perovskite for efficient flexible solar cells, Adv. Energy Mater. 8 (32) (2018) 1801935.

[105] M.J. Carnie, et al., A one-step low temperature processing route for organolead halide perovskite solar cells, Chem. Commun. 49 (72) (2013) 7893–7895.

Metal oxide nanofiber-based electrodes in solid oxide fuel cells

Paola Costamagna[1], Peter Holtappels[2] and Caterina Sanna[1]
[1]DCCI, Department of Chemistry and Industrial Chemistry, University of Genoa, Genoa, Italy, [2]DTU Energy, Department of Energy Conversion and Storage, Technical University of Denmark, Electrovej, Lyngby, Denmark

13.1 Introduction

Solid oxide fuel cells (SOFCs) have been widely investigated during the last decades because of their high energy conversion efficiency, low emission of greenhouse gases, and fuel flexibility [1]. SOFCs are energy systems that convert the fuel chemical energy directly into electrical energy through electrochemical reactions. As represented in Fig. 13.1A, a SOFC consists of a dense electrolyte placed between two porous electrodes, fed respectively with a gaseous fuel, typically hydrogen, and an oxidant, typically oxygen from air. At the latter electrode the oxygen molecules are reduced to oxygen ions (oxygen reduction reaction, ORR):

$$\frac{1}{2}O_2 + 2e^- \rightleftarrows O^{2-} \tag{13.1}$$

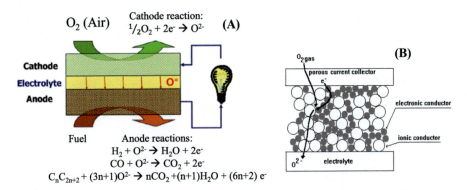

Figure 13.1 Solid oxide fuel cell. (A) Schematic diagram of the working principle [2]. (B) Schematic diagram of a composite cathode [3].
Source: (A) S.P. Jiang, Development of lanthanum strontium cobalt ferrite perovskite electrodes of solid oxide fuel cells—a review. Int. J. Hydrogen Energy 2019;44:7448—7493. https://doi.org/10.1016/j.ijhydene.2019.01.212; (B) P. Costamagna, P. Costa, V. Antonucci, Micro-modelling of solid oxide fuel cell electrodes. Electrochim. Acta 1998;43:375—394. https://doi.org/10.1016/S0013-4686(97)00063-7.

Metal Oxide-Based Nanofibers and Their Applications. DOI: https://doi.org/10.1016/B978-0-12-820629-4.00013-8
© 2022 Elsevier Inc. All rights reserved.

The electrode takes the name of cathode. The oxygen ions migrate through the electrolyte toward the opposite electrode, where they oxidize the fuel. If the fuel is hydrogen, the electrochemical reaction is as follows:

$$H_2 + O^{2-} \rightleftarrows H_2O + 2e^- \tag{13.2}$$

The latter is a hydrogen oxidation reaction (HOR), and the electrode is called an anode. The generated electrons flow through an external electrical circuit toward the cathode, where they are consumed by the reduction reaction. The overall reaction coincides with the combustion of hydrogen:

$$H_2 + \frac{1}{2}O_2 \rightleftarrows H_2O \tag{13.3}$$

The electrochemical reaction route makes possible a direct transformation of a portion of the ΔG of the overall reaction into electrical work, performed by the electrons in the external circuit. This direct transformation does not involve any thermodynamic cycle, so the efficiency limitations related to the Carnot efficiency do not apply. This is the reason for the high energy conversion efficiency, which is one of the attractive features of fuel cells in general.

An important characteristic of SOFCs is to have all solid-state components, including the electrolyte, which avoids the problems posed by liquid electrolytes—such as corrosion and flooding. In SOFCs the electrolyte is a ceramic layer, featuring oxygen ion conductivity. The electrolyte needs to be dense in order to separate fuel and air. This separation is essential to prevent gas leakage, which would cause losses associated with chemical rather than electrochemical reactions, with associated safety issues. The traditional electrolyte material is yttria-stabilized zirconia [YSZ, i.e., $(ZrO_2)_{1-x}(Y_2O_3)_x$, where $x = 0.03-0.1$]. YSZ has a fluorite structure, shown in Fig. 13.2A,

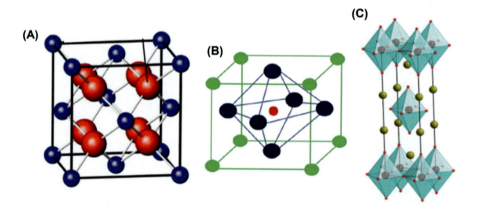

Figure 13.2 Schematic structure of (A) fluorite, (B) cubic perovskite, and (C) Ruddlesden–Popper.

where the coordination numbers for the cation and anion are 8 and 4, respectively. The introduction of lower-valence cations (Y) into the fluorite lattice (ZrO_2) results in the formation of oxygen vacancies, which, at high temperature, are mobile and are the source of oxygen ion conduction [4]. Nevertheless, YSZ needs an operating temperature as high as $950°C-1000°C$ to attain an oxygen ion conductivity of 0.1 S cm^{-1}. For this reason the operating temperature of traditional SOFCs (also called high-temperature SOFCs, HT-SOFCs) is in the range $900°C-1000°C$, which causes several engineering problems related to defective sealing, component degradation, and slow startup. To overcome these problems, ceria-based materials are currently being used as electrolytes, in particular gadolinia-doped ceria [GDC, i.e., $Ce_{1-x}Gd_xO_{1-\delta}$]. For example, the GDC10 ($Ce_{0.8}Gd_{0.1}O_{1.95}$) attains an oxygen ion conductivity of 0.1 S cm^{-1} around 880°C [5]. Furthermore, GDC retains interesting values of oxygen ion conductivity even at 700°C (0.03 S cm^{-1}) and 600°C (0.01 S cm^{-1}). This has led to the development of the so-called intermediate temperature−solid oxide fuel cells (IT-SOFCs), designed for operation in the range $600°C-800°C$. It must be borne in mind that decreasing the operating temperature hinders the kinetics of the electrochemical reactions occurring at the anode and cathode, lowering the cell performance. This has stimulated the design of high-performance electrodes. These are obtained through advanced materials and nano-sized architectures, which must provide simultaneously a high catalytic activity, a high level of conductivity, and a satisfactory porosity, to permit gas diffusion. Furthermore, it must be borne in mind that the electrochemical reaction takes place in a specific region of the electrode, the three-phase boundary (TPB). The name TPB identifies a specific zone where three different phases, that is, the ionic conductor, the electronic conductor, and the gas phase, meet. Since the electrochemical reactions in Eqs. (13.1) and (13.2) involve oxygen ions, electrons, and gaseous species, they can take place only at the TPB. Furthermore, the electrodes must feature both ionic and electronic conductivity inside the electrode bulk so that oxygen ions and electrons can be conveyed to and from the TPB. The term "electrode bulk" is used here to refer to the interior of the electrode thickness. One-dimensional nanomaterials—such as nanofibers (NFs), nanotubes, nanowires, and nanorods—are being intensively studied as promising electrode architectures with a high void degree, large internal surface, and potentially high TPB and ionic and electronic conductivity. The key electrode is the cathode, since the ORR provides the main contribution to the SOFC's internal energy losses.

13.1.1 State-of-the-art architectures and materials for solid oxide fuel cell electrodes

The geometrical features of the TPB and the location and extension of the TPB inside the electrode bulk are essential in determining the SOFC's electrochemical behavior. In addition, to the electrode architecture, the material's catalytic activity is a key factor. Metal oxides are employed mostly as cathode materials. Thus the attention here is focused mainly on cathodes with a short outline about anodes. The term "electrode/electrolyte interface" is used here and defined as the electrolyte surface on which the electrode particles or fibers are attached. The term "particle/fiber bulk" is used as well in referring to the inside of the individual electrode particles or fibers; the term "particle/fiber surface," refers to the particle or fiber's external surface.

13.1.1.1 Electrode architectures

To extend the TPB inside the electrode bulk, conventional SOFC electrodes are prepared as a mixture of ionic and electronic conducting particles. The former are (usually) electrolyte particles, whereas the latter are the "true" electrode material. A schematic of the resulting composite electrode is shown in Fig. 13.1B, where all the particles are schematized as spheres. Fig. 13.1B clarifies that the extension of the TPB inside the electrode bulk is effectively achieved if two conditions are satisfied. The first condition is that there are many contact points between electronic and ionic conducting particles. The second condition is that electrons and ions effectively reach these contact points. In Fig. 13.1B, representing an SOFC cathode, it is shown that the electrons, supplied by the current collector, can migrate throughout the electrode bulk by moving from one electronic conducting particle to another if these particles form conducting clusters spanning throughout the entire electrode. The same consideration applies also the oxygen ions, which need to migrate toward the electrolyte through a continuous path formed by the ionic conducting particles. This is a typical percolation problem, and many theoretical and experimental studied have been developed [3,6−8] demonstrating that for each type of particle (ionic or electronic conducting) there is a threshold, called the percolation threshold, at which cluster connectivity is triggered. In other words, below the percolation threshold, there is no cluster connectivity and the electrode is nonconductive. By contrast, above the percolation threshold, clusters are formed, which easily span the entire electrode and make it conductive. In the case of spherical particles all of the same size, the percolation threshold is about 30%, which means that one type of conductor percolates if the electrode contains at least 30% of particles of that type. Percolation studies [3] have shown that above percolation a further increase in the number of particles of the same type enhances the corresponding type conductivity. On the other hand, the presence of bottlenecks between one particle and another of the same type adversely affects the conductivity. Percolation studies have also addressed the calculation of the TPB, demonstrating that the smaller the particle diameter, the larger the TPB and the better the electrochemical performance. Usually, in the composite electrodes represented in Fig. 13.1B and obtained by powder pressing, particle dimensions are in the range $0.1-1$ μm.

All the latter considerations motivate the interest in NFs. On the one hand, electrospun NFs usually have a diameter as small as 200 nm, and this provides the electrodes with a large internal surface area. For example, about 1.87 m^2 g^{-1} are quoted by Jeon et al. [9]. This is expected to be associated with a high TPB. On the other hand, electrospun NFs are expected to solve percolation problems, guaranteeing continuous conduction paths with no bottlenecks. They have been studied in particular with mixed ionic-electronic conductor (MIEC) materials, in which the NFs

Metal oxide nanofiber-based electrodes in solid oxide fuel cells 305

carry both ionic and electronic pathways simultaneously. In principle, multiple contact points between the ionic and electronic conductive pathways are expected to exist in the material bulk, contributing to the TPB. Nevertheless, this is a debated point [10]. Instead, the electrochemical performance of MIECs has been demonstrated to benefit largely from infiltrations, that is, deposition of nano-sized particles adhering to the MIEC surface. Theories have been developed on the basis of the hypothesis that the infiltrations create a contact between the ionic and electronic conducting pathways, contributing to the TPB formation [11].

13.1.1.2 Cathode materials

In recent years the perovskite metal oxides have attracted much attention as cathode materials for SOFC application. The ideal perovskite structure, shown in Fig. 13.2B, follows the general formula ABO_3, where the A position is occupied by the larger metal cation and the B position is occupied by the smaller metal cation [12]. Perovskite is a versatile metal oxide, since it can be doped in both the A and B sites, leading to a modification of the crystal structure. Furthermore, the A site and B site substitutions are known to affect the oxygen stoichiometry, influencing the oxygen ion transport and oxygen surface exchange kinetics as well as the electronic conductivity [13].

One of the most studied materials for application in HT-SOFC is the Sr-doped $LaMnO_3$, which is an example of A-doped ABO_3 perovskite. The strontium-substituted lanthanum manganite ($La_{1-x}Sr_xMnO_3$, LSM) presents a variable crystal structure depending on the strontium doping level: rhombohedral for $0 < x < 0.5$, tetragonal for $x = 0.5$, and cubic for $x > 0.7$. Both the electrical conductivity and the catalytic activity are enhanced by the substitution of strontium for lanthanum. When the La^{3+} ion at the A site is substituted by a Sr^{2+} ion, the electroneutrality is maintained by the formation of an electric hole on the B site, leading to the oxidation of a manganese ion:

$$LaMnO_3 \xrightarrow{xSrO} La_{1-x}^{3+}Sr_x^{2+}Mn_{1-x}^{3+}Mn_x^{4+}O_3 \tag{13.4}$$

This phenomenon increases the electrical conductivity considerably. Indeed, the electrical conductivity of $LaMnO_3$ is 83 S cm^{-1} at 800°C, whereas $La_{0.6}Sr_{0.4}MnO_3$ reaches 320 S cm^{-1} at the same operating temperature [14].

Lanthanum strontium cobalt ferrite ($La_{1-x}Sr_xCo_{1-y}Fe_yO_{3-\delta}$, LSCF) is one of the most investigated materials for IT-SOFC application. It is an ABO_3 perovskite oxide doped in both the A and B sites. LSCF is an MIEC, since it has a high electronic conductivity and a modest ionic conductivity, which represents its main difference from the pure electronic conductor LSM. Compared to LSM, the electronic and ionic conductivities of LSCF are related to a different behavior of the substituted A and B sites. At the A site, when Sr^{2+} substitutes for La^{3+}, charge compensation occurs by both ionic and electronic defects. It is a combination of valence change of the iron ions, as described in Eq. (13.5), and the formation of doubly ionized oxygen vacancies, as described in Eq. (13.6):

$$2SrO + 2La^{3+} + \frac{1}{2}O_2 + 2Fe^{3+} \xrightarrow{La(Fe,Co)O_3} La_2O_3 + 2Sr^{2+} + 2Fe^{4+} \tag{13.5}$$

$$2SrO + 2La^{3+} + O^{2-} \xrightarrow{La(Fe,Co)O_3} La_2O_3 + 2Sr^{2+} + V_O^{\bullet\bullet} \tag{13.6}$$

where $V_O^{\bullet\bullet}$, according to the Kröger–Vink notation, represents the oxygen vacancy. Furthermore, Co ions on the B site lattice have a smaller binding energy for oxygen than that of Fe ions, increasing the electronic conductivity (due to an increase in the covalence of the Co (3d)-O (2p) bond compared to the Fe (3d)-O (2p) bond) [15]. The electronic conductivity σ_e strongly depends on the Co amount. For example, for $La_{0.6}Sr_{0.4}Co_{0.8}Fe_{0.2}O_{3-\delta}$, $\sigma_e = 1000$ S cm^{-1} at 800°C, whereas for $La_{0.6}Sr_{0.4}Co_{0.2}Fe_{0.8}O_{3-\delta}$, $\sigma_e = 280$ S cm^{-1} at the same operating temperature. The ionic conductivity σ_i slightly depends on the LSCF composition as well: For $La_{0.6}Sr_{0.4}Co_{0.8}Fe_{0.2}O_{3-\delta}$, $\sigma_i = 2 \ 10^{-2}$ S cm^{-1} at 800°C, whereas for $La_{0.6}Sr_{0.4}Co_{0.2}Fe_{0.8}O_{3-\delta}$, $\sigma_i = 1.2 \ 10^{-2}$ in the same conditions. Nevertheless, high concentrations of Co in the LSCF lattice are not recommended, owing to cobalt segregation with consequent loss of conductivity and formation of Co_3O_4 [2].

Recently, innovative materials with $A_{n+1}B_nO_{3n+1}$ crystal structure are being investigated for application in IT-SOFCs because of their relatively large oxygen-ion conductivity coupled with reasonable electronic conductivity. This is the Ruddesdlen–Popper (RP) metal oxide family, and the ideal structure is represented in Fig. 13.2C. RP structures, also written as $AO(ABO_3)_n$, consist in a number n of perovskite layers that are stacked between AO rock salt layers along the crystallographic c-axis. If $n = 1$, then the resulting stoichiometry A_2BO_4 adopts the K_2NiF_4 structure, which is a two-dimensional layered perovskite structure [16]. One example of RP metal oxides is La_2NiO_4, which has an electronic conductivity $\sigma_e = 96$ S cm^{-1} at 800°C and an oxygen ion conductivity $\sigma_i = 0.007$ S cm^{-1} at 600°C, better than that of $La_{0.6}Sr_{0.4}Co_{0.2}Fe_{0.8}O_{3-\delta}$ (0.003 S cm^{-1} at 600°C) [17].

13.1.1.3 Anode materials

Analogous to the cathode represented in Fig. 13.1B, the state-of-the-art SOFC anode is a composite, generally a cermet, formed by a mixture of electronic and ionic conducting particles. Overall, the ionic conducting material is the same as that used for the electrolyte, whereas the electronic conducting material is a metal catalyst. Commonly used metals are Ni, Co, Fe, Pt, Mn, and Ru [18]. Among them, Ni exhibits the highest electrochemical activity for HOR. Furthermore, it is cheaper than Co and noble metals. Another reason for selecting Ni is that it is the state-of-the-art catalyst for the methane steam reforming (SMR) reaction. Since the hydrogen fed to the anode compartment is often produced from natural gas through the SMR reaction, it can contain small quantities of unreacted methane. In this case the presence of Ni in the anode catalyst is beneficial, since it provides in situ conversion of this residual

Metal oxide nanofiber-based electrodes in solid oxide fuel cells   307

methane into hydrogen. For all these reasons, Ni is the state-of-the-art anode material. The electronic conductivity of Ni is $\sigma_e = 138 \times 10^4$ S cm^{-1} at 25°C and $\sigma_e = 2 \times 10^4$ S cm^{-1} at 1000°C. Nevertheless, the use of Ni as the anode material has a number of drawbacks. First, owing to the relatively low melting temperature (1453°C), agglomeration and evaporation occur, with loss of porosity and electrode thinning. Another problem is the thermal expansion coefficient (TEC) mismatch between Ni and the electrolyte, leading to detachment and delamination in passing from the ambient temperature to the operating temperature. All these problems are greatly relieved by mixing Ni and electrolyte particles to form a cermet electrode. As has already been discussed, this also provides an extension of the TPB into the electrode bulk. The Ni/YSZ cermet is the state-of-the-art HT-SOFC anode [4,18,19]. As with cathodes, perovskites are considered good candidates for IT-SOFC anodes [20]. The reason is that even if at intermediate temperature the problems of Ni agglomeration and TEC mismatch are reduced, other Ni problems still hold, such as sulfur poisoning, carbon deposition, and poor oxidation stability, which causes volume instability. Among the perovskites, La-substituted $SrTiO_3$ and Y-substituted $SrTiO_3$ show satisfactory electrical conductivity in reducing atmosphere and good dimensional and chemical stability upon redox cycling. Nevertheless, in some cases the electrocatalytic activity for HOR is very poor. To overcome the poor electrocatalytic properties, the B site is often substituted with Mn or Ga atoms. The $La_4Sr_8Ti_{11}Mn_{1-x}Ga_xO_{38-\delta}$ perovskite reports electrical conductivity values in the range $7.9-6.8$ S cm^{-1} at 900°C in reducing conditions [21], whereas the $Y_{0.09}Sr_{0.91}TiO_3$ shows an electrical conductivity of 73.7 S cm^{-1} at 800°C in forming gas (5 vol.% of hydrogen in argon) [22].

13.2   Nanofiber solid oxide fuel cell electrode preparation through electrospinning

Electrospinning is the main method to produce NF SOFC electrodes, since it makes it possible to obtain different morphologies and to control fiber diameters. Furthermore, it features reproducibility and low effective cost. The working principle is based on the electrostatic attraction, which is achieved by applying a voltage difference between the starting precursor solution, injected through a syringe, and the NFs collector.

13.2.1   Electrode preparation

NF preparation involves three main steps: the preparation of the precursor solution, the electrospinning process, and the thermal treatment of the raw NFs (Table 13.1). The electrospinning precursor solution is generally prepared through a sol−gel process. It needs to contain a solvent or a mixture of solvents, the precursor metal salts, and a carrier polymer. The solvents are essential for the dissolution of both the metal salts and the polymer. It is preferable to choose a solvent with high volatility to induce efficient evaporation during electrospinning. The most used solvents are

Table 13.1 Procedure for preparation of symmetrical cells with electrospun electrodes [23].

1. Preparation of the electrospinning solution
Dissolution of nitrates into water. Addition of PVP (weight ratio PVP powder/overall solutes 0.25). Stirring for 24 hours.

2. Electrospinning process
Regulation of electrospinning parameters.

3. Thermal treatment of the LSCF/PVP fibers
Cutting of the nanofiber tissue in circular shape to obtain the electrodes. Thermal treatment of the electrodes in a furnace, heating ramp 0.5 °C/min, maximum temperature 800 °C.

4. Preparation of the symmetrical cell
Application of the electrodes onto both sides of the electrolyte.

5. Firing of the symmetrical cell
Firing of the symmetrical cells with maximum temperature 950°C for 3 hours.

Source: A. Enrico, W. Zhang, M. Lund Traulsen, E.M. Sala, P. Costamagna, P. Holtappels, $La_{0.6}Sr_{0.4}Co_{0.2}Fe_{0.8}O_{3-\delta}$ nanofiber cathode for intermediate-temperature solid oxide fuel cells by water-based sol−gel electrospinning: Synthesis and electrochemical behaviour. J. Eur. Ceram. Soc. 2018;38:2677−2686. https://doi.org/10.1016/j.jeurceramsoc.2018.01.034.

water, ethanol, methanol, and dimethylformamide. The metal salts have to be soluble in the chosen solvents. Generally, the salts involved in the precursor solution are nitrates, oxides, or hydroxides. Finally, the role of the polymer is to give shape to the NFs. The polymer concentration affects the solution viscosity, which in turns determines the NFs size and morphology. The polymer must be soluble in the chosen solvents and must not interact with metal salts. The most used polymers are polyvinyl pyrrolidone and polyethylene oxide.

The precursor solution is then fed to the syringe of the electrospinning equipment for the extrusion of NFs. There are several parameters that rule the process. The most important are applied voltage, solution flow rate, distance between syringe tip and collector, syringe translation speed, collector rotation speed, environment temperature, and relative humidity. These parameters depend on the specific electrospinning equipment in use, on the solution properties, and on the characteristics of the environment inside the electrospinning chamber (air temperature and relative humidity). As an example, a set of parameters used for electrospinning LSCF cathodes is reported in Table 13.2 [24].

After electrospinning, the raw NF tissue has to undergo thermal treatment to achieve the desired crystal structure. During thermal treatment it is essential to control the temperature range in which the polymer degradation takes place. In this temperature window the temperature increase over time must be kept very slow, of the order of 0.5°C min^{-1}, to avoid damage to the NF morphology by the gas formed by polymer decomposition [24]. The NF tissue can be coupled to the electrolyte as it is (unbroken fibers), or it can undergo gentle disaggregation for some minutes in the sonicator in a terpineol solution, after which a mixture of nanorods is obtained. Alternatively, the NFs can be crushed by using a pestle or ball milling. The length of the nanorods can change depending on the disaggregation method. Then the NF tissue or the nanorod paste has to be anchored on the electrolyte. For the characterization of single electrodes a symmetrical button cell configuration is usually adopted, with identical electrodes on both sides, as shown in Fig. 13.3. Typically, button cells have

Table 13.2 Example of electrospinning parameters for lanthanum strontium cobalt ferrite intermediate temperature-solid oxide fuel cell cathodes [24].

Voltage	Flow rate	Translation speed	Rotation speed	Humidity
17 kV	0.5 mL/h	1 mm/s	750 rpm	30%

Source: A. Enrico, B. Aliakbarian, A. Lagazzo, A. Donazzi, R. Botter, P. Perego, et al. Parameter optimization for the electrospinning of La$_{1-x}$Sr$_x$Co$_{1-y}$Fe$_y$O$_{3-\delta}$ fibers for IT-SOFC electrodes. Fuel Cell 2017;17:415–422. https://doi.org/10.1002/fuce.201600190.

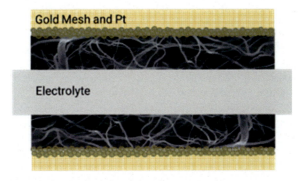

Figure 13.3 Schematic of the experimental symmetrical cell.

diameters of $1-3 \; 10^{-2}$ m. Prior to testing, the symmetrical cells undergo a final thermal treatment to achieve adhesion of the electrode onto the electrolyte.

13.2.2 Typical electrode structures

Electrospinning makes it possible to obtain a variety of NF architectures. Fig. 13.4A shows the morphological characteristics of the LSCF nanorods. Typical nanorod lengths obtained by grinding the unbroken NFs are $0.5-2 \; \mu m$. Fig. 13.4B displays

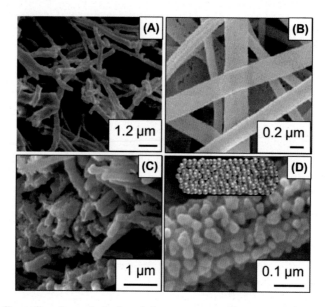

Figure 13.4 Examples of morphologies of electrospun nanofiber cathodes. (A) lanthanum strontium cobalt ferrite nanorods [25]. (B) lanthanum strontium cobalt ferrite unbroken nanofibers [23]. (C) lanthanum strontium cobalt ferrite nanorods infiltrated with gadolinia-doped ceria solution [26]. (D) yttria-stabilized zirconia/lanthanum manganite core–shell nanofibers [9].
Source: (A) P. Costamagna, C. Sanna, A. Campodonico, E.M. Sala, R. Sažinas, P. Holtappels, Electrochemical impedance spectroscopy of electrospun $La_{0.6}Sr_{0.4}Co_{0.2}Fe_{0.8}O_{3-\delta}$ nanorod cathodes for intermediate temperature—solid oxide fuel cells. Fuel Cell 2019;19:472–483. https://doi.org/10.1002/fuce.201800205; (B) A. Enrico, W. Zhang, M. Lund Traulsen, E.M. Sala, P. Costamagna, P. Holtappels, $La_{0.6}Sr_{0.4}Co_{0.2}Fe_{0.8}O_{3-\delta}$ nanofiber cathode for intermediate-temperature solid oxide fuel cells by water-based sol-gel electrospinning: synthesis and electrochemical behaviour. J. Eur. Ceram. Soc. 2018;38:2677–2686. https://doi.org/10.1016/j.jeurceramsoc.2018.01.034; (C) E. Zhao, Z. Jia, L. Zhao, Y. Xiong, C. Sun, M.E. Brito, One dimensional $La_{0.8}Sr_{0.2}Co_{0.2}Fe_{0.8}O_{3-\delta}$/$Ce_{0.8}Gd_{0.2}O19$ nanocomposite cathodes for intermediate temperature solid oxide fuel cells. J. Power Sources 2012;219:133–139. https://doi.org/10.1016/j.jpowsour.2012.07.013; and (D) Y. Jeon, J.H. Myung, S.H. Hyun, Y.G. Shul, J.T.S. Irvine, Corn-cob like nanofibres as cathode catalysts for an effective microstructure design in solid oxide fuel cells. J. Mater. Chem. A 2017;5:3966–3973. https://doi.org/10.1039/c6ta08692f.

typical LSCF unbroken fibers. Typical diameters are in the range 100–500 nm. The unbroken NF electrode has the advantage of ensuring a continuous path for the electrons and oxygen ions, with an improvement in the ionic and electronic electrode conductivity over that of the nanorod electrode.

Both NF and nanorod structures, can be used as backbones for the infiltration technique. An example is shown in Fig. 13.4C, consisting of LSCF NFs infiltrated with GDC nanoparticles. In this case the amount of infiltration is an important parameter, since there is a percolation threshold also for the nanoparticles infiltrated onto the NF surface. Above the percolation threshold the infiltrations contribute to the charge transfer process through the electrode bulk, and this is expected to enhance the electrochemical performance. An even more complex NF structure that can be obtained with the electrospinning technique is the core–shell structure. This consists of two different materials electrospun simultaneously in a core–shell configuration. By using this technique, it is possible to guarantee excellent contact between the two materials. For example, Fig. 13.4D shows NFs with a YSZ core featuring high ionic conductivity and an LSM shell featuring high electronic conductivity [9].

13.3 Overview of electrochemical performance of nanofiber versus conventional solid oxide fuel cell electrodes

The electrochemical performance of SOFC electrodes is expressed in terms of polarization resistance R_p (in Ω cm^2). Conceptually, the electrode polarization resistance is the sum of three contributions, namely, ohmic, activation, and concentration:

$$R_p = R_{ohm} + R_{act} + R_{conc} \tag{13.7}$$

Normally, electrodes have good electronic conductivity, and the resulting ohmic resistance R_{ohm} is therefore practically negligible. The activation resistance R_{act} is intrinsic to the electrochemical reaction. It is associated with the energy barrier to be overcome during the advancement of the reaction. This energy barrier is the sum of a chemical term, that is, the typical chemical activation energy, and an electrical term related to the operating voltage. Typically, the activation resistance R_{act} is a nonlinear function of the operating current and is expressed through the Butler–Volmer equation:

$$j = j_0 \left\{ \exp\left(\eta_{\text{act}} \frac{\alpha nF}{R_{gas}T} \right) - \exp\left[-\eta_{\text{act}} \frac{(1-\alpha)nF}{R_{gas}T} \right] \right\} \tag{13.8}$$

where j is the operating current density (in A cm^{-2}) and η_{act} is the activation voltage loss (in volts), with $R_{act} = \eta_{\text{act}}/j$. The Butler–Volmer equation is the difference between two exponentials, the first representing the forward oxidation current generated at the TPB and the second representing the backward reduction current generated at the same TPB. When there is no net current supplied by the electrode, $j = 0$ and $\eta_{act} = 0$. This is a situation in which the electrode is in dynamic equilibrium, the forward and backward currents both being equal to j_o (in A cm^{-2}), denominated exchange current density, which increases exponentially with temperature. Other parameters that appear in the Butler−Volmer equation are n [−], the number of electrons transferred for each molecule oxidized, and α [−], the charge transfer coefficient. R_{gas} is the gas constant (in J moL^{-1} K) and T is the absolute temperature (in K).

The expression of the Butler−Volmer equation clearly indicates that R_{act} is nonlinear with current. This is typical of IT-SOFCs. Conversely, at small operating currents (where η_{act} is also small) or at the high operating temperatures of HT-SOFCs, the arguments of the exponentials become small. Thus the exponentials can be linearized, obtaining:

$$R_{act} = \frac{R_{gas}T}{j_0 nF} \tag{13.9}$$

Here, R_{act} is constant as a function of current, and the SOFC electrode has linear activation behavior.

Finally, the concentration resistance R_{conc} is related to diffusional limitations occurring inside the electrode [27,28] and in the boundary layer adjacent to the electrode [29]. R_{conc}, usually, is practically constant with temperature, and its contribution to the overall R_p becomes relevant only at high operating temperatures, where R_{act} becomes small.

To complete the picture, the electrode voltage loss is obtained as follows:

$$V_{loss} = R_p j \tag{13.10}$$

In the full SOFC the overall internal voltage loss is the sum of the losses occurring at the anode and cathode plus the ohmic losses associated with the ion flux through the electrolyte. Thus the overall SOFC internal loss is as follows:

$$V_{loss,overall} = V_{loss,anode} + V_{loss,cathode} + V_{loss,electrolyte} \tag{13.11}$$

It must be borne in mind that the operating voltage is:

$$V = V_{Nernst} - V_{loss,overall} \tag{13.12}$$

V_{Nernst} is the theoretical reversible voltage. In the typical operating conditions in HT-SOFCs, $V_{Nernst} \cong 1.07$ V at 1000°C, whereas for IT-SOFCs, $V_{Nernst} \cong 1.12$ V at 700°C. It follows that to have reasonable performance from the fuel cell, electrode losses must be $\ll 0.5$ V. Since reasonable operating currents are of the order of 0.5 A cm^{-2} or larger, the electrode R_p must necessarily be $\ll 1\ \Omega$ cm^2 for real applications. Typically, in SOFCs, R_p is evaluated experimentally through the electrochemical impedance spectroscopy (EIS) technique. This technique is illustrated in Section 13.4.1.

13.3.1 Strontium-doped lanthanum manganite

Fig. 13.5 shows an Arrhenius plot of the R_p^{-1} of the LSM-based cathodes. The conventional HT-SOFC cathodes are prepared by using LSM granular powders, and a significant improvement in electrochemical performance, that is, reduction in R_p, is achieved by reducing the particle size [23]. This is explained by considering that the electrode internal surface and consequently also the TPB are increased by decreasing the particle size. Fig. 13.5 shows that the conventional LSM granular cathode with an average particle size of 1 μm has a polarization resistance $R_p = 9.4\ \Omega\ cm^2$ at 700°C. This value decreases by increasing the operating temperature, reaching 0.79 $\Omega\ cm^2$ at 900°C and 0.38 $\Omega\ cm^2$ at 950°C [30–32]. The pure LSM NF electrode, whose structure is analogous to that reported in Fig. 13.4B, with a NF diameter of 380 nm, shows a polarization resistance of 5.11 $\Omega\ cm^2$ at 700°C, which is lower than that reported above for the conventional powders [32]. This can be ascribed to the continuous conduction paths along the fibers. The small diameter of the fibers can also play a beneficial role.

As has already been discussed, composite cathodes formed by a mixture of LSM and YSZ granular powders exhibit improved performance, owing to the extension of the TPB inside the electrode bulk. For example, LSM/YSZ granular cathodes have demonstrated an $R_p = 2.85\ \Omega\ cm^2$ at 700°C and $R_p = 0.49\ \Omega\ cm^2$ at 900°C [33]. Another option, as already discussed, is provided by infiltration. Polarization

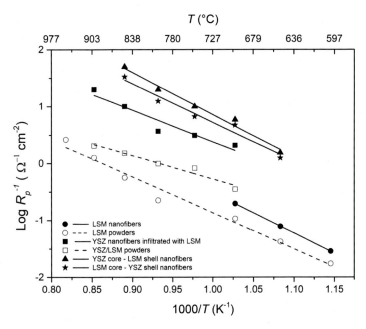

Figure 13.5 Arrhenius plot of the R_p^{-1} of lanthanum manganite. (LSM NFs [32], LSM powders [30–32], YSZ NFs infiltrated with LSM [33], YSZ/LSM powders [33], YSZ core - LSM shell NFs [9], LSM core - YSZ shell NFs [9]).

results are reported for cathodes prepared by using a YSZ NFs scaffold, fiber diameter 200 µm, infiltrated with 50% wt./wt. of LSM nanoparticles. It must be pointed out that the infiltrations are above percolation, which is fundamental for the electrode to achieve electronic conductivity and consequently an effective extension of the TPB throughout the electrode bulk. This electrode has demonstrated a polarization resistance $R_p = 0.32 \ \Omega \ cm^2$ at 750°C [33]. Further improvements are demonstrated by employing a core−shell configuration for the NFs. In particular, with YSZ core−LSM shell NFs (Fig. 13.4D), with a diameter of 350 nm, a polarization resistance as low as $R_p = 0.05 \ \Omega \ cm^2$ has been demonstrated at 800°C. With the same electrode, $R_p = 0.65 \ \Omega \ cm^2$ at 650°C [9].

Fig. 13.5 shows an Arrhenius plot of the R_p^{-1} of the LSM-based cathodes discussed above. The corresponding activation energies are given in Table 13.3., which shows that the NF pure LSM electrode has an activation energy 15 kJ mol^{-1} larger than that of the conventional LSM powder electrode. Fig. 13.5 demonstrates that all the techniques based on coupling LSM with YSZ within the electrode produce an decrease in R_p. This is accompanied by a change in the activation energy E_A (Table 13.3). In more detail, electrodes formed by a mixture of YSZ and LSM

Table 13.3 Activation energies of the intermediate temperature-solid oxide fuel cell cathodes reviewed in this work.

Material	Sample	References	Activation energy E_A (kJmoL^{-1})
LSM	LSM powders	[30−32]	121
	LSM/YSZ powders	[33]	79
	LSM NFs	[32]	136
	YSZ NFs infiltrated with LSM	[33]	107
	YSZ/LSM core−shell NFs	[9]	141
	LSM/YSZ core−shell NFs	[9]	130
Cobalt-based	LSCF powders	[41]	157
	LSCF/GDC powders	[43]	111
	LSCF powders infiltrated with GDC	[41]	160
	LSCF nanorods 50%	[25]	132
	LSCF nanorods 30%	[42]	128
	LSCF NFs 50% porosity	[23]	118
	LSCF NFs 30% porosity	[42]	107
	BSCF powders	[37]	119
	BSCF NFs	[37]	96
	SYC NFs	[38]	180
	SYC NFs infiltrated with GDC	[38]	200
	SSC powders	[39]	162
	SSC NFs	[39]	172
Cobalt-free	NSC powders	[40]	154
	NSC NFs	[40]	168
	PSF powder	[44]	104
	PSF NFs	[44]	116

powders have an activation energy as low as 79 kJ mol^{-1}. This numbers is consistent with the activation energy of oxygen ion conductivity along YSZ, which is around 73 kJ mol^{-1} for temperatures above 650°C [35]. Conversely, the YSZ NFs infiltrated with LSM [33] have an intermediate activation energy (107 kJ mol^{-1}), which could result from a tradeoff between the oxygen ion conduction and charge transfer reaction occurring in the electrode. In this group of materials, YSZ-LSM core–shell NFs demonstrate the best performance and the highest activation energy, 130–141 kJ mol^{-1}, similar to that of the pure LSM NFs (136 kJ mol^{-1}).

13.3.2 Cobalt-based metal oxides

LSCF is the most studied material for IT-SOFC cathodes, since it shows good electrochemical performance in the temperature range 600°C–800°C. An overview of the results obtained from LSCF-based cathodes is shown in Fig. 13.6. The conventional LSCF powder cathode, with 200–300 nm average particle diameter, has a polarization resistance $R_p = 1.7$ Ω cm^2 at 650°C [41], which is about one order of magnitude lower than that of LSM powder cathodes. LSCF NFs are employed at the cathode side by using different arrangements. The LSCF nanorod cathode, with an average NF diameter of 250 nm and 30% porosity, shows a polarization resistance $R_p = 0.45$ Ω cm^2 at 650°C [42]. The LSCF unbroken NF cathode, with the

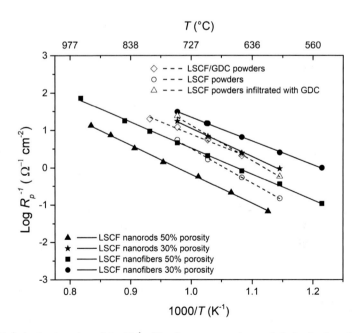

Figure 13.6 Arrhenius plot of the R_p^{-1} of lanthanum strontium cobalt ferrite-based cathodes. (LSCF/GDC powders [43], LSCF powders [41], LSCF powders infiltrated with GDC [41], LSCF nanorods 50% porosity [25], LSCF nanorods 30% porosity [42], LSCF NFs 50% porosity [23], LSCF NFs 30% porosity [42])

same NF diameter and electrode porosity, shows a polarization resistance of $0.2 \, \Omega \, cm^2$ at 650°C [42]. The performance of the unbroken LSCF NF is better than both the nanorods and the conventional powder electrodes [23]; again, this can be ascribed to the continuous path guaranteed by the unbroken morphology. Nevertheless, the electrode porosity highly affects the electrochemical performance. Indeed, the LSCF unbroken NF cathode with an average NF diameter of 250 nm and a porosity of 50%, reports a polarization resistance of $1 \, \Omega \, cm^2$ at 650°C [23,34], and LSCF nanorods with an average NFs diameter of 300 nm and a porosity of 50% show a polarization resistance $R_p = 3.9 \, \Omega \, cm^2$ at 660°C [25].

As with the LSM cathode, the electrochemical performance is improved by adding materials with high oxygen ion conductivity. The composite obtained by mixing LSCF and GDC powders is one of the most popular IT-SOFC cathodes [2]. This composite cathode, featuring an average particle diameter of 470 and 160 nm for LSCF and GDC, respectively, shows a polarization resistance of $R_p = 0.52 \, \Omega \, cm^2$ at 650°C [43]. The infiltration technique has been widely investigated as well. For example, LSCF powder electrodes with an average particle diameter of 200–300 nm, infiltrated with GDC nanoparticles with an average particle diameter of 10–40 nm, give a polarization resistance $R_p = 0.45 \, \Omega \, cm^2$ at 650°C [41]. The infiltration technique has been demonstrated also with LSCF NF cathodes. For example, an electrode consisting of LSCF nanorods with an average diameter of 200–300 nm and infiltrated with GDC nanoparticles deposited from a $0.25 \, mol \, L^{-1}$ GDC solution has demonstrated a polarization resistance as low as $0.10 \, \Omega \, cm^2$ at 650°C [36]. On the other hand, at present, no results have been reported in the literature for infiltrated unbroken LSCF NFs.

The activation energies associated with the LSCF-based cathodes discussed above are given in Table 13.3. Here, the highest activation energies are reported for the pure LSCF powder and for the LSCF powders infiltrated with GDC ($157-160 \, kJ \, mol^{-1}$). Also here, electrodes formed by a mixture of YSZ and LSM powders have a much lower activation energy ($111 \, kJ \, mol^{-1}$). Nevertheless, this is significantly higher than the activation energy for oxygen ion conduction along GDC ($70 \, kJ \, mol^{-1}$ for temperatures above 400°C [5]). Pure LSCF nanorods and NFs show activation energies in the ranges $128-132 \, kJ \, mol^{-1}$ and $107-118 \, kJ \, mol^{-1}$ respectively. Interestingly, different laboratories have reported consistent activation energies for nanorods and NFs [23,25,34,42]. The former are larger than the latter, and both are significantly lower than the activation energies of the pure LSCF powder electrodes.

Besides LSCF, other Co-based metal oxide materials have been used to prepare NFs for application as IT-SOFCs cathodes. The electrochemical results are reported in Fig. 13.7. $Ba_{1-x}Sr_xCo_{1-y}Fe_yO_{3-d}$ (BSCF), very similar to LSCF, has a high oxygen ion conductivity, due to its oxygen-deficient perovskite structure, and exhibits a moderately high electronic conductivity, owing to the mixed valence of the transition metal B site cations. The conventional BSCF powder cathode has $R_p = 0.468 \, \Omega \, cm^2$ at 600°C, whereas the BSCF unbroken NF cathode, with an average NF diameter of 100–200 nm, shows an $R_p = 0.094 \, \Omega \, cm^2$ under the same operating conditions [37]. Another Co-based material potentially used in

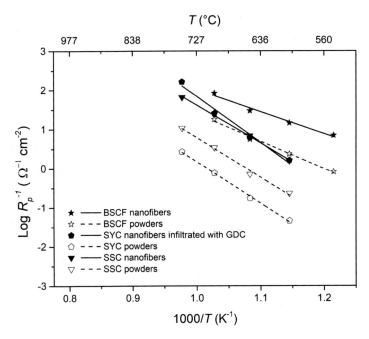

Figure 13.7 Arrhenius plot of the R_p^{-1} of other cobalt-based cathodes. (BSCF NFs [37], BSCF powders [37], SYC NFs infiltrated with GDC [38], SYC powders [38], SSC NFs [39], SSC powders [39])

IT-SOFC cathodes is $Sr_{1-x}Y_xCoO_{2.65-d}$ (SYC). Compared to LSCF and BSCF, the SYC perovskite has recently attracted attention for its promising ionic and electronic conductivities. The polarization resistance of the SYC NF cathode is $R_p = 4.73\ \Omega\ cm^2$ at 650°C, which is significantly larger than the results reported above. Nevertheless, the polarization resistance drops drastically by infiltrating the electrode with GDC. The SYC NF cathode infiltrated with a GDC solution in mass ratio 1:0.44 shows a polarization resistance $R_p = 0.2\ \Omega\ cm^2$ at 650°C [38]. Another promising material for IT-SOFC cathode applications is $Sm_{1-x}Sr_xCoO_{3-d}$ (SSC). Again, it is a perovskite structure with high ionic and electronic conductivity. The polarization resistance at 650°C is $R_p = 0.13\ \Omega\ cm^2$ for the conventional SSC powder cathode and only $R_p = 0.024\ \Omega\ cm2$ for the SSC NF cathode, with NF diameters in the range 80–300 nm [39]. It is interesting to notice that for all these materials the NF electrode gives better performance than the conventional powder electrode, as represented in Fig. 13.7. Table 13.3 shows that with BSCF the NF electrode has an associated activation energy 23 kJ mol^{-1} lower than that of the conventional powder electrode. This is in agreement with the findings previously reported for LSCF. Instead, with SSC the behavior is similar to that of LSM, since the NF electrode has an associated activation energy 10 kJ mol^{-1} higher than that of the conventional powder electrode.

13.3.3 Cobalt-free metal oxides
13.3.3.1 Cathodes

Co-based perovskite oxides are widely used as cathode materials for IT-SOFCs. Nevertheless, the use of Co has some drawbacks. At high temperatures, Co is not stable, because the reduction of Co associated with the formation of oxygen vacancies leads to lattice expansion. Furthermore, Co is highly toxic [2], and the Co veins are all located in a part of the world that is afflicted by geopolitical problems. All these implications motivate interest in Co-free metal oxides, in particular the new emergent RP class of materials. Also with RP materials, NF electrodes have been developed to improve the electrochemical performance at the IT-SOFC operating temperature. Results are reported in Fig. 13.8. Among the RP materials for cathode application, $Nd_{2-x}Sr_xCuO_4$ (NSC) has attracted attention because of its good catalytic activity for the ORR. The conventional NSC powder cathode shows a polarization resistance $R_p = 0.39$ Ω cm^2 at 700°C. An even lower polarization resistance is achieved by employing the NSC NF cathode, with NF diameters in the range of 100–200 nm. For the latter an $R_p = 0.26$ Ω cm^2 is reported at the same temperature of 700°C [40].

Besides the RP structures, other Co-free, single perovskites are under development. An example is the composite $Pr_{0.6}Sr_{0.4}FeO_{3\delta}$-$Ce_{0.9}Pr_{0.1}O_{2\delta}$ (PSF-CP), which shows higher oxygen permeation flux compared to other materials under the same operating conditions. This feature makes PSF-CP an interesting candidate for

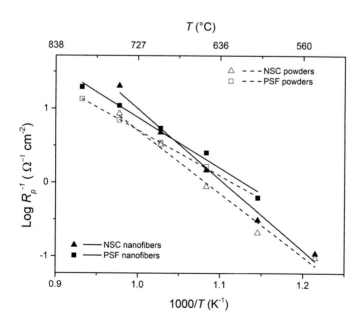

Figure 13.8 Arrhenius plot of the R_p^{-1} of Co-free cathodes. (NSC NFs [40], NSC powders [40], PSF NFs [44], PSF powders [44])

IT-SOFC applications. The polarization resistance of a conventional PSF-CP granular powder cathode is $R_p = 0.35\ \Omega\ cm^2$ at 700°C, which decreases to $R_p = 0.25\ \Omega\ cm^2$ with PSF-CP NFs [44]. It is interesting to notice that for all the Co-free materials, the NF electrode gives better performance than the conventional powder electrode, as shown in Fig. 13.8. Table 13.3 shows that the NF-based electrodes have activation energies $10-15\ kJ\ mol^{-1}$ larger than those of conventional powder electrodes.

13.3.3.2 Anodes

As was reported previously, La-substituted $SrTiO_3$ (LST) is considered a promising material for SOFC anode application. To enhance the poor electrocatalytic activity for the HOR, LST NFs infiltrated with GDC are employed. The polarization resistance is $R_p = 0.95\ \Omega\ cm^2$ at 800°C in an atmosphere composed of 97% H_2 and 3% H_2O [45].

13.4 Understanding the structure-performance relationship in nanofiber solid oxide fuel cell electrodes: experimental characterization and numerical modeling

In the previous Section the relationship between structure and performance was discussed on a qualitative basis. Experimental techniques coupled to modeling pave the way to a quantitative analysis.

13.4.1 Electrochemical impedance spectroscopy: experimental characterization and equivalent circuit modeling

EIS [46,47] is a useful tool for investigating novel SOFC electrode materials and architectures. Simplifying, it is an application of the typical impedance measurement approach developed in electrical engineering (Fig. 13.9A), and it is especially used for solid-state electrochemical cells such as solid oxide cells, including both SOFCs and solid oxide electrolysis cells. This is the technique that is used to measure experimentally the electrode polarization resistance R_p.

The acquisition of the experimental data is typically performed on symmetrical button cells, as described in Section 13.2.1. Prior to measurement the symmetrical button cells are mounted in a setup in which each cell is fixed between two meshes, usually in gold or platinum, which act as current collectors (see Fig. 13.9B). Then the setup is placed in a furnace, sealed, and connected to the gas supply system and to the impedance measurement equipment.

It is well known that in EIS studies, the limiting step is the analysis of the experimental data rather than the acquisition of the data themselves [48]. Impedance modeling is therefore an essential step in the EIS methodology. As far as metal oxide NF-based SOFC electrodes are concerned, the type of modeling that is carried

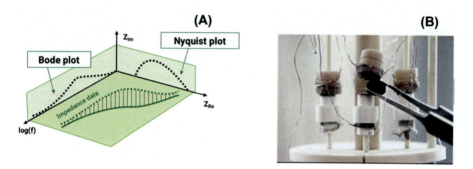

Figure 13.9 Electrochemical impedance spectroscopy experimental technique. (A) Bode and Nyquist visualization of impedance data. (B) The experimental test rig.
Source: Courtesy of DTU Energy, Denmark.

Table 13.4 Main impedance circuit elements used in the equivalent circuit modeling of metal oxide nanofiber-based electrodes for solid oxide fuel cells.

Name	Chemical–physical phenomenon	Equation	Parameters
L	Inductance	$Z(\omega) = i\omega L$	L = inductance
R	Resistance	$Z(\omega) = R$	R = resistance
CPE	Double-layer charge/discharge	$Z(\omega) = \frac{1}{Q(i\omega)^\gamma}$	Q = double-layer capacitance[a]; γ = fitting parameter
RQ	Electrochemical reaction through the electrochemical double layer[a]	$Z(\omega) = \frac{1}{R^{-1} + Q(i\omega)^\gamma}$	Q, γ same as above; R = resistance associated with charge transfer process
Gerischer (G)	Charge transport coupled to distributed electrochemical reaction	$Z(\omega) = \frac{1}{Y\sqrt{k + i\omega}}$	k = oxygen surface exchange coefficient; Y = structural parameters and solid phase oxygen diffusion coefficient
Finite-length Warburg (FLW)	Gas phase diffusion	$Z(\omega) = \frac{\tanh(\sqrt{Bi\omega})}{\sqrt{Y_1 i\omega}}$	$B = \delta^2/D$, where δ = thickness of gas diffusion layer and D = gas phase oxygen diffusion coefficient; Y_1 = phenomenological parameter

[a]The electrochemical double-layer is the basic schematization of the electrochemical phenomena occurring at the TPB [50].

out to give an interpretation to the EIS experimental results is the so-called equivalent circuit (EC) modeling. The analysis of experimental data is carried out through an iterative process based on translating a physical model into an EC and fitting the experimental data through nonlinear least squares algorithms [49]. The main circuit elements used in the EC modeling of metal oxide NF-based electrodes for SOFCs are summarized in Table 13.4. Knowing the physical meaning of the EC parameters is essential, since this allows the drawing of quantitative information from the fitting parameters and thus the assessment of the appropriateness of the model. Several electrochemical processes have already been described and analyzed in terms of their similarity to ECs [47,51].

13.4.1.1 Equivalent circuit modeling of metal oxide nanofiber-based electrodes

It has been demonstrated that the EC model L-R_s-RQ-G can be used to fit the EIS experimental data obtained from metal oxide NF-based electrodes for SOFC applications [34]. The element L accounts for the inductance of the experimental apparatus. By increasing the operating temperature, the FLW element is added to account for gas phase diffusion, and the EC becomes L-R_s-RQ-G-FLW [34]. The fittings allow identification of a resistance associated with each EC element, which is named after the element itself. Thus, for example, R_G is the resistance associated with the Gerischer element, and R_{RQ} is the resistance associated with the RQ element. Furthermore, if the EIS spectrum features two arcs, R_{HF} is the resistance associated with the high-frequency arc, and R_{LF} is the resistance associated with the low-frequency arc. The following relationship holds:

$$R_p = R_{HF} + R_{LF} \tag{13.13}$$

where, in the case of metal oxide NF-based electrodes for SOFC applications [34],

$$R_{HF} = R_{RQ} + R_G \tag{13.14}$$

$$R_{LF} = R_{FLW} \tag{13.15}$$

with R_{LF} not being visible at temperatures lower than 750°C.

In Fig. 13.10 a complete set of EIS experimental results obtained from LSCF NF SOFC cathodes is shown, obtained in the range 550°C–950°C [34]. The experimental results are plotted in the form of Nyquist and Bode plots, together with the fittings. The software Elchemea [52] was used to implement the EC model adopted and to fit it to the EIS experimental data. The fitting value obtained for the L element was subtracted from both the experimental and EC modeling results. In Fig. 13.10 the Nyquist plot of the EIS experimental data at 550°C shows a single arc, which at high frequency becomes a straight line with a 45-degree slope with respect to the real axis. This is usually referred to as Gerischer behavior [53–55]. Fig. 13.10 shows that with increasing temperature (600°C–750°C) this arc gradually shrinks and deforms, with the high-frequency straight line becoming remarkably narrower in range and reduced in slope, resulting in an overall arc depression. When the temperature is increased above 750°C, the impedance arc shrinks further, and the high-frequency straight line typical of the Gerischer behavior tends to bend, so at 950°C the arc has the shape of a depressed semicircle. When the temperature is increased above 750°C, the shrinkage of the high-frequency arc reveals a second, low-frequency arc. The dimensions and shape of this low-frequency arc apparently

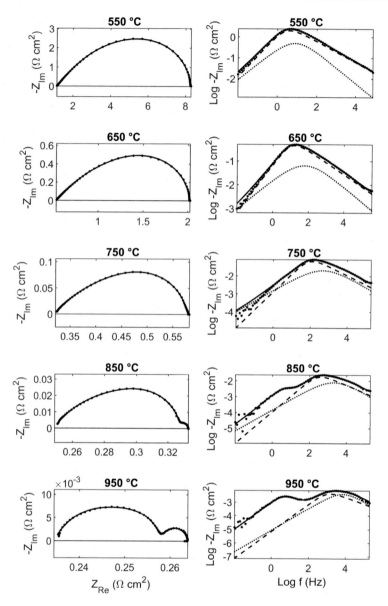

Figure 13.10 Electrochemical impedance spectroscopy results obtained from lanthanum strontium cobalt ferrite nanofiber cathodes. * = experimental data; – – – = fittings through the L-R$_s$-RQ-G-FLW equivalent circuits. (Left) Nyquist plots. (Rigth) Bode plots, displaying also the individual contributions of the Gerischer (– – –) and RQ (· · · ·) elements [34].
Source: P. Costamagna, E.M. Sala, W. Zhang, M. Lund Traulsen, P. Holtappels, Electrochemical impedance spectroscopy of La$_{0.6}$Sr$_{0.4}$Co$_{0.2}$Fe$_{0.8}$O$_{3-\delta}$ nanofiber cathodes for intermediate temperature-solid oxide fuel cell applications: A case study for the 'depressed' or 'fractal' Gerischer element. Electrochim. Acta 2019;319:657–671. https://doi.org/10.1016/j.electacta.2019.06.068.

do not change significantly with temperature, which suggests that this arc is related to the physical process of gas phase diffusion [23,28,56].

Correspondingly, in the Bode Z_{Im} plots, the experimental data display, at low temperature (550°C), a single process with a well-defined sharp maximum, clearly individuating one single time constant. When the temperature is increased, the maximum decreases, and simultaneously it shifts toward higher frequencies, indicating that the process is thermally activated. At high temperatures (800°C−950°C) the high-frequency maximum is no longer sharp but rather is broad, and simultaneously a curvature appears in the high-frequency part of the spectrum, which suggests that more than one process is reflected. Still referring to the Bode Z_{Im} plots, at temperatures of 750°C−800°C and above, a second maximum becomes gradually visible in the low-frequency region, corresponding to the diffusive low-frequency arc already individuated in the Nyquist plot. By increasing the temperature to 950°C, the frequency of this second maximum and the impedance do not change, indicating that this second phenomenon is not thermally activated, which supports the hypothesis of an originating diffusive process.

In Fig. 13.10 the Bode Z_{Im} plots also display the individual contributions of the G and RQ components of the Rs-RQ-G circuits. At all temperatures, both the G and RQ processes contribute in the same range of frequencies. It must be borne in mind that the G element represents the electrochemical reaction coupled to charge transfer occurring along the fibers within the electrode bulk, whereas the RQ element represents the electrochemical phenomenon occurring at the electrode-electrolyte interface. In light of this it is not surprising that the G and RQ processes contribute in the same range of frequencies, since the ORR occurs in proximity to the LSCF NF in both cases, and therefore the processes display some similarities. At all temperatures the peak frequency of the G process is lower than that of the RQ process. This is consistent with the fact that the G process, in addition, to the charge transfer reaction, embeds the oxygen ion conduction along the electrode thickness, which is a slow process compared to the electrochemical reaction typically represented by the RQ element. With increasing temperature, both maxima shift toward higher frequencies, since both processes are thermally activated.

Still concerning the Bode Z_{Im} plots, it is clear that at low temperature the G element accounts almost completely for the impedance of the sample, the RQ contribution being practically negligible. Correspondingly, the experimental data show one single maximum, which is rather sharp in shape. When the temperature is increased, the relative importance of the RQ contribution tends to increase, and at 950°C the G and RQ contributions are comparable, the maxima being separated by one order of magnitude in frequency. Because of this, the G and RQ contributions merge into one single high-frequency impedance maximum, which is rather broad in shape, as has already been mentioned.

13.4.1.2 *Electrochemical behavior of metal oxide nanofiber-based electrodes*

Nielsen et al. studied LSCF/GDC conventional composite SOFC cathodes with granule size of the order of 0.5−1 μm [29]. The penetration of the electrochemical reaction, starting from the electrode-electrolyte interface toward the electrode bulk, was shown to be of the order of 40 μm at a temperature of 550°C and about 10 μm at a temperature of 850°C. An analogous shrinkage of the electrochemically active electrode thickness with increasing temperature has been demonstrated for LSM/YSZ granular composite cathodes [57]. With LSCF NF-based electrodes the EIS studies reported in Section 4.1.1 [34] suggest that the phenomenon is analogous. At low temperature the electrochemical reaction penetrates the electrode in depth, and this is accompanied by charge transport, in particular electrons and oxygen ions, through the electrode thickness. Simultaneously, the electrochemical reaction also occurs at the electrode-electrolyte interface. When the temperature is increased, the electrochemical reactions tend to develop more at the electrode-electrolyte interface rather than penetrating the electrode thickness.

13.4.2 One-dimensional pseudohomogeneous model of infiltrated mixed ionic-electronic conductor nanofiber electrodes

In Section 13.1.1.1 it was underlined that, especially with MIEC electrodes, infiltrations improve the electrochemical performance. This takes place with both conventional granular electrodes and NF-based electrodes. For NF-based electrodes, studies attempting to explain this effect have been developed in the literature [11], based on the consideration that in the pristine MIEC NF electrodes there are conducting paths for electrons and for oxygen ions, which are separated from each other, that is, there are no contact points between them. As shown in Fig. 13.11, the

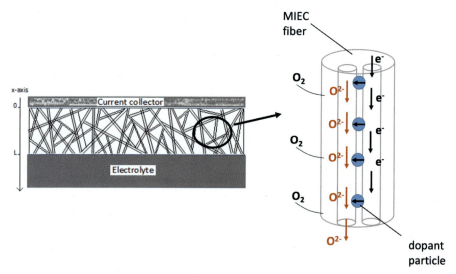

Figure 13.11 (Right) Schematic of a nanofiber-based cathode for application in solid oxide fuel cells. (Left) Mechanism of charge transport along the nanofibers, and infiltration effect [11].
Source: A. Enrico, P. Costamagna, Model of infiltrated $La_{1-x}Sr_xCo_{1-y}Fe_yO_{3-\delta}$ cathodes for intermediate temperature solid oxide fuel cells. J. Power Sources 2014;272:1106–1121. https://doi.org/10.1016/j.jpowsour.2014.08.022.

Metal oxide nanofiber-based electrodes in solid oxide fuel cells

model is based on the idea that the infiltrating particles adhere to the surface of the NFs, creating contact points between the ionic and electronic conductive paths. In this way, the infiltrated particles make it possible the exchange of electrical charges between the two conducting paths, extending the TPB deeply into the electrode thickness, away from the electrode/electrolyte interface, with a consequent improvement of electrochemical performance.

13.4.2.1 Model development

The main hypotheses of the model are as follows [11]:

- It has DC steady-state operating conditions.
- The model is pseudohomogeneous: the scaffold electrode structure, formed by straight fibers randomly distributed within the electrode, is considered as continuous and homogeneous.
- The model is one-dimensional along the x-coordinate (see Fig. 13.11).
- The temperature and pressure are uniform throughout the electrode.
- The dopant particles are regularly distributed along the electrode thickness.
- The whole interface between the dopant particles and the MIEC fibers is a TPB.
- At the electrode-electrolyte boundary, the whole interface between the electrolyte and the electronic conducting paths of the electrode is a TPB.
- Mass transport limitations are not taken into account.

The model is based on Ohm's law, written in differential form, for the transport of oxygen ions and electrons, with j_{io} and j_{el} being the ionic and electronic current densities, respectively (in A cm^{-2}). V_{el} and V_{io} represent the voltage along the electronic and ionic conducting rails, respectively, in the NFs (in V), and σ_{el}^{eff} and σ_{io}^{eff} are effective electronic and ionic conductivities, respectively, along the NFs (in S cm^{-1}). The electrochemical reaction rate is expressed through the Butler–Volmer equation [Eq. (13.4)], where j is the current density referred to the TPB area (S_{TPB}, TPB area per unit volume of the electrode per centimeter):

$$
\begin{cases}
\dfrac{dV_{io}}{dx} = -\dfrac{j_{io}}{\sigma_{io}^{\text{eff}}} \\[2mm]
\dfrac{dV_{el}}{dx} = -\dfrac{j_{el}}{\sigma_{el}^{\text{eff}}} \\[2mm]
j = j_0 \left\{ -\exp\left(\eta\dfrac{\alpha nF}{R_{gas}T}\right) + \exp\left[-\eta\dfrac{(1-\alpha)nF}{R_{gas}T}\right] \right\} \\[2mm]
\dfrac{di_{io}}{dx} = -\dfrac{di_{el}}{dx} = S_{\text{TPB}}j \\[2mm]
\eta = (V_{el} - V_{io})^{\text{th}} - (V_{el} - V_{io})
\end{cases}
\tag{13.16}
$$

The electrochemical reaction results in a transfer of charges between the ionic and electronic conducting rails that are present in the fibers. In Eq. (13.10), this exchange of charges is expressed through a charge balance equation. Finally, in

Eq. (13.10), the definition of overpotential η is recalled. The other terms that appear in Eq. (13.10) were defined in Section 13.3. Equation (13.10) contains a number of parameters, which are evaluated in detail on the basis of literature experimental data and on the basis of the NF morphology. In particular, S_{TPB} is evaluated as a function of the infiltration level, expressed as n_p/L (number of particles per unit thickness (cm^{-1}) of the electrode). Boundary condition for the system of equations Eq. (13.16) are expressed through the overall operating electrode current density J_{tot} (in A cm^{-2}). The model equations are integrated in MATLAB®.

13.4.2.2 Model results

The model makes it possible to evaluate the extension of the TPB inside the bulk of the electrode, and then the distribution of current density and voltage in the electrode bulk along the x-coordinate, and finally the overall electrode polarization resistance R_p. The results show an increased penetration of the electrochemical reaction into the electrode bulk [11]. The results shown in Fig. 13.12 refer to GDC-infiltrated LSCF cathodes when the infiltration load n_p/L was varied from 0 to 170 particles μm^{-1}. Fig. 13.12 shows simulation results [11] together with literature experimental data [26] for R_p^{-1}, demonstrating reasonably good agreement. The results show the improvement of performance due to infiltrations. For infiltration loads between 50 and 150 particles μm^{-1}, the simulated value of R_p^{-1} is slightly below the experimental data; this is possibly due to percolation occurring among the infiltrations, which enhances the ionic conductivity and thus the R_p^{-1} of the electrode and is not accounted for by the model. On the other hand, for infiltration loads

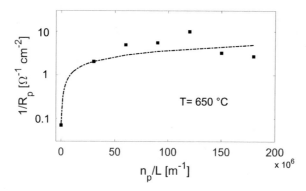

Figure 13.12 R_p^{-1} of lanthanum strontium cobalt ferrite nanofiber cathodes infiltrated with various levels of gadolinia-doped ceria [11].
Source: Symbols: experimental data, taken from E. Zhao, Z. Jia, L. Zhao, Y. Xiong, C. Sun, M.E. Brito, One dimensional La$_{0.8}$Sr$_{0.2}$Co$_{0.2}$Fe$_{0.8}$O$_{3-\delta}$/Ce$_{0.8}$Gd$_{0.2}$O$_{1.9}$ nanocomposite cathodes for intermediate temperature solid oxide fuel cells. J. Power Sources 2012;219:133−139. https://doi.org/10.1016/j.jpowsour.2012.07.013. Lines: simulations, taken from A. Enrico, P. Costamagna, Model of infiltrated La$_{1-x}$Sr$_x$Co$_{1-y}$Fe$_y$O$_{3-\delta}$ cathodes for intermediate temperature solid oxide fuel cells. J. Power Sources 2014;272:1106−1121. https://doi.org/10.1016/j.jpowsour.2014.08.022.

$n_p/L > 150$, gas diffusion limitations are expected to occur in the experimental electrode, which may explain the experimental loss of performance.

13.5 Summary

In SOFCs the most critical component is the air electrode (cathode), which accounts for the largest internal loss. Metal oxides are the state-of-the-art materials for SOFC cathodes. The conventional electrode architecture is obtained by powder pressing. A number of strategies have been proposed to enhance the cathode performance, especially to reduce the SOFC operating temperature. New metal oxide materials are being investigated as well as new electrode manufacturing techniques. Electrospinning is currently at the forefront of research, as it is inexpensive and simple and provides reproducible results. In this chapter a number of metal oxide cathodes for SOFC application were reviewed, in particular the electrochemical performance of conventional powder cathodes is compared to that obtained from electrospun fiber-based cathodes. The results show that, on the whole, the materials in the form of fibers provide better electrochemical performance than powders. Nevertheless, it has been shown that too high a void degree can compromise this result. Furthermore, the fibers feature an increase in activation energy of $10-20 \text{ kJ mol}^{-1}$ compared to the powders, with two exceptions: the LSCF and BSCF materials. Further experimental evidence is needed to confirm this effect.

Electrode modeling, carried out through different approaches, makes it possible to assess the penetration of the electrochemical reaction into the electrode bulk. This penetration typically shrinks by increasing the operating temperature. Finally, infiltrating the electrodes with dopants improves the electrochemical performance, as has been demonstrated by both experimental and modeling results.

References

[1] K. Huang, J.B. Goodenough, Solid Oxide Fuel Cell Technology, Woodhead Publishing Limited, Cambridge, UK, 2009.

[2] S.P. Jiang, Development of lanthanum strontium cobalt ferrite perovskite electrodes of solid oxide fuel cells—a review, Int. J. Hydrogen Energy 44 (2019) 7448−7493. Available from: https://doi.org/10.1016/j.ijhydene.2019.01.212.

[3] P. Costamagna, P. Costa, V. Antonucci, Micro-modelling of solid oxide fuel cell electrodes, Electrochim. Acta 43 (1998) 375−394. Available from: https://doi.org/10.1016/S0013-4686(97)00063-7.

[4] S. Tao, J.T.S. Irvine, Discovery and characterization of novel oxide anodes for solid oxide fuel cells, J. Chem. Soc. Jpn. (2004) 83−95. Available from: https://doi.org/10.1002/tcr.20003.

[5] E. Jud, L.J. Gauckler, The effect of cobalt oxide addition on the conductivity of $Ce_{0.9}Gd_{0.1}O_{1.95}$, J. Electroceram. 15 (2005) 159−166. Available from: https://doi.org/10.1007/s10832-005-2193-3.

[6] P. Costamagna, M. Panizza, G. Cerisola, A. Barbucci, Effect of composition on the performance of cermet electrodes. Experimental and Theoretical Approach, Electrochim. Acta 47 (2002) 1079−1089. Available from: https://doi.org/10.1016/S0013-4686(01)00830-1.

[7] A. Barbucci, R. Bozzo, G. Cerisola, P. Costamagna, Characterisation of composite SOFC cathodes using electrochemical impedance spectroscopy. Analysis of Pt/YSZ and LSM/YSZ Electrodes, Electrochim. Acta 47 (2002) 2183−2188. Available from: https://doi.org/10.1016/s0013-4686(02)00095-6.

[8] P. Costamagna, P. Costa, E. Arato, Some more considerations on the optimization of cermet solid oxide fuel celle electrodes, Electrochim. Acta 43 (1998) 967−972. Available from: https://doi.org/10.1016/s0013-4686(97)00262-4.

[9] Y. Jeon, J. ha Myung, S. hoon Hyun, Y. gun Shul, J.T.S. Irvine, Corn-cob like nanofibres as cathode catalysts for an effective microstructure design in solid oxide fuel cells, J. Mater. Chem. A 5 (2017) 3966−3973. Available from: https://doi.org/10.1039/c6ta08692f.

[10] Z. Lu, J. Hardy, J. Templeton, J. Stevenson, Extended reaction zone of $La_{0.6}Sr_{0.4}Co_{0.2}Fe_{0.8}O_3$ cathode for solid oxide fuel cell, J. Power Sources 198 (2012) 90−94. Available from: https://doi.org/10.1016/j.jpowsour.2011.09.020.

[11] A. Enrico, P. Costamagna, Model of infiltrated $La_{1-x}Sr_xCo_{1-y}Fe_yO_{3-\delta}$ cathodes for intermediate temperature solid oxide fuel cells, J. Power Sources 272 (2014) 1106−1121. Available from: https://doi.org/10.1016/j.jpowsour.2014.08.022.

[12] P. Kaur, K. Singh, Review of perovskite-structure related cathode materials for solid oxide fuel cells, Ceram. Int. 46 (2020) 5521−5535. Available from: https://doi.org/10.1016/j.ceramint.2019.11.066.

[13] G. Yang, W. Jung, S.J. Ahn, D. Lee, Controlling the oxygen electrocatalysis on perovskite and layered oxide thin films for solid oxide fuel cell cathodes, Appl. Sci. (2019) 9. Available from: https://doi.org/10.3390/app9051030.

[14] S.P. Jiang, Development of lanthanum strontium manganite perovskite cathode materials of solid oxide fuel cells: a review, J. Mater. Sci 43 (2008) 6799−6833. Available from: https://doi.org/10.1007/s10853-008-2966-6.

[15] F. Prado, T. Armstrong, A. Caneiro, A. Manthiram, Structural Stability and Oxygen Permeation Properties of $Sr_{3-x} La_x Fe_{2-y} Co_y O_{7-\delta}$ ($0 \le x \le 0.3$ and $0 \le y \le 1.0$), J. Electrochem. Soc. 148 (2001) J7−J14. Available from: https://doi.org/10.1149/1.1354605.

[16] G. Nirala, D. Yadav, S. Upadhyay, Ruddlesden−Popper phase A_2BO_4 oxides: recent studies on structure, electrical, dielectric, and optical properties, J. Adv. Ceram. 9 (2020) 129−148. Available from: https://doi.org/10.1007/s40145-020-0365-x.

[17] M. Ghamarinia, A. Babaei, C. Zamani, Electrochemical characterization of La_2NiO_4-infiltrated $La_{0.6}Sr_{0.4}Co_{0.2}Fe_{0.8}O_{3-d}$ by analysis of distribution of relax times, Electrochim. Acta 353 (2020) 1−9. Available from: https://doi.org/10.1016/j.electacta.2020.136520.

[18] B. Shri Prakash, S. Senthil Kumar, S.T. Aruna, Properties and development of Ni/YSZ as an anode material in solid oxide fuel cell: a review, Renew. Sustain. Energy Rev. 36 (2014) 149−179. Available from: https://doi.org/10.1016/j.rser.2014.04.043.

[19] L.J. Gauckler, D. Beckel, B.E. Buergler, E. Jud, U.P. Muecke, M. Prestat, et al., Solid oxide fuel cells: systems and materials, Chimia (Aarau) 58 (2004) 837−850. Available from: https://doi.org/10.2533/000942904777677047.

[20] C.D. Savaniu, J.T.S. Irvine, La-doped $SrTiO_3$ as anode material for IT-SOFC, Solid. State Ion. 192 (2011) 491−493. Available from: https://doi.org/10.1016/j.ssi.2010.02.010.

[21] M.J. Escudero, J.T.S. Irvine, L. Daza, Development of anode material based on La-substituted $SrTiO_3$ perovskites doped with manganese and/or gallium for SOFC, J. Power Sources 192 (2009) 43−50. Available from: https://doi.org/10.1016/j.jpowsour.2008.11.132.

[22] X. Li, H. Zhao, W. Shen, F. Gao, X. Huang, Y. Li, et al., Synthesis and properties of Y-doped $SrTiO_3$ as an anode material for SOFCs, J. Power Sources 166 (2007) 47−52. Available from: https://doi.org/10.1016/j.jpowsour.2007.01.008.

[23] A. Enrico, W. Zhang, M. Lund Traulsen, E.M. Sala, P. Costamagna, P. Holtappels, $La_{0.6}Sr_{0.4}Co_{0.2}Fe_{0.8}O_{3-\delta}$ nanofiber cathode for intermediate-temperature solid oxide fuel cells by water-based sol-gel electrospinning: synthesis and electrochemical behaviour, J. Eur. Ceram. Soc. 38 (2018) 2677−2686. Available from: https://doi.org/10.1016/j.jeurceramsoc.2018.01.034.

[24] A. Enrico, B. Aliakbarian, A. Lagazzo, A. Donazzi, R. Botter, P. Perego, et al., Parameter optimization for the electrospinning of $La_{1-x}Sr_xCo_{1-y}Fe_yO_{3-\delta}$ fibers for IT-SOFC electrodes, Fuel Cells 17 (2017) 415−422. Available from: https://doi.org/10.1002/fuce.201600190.

[25] P. Costamagna, C. Sanna, A. Campodonico, E.M. Sala, R. Sažinas, P. Holtappels, Electrochemical impedance spectroscopy of electrospun $La_{0.6}Sr_{0.4}Co_{0.2}Fe_{0.8}O_{3-\delta}$ nanorod cathodes for intermediate temperature—solid oxide fuel cells, Fuel Cells 19 (2019) 472−483. Available from: https://doi.org/10.1002/fuce.201800205.

[26] E. Zhao, Z. Jia, L. Zhao, Y. Xiong, C. Sun, M.E. Brito, One dimensional $La_{0.8}Sr_{0.2}Co_{0.2}Fe_{0.8}O_{3-\delta}/Ce_{0.8}Gd_{0.2}O_{1.9}$ nanocomposite cathodes for intermediate temperature solid oxide fuel cells, J. Power Sources 219 (2012) 133−139. Available from: https://doi.org/10.1016/j.jpowsour.2012.07.013.

[27] M. Cannarozzo, A. Del Borghi, P. Costamagna, Simulation of mass transport in SOFC composite electrodes, J. Appl. Electrochem. 38 (2008) 1011−1018. Available from: https://doi.org/10.1007/s10800-008-9527-1.

[28] M. Cannarozzo, S. Grosso, G. Agnew, A. Del Borghi, P. Costamagna, Effects of mass transport on the performance of solid oxide fuel cells composite electrodes, J. Fuel Cell Sci. Technol. 4 (2007) 99−106. Available from: https://doi.org/10.1115/1.2393311.

[29] J. Nielsen, T. Jacobsen, M. Wandel, Impedance of porous IT-SOFC LSCF:CGO composite cathodes, Electrochim. Acta 56 (2011) 7963−7974. Available from: https://doi.org/10.1016/j.electacta.2011.05.042.

[30] S. Wang, Y. Jiang, Y. Zhang, J. Yan, W. Li, Promoting effect of YSZ on the electrochemical performance of YSZ + LSM composite electrodes, Solid. State Ion. 113−115 (1998) 291−303. Available from: https://doi.org/10.1016/s0167-2738(98)00379-8.

[31] S. Wang, X. Lu, M. Liu, Electrocatalytic properties of $La_{0.9}Sr_{0.1}MnO_3$-based electrodes for oxygen reduction, J. Solid. State Electrochem. 6 (2002) 384−390. Available from: https://doi.org/10.1007/s10008-001-0250-7.

[32] J. Parbey, Q. Wang, J. Lei, M. Espinoza-Andaluz, F. Hao, Y. Xiang, et al., High-performance solid oxide fuel cells with fiber-based cathodes for low-temperature operation, Int. J. Hydrog. Energy 45 (2020) 6949−6957. Available from: https://doi.org/10.1016/j.ijhydene.2019.12.125.

[33] M. Zhi, N. Mariani, R. Gemmen, K. Gerdes, N. Wu, Nanofiber scaffold for cathode of solid oxide fuel cell, Energy Env. Sci. 4 (2011) 417−420. Available from: https://doi.org/10.1039/c0ee00358a.

[34] P. Costamagna, E.M. Sala, W. Zhang, M. Lund Traulsen, P. Holtappels, Electrochemical impedance spectroscopy of $La_{0.6}Sr_{0.4}Co_{0.2}Fe_{0.8}O_{3-\delta}$ nanofiber cathodes for intermediate temperature-solid oxide fuel cell applications: A case study for the 'depressed' or 'fractal' Gerischer element, Electrochim. Acta 319 (2019) 657−671. Available from: https://doi.org/10.1016/j.electacta.2019.06.068.

[35] C. Ahamer, A.K. Opitz, G.M. Rupp, J. Fleig, Revisiting the temperature dependent ionic conductivity of yttria stabilized zirconia (YSZ), J. Electrochem. Soc. 164 (2017) F790−F803. Available from: https://doi.org/10.1149/2.0641707jes.

[36] E. Zhao, X. Liu, L. Liu, H. Huo, Y. Xiong, Effect of $La_{0.8}Sr_{0.2}Co_{0.2}Fe_{0.8}O_{3-\delta}$ morphology on the performance of composite cathodes, Prog. Nat. Sci. Mater. Int. 24 (2014) 24−30. Available from: https://doi.org/10.1016/j.pnsc.2014.01.008.

[37] N.T. Hieu, J. Park, B. Tae, Synthesis and characterization of nanofiber-structured $Ba_{0.5}Sr_{0.5}Co_{0.8}Fe_{0.2}O_{3-\delta}$ perovskite oxide used as a cathode material for low-temperature solid oxide fuel cells, Mater. Sci. Eng. B-Solid State Mater Adv. Technol. 177 (2012) 205−209. Available from: https://doi.org/10.1016/j.mseb.2011.12.018.

[38] L. Fan, L. Liu, Y. Wang, H. Huo, Y. Xiong, One-dimensional $Sr_{0.7}Y_{0.3}CoO_{2.65-\delta}$ nanofibers as cathode material for IT-SOFCs, Int. J. Hydrog. Energy 39 (2014) 14428−14433. Available from: https://doi.org/10.1016/j.ijhydene.2014.02.030.

[39] M. Ahn, J. Lee, W. Lee, Nanofiber-based composite cathodes for intermediate temperature solid oxide fuel cells, J. Power Sources 353 (2017) 176−182. Available from: https://doi.org/10.1016/j.jpowsour.2017.03.151.

[40] L.P. Sun, Q. Li, H. Zhao, J.H. Hao, L.H. Huo, G. Pang, et al., Electrochemical performance of $Nd_{1.93}Sr_{0.07}CuO_4$ nanofiber as cathode material for SOFC, Int. J. Hydrogen. Energy 37 (2012) 11955−11962. Available from: https://doi.org/10.1016/j.ijhydene.2012.05.112.

[41] J. Chen, F. Liang, B. Chi, J. Pu, S.P. Jiang, L. Jian, Palladium and ceria infiltrated $La_{0.8}Sr_{0.2}Co_{0.5}Fe_{0.5}O_{3-\delta}$ cathodes of solid oxide fuel cells, J. Power Sources 194 (2009) 275−280. Available from: https://doi.org/10.1016/j.jpowsour.2009.04.041.

[42] Y. Chen, Y. Bu, Y. Zhang, R. Yan, D. Ding, B. Zhao, et al., A highly efficient and robust nanofiber cathode for solid oxide fuel cells, Adv. Energy Mater. 7 (2017) 1−7. Available from: https://doi.org/10.1002/aenm.201601890.

[43] F. Qiang, K.N. Sun, N.Q. Zhang, X.D. Zhu, S.R. Le, D.R. Zhou, Characterization of electrical properties of GDC doped A-site deficient LSCF based composite cathode using impedance spectroscopy, J. Power Sources 168 (2007) 338−345. Available from: https://doi.org/10.1016/j.jpowsour.2007.03.040.

[44] C. Jin, Y. Mao, D.W. Rooney, N. Zhang, K. Sun, Preparation and characterization of $Pr_{0.6}Sr_{0.4}FeO_{3-\delta}$-$Ce_{0.9}Pr_{0.1}O_{2-\delta}$ nanofiber structured composite cathode for IT-SOFCs, Ceram. Int. 42 (2016) 9311−9314. Available from: https://doi.org/10.1016/j.ceramint.2016.02.121.

[45] L. Fan, Y. Xiong, L. Liu, Y. Wang, H. Kishimoto, K. Yamaji, et al., Performance of $Gd_{0.2}Ce_{0.8}O_{1.9}$ infiltrated $La_{0.2}Sr_{0.8}TiO_3$ nanofiber scaffolds as anodes for solid oxide fuel cells, J. Power Sources 265 (2014) 125−131. Available from: https://doi.org/10.1016/j.jpowsour.2014.04.109.

[46] M.E. Orazem, B. Tribollet, Electrochemical Impedance Spectroscopy, 2nd edition, John Wiley & Sons, Hoboken, New Jersey, 2017.

[47] M. Sluyters-Rehbach, S.H. Sluyters, Electroanalytical Chemistry, A Series of Advances, Marcel Dekker, New York, 1970.

[48] J.P. Diard, C. Montella, Non-intuitive features of equivalent circuits for analysis of EIS data. The example of EE reaction, J. Electroanal. Chem. 735 (2014) 99−110. Available from: https://doi.org/10.1016/j.jelechem.2014.08.029.

[49] B.A. Boukamp, A nonlinear least squares fit procedure for analysis of immittance data of electrochemical systems, Solid. State Ion. 20 (1986) 31−44. Available from: https://doi.org/10.1016/0167-2738(86)90031-7.

[50] M. Chatti, J.L. Gardiner, M. Fournier, B. Johannessen, T. Williams, T.R. Gengenbach, et al., Intrinsically stable situ generated electrocatalyst long-term oxidation of acidic water at up to 80°C, Nat. Cat. 2 (2019) 457–465. Available from: https://doi.org/10.1038/s41929-019-0277-8.

[51] E. Barsoukov, J.R. Macdonald, Second ed, Impedance Spectroscopy: Theory, Experiment, and Applications, 177, John Wiley & Sons, Hoboken, New Jersey, 2005.

[52] S. Koch, K.V. Hansen, B.S. Johansen, Elchemea (2005). Available from: https://www.elchemea.com/. (accessed 12.08.21).

[53] S.B. Adler, J.A. Lane, B.C.H. Steele, Electrode kinetics of porous mixed-conducting oxygen electrodes, J. Electrochem. Soc. 143 (1996) 3554–3564. Available from: https://doi.org/10.1149/1.1837252.

[54] S.B. Adler, Mechanism and kinetics of oxygen reduction on porous $La_{1-x}Sr_xCoO_{3-\delta}$ electrodes, Solid. State Ion. 111 (1998) 125–134. Available from: https://doi.org/10.1016/S0167-2738(98)00179-9.

[55] P. Hjalmarsson, M. Søgaard, M. Mogensen, Oxygen transport properties of dense and porous $(La_{0.8}Sr_{0.2})_{0.99}Co_{0.8}Ni_{0.2}O_{3-\delta}$, Solid. State Ion. 180 (2009) 1290–1297. Available from: https://doi.org/10.1016/j.ssi.2009.07.012.

[56] T. Jacobsen, P.V. Hendriksen, S. Koch, Diffusion and conversion impedance in solid oxide fuel cells, Electrochim. Acta 53 (2008) 7500–7508. Available from: https://doi.org/10.1016/j.electacta.2008.02.019.

[57] J. Nielsen, J. Hjelm, Impedance of SOFC electrodes: a review and a comprehensive case study on the impedance of LSM:YSZ cathodes, Electrochim. Acta 115 (2014) 31–45. Available from: https://doi.org/10.1016/j.electacta.2013.10.053.

Synthesis of one-dimensional metal oxide–based crystals as energy storage materials

14

Andrea La Monaca, Daniele Campanella and Andrea Paolella
Centre d'Excellence en Électrification Des Transports et Stockage d'Énergie, Hydro-Québec, Varennes, QC, Canada

14.1 Introduction

Since their first appearance in 1991, lithium ion (Li-ion) batteries have steadily become the most employed energy storage systems in the fields of portable electronics and electric vehicles, owing to their high specific energy and low fabrication cost. Li-ion batteries are composed of an anode, a cathode, a separator, and an electrolyte (in case of all-solid batteries the separator coincides with the electrolyte). During battery charge, an electrochemical reaction based on electron transfer occurs (through the external circuit), and ions transfer (through the electrolyte) from the positive to the negative electrode, while in the discharge process, electrons spontaneously flow in the opposite direction (from negative to positive). Other metal ion batteries based on same principles are possible, including sodium ion, potassium ion, aluminum ion, and zinc ion batteries. One-dimensional materials are becoming essential components of batteries [1] owing to the combination of mechanical strength improvement and high porosity (good electrode-electrolyte contact). Because of the conduction of the oriented metal ions along a micron-long dimension, nanofibers (NFs) and nanowires (NWs) have been considered as potential materials for solid-state electrolytes (mainly as filler in polymer-based electrolytes), although the conductive mechanism is still under investigation [2].

These materials can be obtained by several scalable techniques, such as spinning and wet chemical methods (Fig. 14.1). When spinning methods are used (e.g., electrospinning, solution blow spinning, centrifugal solution spinning), it is possible to obtain NFs composed of interconnected metal oxides nanocrystals. The fibers may be hollow, and their morphology could present a high surface area, increasing their reactivity toward alkali metal intercalation and deintercalation. For example, in a standard electrospinning process, metal precursors are dissolved in a solvent with a polymer, and the polymeric solution is subsequently spun through the application of high voltage on the syringe needle while grounding a collector. Thus a droplet that is formed on the tip of the needle is attracted to the collector, losing solvent by evaporation and leaving a solid nonwoven fiber residual. Typically, a sintering step

---

Metal Oxide-Based Nanofibers and Their Applications. DOI: https://doi.org/10.1016/B978-0-12-820629-4.00014-X
© 2022 Elsevier Inc. All rights reserved.

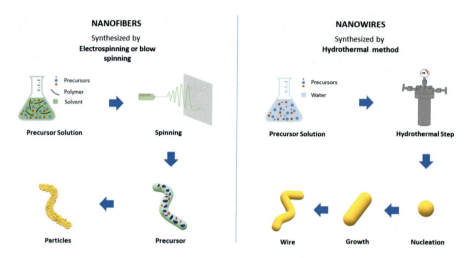

Figure 14.1 Schematics of nanofiber and nanowire synthesis.

is required to burn the polymer matrix and to convert metal precursors into more stable crystalline oxides.

On the other hand, by using a wet synthetic method (hydrothermal, sol–gel), elongated crystals with nanometer-size diameters are obtained (NWs), usually with inferior surface roughness and a higher degree of brittleness. In a standard hydrothermal method, metal precursors are dissolved in water, and the solution is then heated in a Teflon liner in a temperature range of 150°C–250°C for a few hours. After purification of the product a further annealing step is usually required to improve the crystallinity of the final particles. The chapter shows methods for synthesis of metal oxide NFs and NWs, in alphabetical order.

14.2 Aluminum oxide

Beta-alumina NFs may be used as filler in polymer electrolytes for sodium-metal batteries [3]. The fibers are synthesized by electrospinning a solution composed of a mixture of Al precursors (aluminum nitrate and aluminum isopropoxide), lithium nitrate, nitric acid, and polyvinyl pyrrolidone (PVP) in water. After sintering at 600°C, the alumina membrane is dip-coated by polyvinyl difluoride polymer and then immersed in 1 M NaClO$_4$ in ethylene carbonate and diethyl carbonate with 5% fluoroethylene carbonate. The starting solution was composed of aluminum nitrate and polyacrilonitrile in dimethylformamide. After a heat treatment process at 300°C under air atmosphere, electrospun Al$_2$O$_3$ NWs were used as interlayer in Al-air batteries for devices with a low current application, such as portable electronics and biosensors. Another interesting method to produce Al$_2$O$_3$ NFs was proposed by Tian et al. [4], in which cotton was sputtered with Al$_2$O$_3$ and NFs were obtained after a calcination step at 700°C in air for 4 h (Fig. 14.2A–H). The final

Synthesis of one-dimensional metal oxide–based crystals as energy storage materials 335

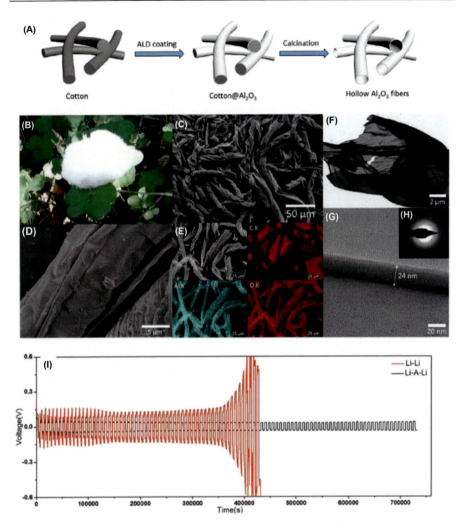

Figure 14.2 Fabrication and characterization of hollow Al_2O_3 fibers. (A) Illustration of the synthesis of the hollow Al_2O_3 fibers. (B) Digital photograph. (C, D) Scanning electron microscope images. (E) Energy-dispersive X-ray spectroscopy mapping. (F, G) Transmission electron microscope images. (H) The selected-area electron diffraction pattern of the hollow Al_2O_3 fibers. (I) Electrochemical performance of the symmetric cells. Li–A–Li cells (black) and Li–Li cells (red). The current density was 1 mA cm^{-2} with a stripping/plating capacity of 1 mAh cm^{-2}.
Source: R. Tian, X. Feng, H. Duan, P. Zhang, H. Li, H. Liu, Low-weight 3D Al2O3 network as an artificial layer to stabilize lithium deposition. ChemSusChem. 2018;2:3243–3252.

three-dimensional (3D) Al_2O_3 NF layer was placed in contact with lithium metal and cycled in symmetrical cells as shown in Fig. 14.2I. Li−Li protected with 3D Al_2O_3 NFs showed much higher stability than cells composed of bare Li−Li. On the basis of a similar idea, Zuo et al. prepared electrospun Al_2O_3 NFs to protect the aluminum metal anode from corrosion in an Al-air battery [5].

14.3 Copper oxide

Copper oxide (CuO) NFs [6] are synthesized on copper foil surface by a simple chemical route. After treatment with NaOH and $K_2S_2O_8$ (see Fig. 14.3A−D) a network of CuO is formed. The copper plate having a NF network on its surface was tested as anode material in a Li-ion battery; the material released a capacity of 300 mAh g^{-1} at a C rate of 0.3 C. CuO NFs were made by an electrospinning process by Sahay et al. [7]. The authors used copper acetate hydrate as the precursor by dissolving it in a 10% aqueous polyvinyl alcohol (PVA) solution. After sintering at 800°C, the final fibers were mainly composed of nano-sized CuO particles as shown in Fig. 14.3E−H; thus the material was tested as the anode in Li-ion batteries delivering 400 mAh g^{-1} at a current of 100 mA g^{-1} for 100 cycles (Fig. 14.3I and L).

An alternative way to synthesize CuO NWs for Li-ion batteries involves the use of a very simple method [8]: $CuSO_4 \cdot 5H_2O$ was dissolved in deionized water under constant stirring, and then 1 M KOH and ammonia solutions were dropped into the previous mixture. CuO was formed after slow conversion of blue $Cu(OH)_2$ NWs into dark solution. The CuO NWs exhibited excellent cycling performances, with a stable capacity of 650 mAh g^{-1} over 100 cycles at 0.5 C. The possibility of growing Co_3O_4/CuO nanowire heterostructures on Ni foam has also been shown; first, a Cu layer was deposited by e-beam evaporation followed by an annealing step at 400°C under air atmosphere. CuO NWs were later decorated by Co_3O_4 to boost the electrochemical performance of the material as a negative electrode in a Li-ion battery [9]. Co_3O_4/CuO heterostructures revealed a high reversible capacity of 1191 mAh g^{-1} after 200 cycles at a current density of 200 mA g^{-1}, with 90.9% capacity retention.

14.4 Iron oxide

An interesting anode material for Li-ion batteries is represented by hollow fibers of α-Fe_2O_3. Chaudhari el al. [10] performed the synthesis via an electrospinning technique by using a solution based on iron acetylacetonate and PVP. Fig. 14.4A−C show that hollow fibers annealed at 500°C were composed of interconnected Fe_2O_3 nanocrystals. The discharge mechanism suggested by cyclic voltammetry tests revealed the initial insertion of lithium in the structure, with the formation of hexagonal α-$Li_xFe_2O_3$. The next step shows the irreversible conversion of the structure into cubic $Li_2Fe_2O_3$ (disappearance of reduction peak II in Fig. 14.4D). As shown in Fig. 14.4E, the α-Fe_2O_3 hollow fibers electrode demonstrates a stable capacity of 1293 mAh g^{-1}

Synthesis of one-dimensional metal oxide−based crystals as energy storage materials 337

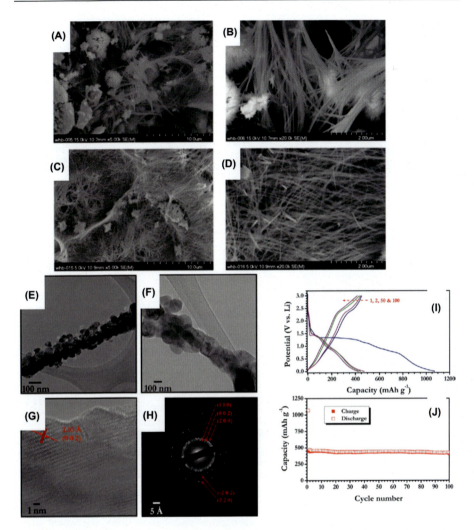

Figure 14.3 Scanning electron microscopy images of the copper foils immersion in 0.25 M NaOH containing 17 mM safety date sheet and 9 mM $K_2S_2O_8$ for (A, B) 12 h and (C, D) 24 h. B and C are images at higher magnifications. Transmission electron microscope pictures of (E, F) CuO nanofibers sintered at 800°C for 1 h with different magnifications. High resolution transmission electro microscope photograph of (G) CuO nanofibers and (H) selected-area electron diffraction pattern. (I) Typical galvanostatic charge-discharge curves of Li/CuO nanofiber half-cell cycled between 0.005 and 3 V versus Li at constant current density of 100 mA g^{-1} at room temperature. (J) Plot of capacity versus cycle number
Source: H. Wang, Q. Pan, J. Zhao, G. Yin, P. Zuo, Fabrication of CuO film with network-like architectures through solution-immersion and their application in lithium ion batteries. J. Power Sources. 2007;167:206−211; R. Sahay, P.S. Kumar, V. Aravindan, J. Sundaramurthy, W.C. Ling, S.G. Mhaisalkar, et al. High aspect ratio electrospun CuO nanofibers as anode material for lithium-ion batteries with superior cycleability. J. Phys. Chem. C. 2012;116:18087−18092.

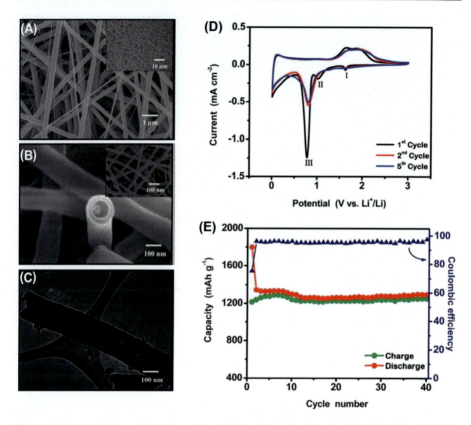

Figure 14.4 (A) Field emission scanning electron microscopy images of Fe(acac)$_3$-polyvinyl pyrrolidone composite fibers at higher magnification (lower magnification in the inset). (B) Field emission scanning electron microscope images of α-Fe$_2$O$_3$ hollow fibers at higher magnification (lower magnification in the inset). (C) Transmission electron microscope image of α-Fe$_2$O$_3$ hollow fibers. (D) Cyclic voltammograms of the electrode made from α-Fe$_2$O$_3$ hollow fibers (scan rate: 0.1 mV s^{-1} over the voltage range of 0.005–3.0 V). (E) Cycling performance of the electrode made from α-Fe$_2$O$_3$ hollow fibers at a current density of 60 mA g^{-1} (0.06 C).
Source: S. Chaudhari, M. Srinivasan, 1D hollow a-Fe$_2$O$_3$ electrospun nanofibers as high performance anode. J. Mater. Chem. 2012;22:23049–23056.

after 40 cycles. Jayraman et al. [11] proposed a full battery composed by electrospun α-Fe$_2$O$_3$ NWs and a LiMn$_2$O$_4$ cathode. The cell exhibited a capacity of 119 mAh g^{-1} at a C rate of 1 C and a capacity retention of about 70% after 4000 cycles.

14.5 Manganese oxide

Interconnected manganese oxide (MnO$_2$) NWs with a diameter of 10–12 nm have been synthesized by Wu et al. [12] via electrodeposition from an aqueous solution of manganese acetate (see Fig. 14.5A and B). After annealing in a temperature

Synthesis of one-dimensional metal oxide−based crystals as energy storage materials

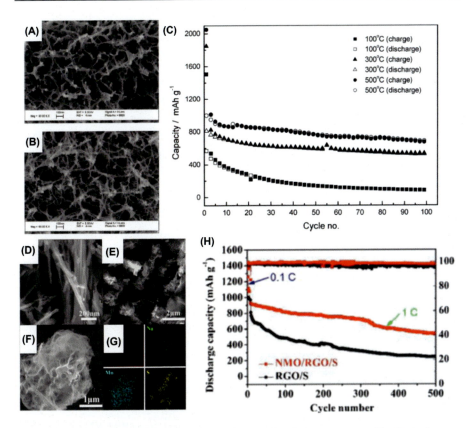

Figure 14.5 Scanning electron microscope micrographs of a manganese oxide electrode after different annealing temperatures: (A) 100°C and (B) 300°C. (C) The relationship between specific capacity and cycle number of a manganese oxide electrode treated at 100°C, 300°C, and 500°C. The morphology and structure of the Na$_4$Mn$_9$O$_{18}$ (NMO) nanorods and Na$_4$M$_9$O$_{18}$/reduced graphene oxide (RGO)/sulfur (S) microspheres. (D) Scanning electron microscope image of Na$_4$Mn$_9$O$_{18}$. (E-F) Scanning electron microscope image and (G) Energy-dispersive X-ray spectroscopy elemental mapping of Na$_4$Mn$_9$O$_{18}$/reduced graphene oxide/sulfur. (H) Cycling performances of Li-S battery with RGO/S and NMO/RGO/S composites.
Source: M. Wu, P.J. Chiang, J, Lee, J. Lin, Synthesis of manganese oxide electrodes with interconnected nanowire structure as an anode material for rechargeable lithium ion batteries. J. Phys. Chem. B. 2005;109:23279−23284; X. Wang, Z. Sun, Y. Zhao, J. Li, Na$_4$Mn$_9$O$_{18}$ nanowires wrapped by reduced graphene oxide as efficient sulfur host material for lithium/sulfur batteries. J. Solid. State Electrochem. 2020;24:111−119.

range between 100°C and 500°C, the material was tested as anode in the voltage range of 0−3 V; the electrode that was annealed at 300°C manifested the highest discharge capacity (about 800 mAh g^{-1}). By contrast, the sample that was annealed at lower temperatures presented a strong capacity that faded in the first cycles. The high presence of residual water content affected the global cell resistance, with detrimental effects on the cycle-life performance (see Fig. 14.5C).

Wang et al. presented a simple hydrothermal route for obtaining $Na_4Mn_9O_{18}$ (NMO) NWs [13]: $KMnO_4$, NaOH, and $MnSO_4$ were dissolved in deionized water and heated at 180°C for 20 h. To limit the dissolution of polysulfides, a NMO/reduced graphene oxide (RGO)/sulfur composite was tested against lithium metal; the electrolyte that was used was 1.0 M lithium bis(fluorosulfonyl)imide (LiTFSI) in DOL/DME (1:1 by volume) with the addition of 1% $LiNO_3$. The NMO/RGO/S electrode delivered augmented capacity compared with the cathode without NMO (RGO/S), especially at high current density. The immobilization affinity of NMO toward polysulfides Li_2S_x strongly influences the cycling performances of Lithium/sulfur (Li/S) batteries. In a long-term cycling test at 1 C, NMO/RGO/S revealed an initial capacity of 952.8 mAh g^{-1} and a reversible specific capacity of 520 mAh g^{-1} after 500 cycles, while the RGO/S electrode showed a globally lower capacity and a remaining capacity of 240 mAh g^{-1} (see Fig. 14.5D−H).

Centrifugal spinning technology allowed the preparation of layered transition metal $Na_{0.8}Li_{0.4}Ni_{0.15}Mn_{0.55}Co_{0.1}O_2$ [14]. To fabricate the precursor fibers, different solutions of polyacrylonitrile into N,N-dimethylformamide (DMF) and stoichiometric amounts of metal nitrates were prepared. The precursor solutions were loaded into the spinneret through the top opening, and then high-speed rotation (4000 rpm) was applied. The final Li-substituted, Na-layered fiber architecture was composed of interconnected nanoparticles with a large electrode-electrolyte contact area and improved wettability. Electrochemical performances in the Na-ion cell revealed for the cathode fibers the highest capacities of 138, 113, and 94 mAh g^{-1} at current densities of 15, 75, and 300 mA g^{-1}, respectively.

Electrospun Mn_3O_4 NFs were successfully proposed by Fan et al. in 2007 [15]. Manganese acetate tetrahydrate was added until saturation in a solution of poly (methyl methacrylate) (PMMA) in DMF:chloroform 1:1 v/v. After the electrospinning process and subsequent annealing at 300°C to remove the organic residuum, the final diameters of the NFs ranged from under 100 to over 300 nm, depending on the concentration of PMMA in the initial solution. The NFs showed a retained capacity of 400 mAh g^{-1} after 50 cycles at a current density of 0.5 mA cm^{-2}.

Wang et al. [16] reported $K_{0.7}Fe_{0.5}Mn_{0.5}O_2$ NWs as a promising material for K-ion battery. A 3D network composed of nanocrystals encapsulated by soft carbon layers provided a stable crystal structure during the K-ion insertion and extraction processes. The remarkable potassium storage performances at high current rate (a capacity of 119 mAh g^{-1} at a current density of 20 mA g^{-1}) of the interconnected $K_{0.7}Fe_{0.5}Mn_{0.5}O_2$ NWs is attributed to their peculiar crystal framework and the abundant presence of pores.

14.6 Nickel oxide

Aravindan et al. [17] described the preparation of nickel oxide (NiO) NFs by electrospinning. In the synthesis a homogeneous solution of nickel acetate as Ni precursor was prepared in DMF, with the addition of polyvinyl acetate and acetic acid.

After sintering at 800°C for 1 h, the NFs were used as the anode in a Li-ion battery, delivering good performance at high current densities; in particular, the cell presented a good reversible capacity around 409 mAh g^{-1} at 2 A g^{-1}, with a Coulombic efficiency above 98%. NiO-zinc oxide (ZnO) electrospun fibers [18] are a promising anode material. The synthesis involved the mixture of nickel nitrate and zinc nitrate in PVP followed by an annealing step at 500°C in air. The good electrochemical performance relates strictly to the one-dimensional structure, which is characterized by assembled primary nanocrystals and abundant nanopores. This offers several advantages, including a large electrode-electrolyte area ratio, a short Li-ion diffusion path, and an accommodation strain induced by the volume change during the processes of electrochemical reactions. Copper-doped electrospun NiO NFs may be prepared on indium tin oxide (ITO) glass support and used in electrochromic devices [19]. Compared to the undoped NiO NFs, $Ni_{0.97}Cu_{0.03}O$ NFs manifested excellent electrochromic performance, including short response times (coloring in 1.6 s and bleaching in 0.9 s), a large transmittance modulation (73% at 550 nm), high coloration efficiency (77.9 cm^2 C^{-1}), and good cycling durability (80% after 2000 cycles). Fibrous $NiCo_2O_4$ has been prepared by Zhao et al. [20] by a simple hydrothermal method: Carbon tubes bundles (CTBs) were mixed with cobalt nitrate, nickel nitrate, and urea into deionized water; the resulting blend was put in a Teflon-lined autoclave at 120°C for 6 h. Subsequently, the synthesized material underwent annealing at 350°C for 1 h and carbonization at 600°C for 1 h in argon. The sodium storage properties of $NiCo_2O_4$/CTBs NWs were explored; the material exhibited a high capacity of 236 mAh g^{-1} at the higher current density (5 A g^{-1}). Lowering the current density to 0.1 A g^{-1} permitted recovery of 98.1% of the initial capacity (464 mAh g^{-1}). A sodium hybrid capacitor was fabricated with $NiCo_2O_4$/CTBs and commercial activated carbon (AC) as both negative and positive electrodes ($NiCo_2O_4$/CTBs/AC). The system was able to deliver around 100% of capacity retention for 5000 cycles. Moreover, electrospun $NiCo_2O_4$ NFs were used by Liu et al. [21] with sulfur and carbon nanotubes to create a stable composite cathode in a Li-S battery. One-dimensional $NiCo_2O_4$ NFs could provide good electronic conductivity, an ion diffusion path during the charge-discharge processes of the sulfur cathode, and enough active sites to control the polysulfide dissolution. It was possible to synthesize NiO, CeO_2, and NiO-CeO_2 fibers by the solution blow spinning method [22], in which nickel nitrate and cerium nitrate were dissolved in ethanol/DMF, (1:1 v/v). Afterwards, the viscosity of each respective solution was adjusted with the addition of PVP. The solutions were subsequently blown with a gas pressure of 0.41 MPa and an injection rate of 3 mL h^{-1} onto a preheated aluminum foil at a distance of 60 cm. The final precursor fibers were calcined at 800°C for 1 h in air. Fig. 14.6 shows the morphology and energy-dispersive X-ray spectroscopy (EDS) mapping of the final product. In K-ion batteries, NiO hollow fibers exhibited the best specific capacity, followed by CeO_2 and NiO-CeO_2 electrodes because of the low electrical conductivity of CeO_2. Huang et al. [23] explored the possibility of synthesizing $Na_3Ni_2SbO_6$ and $Na_3Ni_2BiO_6$ NFs by mixing metal precursors with polyacrilonitrile; the exerted strategy contemplated the insertion of diisopropyl azodiformate as an additional

342   Metal Oxide-Based Nanofibers and Their Applications

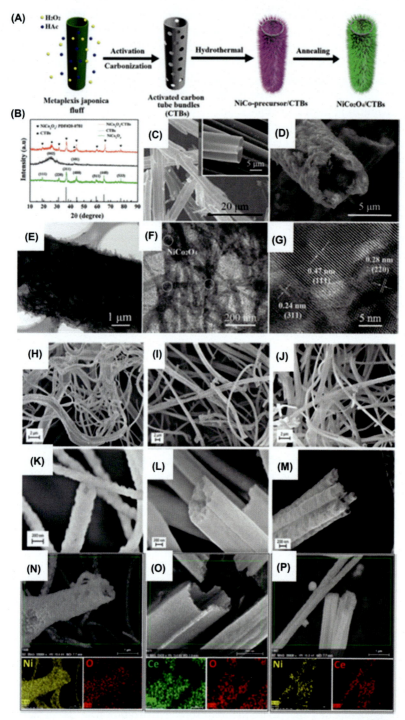

(*Continued*)

Synthesis of one-dimensional metal oxide—based crystals as energy storage materials 343

foam agent to control the microstructure and morphology of the fibers. The NFs with the lowest crystallinity and highest porous structure showed the best cycling performances in Na-ion batteries because of the high number of voids present in the modified fibers. At the same time, though, the presence of large macropores had the effect of diminishing the active area and increasing useless space within the fibers, leading to an eventual deterioration of ion transport properties and electro-chemical performances.

14.7 Silicon oxide and silicates

Silicon@silicon oxide core-shell NWs were synthesized in Ar/H_2 atmosphere at 1600°C with titanium powder as reducing agent [24]. Structural analysis revealed a catalytic effect of the spherical silicon nanoparticles with an oxide shell on nanowire growth along the crystallographic (111) direction. The material showed good perfor-mances in Li-ion batteries, with a reversible capacity of about 1640 mAh g − 1 after 100 cycles at a current density of $1 A g^{-1}$. Jia et al. [25] proposed a battery separator composed of $Ca_6Si_6O_{17}(OH)_2$ xonotlite NWs. The material was prepared by dropping $Na_2SiO_3 \cdot 9H_2O$ aqueous solution into $CaCl_2 \cdot 2H_2O$ solution under continuous stirring. The resulting solution was then added to a 100-mL Teflon-lined autoclave and heated at 200°C for 12 h. Xonolite NWs were tested as a separator in a $LiFePO_4$/lithium metal battery and displayed enhanced thermal and electrochemical stability with respect to a commercial polypropylene separator. Palygorskite $[(Mg,Al)_2Si_4O_{10}(OH)]$ NWs are commercial nanomaterials with a one-dimensional fiber structure (~ 50 nm in diameter and $\sim 1\,\mu m$ in length). Yao et al. [26] used them as ceramic fillers in polyvinylidene fuuoride (PVDF):$LiClO_4$ polymer electrolyte, obtaining a high ionic conductivity at room temperature ($1.2 \times 10^{-4}\,S\,cm^{-1}$). The material displayed a high electrochemical stability up to 4 V, which rendered it compatible with high-voltage cathodes. Electrochemical tests in a $Li(Ni_{1/3}Mn_{1/3}Co1_{/3})O_2$ (NMC111) cathode//

Figure 14.6 (A) Preparation procedure of the $NiCo_2O_4$/carbon tubular bundles composite. (B) X-ray diffraction spectra of the samples. (C) Scanning electron microscope image of the carbon tubular bundles.The inset presents the scanning electro microscope image of the carbon tubular bandles with a tubular cross section. (D) Scanning electron microscope image of the $NiCo_2O_4$/carbon tubular bandles composite. (E—G) Transmission electron microscope images of the $NiCo_2O_4$/CTBs composite. Field emission scanning electron microscope field emission scanning electron microscope images and energy-dispersive X-ray spectroscopy mapping of the (H, M, P) NiO, (I, Q) CeO_2, and (L, O, R) $NiO\text{-}CeO_2$ hollow fibers.
Source: J. Zhao, C. Zhou, Y. Li, K. Cheng, K. Zhu, K. Ye, et al. Nickel cobalt oxide nanowires-modified hollow carbon tubular bundles for high-performance sodium-ion hybrid capacitors. Int. J. Energy Res. [Internet]. 2020;(October 2019):3883−3892. Available from: https://doi.org/10.1002/er.5185; V.D. Silva, L.S. Ferreira, A.J.M. Araújo, T.A. Sim, P.F. Grilo, M. Tahir, et al. Ni and Ce oxide-based hollow fi bers as battery-like electrodes. J. Alloy. Compd. 2020;830:154633.

lithium metal battery revealed good electrochemical performance for the modified solid electrolyte. The discharge capacity increased from 117.6 to 121.4 mAh g^{-1} in the first five cycles, with a retained capacity of 118.1 mAh g^{-1} after 200 cycles at 0.3 C.

14.8 Tin oxide

The synthesis of porous SnO$_2$ nanotubes has been described by Li et al. [27]. After the preparation of a solution of PVP and SnCl$_4$ in ethanol, SnO$_2$ fibers, shown in Fig. 14.7A−D, were produced by electrospinning followed by calcination at 500°C under air atmosphere. The resulting hollow fibers were tested as an anode in a Li-ion battery and showed a capacity of 300 mAh g − 1 at a C rate of 2.4 C. Interestingly, Bonino et al. [28] employed SnSO$_4$ as Sn precursor in the electrospinning synthesis of SnO$_2$ with polyacrylonitrile polymer. At 600°C, SnSO$_4$ underwent decomposition with formation of SnO$_2$ and SO$_2$, allowing the synthesis of NFs. The good mechanical stability and electrical conductivity of the NFs allowed the elimination of polymer binders and the direct use as a Li-ion battery anode at high current density such as 500 mA g^{-1}. In a plasma reactor it is possible to prepare SnO$_2$ NWs with high yield after reacting Sn wires with oxygen gas [29]. After a treatment with H$_2$ the surface of SnO$_2$ fibers was covered by metallic Sn to help the formation of Li−Sn alloy during battery discharge. The anode materials show high stability with a discharge capacity of about 800 mAh g^{-1}.

14.9 Titanium oxides and titanates

Anatase titanium oxide (TiO$_2$)−amorphous SiO$_2$ NFs have been synthesized by electrospinning followed by heat treatment at 500°C [30]. Sol−gel TiO$_2$ nanoparticles were mixed with NFs and tested in half-cell. The presence of TiO$_2$ NFs strongly enhanced the specific capacity at a fast charge-discharge rate of 1 A g^{-1} (80 mAh g^{-1} with TiO$_2$ NFs versus 20 mAh g^{-1} for the sample without NFs). Hoshide et al. [31] demonstrated that titania nanosheets and graphene oxide (GO) could be assembled into macroscopic fibers by a wet spinning technique as shown in Fig. 14.7E. Consequently, the authors were able to build a full flexible titania/LiMn$_2$O$_4$ fiber battery that could release a capacity of 100 mAh g^{-1} at a current of 0.0425 mA.

Sodium-doped titanium oxide NWs may be synthesized by a hydrothermal route as suggested by Huang et al. [32]. In their procedure, tetrabutyl titanate was added in a NaOH alkaline solution and then heated in a Teflon-lined autoclave to 190°C for 8−18 h. Crystals composed of hydrogenated sodium titanate were produced upon drying at 60°C for 24 h. The final sodium-containing TiO$_2$ NWs were eventually prepared by thermal treatment of hydrogenated sodium titanate precursors at 550°C for 10 h under argon atmosphere in a tube furnace. To control polysulfide dissolution in a Li-S battery, a polypropylene separator was modified with a

Synthesis of one-dimensional metal oxide–based crystals as energy storage materials 345

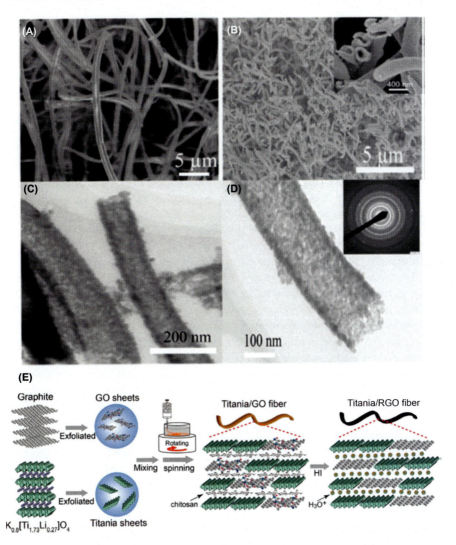

Figure 14.7 Scanning electron microscope images of (A) electrospun precursor fibers after vacuum drying and (B) SnO$_2$ nanotubes at low magnification and high magnification (inset). Transmission electron microscope images of SnO$_2$ nanotubes at (C) low magnification and (D) high magnification (inset is the corresponding selected-area electron diffraction pattern). (E) Schematic illustration of the fabrication process for the hybrid fiber of titania/reduced graphene oxide, which could be used for assembling fiber batteries. As both titania sheets and graphene oxide carry surface negative charges and have high stability in aqueous media, they could be homogeneously mixed. The wet spinning of the mixture produced a hybrid fiber with indiscriminate stacking of titania and graphene oxide sheets. The subsequent treatment in H$_I$ could reduce the oxidizing species on the graphene plane and simultaneously eliminate the polymer chitosan glue between the sheets.

(*Continued*)

346   Metal Oxide-Based Nanofibers and Their Applications

nanowire/PVP interlayer to face the sulfur cathode. The battery using a modified NWs/PP separator exhibited an initial discharge capacity of 813 mAh g^{-1} and a retained capacity of ~ 541 mAh g^{-1} after 500 cycles at a C rate of 1 C, with a capacity fading rate of about 0.067% for each cycle.

Oxygen may be partially replaced by nitrogen to form titanium oxynitride as reported by Lee et al. [33]. Hydrogen titanate ($H_2Ti_3O_7$) NWs 100 nm in diameter and several micrometers in length were treated in NH_3 atmosphere at 700°C, and after a first phase transition to an anatase TiO_2 intermediate, cubic Ti(O,N) NWs were obtained (see Fig. 14.8A). This result suggests that metal oxide NWs may be used as template for other materials. Ti(O,N) was applied as cathode material with a carbon anode and manifested a capacity of 500 C g^{-1} at a current of 1 A g^{-1} after 500 cycles with an interesting pseudocapacitive behavior. Park et al. [34] proposed a similar method of nitridation process (annealing in NH_3 atmosphere) for electrospun spinel lithium titanate $Li_4Ti_5O_{12}$ (LTO). The distribution of conducting TiN/TiO_xN_y on the surface of LTO NFs improved the cycling performances at high C rates (> 1 C).

Furthermore, titanium can be used to prepare $Li_{0.33}La_{0.557}TiO_3$ (LLTO) solid electrolyte via electrospinning [35]. At first, $Ti(OC_4H_9)_4$, $La(NO_3)_3 \cdot 6H_2O$, and $LiNO_3$ of various corresponding molar mass were dissolved in a transparent solution by using DMF as a solvent with acetic acid and PVP. Fig. 14.8H displays the morphology of electrospun fibers after a calcination step at 800°C for 2 h. LLTO NWs were used as ceramic filler to hinder the crystallization of polymer electrolyte based on polyethylene oxide (PEO):LiTFSI. In addition, LLTO ceramic was able to create an ion transport pathway at the interface between the filler and the polymer, with the effect of increasing the ionic conductivity. The discharge capacity of $LiFePO_4$ by using PEO:LiTFSI as an electrolyte was strongly improved after the addition of LLTO NWs as filler (Fig. 14.8N). In a later study, Zhu et al. [36] presented a composite electrolyte based on electrospun LLTO NWs, PEO, polypropylene carbonate, and LiTFSI. A solid-state $LiFePO_4$/Li battery employing the new composite as electrolyte exhibited a discharge capacity of 130 mAh g^{-1} at 60°C after 100 cycles at 0.5 C. NASICON $NaTi_2(PO_4)_3$ (NTP) NWs were synthesized by a hydrothermal method [37], in which titanium (IV) oxide acetylacetonate was slowly dissolved in isopropyl alcohol, and the addition of phosphoric acid then produced a yellow suspension. Subsequently, an aqueous solution of sodium acetate was added, and the obtained colloidal solution was transferred into a Teflon-lined stainless steel autoclave and treated at 180°C for 12 h. NTP NWs were used as cathode material in a Na-Mg hybrid battery and revealed a capacity of about 125 mAh g^{-1} for 50 cycles at a current of 1 C.

Source: L. Li, X. Yin, S. Liu, Y. Wang, L. Chen, T. Wang, Electrospun porous SnO2 nanotubes as high capacity anode materials for lithium ion batteries. Electrochem. commun. [Internet]. 2020;12(10):1383−1386. Available from: http://doi.org/10.1016/j. elecom.2010.07.026; T. Hoshide, Y. Zheng, J. Hou, Z. Wang, Q. Li, Z. Zhao, et al. Flexible lithium-ion fiber battery by the regular stacking of two-dimensional titanium oxide nanosheets hybridized with reduced graphene oxide. Nano Lett. 2017;17:3543−3549.

Synthesis of one-dimensional metal oxide—based crystals as energy storage materials 347

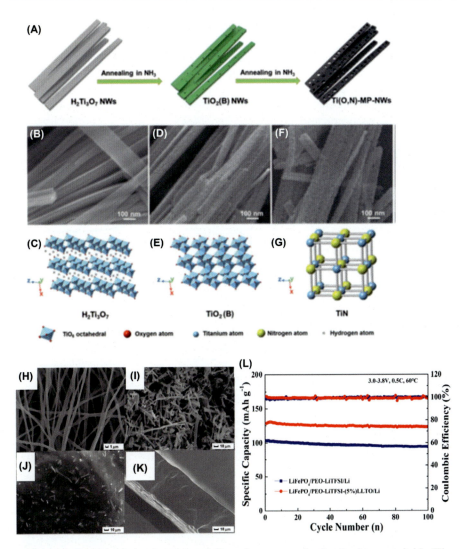

Figure 14.8 (A) Schematic of a proposed formation process for the titanium oxynitride (Ti(O,N)) mesoporous nanowires. (B) Scanning electron microscope image and (C) crystal structure of $H_2Ti_3O_7$ nanowires. (D) Scanning electron microscope image and (E) crystal structure of titanium oxide (B) mesoporous nanowires. (F) Scanning electron microscope image of Ti(O,N) mesoporous nanowires and (G) crystal structure of TiN. Scanning electron microscope images of (H) electrospun fibers, (I) $Li_{0.33}La_{0.557}TiO_3$ nanowires, and (L) the surface and (M) cross section of prepared composite electrolyte membrane. (N) The cycle performance of the solid-state cell at 0.5 C under 60°C

Source: H. Park, T. Song, T. Han, U. Paik, Electrospun Li4Ti5O12 nanofibers sheathed with conductive TiN/TiOxNy layer as an anode material for high power Li-ion batteries. J. Power Sources. 2013;244:726–730; L. Zhu, P. Zhu, Q. Fang, M. Jing, X. Shen, A novel solid PEO/LLTO-nanowires polymer composite electrolyte for solid-state lithium-ion battery. Electrochim. Acta [Internet]. 2018;292:718–726. Available from: https://doi.org/10.1016/j.electacta.2018.10.005.

14.10 Tungsten oxide and tungstates

Electrospun WO$_x$-C NFs were fabricated by using PVP and ammonium metatungstate hydrate as tungsten precursors in dimethylformamide [38]. The as-spun material was annealed at 700°C and was found to maintain good flexibility. High resolution transmission electron microscope (TEM) imaging showed that after carbonization of PVP the small WO$_x$ nanoparticles (<3 nm; see Fig. 14.9A−D) were uniformly dispersed

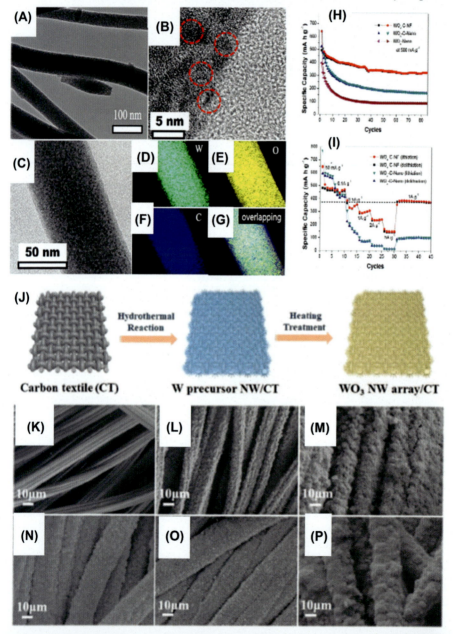

(Continued)

inside the fibers. Because of this aspect, the material was directly applied as a positive electrode against metallic lithium in a half-cell. As a reference, the authors compared the cycling performances of the electrode containing fibers with those of electrodes made with WO_3 nanoparticles dispersed in a slurry with PVDF binder on copper foil. The WO_3 NF—based battery had the best cycling performances, exhibiting stable capacities (~ 350 mAh g^{-1}) at a current of 500 mA g^{-1} (see Fig. 14.9E—F). Yu et al. [39] proposed a new 3D nanocomposite based on hydrothermal WO_3 NWs distributed on a graphene matrix. After the preparation of GO by Hammer's method, GO was dispersed in water and heated with Na_2WO_4 and NaCl in a Teflon liner at 180°C for 20 h. The resulting composite was tested in a Li-ion battery. The electrochemical performances revealed an initial discharge capacity of 923 mAh g^{-1} and a charge capacity of 622 mAh g^{-1}, with a Coulombic efficiency of 67.4% at the first cycle, demonstrating a superior lithium storage capacity. Then the reversible capacity value was nearly 100% after 100 cycles with an applied current of 100 mA g^{-1}, exhibiting an excellent capacity retention. On carbon textile, after hydrothermal treatment with Na_2WO_4, it was possible to grow an in situ needle-like WO_3 nanowire array as a photocatalyst for Li—O_2 batteries [40]. The hydrothermal reaction time played a key role in controlling the morphology of the WO_3; after 20 h of stirring, a higher amount of W precursor was deposited, blocking electrolyte permeability and electron conductivity (Fig. 14.9L—R). The WO_3 array was able to reduce the oxidation voltage of Li_2O_2 from 4.4 to 3.55 V in a photo-assisted Li—O_2 battery.

14.11 Vanadium oxide

Mai et al. [41] proposed ultralong hierarchical vanadium oxide NWs with diameters of 100—200 nm and lengths up to several millimeters that were obtained by using low-cost starting materials (polyvinyl alcohol and ammonium metavanadate) in an

Figure 14.9 (A) Transmission electron microscope image of WO_x-C nanofibers. (B) High resolution transmission electron microscope (high resolution-transmission electron microscope) image of WO_x-C nanofibers. (C) high resolution-transmission electron microscope image of the WO_x-C nanofibers and corresponding electron energy loss spectroscopy mapping images of the (D) W, (E) O, (F) C, and (G) overlapping of the W, O, and C contents. (H) Capacity retention profiles of WO_x-C nanofibers, WO_x-C nanofibers, and WO_x nanoelectrodes. The charge-discharge rate was fixed at 500 mA g^{-1}. (I) Capacity changes of WO_x-C nanofibers and WO_x-C nanoelectrodes under different current density conditions (0.05—5 A h^{-1}). (L) Schematic for the growth of WO_3 on carbon textiles. (M—R) Growth of WO_3 on carbon textiles at different hydrothermal times: (M) 0 h, (N) 4 h, (O) 8 h, (P) 12 h, (Q) 16 h, (R) 20 h.
Source: J. Lee, C. Jo, B. Park, W. Hwang, H.I. Lee, S. Yoon, et al. Simple fabrication of flexible electrode with high metal-oxide content: electrospun reduced tungsten oxide/carbon nanofibers for lithium ion battery applications. Nanoscale. 2012;1:1—3; Y. Feng, H. Xue, T. Wang, H. Gong, B. Gao, W. Xia, et al. Enhanced Li2O2 decomposition in rechargeable Li — O2 battery by incorporating WO3 nanowire array photocatalyst. ACS Sustian. Chem. Eng. 2019;7:5931—5939.

electrospinning process coupled with thermal annealing. The hierarchical NWs were constructed from attached vanadium oxide nanorods with diameters around 50 and lengths of 100 nm. Fig. 14.10 shows the final electrospun fibers as aggregation of vanadium oxide nanorods. This interesting nanostructure exhibited high performance as a cathode material for Li-ion batteries, with a high discharge capacity up to 400 mAh g^{-1}. Zhao et al. [42] investigated the influence on capacity retention

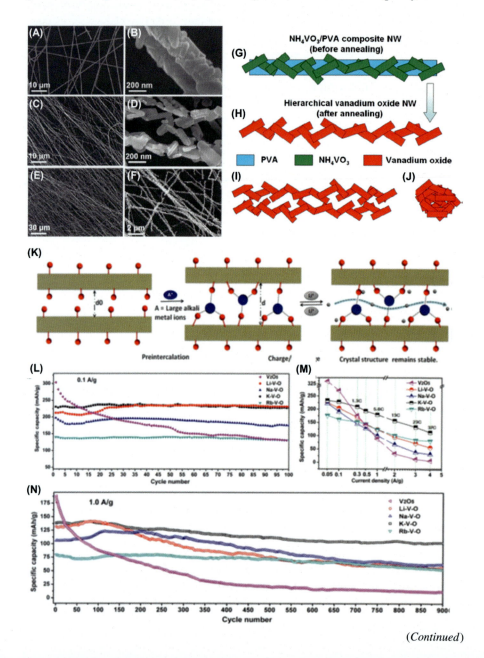

(Continued)

Synthesis of one-dimensional metal oxide−based crystals as energy storage materials 351

of the intercalation of lithium ions in vanadium oxide (V_2O_5) and preintercalated vanadium oxide NWs, A−V−O (A = alkali metal cation such as Na^+, K^+, and Rb^+; see Fig. 14.10A). After 100 cycles at a C rate of 0.1 A g^{-1} (Fig. 14.10B) the capacity retention significantly increased from 37.6% for baseline V_2O_5 to over 95% for preintercalated A−V−O, although the preintercalated compounds presented lower initial capacities than V_2O_5. The K−V−O electrodes exhibited the highest capacity retention; after 900 cycles, the K−V−O electrode retained 76% of 139 mAh g^{-1} compared with 39% for Li−V−O and 50% for Na−V−O under the same working conditions. Alkali metals can enlarge the diffusion channel in V_2O_5 NWs. Sodium-stabilized V_2O_5 with a $Na_{0.33}V_2O_5$ structure in a Zn-ion battery [43] showed a high capacity (367.1 mAh g^{-1} at a current density of 0.1 A g^{-1}) and a long cycling life (with a capacity retention of 93% after 1000 cycles). $Na_2V_6O_{16}•1.63H_2O$ NWs were also easily synthesized by a hydrothermal method [44]. V_2O_5 and NaOH were dissolved in distilled water, and then the solution was transferred into an autoclave and heated at 180°C for 24 h. The NWs were tested as a cathode in a Zn-metal battery, delivering a high specific capacity of 352 mAh g^{-1} at a current of 50 mA g^{-1} and long-term cycle stability, with a capacity retention of 90% after 6000 cycles.

Recently, Lin et al. [45] proposed a micron-scale VO_x NF−graphene nanoscroll composite synthesized via a hydrothermal method. After the dispersion of graphene in water, a vanadate ($V_6O_{13-\delta}$) solution was added, and the mixture reacted at 180°C in an autoclave for 48 h. The hierarchal architecture shown in Fig. 14.11 was obtained by calcination step at 450°C in inert atmosphere and then retreated in a reducing atmosphere (H_2/N_2 mixture) at 200°C to obtain the final product. NFS@graphene was used as the cathode in an aqueous Zn battery and showed stable cycling performance upto 16 A g^{-1}.

Figure 14.10 (A, B) Field emission scanning electron microscopy images of electrospun NH_4VO_3/polyvinyil alcohol composite nanowires. (C−F) field emission scanning electron microscope images of the ultralong hierarchical vanadium oxide nanowires after annealing. (G, H) Schematic illustration of formation of the ultralong hierarchical vanadium oxide nanowires during annealing. (I) Side view of two ultralong hierarchical vanadium oxide nanowires near each other. (L) Self-aggregation of short vanadium oxide nanorods. Schematic representation and electrochemical properties of large alkali metal ion intercalation. (M) Schematic representation of large alkali metal ion intercalation. (N) Cycling performance of A−V−O nanowires formed by preintercalating large alkali metal ions into vanadium oxides with a charge-discharge rate of 0.1 A g^{-1}. (O) Rate performance of A−V−O nanowires. A−V−O nanowires are cycled at various rates from 0.05 to 4.0 A g^{-1}. Here, nC denotes the rate at which a full charge or discharge takes $1/n$ h. (P) Cycling performance of A−V−O nanowires at the charge-discharge rate of 1.0 A g^{-1}. *Source*: L. Mai, L. Xu, C. Han, X. Xu, Y. Luo, S. Zhao, et al. Electrospun ultralong hierarchical vanadium oxide nanowires with high performance for lithium ion batteries. Nanoletters. 2010;10:4750−4755; Y. Zhao, C. Han, J. Yang, J. Su, X. Xu, S. Li, et al. stable alkali metal ion intercalation compounds as optimized metal oxide nanowire cathodes for lithium batteries. Nanoletters. 2015;15:2180−2185.

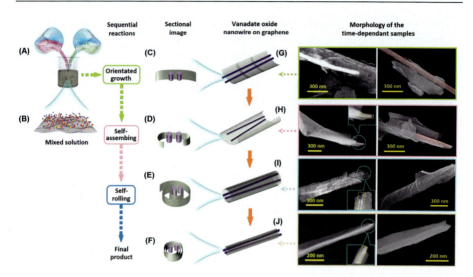

Figure 14.11 (A, B) Schematic illustrations of the mixed solution composed of graphene and the vanadium ions. (C–J) Schematic illustrations, (K–N) scanning electron microscope, and (O–R) brightfield transmission electron microscope images of the intermediate products for the digital negativities sample with 5 h (C, G, K, O) 10 h (D, H, L, P) 18 h (E, I, M, Q), and 24 h (F, J, N, R) of hydrothermal treatments.
Source: Y. Lin, F. Zhou, M. Chen, S. Zhang, C. Deng, Building defect-rich oxide nanowires @ graphene coaxial scrolls to boost high-rate capability, cycling durability and energy density for flexible Zn-ion batteries. Chem. Eng. J. [Internet]. 2020;396(April):125259. Available from: https://doi.org/10.1016/j.cej.2020.125259.

14.12 Zinc oxide

Yin et al. [46] prepared electrospun ZnO NFs by mixing zinc acetate dissolved in acetic acid and a PVA-water solution. A few drops of Triton X were added to the mixture before the electrospinning step. ZnO NFs were mixed with graphene nanoribbons showing a high specific capacity of 6300 mAh g^{-1} carbon in a Li–O$_2$ battery at a rate of 210 mA g^{-1}. Zhao et al. [47] investigated the possibility of growing ZnO NWs on nickel foam by a chemical bath deposition method (see Fig. 14.11). First, nickel foam was wetted with several droplets of zinc acetate dihydrate in ethanol (for three times), and after annealing at 300°C, a layer of ZnO nanoseeds was formed. Afterwards, the sample was treated with zinc nitrate, hexamethylenetetramine, ammonium hydroxide, and polyethylenimine before heating at 95°C for 24 h under air atmosphere. Subsequently, ZnO NWs were used as an interlayer in a Li-S battery with a sulfur loading of 3 mg cm^{-2}. The material could release a reversible capacity of 776 mAh g^{-1} after 200 cycles at a current of 1 C, corresponding to a capacity loss of 0.05% per cycle.

ZnO NWs may also be prepared by an easy hydrothermal route [48]. ZnCl$_2$ and Na$_2$CO$_3$ were dissolved in water and successively put into a 100-mL Teflon-lined stainless steel autoclave, which then was sealed and maintained at 120°C for 14 h. ZnO NWs

Synthesis of one-dimensional metal oxide–based crystals as energy storage materials 353

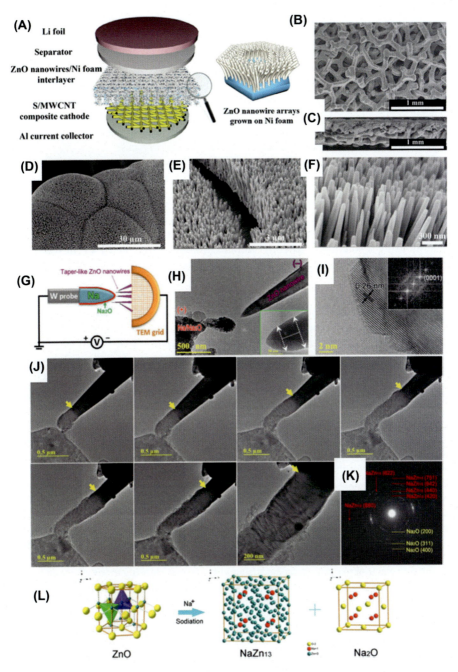

Figure 14.12 (A) Illustration of the assembled cell model, including a conceptual drawing of the zinc oxide nanowire arrays on the surface of Ni foam. (B) Top view and (C) cross-sectional low-magnification scanning electron microscope images of the hybrid zinc oxide nanowires/Ni

(*Continued*)

were later used to prepare an electrode with acetylene black, carboxymethylcellulose, and polytetrafluoroethylene on nickel foam as an electrode for a Ni/Zn battery. The NWs exhibited an average discharge capacity of 609 mAh g^{-1} at 0.2 C after 30 cycles. The ZnO nanowire sodiation mechanism was elucidated by Xu et al. [49] by using in situ transmission electron microscopy, as shown in Fig. 14.12. The sodiation mechanism could be summarized as a displacement reaction that results in $NaZn_{13}$ nanograin intermediate: $xZnO + (2x + 1)Na_2O + (2x + 1)e^- \rightarrow xNa_2O + NaZn_x$. Moreover, the authors observed an unexpected fast sodium insertion in the NWs due to abundant ion transport pathways through the channels of interconnected ZnO nanocrystals.

14.13 Zirconate fibers

Liu et al. reported a synthesis of $Li_6Zr_2O_7$ NFs by a simple electrospinning technique by using a solution of zirconyl nitrate, lithium acetate, and PVP in water and ethanol [50]. After sintering at 750°C for 1 h, as-prepared $Li_6Zr_2O_7$ NFs exhibited high conductivity: 1.27×10^{-5} and 1.16×10^{-3} S cm^{-1} at 200°C and 400°C, respectively. $Li_7La_2Zr_3O_{12}$ structure and lithium-ion dynamics in fluoride-doped cubic (LLZO) NFs as a solid electrolyte conductor for all-solid-state batteries can be prepared by electrospinning as well [51]. The synthesis procedure started with the preparation of stable colloidal solutions containing PVP template, lithium nitrate, aluminum nitrate, lanthanum nitrate, and zirconium acetate. After electrospinning, dense ceramic NFs were formed with a calcination step at 800°C in air (Fig. 14.13). The aligned LLZO NFs (75% w/w) were then combined with PVDF

foam interlayer. (D−F) High-magnification scanning electron microscope images of the hybrid zinc oxide nanowires/Ni foam interlayer. (G) Schematic illustration of the experimental setup for in situ electrochemical sodiation of nanosilicon borides. (H) Panoramic Transmission electron microscope image of the nanosilicon borides constructed with a taper-like zinc oxide nanowire inside a transmission electron microscope; inset shows the taper-like morphology of the zinc oxide nanowire tip with a taper degree of 1:3. (I) High resolution -transmission electron microscope image of the zinc oxide nanowire tip shows its single-crystalline nature; inset is the fast fourier transform pattern corresponding to the high resolution-transmission electron microscope image. (L) Time-resolved transmission electron microscope images from video frames showing morphology and structure evolution as a function of sodiation time. (M) Electron diffraction patterns of sodiated products demonstrating the formation of the $NaZn_{13}$ and Na_2O phases. (N) Schematic illustration of the crystal phase transformation of the zinc oxide anode during the electrochemical sodiation process.

Source: B.P.A. Brush, T. Zhao, Y. Ye, X. Peng, G. Divitini, H. Kim, et al. Advanced lithium−sulfur batteries enabled by a bio-insipred polysufide dsorptive bursh. Adv. Funct. Mater. 2016;26:8418−8426; F. Xu, Z. Li, L. Wu, Q. Meng, H.L. Xin, J. Sun, et al. In situ TEM probing of crystallization form-dependent sodiation behavior in ZnO nanowires for sodium-ion batteries. Nano Energy [Internet]. 2016;30:771−779. Available from: http://doi.org/10.1016/j.nanoen.2016.09.020.

Synthesis of one-dimensional metal oxide—based crystals as energy storage materials

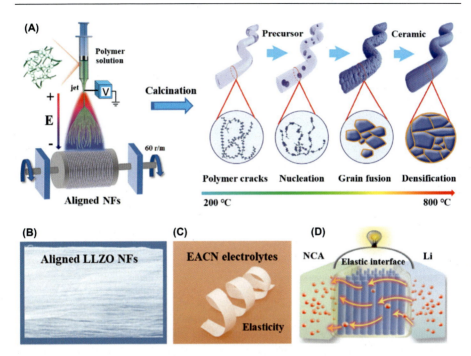

Figure 14.13 Fabrication procedure, mechanical properties, and ionic conductions of the elastic and well-aligned ceramic nanofiber based electrolytes. (A) A general picture of using a sol—gel electrospinning method followed by calcination to fabricate well-aligned ceramic structure and lithium-ion dynamics in fluoride-doped cubic nanofibers. (B) Digital image of the aligned structure and lithium-ion fluoride-doped cubic nanofiber membrane. (C) Digital image of the elastic ans well-aligned ceramic nanofiber electrolyte membrane showing the elasticity and softness. (D) Schematic illustration of the ion conduction and elastic interface of elastic and well-aligned ceramic nanofiber electrodes in a solid-state lithium nickel colbalt aliuminium oxides/elastic and well-aligned ceramic nanofiber/lithium battery.
Source: Y. Zhao, J. Yan, W. Cai, Y. Lai, J. Song, J. Yu, et al. Elastic and well-aligned ceramic LLZO nano fiber based electrolytes for solid-state lithium batteries. Energy Storage Mater. [Internet]. 2019;23(May):306—313. Available from: https://doi.org/10.1016/j.ensm.2019.04.043.

and LiClO$_4$ to form a new hybrid electrolyte with an ionic conductivity of 1.16×10^{-4} S cm^{-1} at 30°C. This electrolyte was used in a Li metal battery with a LiNi$_{0.8}$Co$_{0.15}$Al$_{0.05}$O$_2$ lithium nickel cobalt aliuminium oxides (NCA) cathode and displayed a capacity of around 150 mAh g^{-1} after 15 cycles.

Wan et al. [52] proposed new electrospun LLZO NWs as ceramic fillers in PEO: LiTFSI polymer electrolyte. The new composite showed an ionic conductivity of 2.39×10^{-4} S cm^{-1} at room temperature and 1.53×10^{-3} S cm^{-1} at 60°C. With the application of the LLZO nanowire—PEO composite electrolyte, a symmetrical Li/Li cell showed very low polarization, that is, less than 0.2 V at a current of 0.8 mA cm^{-2}. Furthermore, the LiFePO$_4$/Li metal-solid polymer electrolyte cell presented a specific capacity of 158.8 mAh g^{-1} after 70 cycles at 0.5°C and 60°C and a specific capacity of 158.7 mAh g^{-1} after 80 cycles at 0.1°C and 45°C. Yang et al. [53] observed a strong increment of the values of ionic conductivity in polyacrilonitrile:LiClO$_4$ polymer electrolyte (from 4.06×10^{-7} to about 10^{-4} S cm^{-1}) at room temperature when electrospun LLZO NWs were used as ceramic filler (5% w/w) with undoped c-LLZO, Al-LLZO, and Ta-LLZO compositions. Yang et al. [54] identified the formation of La$_2$Zr$_2$O$_7$ and amorphous Li compound as the first intermediates during sintering at 700°C of electrospun LLZO.

References

[1] A. Kilic, E. Serife, E. Stojanovska, A review of nanofibrous structures in lithium ion batteries, J. Power Sources 300 (2015) 199−215.

[2] A. La, Monaca, A. Paolella, A. Guerfi, F. Rosei, K. Zaghib, Electrospun ceramic nanofibers as 1D solid electrolytes for lithium batteries, Electrochem. commun. [Internet] 104 (June) (2019) 106483. Available from: https://doi.org/10.1016/j.elecom.2019.106483.

[3] D. Lei, Y. He, H. Huang, Y. Yuan, G. Zhong, Q. Zhao, et al., Cross-linked beta alumina nanowires with compact gel polymer electrolyte coating for ultra-stable sodium metal battery, Nat. Commun. [Internet] 10 (2019) 4244. Available from: https://doi.org/10.1038/s41467-019-11960-w. Available from.

[4] R. Tian, X. Feng, H. Duan, P. Zhang, H. Li, H. Liu, Low-weight 3D Al$_2$O$_3$ network as an artificial layer to stabilize lithium deposition, ChemSusChem. 2 (2018) 3243−3252.

[5] Y. Zuo, Y. Yu, H. Liu, Z. Gu, Q. Cao, C. Zuo, Electrospun Al$_2$O$_3$ film as inhibiting corrosion interlayer of anode for solid aluminum—air batteries, Batteries. 6 (2020) 19.

[6] H. Wang, Q. Pan, J. Zhao, G. Yin, P. Zuo, Fabrication of CuO film with network-like architectures through solution-immersion and their application in lithium ion batteries, J. Power Sources 167 (2007) 206−211.

[7] R. Sahay, P.S. Kumar, V. Aravindan, J. Sundaramurthy, W.C. Ling, S.G. Mhaisalkar, et al., High aspect ratio electrospun CuO nano fibers as anode material for lithium-ion batteries with superior cycleability, J. Phys. Chem. C. 116 (2012) 18087−18092.

[8] L.B. Chen, N. Lu, C.M. Xu, H.C. Yu, T.H. Wang, Electrochemical performance of polycrystalline CuO nanowires as anode material for Li ion batteries, Electrochim. Acta 54 (2009) 4198−4201.

[9] J. Wang, Q. Zhang, X. Li, D. Xu, Z. Wang, H. Guo, et al., Three-dimensional hierarchical Co$_3$O$_4$/CuO nanowire heterostructure arrays on nickel foam for high-performance lithium ion batteries, Nano Energy [Internet] 6 (2014) 19−26. Available from: https://doi.org/10.1016/j.nanoen.2014.02.012. Available from.

[10] S. Chaudhari, M. Srinivasan, 1D hollow a-Fe$_2$O$_3$ electrospun nanofibers as high performance anode, J. Mater. Chem. 22 (2012) 23049−23056.

[11] S. Jayaraman, V. Aravindan, M. Ulaganathan, W.C. Ling, Ultralong durability of porous α-Fe$_2$O$_3$ nanofi bers in practical Li-ion confi guration with LiMn$_2$O$_4$ cathode, Adv. Sci. 2 (2015) 1−5.

[12] M. Wu, P.J. Chiang, J. Lee, J. Lin, Synthesis of manganese oxide electrodes with interconnected nanowire structure as an anode material for rechargeable lithium ion batteries, J. Phys. Chem. B 109 (2005) 23279−23284.

[13] X. Wang, Z. Sun, Y. Zhao, J. Li, $Na_4Mn_9O_{18}$ nanowires wrapped by reduced graphene oxide as efficient sulfur host material for lithium/sulfur batteries, J. Solid. State Electrochem. 24 (2020) 111−119.

[14] Y. Lu, M. Yanilmaz, C. Chen, Y. Ge, M. Dirican, J. Zhu, et al., Lithium-substituted sodium layered transition metal oxide fi bers as cathodes for sodium-ion batteries, Energy Storage Mater. [Internet] 1 (2015) 74−81. Available from: https://doi.org/10.1016/j.ensm.2015.09.005. Available from.

[15] Q. Fan, M.S. Whittingham, E.S. Lett, Q. Fan, M.S. Whittingham, Electrospun manganese oxide nanofibers as anodes for lithium-ion batteries, Electrochem. Solid. State Lett. 10 (3) (2007) A48−A51.

[16] X. Wang, X. Xu, C. Niu, J. Meng, M. Huang, X. Liu, et al., Earth abundant Fe/Mn-based layered oxide interconnected nanowires for advanced K-ion full batteries, Nanoletters 17 (2017) 544−550.

[17] V. Aravindan, P. Suresh, J. Sundaramurthy, W. Chui, S. Ramakrishna, S. Madhavi, Electrospun NiO nanofibers as high performance anode material for Li-ion batteries, J. Power Sources [Internet] 227 (2013) 284−290. Available from: https://doi.org/10.1016/j.jpowsour.2012.11.050. Available from.

[18] L. Qiao, L. Qiao, X. Li, Y. Zheng, D. He, Single electrospun porous NiO−ZnO hybrid nanofibers as anode materials for advanced lithium-ion batteries, Nanoscale. (2013) 3037−3042.

[19] Y. Li, Y. Cui, Z. Yao, G. Liu, F. Shan, Fast electrochromic switching of electrospun Cu-doped NiO nanofibers, Scr. Mater. 178 (2020) 472−476.

[20] J. Zhao, C. Zhou, Y. Li, K. Cheng, K. Zhu, K. Ye, et al. Nickel cobalt oxide nanowires-modified hollow carbon tubular bundles for high-performance sodium-ion hybrid capacitors. Int. J. Energy Res. [Internet]. 2020;(October 2019):3883−3892. Available from: https://doi.org/10.1002/er.5185

[21] Y. Liu, D. Han, L. Wang, G. Li, S. Liu, X. Gao, $NiCO_2O_4$ Nanofibers as carbon-free sulfur immobilizer to fabricate sulfur-based composite with high volumetric capacity for lithium−sulfur battery, Adv. Energy Mater. 1803477 (2019) 1−10.

[22] V.D. Silva, L.S. Ferreira, A.J.M. Araújo, T.A. Sim, P.F. Grilo, M. Tahir, et al., Ni and Ce oxide-based hollow fi bers as battery-like electrodes, J. Alloy. Compd. 830 (2020) 154633.

[23] B. Huang, H. Wang, S. Zhang, C. Deng, Building 1D nano fibers with controlled porosity and crystallinity for honeycomb-layered oxide to achieve fast ion kinetics and superior sodium storage performance, Electrochim. Acta [Internet] 334 (2020) 135644. Available from: https://doi.org/10.1016/j.electacta.2020.135644. Available from.

[24] C. Zhang, L. Gu, N. Kaskhedikar, G. Cui, J. Maier, Preparation of silicon@silicon oxide core − shell nanowires from a silica precursor toward a high energy density li-ion battery anode, ACS Appl. Mater. Interfaces 5 (2013) 12340−12345.

[25] S. Jia, S. Yang, M. Zhang, K. Huang, J. Long, J. Xiao, Eco-friendly xonotlite nanowires / wood pulp fibers ceramic hybrid separators through a simple papermaking process for lithium ion battery, J. Memb. Sci. [Internet] 597 (December 2019) (2020) 117725. Available from: https://doi.org/10.1016/j.memsci.2019.117725. Available from.

[26] P. Yao, B. Zhu, H. Zhai, X. Liao, Y. Zhu, W. Xu, PVDF/palygorskite nanowire composite electrolyte for 4 V rechargeable lithium batteries with high energy density, Nano Lett. 18 (2018) 6113−6120.

[27] L. Li, X. Yin, S. Liu, Y. Wang, L. Chen, T. Wang, Electrospun porous SnO_2 nanotubes as high capacity anode materials for lithium ion batteries, Electrochem. commun. [Internet] 12 (10) (2020) 1383−1386. Available from: https://doi.org/10.1016/j.elecom.2010.07.026. Available from.

[28] C.A. Bonino, L. Ji, Z. Lin, X. Zhang, S.A. Khan, Electrospun carbon-tin oxide composite nanofibers for use as lithium ion battery anodes, ACS Appl. Mater. Interfaces 3 (2011) 2534−2542.

[29] P. Meduri, C. Pendyala, V. Kumar, G.U. Sumanasekera, M.K. Sunkara, Hybrid tin oxide nanowires as stable and high capacity anodes for Li-ion batteries 2009, Nanoletters 9 (2) (2009) 612−616.

[30] X. Wang, M. Xi, X. Wang, H. Fong, Z. Zhu, Flexible composite felt of electrospun TiO_2 and SiO_2 nano fibers infused with TiO_2 nanoparticles for lithium ion battery anode, Electrochim. Acta 190 (2016) 811−816.

[31] T. Hoshide, Y. Zheng, J. Hou, Z. Wang, Q. Li, Z. Zhao, et al., Flexible lithium-ion fiber battery by the regular stacking of two- dimensional titanium oxide nanosheets hybridized with reduced graphene oxide, Nano Lett. 17 (2017) 3543−3549.

[32] Z. Huang, M. Yang, J. Qi, P. Zhang, L. Lei, Q. Du, Mitigating the polysulfides "shuttling" with TiO_2 nanowires/nanosheets hybrid modi fi ed separators for robust lithium-sulfur batteries, Chem. Eng. J. 387 (December 2019) (2020) 1−10.

[33] J. Dong, Y. Jiang, Q. Li, Q. Wei, W. Yang, S. Tan, et al., Pseudocapacitive titanium oxynitride mesoporous nanowires with iso-oriented nanocrystals for ultrahigh-rate sodium ion hybrid capacitors †, J. Mater. Chem. A Mater Energy Sustain 5 (2017) 10827−10835.

[34] H. Park, T. Song, H. Han, U. Paik, Electrospun $Li_4Ti_5O_{12}$ nanofibers sheathed with conductive TiN/TiO_xN_y layer as an anode material for high power Li-ion batteries, J. Power Sources 244 (2013) 726−730.

[35] L. Zhu, P. Zhu, Q. Fang, M. Jing, X. Shen, A novel solid PEO/LLTO-nanowires polymer composite electrolyte for solid-state lithium-ion battery, Electrochim. Acta [Internet] 292 (2018) 718−726. Available from: https://doi.org/10.1016/j.electacta.2018.10.005. Available from.

[36] L. Zhu, P. Zhu, S. Yao, X. Shen, F. Tu, High - performance solid PEO/PPC/LLTO—nanowires polymer composite electrolyte for solid-state lithium battery, Int. J. Energy Res. May (2019) 4854−4866.

[37] Y. Xu, W. Cao, Y. Yin, J. Sheng, Q. An, Q. Wei, et al., Novel $NaTi_2(PO_4)_3$ nanowire clusters as high performance cathodes for Mg-Na hybrid-ion batteries, Nano Energy [Internet] 55 (September 2018) (2019) 526−533. Available from: https://doi.org/10.1016/j.nanoen.2018.10.064. Available from.

[38] J. Lee, C. Jo, B. Park, W. Hwang, H.I. Lee, S. Yoon, et al., Simple fabrication of flexible electrode with high metal-oxide content: electrospun reduced tungsten oxide/carbon nanofibers for lithium ion battery applications, Nanoscale. 1 (2012) 1−3.

[39] M. Yu, H. Sun, X. Sun, F. Lu, T. Hu, G. Wang, et al., 3D WO_3 nanowires/graphene nanocomposite with improved reversible capacity and cyclic stability for lithium ion batteries, Mater. Lett. [Internet] 108 (2013) 29−32. Available from: https://doi.org/10.1016/j.matlet.2013.06.067. Available from.

[40] Y. Feng, H. Xue, T. Wang, H. Gong, B. Gao, W. Xia, et al., Enhanced Li_2O_2 decomposition in rechargeable $Li − O_2$ battery by incorporating WO_3 nanowire array photocatalyst, ACS Sustain. Chem. Eng 7 (2019) 5931−5939.

[41] L. Mai, L. Xu, C. Han, X. Xu, S. Luo, S. Zhao, et al., Electrospun ultralong hierarchical vanadium oxide nanowires with high performance for lithium ion batteries, Nanoletters 10 (2010) 4750−4755.

[42] Y. Zhao, C. Han, J. Yang, J. Su, X. Xu, S. Li, et al., stable alkali metal ion intercalation compounds as optimized metal oxide nanowire cathodes for lithium batteries, Nanoletters 15 (2015) 2180−2185.

[43] P. He, G. Zhang, X. Liao, M. Yan, X. Xu, Q. An, et al., Sodium Ion stabilized vanadium oxide nanowire cathode for high-performance zinc-ion batteries, Adv. Energy Mater. 1702463 (2018) 1−6.

[44] P. Hu, T. Zhu, X. Wang, X. Wei, M. Yan, J. Li, et al., Highly durable $Na_2V_6O_{16}\cdot1.63H_2O$ nanowire cathode for aqueous zinc-ion battery, Nano Lett. (2018) 6−11.

[45] Y. Lin, F. Zhou, M. Chen, S. Zhang, C. Deng, Building defect-rich oxide nanowires @ graphene coaxial scrolls to boost high-rate capability, cycling durability and energy density for flexible Zn-ion batteries, Chem. Eng. J. [Internet] 396 (April) (2020) 125259. Available from: https://doi.org/10.1016/j.cej.2020.125259. Available from.

[46] J. Yin, J.M. Carlin, J. Kim, Z. Li, J.H. Park, B. Patel, et al., Synergy between metal oxide nanofibers and graphene nanoribbons for rechargeable lithium-oxygen battery cathodes, Adv. Energy Mater. 5 (2015) 1401412.

[47] B.P.A. Brush, T. Zhao, Y. Ye, X. Peng, G. Divitini, H. Kim, et al., Advanced lithium−sulfur batteries enabled by a bio-insipred polysufide dsorptive bursh, Adv. Funct. Mater. 26 (2016) 8418−8426.

[48] J.L. Yang, Y.F. Yuan, H.M. Wu, Y. Li, Y.B. Chen, S.Y. Guo, Preparation and electrochemical performances of ZnO nanowires as anode materials for Ni/Zn secondary battery, Electrochim. Acta 55 (2010) 7050−7054.

[49] F. Xu, Z. Li, L. Wu, Q. Meng, H.L. Xin, J. Sun, et al., In situ TEM probing of crystallization form-dependent sodiation behavior in ZnO nanowires for sodium-ion batteries, Nano Energy [Internet] 30 (2016) 771−779. Available from: https://doi.org/10.1016/j.nanoen.2016.09.020. Available from.

[50] Y. Liu, X. Hua, Preparation of $Li_6Zr_2O_7$ nanofibers with high Li-ion conductivity by electrospinning, Int. J. Appl. Ceram. Technol. 583 (3) (2016) 579−583.

[51] Y. Zhao, J. Yan, W. Cai, Y. Lai, J. Song, J. Yu, et al., Elastic and well-aligned ceramic LLZO nano fiber based electrolytes for solid-state lithium batteries, Energy Storage Mater. [Internet] 23 (May) (2019) 306−313. Available from: https://doi.org/10.1016/j.ensm.2019.04.043. Available from.

[52] Z. Wan, D. Lei, W. Yang, C. Liu, K. Shi, X. Hao, et al., Low resistance-integrated all-solid-state battery achieved by $Li_7La_3Zr_2O_{12}$ nanowire upgrading polyethylene oxide (PEO) composite electrolyte and PEO cathode binder, Adv. Funct. Mater. 1805301 (2019) 1−10.

[53] T. Yang, J. Zheng, Q. Cheng, Y. Hu, C.K. Chan, Composite polymer electrolytes with $Li_7La_3Zr_2O_{12}$ garnet-type nanowires as ceramic fillers: mechanism of conductivity enhancement and role of doping and morphology, ACS Appl. Mater. Interfaces 9 (2017) 21773−21780.

[54] T. Yang, Z.D. Gordon, Y. Li, C.K. Chan, Nanostructured garnet-type solid electrolytes for lithium batteries: electrospinning synthesis of $Li_7La_3Zr_2O_{12}$ nanowires and particle size-dependent phase transformation, J. Phys. Chem. C. 119 (27) (2015) 14947−14953.

Supercapacitors based on electrospun metal oxide nanofibers

15

Di Tian, Ce Wang and Xiaofeng Lu
Alan G. MacDiarmid Institute, College of Chemistry, Jilin University, Changchun, P.R. China

15.1 Introduction

High dependence on fossil fuels in human society has resulted in the dual crisis of energy and the environment, which promote the development of efficient and green energy storage and conversion devices. Supercapacitors have become one of the most attractive devices because of their advantages, such as a fast charge-discharge process, a long cycle life, and high power density [1]. According to the energy storage mechanism, there are two kinds of supercapacitors. One is electric double-layer capacitors (EDLCs), which realize charge storage by forming a double electron layer through a physical adsorption process. The other is pseudocapacitors, which are associated with the redox reaction between active materials and electrolyte to storage charges [2,3].

The electrochemical performance of supercapacitors is closely related to the composition and architecture of the electrode materials. From the structure point of view, many types of electrode materials from zero to three dimensions have been designed [4−6]. Among them, electrospun one-dimensional (1D) nanomaterials have bright prospects as supercapacitor electrode materials because of their high aspect ratio for efficient electron and mass transport capability, easy preparation, and strong plasticity. In terms of compositions, electrode materials can be classified into two categories: EDLC-based materials and pseudocapacitor-based materials. Carbon materials are of the former type, which possess good conductivity and stability. Materials of the latter type, which have high electron storage capacities, include metal oxides, sulfides, nitrides, phosphides, and selenides; conducting polymers; MXenes; metal organic frameworks (MOF); and covalent organic frameworks (COF) [7−11]. Especially with the deepening of research, according to the specific electrochemical results, some pseudocapacitive active materials are distinguished from the typical ones and are defined as battery-type electrode materials, whose electron storage capabilities are usually evaluated by specific capacity ($C\ g^{-1}$ or $mA\ h\ g^{-1}$) instead of specific capacitance ($F\ g^{-1}$) [12,13].

Metal oxides are rich in types and resources and exhibit high redox activities, serving as promising electrode materials for supercapacitors. Electrospun metal

---

Metal Oxide-Based Nanofibers and Their Applications. DOI: https://doi.org/10.1016/B978-0-12-820629-4.00012-6
© 2022 Elsevier Inc. All rights reserved.

oxide nanofiber (EMONF)—based materials not only provide distinct 1D structures that are beneficial for electron transport and ion diffusion, but also have great potential for storing electrons. To fabricate high-performance electrode materials, a large number of MONFs and MONF—based composites with various morphologies and compositions have been prepared via electrospinning technologies combined with other approaches. These materials displayed favorable electrochemical behaviors and brilliant prospects for energy storage.

15.2 Electrospun metal oxide nanofibers

Generally, EMON have been fabricated via an electrospinning and calcination process. It has been found that the electrospinning parameters, including voltage, spinning rate, the distance between the spinneret and the collector, the concentration and types of spinning solution, and the calcination procedure, have great influence over the composition and morphology of EMON. Common polymers involved in the spinning solution preparation include polyacrylonitrile (PAN), polyvinyl pyrrolidone (PVP), polymethyl methacrylate (PMMA), polyvinyl acetate (PVAc), and polyvinyl alcohol (PVA), which can be combined with various metal salts to produce single metal oxides, bimetallic oxides, and even polymetallic oxides.

15.2.1 Single metal oxides

Ruthenium oxide (RuO_2), involving three oxidation states within 1.4 V and reversible redox chemistry, offers great promise as a supercapacitor electrode material. Electrospinning techniques can endow RuO_2 with the characteristics of good conductivity, fast proton transport, and large specific surface area, leading to its highest specific capacitance among a large number of electrospun transition metal oxides. Hyun and coworkers prepared electrospun RuO_2 NF mats using ruthenium(III) chloride hydrate ($RuCl_3 \cdot xH_2O$)/PVAc as precursor [14]. Its superb electrical conductivity (288 S cm^{-1}), unique morphology, and large specific surface area contribute to the good rate performance, but the dense crystalline lattice structure limits the transport of protons (H^+). Therefore a thin hydrous RuO_2 overlayer with excellent proton transport was covered onto the crystalline RuO_2 NF mats via cyclic voltammetric (CV) deposition, resulting in an improved specific capacitance of 104.3 F g^{-1} and excellent cycle life (87.2% capacitance retention, 30,000 cycles).

Although RuO_2 has attracted increasing attention for supercapacitor electrodes, its properties, such as exceedingly high cost and environmentally poisonous characteristic, become the main hindrance in its commercialization. Fortunately, there have been some cheap and high-performance transition metal oxide electrode materials in the supercapacitor field. For example, Mn-based oxides serve as promising electrode materials for supercapacitors, owing to their various redox states, abundant natural resources, and low cost [15]. The structure and components of electrospun Mn-based oxides can be tuned via controlling the mixture ratio or posttreatments. The architectures of materials are related to their specific surface area, pore size and distribution, and aspect ratio, further affecting the electrochemical performance. Mn_3O_4 with various nanostructures (nanoparticles, nanorods, and NFs) were derived from manganese acetate tetrahydrate $(Mn(CH_3COO)_2 \cdot 4H_2O)$/ PVP with different ratios though an electrospinning and calcination process [16]. Mn_3O_4 NFs derived from a $Mn(CH_3COO)_2 \cdot 4H_2O$/PVP ratio of 1:1 exhibited a specific capacitance of $210 \, F \, g^{-1}$ and $155 \, F \, g^{-1}$ at $0.3 \, A \, g^{-1}$ in 1 M KCl and 1 M Na_2SO_4, respectively, which were much higher than those of Mn_3O_4 nanoparticles $(58 \, F \, g^{-1})$, nanorods $(65 \, F \, g^{-1})$, and NFs composed of dense nanoparticles $(86 \, F \, g^{-1})$. The smaller contact and interparticle resistance in a unique NF structure consisting of the pores and their homogenous distribution could be responsible for these results. Recent research has revealed that the limitation of electrochemical performance for nanoparticles was caused mainly by their inherent agglomeration. Thereby, a combination of electrospinning, frozen section, and calcination processes was applied to fabricate well-dispersed MnO_2 nanoparticles using PVP/Mn $(CH_3COO)_2 \cdot 4H_2O$ aligned NFs as precursors, which showed a high specific capacitance $(420 \, F \, g^{-1}$ at $0.1 \, A \, g^{-1})$ and good cycling stability (more than 92%, 4000 charge-discharge cycles) due to the nanometer size effect and the microminiaturization of the resulting devices (Fig. 15.1) [17]. From the chemical composition of electrospun Mn-based oxides, the low oxide state of Mn may have a negative impact on their electrochemical behaviors. Electrospun manganese(III)-acetylacetonate/PVP NFs were calcined in air and Ar to produce Mn_3O_4 and Mn_2O_3 NFs, respectively, which were converted into MnO_2 via a galvanostatic oxidation process [18]. As the valence rises, the charge storage capacities of MnO_x were improved. The galvanostatically oxidized Mn_3O_4 illustrated an outstanding specific capacitance of $380 \, F \, g^{-1}$ under a mass loading of $1.2 \, mg \, cm^{-2}$. In addition, the variable compositional ratios of electrospun Mn_3O_4/Mn_2O_3 hybrids can be achieved by controlling the calcination temperatures [19]. Interestingly, when the sample was calcined at 500°C, the highest capacitance of $360.7 \, F \, g^{-1}$ was obtained at $1 \, A \, g^{-1}$, reflecting that the mixed-phase were conductive to reducing the interfacial resistance within MnO_x NFs and electrolyte.

V_2O_5 is an attractive supercapacitor electrode material, owing to the multiple oxidation states of vanadium, layered structure, and redox reactions existing in surface and bulk. Surprisingly, electrospun V_2O_5 NFs exhibit good electrochemical performance as both cathode materials and anode materials in a supercapacitor device [20]. However, the operating potential windows are limited, owing to the presence of water decomposition within an electrochemical window of 1.23 V, which is adverse to both the capacitance of V_2O_5 and the energy density of the device. These problems can be solved by employing an organic electrolyte or building an asymmetrical supercapacitor device. For example, electrospun V_2O_5 NFs yielded $250 \, F \, g^{-1}$ in organic electrolyte (1 M $LiClO_4$ in propylene carbonate) within a potential range of 3 V, which was higher than that in 2 M KCl $(190 \, F \, g^{-1})$, 1 M H_2SO_4 $(106 \, F \, g^{-1})$, and 2 M KOH $(8 \, F \, g^{-1})$ with a potential window of 0.9 V. An asymmetrical supercapacitor based on electrospun V_2O_5 NFs (negative electrode material) and polyaniline NFs (positive electrode material) in

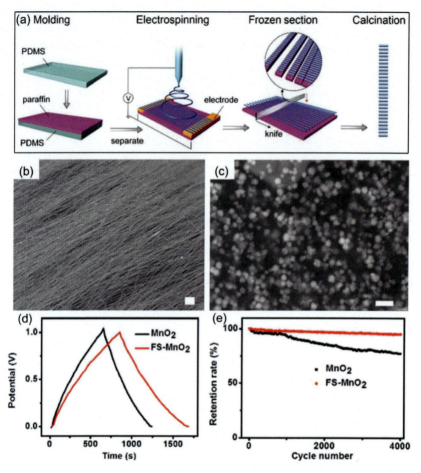

Figure 15.1 (A) Schematic illustration of the preparation of MnO$_2$ nanoparticles. Scanning electron microscope images of (B) aligned nanofibers and (C) monodispersed MnO$_2$ nanoparticles after ultrasonic treatment. (D) Galvanostatic charge-discharge curves at 0.5 A g^{-1} and (E) cycling stability of samples.
[17] Copyright 2015, Royal Society of Chemistry.

aqueous electrolyte was designed by Mak and coworkers, providing a wider operating potential up to 2.0 V [21]. It delivered an energy density of 26.7 Wh kg^{-1} at a power density of 0.22 kW kg^{-1} compared to that of a symmetrical cell (5.2 Wh kg^{-1} at 0.22 kW kg^{-1}). In addition, this cell retained 73% of its specific capacitance after 2000 cycles. In another report, the electrospinning technique makes it possible for an asymmetrical supercapacitor device to be assembled by using a flexible V$_2$O$_5$ NF membrane, reaching a high energy density, up to 32.2 W h kg^{-1} at an average power density of 128.7 W kg^{-1}, exhibiting an excellent cycling stability [22]. On the other hand, Aravindan and coworkers fabricated a high energy density hybrid supercapacitor using electrospun V$_2$O$_5$ NFs as cathode materials [23]. Different from conventional

NFs, continuous tubular V_2O_5 NFs were achieved via an electrospinning and controlled heat treatment [24]. Compared with the conventional electrospun V_2O_5 fibers and the core-shell $C-V_2O_5$ NFs, the prepared tubular V_2O_5 materials showed a superior charge storage ability, which is attributed to its larger specific surface area.

In addition to the typical pseudocapacitive materials, some metal oxide-based pseudocapacitive materials can be found in previous work, which were classified as battery materials because their cyclic voltammetry (faradaic redox peaks, often with rather large voltage separation) and constant current charge-discharge curves (voltage plateau) are obviously different from the typical ones. Nickel(II) oxide (NiO) is one of such electrode materials with a theoretical capacitance of 2584 F g^{-1} within a voltage window of 0.5 V [25]. Its pseudocapacitive characteristic is reflected in the redox peaks of CV curves, which arise from the mutual transformation between Ni(II) and Ni(III). The capacitance of NiO depends largely on its specific surface area, which can be regulated by using various mixtures as precursors. In a typical example, hollow NiO NFs were fabricated by electrospinning a precursor (PVP, nickel nitrate hexahydrate $(Ni(NO_3)_2 \cdot 6H_2O)$, citric acid (CA), Pluronic P123, water, and ethanol) followed by calcination (500°C for 3 h in air) [26]. The formation of hollow structures depends mainly on the introduction of citric acid. In detail, the nickel citrate, deriving from CA and $Ni(NO_3)_2 \cdot 6H_2O$, is linked with surfactant via a hydrogen bond on the surface of the P123 micellar rod. Owing to the larger space steric hindrance and the limitation of molecular geometry, the lamellar structure was formed. After calcination the removal of PVP leads to the hollow NiO nanotube comprising many NiO sheets (NiO/CA). Compared with solid NiO NFs (NiON), NiO/CA shows a larger specific surface area (212.1 m^2 g^{-1}) and specific capacitance (336 F g^{-1}, 2.5 times of NiON electrodes). The capacitance remained 87% of the initial value after 1000 cycles, demonstrating its excellent cycle stability. The better electrochemical performance is attributed to the hollow structure, which not only can promote the insertion and extrusion of OH$^-$ but also offer more active surface area for faradaic reactions. In addition, Vidhyadharan and coworkers proved that NiO nanowires (NWs) produced by using an aqueous polymer/nickel acetate tetrahydrate $(Ni(CH_3COO)_2 \cdot 4H_2O)$ solution have smaller diameters and a higher degree of crystallinity compared to the ones obtained by using organic solvents [27]. A supercapacitor device based on the NiO NWs was fabricated, delivering a capacity retention of 100% and high Coulombic efficiency (98%) during 1000 charge-discharge cycles. Apart from the morphologies, the compositions also have an important effect on the electrochemical performance of the NiO NFs. For example, La doping can contribute to the improved electrochemical performance of NiO NFs. The La-doped NiO NFs with the molar ratios of La/Ni = 1.5% showed a high specific capacitance (94.85 F g^{-1}), which is 5.3 times higher than the capacitors with pure NiO NFs as the positive electrode materials in an asymmetrical supercapacitor [28]. Another asymmetrical supercapacitor was assembled by using electrospun NiO NFs composed of densely packed hexagonal nanoparticles as cathode materials [29], showing a gravimetric capacitance of 141 F g^{-1} and an energy density of 43.75 Wh kg^{-1}. For fabrication of the above electrodes, NiO NFs mixed with binder and carbon black were pressed onto a current collector followed by thermal treatment, leading to the reduction of effective contact area between the electrolyte and the electrode materials. Therefore porous NiO NFs on nickel foam were also achieved for binder-free electrodes via a simple electrospinning technique [30]. The porous nature of both the NiO NFs and Ni foam current collector and the binder-free feature of the electrode greatly facilitate the ion and electron transport, contributing to its excellent rate capability.

Co_3O_4 is another ideal battery-type electrode material, owing to its nontoxicity, natural abundance, controllable size and shape, and good redox property. Over the past few years, a large number of Co_3O_4 nanomaterials have been designed with the various morphologies of nanoparticles, nanowires, nanosheets, and nanotubes—through a series of synthetic strategies such as coprecipitation; thermal decomposition; hydrothermal and microwave hydrothermal reactions; sputtering; electrospinning; sol−gel; solvothermal reaction; plasma, chemical bath, electrophoretic, and electrochemical deposition; combustion; as well as cryochemical techniques. Among them, electrospun nanomaterials usually yield higher charge storage capacities, owing to their larger specific surface areas and distinct electron and mass transport capability. Electrospun PVP/cobalt acetate tetrahydrate $(Co(CH_3COO)_2 \cdot 4H_2O)$ mats were calcined at 475°C for 5 h to produce Co_3O_4 NFs [31]. It delivered a specific capacitance of 407 F g^{-1} at 5 mV s^{-1}, which is higher than that of many Co_3O_4 nanomaterials synthesized by many other methods, such as water-controlled precipitation (102.5 F g^{-1}, 1 A g^{-1}) [32], hydrothermal reaction (323.0 F g^{-1}, 2 A g^{-1}) [33], hard template (102.0 F g^{-1}, 3 mV s^{-1}) [34], electrodeposition (310.0 F g^{-1}, 5 mV s^{-1}) [35], and plasma spray (162.0 6 F g^{-1}, 2.7 A g^{-1}) [36]. Furthermore, to increase the specific surface areas of electrospun NFs, surfactant is usually added to the spinning solution. For example, the addition of P123 can effectively inhibit the grain growth of Co_3O_4, resulting in a specific surface area of 20.89 m^2 g^{-1}, which is nearly double that of NFs produced without P123 [37]. Interestingly, Niu and coworkers designed a gradient electrospinning and controlled pyrolysis route to prepare Co_3O_4 mesoporous nanotubes [38]. In this system, low-, middle-, and high-molecular-weight PVA at a weight ratio of 3:2:1 were dissolved into water to produce the $PVA/Co(CH_3COO)_2 \cdot 4H_2O$ precursor, which played a key role in the formation of tubular structure. In detail, under a strong electrostatic tension force, low-, middle- and high-molecular-weight PVA corresponds to the center, middle layer, and outer layer, respectively, of nanowire. During sintering in air, low- and medium-weight PVA pyrolyzes and moves toward their respective boundary in conjunction with inorganic materials, leading to the Co_3O_4 mesoporous nanotubes. From the scanning electron microscope (SEM) and transmission electron microscope (TEM) images, Co_3O_4 mesoporous nanotubes with a diameter of 50 nm were composed of 5 nm nanoparticles. A microsupercapacitor device was built to reveal the good application potential of this material, delivering a specific capacity of 25 F cm^{-3} at 0.01 mV s^{-1} and a capacity retention of 98% after 10,000 cycles.

Compared with NiO and Co_3O_4, Fe_2O_3 possesses lower theoretical capacitances, but its features, such as low cost, environmental friendliness and corrosion resistance, have attracted huge attention. The crystallinity, phase purity, and morphology of Fe_2O_3 play a vital role in its electrochemical performance as a supercapacitor electrode material. For instance, PVAc and PVP were mixed separately with ferric acetyl acetonate

(Fe(acac)$_3$) to produce α-Fe$_2$O$_3$ with two distinct morphologies: nanograins and porous fibers (PFs), which stem from the chemical interactions between the precursor of metal oxide and the polymers [39]. The corresponding morphology and electrochemical performance are exhibited in Fig. 15.2. The unique nanostructure of Fe$_2$O$_3$ PFs contributes to the fast electron mobility and the enhanced interaction between electrode and electrolyte, resulting in an excellent cycling performance. Surprisingly, Fe$_2$O$_3$ exhibited good capacitance not only in the positive windows but also in the negative windows. Recently, Jiang and coworkers fabricated a flexible Fe$_2$O$_3$ membrane as negative electrodes for asymmetrical supercapacitor (ASC) devices, yielding a high specific capacitance (255 F g^{-1} at 2 mV s^{-1}) in neutral electrolytes and excellent rate capabilities [22]. Copper(II)oxide (CuO) also shows low theoretical capacitances, however, it displays a lot of advantages, such as low toxicity, large abundance, low cost, and environmental stability. CuO nanowires were prepared via an electrospinning process and were gathered by tens of dense cuboidal particles (10 nm), presenting high specific capacitances in both potassium hydroxide (KOH) and lithium hydroxide (LiOH) electrolytes at 2 A g^{-1} [40]. Moreover, the superior crystallinity of CuO leads to its excellent

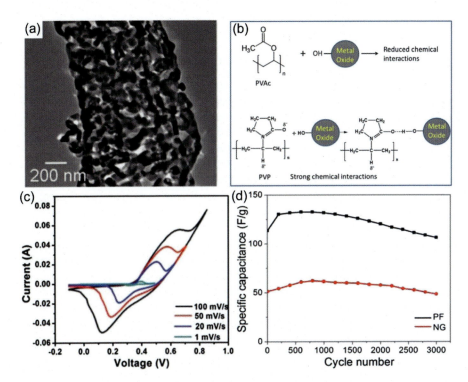

Figure 15.2 (A) Transmission electron microscope image of a single Fe$_2$O$_3$ porous fiber. (B) A schematic demonstration of the polymer polyvinyl pyrrolidone−metal oxide and polyvinyl acetale−metal oxide interactions. (C) cuclic voltammetric of Fe$_2$O$_3$ porous fibers at diverse scan rates and (D) cycling stabilities of Fe$_2$O$_3$ porous fiber and Fe$_2$O$_3$ nanograin. [39] Copyright 2013, Royal Society of Chemistry.

cycling stability, which is much higher than that of some reported CuO materials with different morphologies or synthetic approaches, such as CuO nanosheets (template growth method) [41], CuO nanoflowers (simple chemical precipitation) [42], CuO nanoflakes (oxidation by NaOH) [43], CuO with lotus-like nanostructure (liquid solid reaction) [44], CuO cauliflower (potential dynamic) [45], and CuO nanowire (anodization) [46]. Electrospun CuO nanowires were also used as anode in an asymmetrical supercapacitor [47], which delivered a maximum specific energy density of $29.5\ Wh\ kg^{-1}$ at the specific power density of $800\ W\ kg^{-1}$.

Titanium dioxide (TiO_2) was extensively studied as an electrode material in energy storage devices because of its pseudocapacitive behavior, high chemical stability, excellent functionality, nontoxicity, natural abundance, and cost effectiveness. Electrospun TiO_2 NFs were fabricated and used in a flexible supercapacitor, reaching a specific capacitance of $310\ F\ g^{-1}$ with a maximum energy density of $43.05\ Wh\ kg^{-1}$ [48]. To further improve the electrochemical performance of TiO_2, other metal ions (Nb, Zr, and Ta) were doped into TiO_2 via an electrospinning process [49,50]. Among them, Nb-doped TiO_2 showed the greatest improvement of energy storage capacity (from 40 to $280\ F\ g^{-1}$), and both Nb and Ta doping brought about the excellent cycling stability of TiO_2-based materials.

15.2.2 Bimetallic or polymetallic oxides

Compared with single transition metal oxides, bimetallic or polymetallic oxides usually present richer redox reactions, higher electrochemical activity, and even better electrical conductivity, which have been used as efficient electrodes for supercapacitors [51]. Perovskite-type metal oxides (ABO_3) are good candidates for supercapacitor electrodes, owing to their special electric conductivity, large operating potential window, and good electrochemical activity. ABO_3 contains two types of charge storage mechanisms [52]. On the one hand, the multiple transition metal ions, following cation intercalation, store charges by rapid surface redox reaction. On the other hand, their characteristics, including the distinct structural specificity and oxygen vacancy, provide anion intercalated energy storage capacity in the supercapacitor. Therefore the electrochemical performance of ABO_3 can be controlled by substituting the ions at both the A and B sites. The doping amount, doping sites, and metal species will directly determine the specific areas and electrochemical activity of ABO_3. Compared with bulk ones (limited surface area) or nanoparticles (easy agglomeration), electrospun ABO_3 NFs have attracted greater attentions.

Up to now, a series of $La_xSr_{1-x}BO_{3-\delta}$ (B = Ni, Co, and Fe) NFs have been reported and presented improved electrochemical performance through the Sr doping for La sites [53−55]. Among them, $La_xSr_{1-x}NiO_{3-\delta}$ NFs exhibited the maximum specific capacitance value of $719\ F\ g^{-1}$ at $2\ A\ g^{-1}$ and excellent rate performance, which are attributed mainly to the rich oxygen vacancies caused by Sr doping. The corresponding capacitor device can operate at a potential window as high as 2 V, leading to a high energy density of $81.4\ Wh\ kg^{-1}$ at a power density of $500\ W\ kg^{-1}$. In addition to single-site doping, two-site doping was also applied to perovskite electrode materials, such as $La_xSr_{1-x}CoMn_{0.9}O_{3-\delta}$ $(0.3 \leq x \leq 1)$ and

La$_x$Sr$_{1-x}$CuMn0.9O$_{3-\delta}$ ($0.1 \leq x \leq 1$) materials [56,57]. Similarly, George and coworkers designed and prepared high-crystalline SrMnO$_3$ perovskite oxide NFs via sol−gel electrospinning followed by a calcination process at different temperatures [58]. The effects of doping Ba/Ca on Sr and Co/Fe/Ni on Mn on the specific capacitance of SrMnO$_3$ NFs were studied [59]. The results reflected that the SrMnO$_3$ NF electrode materials that were calcined at 700°C displayed an electrochemical capacitance of 321.7 F g^{-1} at 0.5 A g^{-1}, and meanwhile 20 mol% Ba loading showed the best performance as the supercapacitor electrode materials with a specific capacitance of 446.8 F g^{-1} at 0.5 A g^{-1}. In addition, this material retained 87% of the initial capacitance after 5000 charge-discharge process. The enhancement in specific capacitance by heteroatom doping is explained as the increase in oxygen vacancies because of lattice distortions.

The NF materials with spinel structure (AB$_2$O$_4$) have aroused interest in the supercapacitor field because the voids in crystal structures facilitate the intercalation and delamination of electrolyte, contributing to easy redox reactions and enhanced electrochemical performance. Up to now, many kinds of 1D spinel transition metal oxides have been produced through the combination of electrospinning and calcination processes. They displayed various morphologies and were used as supercapacitor electrode materials. For instance, NiCo$_2$O$_4$ NFs, nanotubes, and nanobelts were successfully obtained by adjusting the ratio of metal salt to polymer for supercapacitor electrodes [60]. Among them the porous NiCo$_2$O$_4$ nanotubes possessed the highest Brunauer, Emmett and Teller (BET) surface area of 36.9 m^2 g^{-1} and the largest specific capacitance of 1647.6 F g^{-1}, which is superior to that of previously reported 1D NiCo$_2$O$_4$ nanostructures—such as nanoneedles, porous NWs, and NWs [61−63]. Meanwhile, its energy density (38.5 Wh kg^{-1}) is approximately 1.6 and 2.0 times as high as that of NiCo$_2$O$_4$ NFs and NiCo$_2$O$_4$ nanobelts, respectively. Similarly, NiCo$_2$O$_4$ microbelts with magnetic and electrochemical properties have been fabricated [64], delivering a specific capacitance of 245 F g^{-1} in 1 M KOH electrolyte. In particular, a novel tube-in-tube structure was achieved for NiCo$_2$O$_4$ by tuning the heating rates [65]. The assembled NiCo$_2$O$_4$ tube-in-tube−activated carbon (AC) ASC devices exhibited good rate properties and excellent cycling stability (about 13% loss, 10,000 cycles). Apart from nickel, the substitution of Zn in place of one Co cation at a tetrahedral and/or octahedral position in spinel structures of Co$_3$O$_4$ will produce ZnCo$_2$O$_4$ [66]. The obtained ZnCo$_2$O$_4$ nanotubes with hollow structures not only shorten the distance between the electrolyte and internal electrode interface, but also alleviate the volume change caused by the repeated insertion and extraction of OH$^-$ ions, achieving the higher specific capacitance and better cycling behavior compared to ZnCo$_2$O$_4$ nanoparticles. Similarly, Mn occupying the octahedral sites in Mn$_3$O$_4$ was replaced by Zn to achieve ZnMn$_2$O$_4$ with a spinel structure [67]. For example, the prepared ZnMn$_2$O$_4$ NFs composed of interconnected nanoparticles were prepared, showing a specific capacitance of 240 ($\pm$5) F g^{-1} at 1 A g^{-1} in 1 M Na$_2$SO$_4$. Recently, the spinel-NiMn$_2$O$_4$ NFs with surface areas of 50 m^2 g^{-1} were also prepared as a supercapacitor electrode [68] which illustrated a high specific capacitance of 410 ($\pm$5) F g^{-1} at 1 A g^{-1}. The prepared NiMn$_2$O$_4$ NF−based solid-state symmetric supercapacitor yielded an energy density of $\sim$95 Wh kg^{-1} and a power density of 1030 W kg^{-1}

within a wide potential window of 2.0 V, which could support a red-colored LED (1.8 V, 20 mA) for at least 5 min. In other reports, metal oxides with ABO_4 structure have also attracted attentions for supercapacitor electrodes. In detail, electrospun porous $MnMoO_4$ nanotubes with a diameter of 120 nm were fabricated and delivered a high specific capacitance (620 F g^{-1}, 1 A g^{-1}), excellent rate capability (460 F g^{-1}, 60 A g^{-1}), and outstanding cycling stability (no decay in specific capacitance for 10,000 cycles) [69]. The specific capacitance is higher than that of MoO_3 NFs, MnO_x nanorods, and $MnMoO_4$ nanochains and is associated with both the hollow structure and the richer redox chemistry reactions provided by manganese and molybdenum ions. Moreover, $NiMoO_4$ with four different mesoporous hierarchical structures (solid NFs, porous NFs, hollow NFs, and microplates) as supercapacitor electrode materials was reported by Sudheendra Budhiraju and coworkers [70]. The highest specific capacity (214 mAh g^{-1}) was obtained for the hollow $NiMoO_4$ NFs. Meanwhile, an asymmetrical supercapacitor ($NiMoO_4$//AC) delivered a larger energy density of 39 Wh kg^{-1} than many $NiMoO_4$-based supercapacitor devices and long cycling performance (capacity retention of 97% after 5000 cycles), benefitting from the high specific surface area ($\sim$105 m^2 g^{-1}), optimum pore size (3 nm), and adequate pore volume (0.3631 cm^3 g^{-1}) of the material.

Recently, lithium ion hybrid supercapacitors, consisting of the supercapacitor active materials and Li-ion insertion and extraction materials, have become promising energy storage systems, owing to the increased power density during the reversible nonfaradaic course and large energy density supplied by the faradaic behavior. Therefore a series of Li-based transition metal oxides were investigated as active insertion materials. For instance, two hybrid supercapacitors were constructed by using two types of electrospun $LiNi_{0.5}Mn_{1.5}O_4$ spinel materials with different morphologies as the cathodes and nitrogen-doped graphene or active carbon as anodes, which delivered a maximum energy density of 15 and 19 Wh kg^{-1}, respectively [71,72]. Another example has showed that a significantly higher energy density of 180.2 Wh kg^{-1} at the power density of 248.0 W kg^{-1} was achieved for the $LiNi_{0.4}Co_{0.6}O_2$-based asymmetrical device ($LiNi_{0.4}Co_{0.6}O_2$ as cathode materials and AC as anode materials) with a wider potential window of 3.0 V [73]. Because of high reversibility, easy synthetic processes, low cost, and high theoretical capacity (about 388 mA h g^{-1}), $TiNb_2O_7$ NFs prepared by a scalable electrospinning technique were also applied to a lithium ion capacitor [74]. The reversible intercalation of lithium (3.45 mol) is feasible with good capacity retention characteristics, which against the theoretical limitation of five moles. $TiNb_2O_7$/AC delivered a maximum energy density of about 43 Wh kg^{-1}.

15.3 Electrospun metal oxide nanofiber–based composites

The electrochemical behaviors of EMONFs have been improved by regulating their morphology and composition. However, metal oxides have limitations in some

Supercapacitors based on electrospun metal oxide nanofibers

aspects, such as poor conductivity and stability. With the continuous development of electrode materials, the combination of metal oxide and other materials has become an effective treatment to further enhance the electrochemical performance of electrode materials. To date, EMONF−based composites can be divided into four categories: metal oxide/metal oxide (metal), metal oxide/carbon material, metal oxide/conducting polymer, and other composites.

15.3.1 Metal oxide/metal oxide (metal hydroxide, metal) composites

There are two synthetic routes for the fabrication of metal oxide/metal oxide (metal) composites. One is electrospinning of mixed metal precursors followed by a calcination process to obtain the metal oxide/metal oxide composites. For example, $RuO_2−Ag_2O$ NWs, $RuO_2−Mn_3O_4$ NFs, and $SrRuO_3$-RuO_2 NFs were prepared through an electrospinning and calcination process using $RuCl_3 \cdot xH_2O/AgNO_3/$ PVP, $RuCl_3 \cdot xH_2O/$strontium (II) chloride hexahydrate/PVAc, and $RuCl_3$/manganese acetylacetonate/PVAc, respectively, as precursors, [75−77]. A coelectrospinning method was applied to fabricate MnO_x-RuO_2 NFs [78], exhibiting a specific capacitance of 208.7 F g^{-1} at 10 mV s^{-1}, which is lower than the specific capacitance of RuO_2/Mn_3O_4 NFs (293 F g^{-1}) and higher than that of both $RuO_2−Ag_2O$ nanowires (192 F g^{-1}) and RuO_2-Mn_3O_4 NFs (173.25 F g^{-1}). In addition, a series of $LaNiO_3$/NiO NFs were obtained by tuning the molar ratios of La:Ni = 0:1, 1:1, 1:2, and 1:4 [79]. The La:Ni = 1:2 hollow NFs with mesoporous wall showed the largest specific capacitance and excellent cycling performance (10% capacitance loss after 1000 cycles). Similarly, the electrochemical performance of $NiO@Co_3O_4$ NFs with different mass ratios and different additives was investigated [80]. Among them, $NiO@Co_3O_4$ NFs composed of many NiO and Co_3O_4 sheets possessed the highest specific capacitance (788 F g^{-1} at 5 mA cm^{-2}), which is attributed to the large capacitance and redox activity of NiO along with the high specific surface area and fast ion insert-emergence rate of Co_3O_4. Our group has also prepared V_2O_5 doped α-Fe_2O_3 NFs via a similar one-step electrospinning technique followed by calcination [81]. Compared with pristine α-Fe_2O_3, the V_2O_5/α-Fe_2O_3 NFs exhibited features such as perfect reversibility, better capacitance, and better cycling stability. Similarly, an electrospun $Mn_2O_3/ZnMn_2O_4$ composite showed an improved electrochemical performance compared with pure Mn_2O_3 and $ZnMn_2O_4$ [82]. MnO_x/SnO_2 fibers exhibited a higher specific capacitance than MnO_x and $MnO_x/$ CNT in previous reports [83]. $NiCo_2O_4@Au$ nanotubes with a high specific capacitance (1013.5 F g^{-1}) and excellent cycling stability (85.13% after 10,000 cycles) were used as the positive electrode to fabricate an asymmetrical supercapacitor device within a maximum voltage of 1.45 V, which delivered a high energy density of 19.56 Wh kg^{-1} [84].

The other synthetic route to prepare metal oxide/metal oxide (metal hydroxide) is carried out through an electrospinning followed by calcination and posttreatment, such as electrochemical deposition and hydrothermal reaction. As an example, electrospun platinum NF mats were selected as the conducting core for the electrochemical deposition of hydrous RuO_2 overlayers to construct a supercapacitor with a good rate performance (only a 21.4% decrease from 10 to 1000 mV s^{-1}), owing to the high electrical conductivity (412 S cm^{-1}) and porous morphologies of the Pt NFs [85]. The use of TiO_2 as substrate to support RuO_2 has also been reported via electrochemical deposition in previous work [86]. Heat treatment at 800°C brings about enhanced electrical conductivity for TiO_2, leading to an improved specific capacitance and rate performance (a capacity loss of 33% from 10 to 1000 mV s^{-1}) of the RuO_2/TiO_2 electrode, which is more than 20 over that of the pristine TiO_2. Similarly, nanonetworks of MnO_2 shell/Ni current collector core were prepared via an electrospinning, calcination, and electrochemical deposition process [87]. The unique architectures contribute to the increase in surface area and the reduction in charge-transfer resistance, delivering a fast charge-discharge rate and a higher specific capacitance (214 F g^{-1}) than that of MnO_2 on Ni foam. In addition to the electrochemical deposition, hydrothermal synthesis is also a typical method to grow MnO_2 on a variety of substrates. Up to now, TiO_2/MnO_2, $La_{0.7}Sr_{0.3}CoO_{3-\delta}$/$MnO_2$ nanorods, and α-MoO_3/MnO_2 composites have been investigated as supercapacitor electrode materials [88−91]. Among them, $La_{0.7}Sr_{0.3}CoO_{3-\delta}$ lanthanum strontium colbaltite (LSC)@MnO_2 core-shell nanorods displayed the largest specific capacitance of 570 F g^{-1} at 1 A g^{-1} and long cycling life (capacitance retention remained at 97.2% after 5000 cycles) due to the excellent stability and ion-electron double conductivity provided by LSC and the increased effective contact areas supplied by gridlike MnO_2 nanosheets. In addition, three-dimensional (3D) ZnO NF @$Ni(OH)_2$ nanoflake with ultrahigh specific capacitance (2218 F g^{-1} at 2 mV s^{-1}) was fabricated, which is much higher than that of pure $Ni(OH)_2$ (1604 F g^{-1}) and ZnO (112 F g^{-1}) (Fig. 15.3) [92]. An asymmetrical supercapacitor assembled by ZnO@$Ni(OH)_2$ hybrids (positive electrode) and porous carbon microfibers (negative electrode) yielded a high energy density of 57.6 Wh kg^{-1} with a power density of 129.7 W kg^{-1}. A novel Co_3O_4@$NiMoO_4$ composite with a tubelike yolk-shell structure was synthesized by Dong and coworkers [93]. The interspaces in the yolk-shell structure and the multiple transition metals of the Co_3O_4@$NiMoO_4$ composite give better electrochemical behavior (913.25 F g^{-1} and large capacitance retention of 88% at high current density) than bare $NiMoO_4$. The advantages of composites for supercapacitor electrodes are also reflected in the combination of $ZnCo_2O_4$ and $NiCo_2O_4$ [94]. In comparison with the individual components of either $ZnCo_2O_4$ NWs or $NiCo_2O_4$ nanosheets, the $ZnCo_2O_4$@$NiCo_2O_4$ composite showed a higher specific capacitance (1476 F g^{-1}, 1 A g^{-1}) and better rate capability (942 F g^{-1}, 20 A g^{-1}), and meanwhile maintained 98.9% capacity after 2000 cycles at 10 A g^{-1}.

15.3.2 Metal oxide/carbon-based composites

Generally, carbon materials with superior conductivity and cycling stability can be hybridization with metal oxides to realize a synergistic effect for pseudocapacitors and double-layer capacitors. The hybridization of metal oxides and carbon materials includes two categories. One is metal oxides embedded in carbon NFs, which are generally achieved through an electrospinning followed by thermal treatment. This

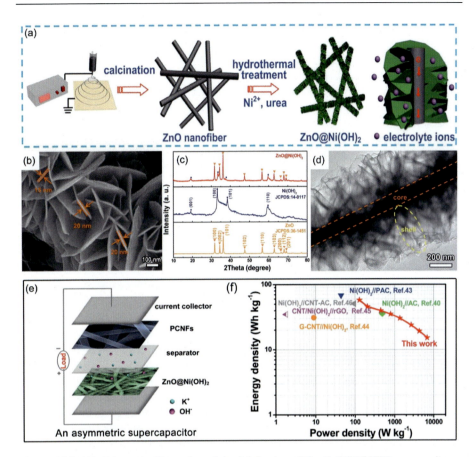

Figure 15.3 (A) Schematic illustration of the fabrication of the ZnO@Ni(OH)$_2$ composite. (B) Scanning electron microscope images of ZnO@Ni(OH)$_2$ hybrid. (C) X-ray diffraction patterns of the products. (D) Transmission electron microscope images of ZnO@Ni(OH)$_2$ nanostructures. (E) Schematic demonstration and (F) Ragone plot of the assembled ZnO@Ni(OH)$_2$//PCNF asymmetrical supercapacitor in comparison with Ni(OH)$_2$-based devices. [92] Copyright 2015, Royal Society of Chemistry.

method can make two materials contact more closely. During electrospinning, the nanostructures of mixed precursors (metal salts and polymers) are preliminarily constructed. During the course of thermal treatment, metal salts are converted into their corresponding metal oxides, and the released gas is favorable for the pore formation of materials, providing a large number of electrochemical active sites and larger specific surface areas for the whole electrode materials. On the other hand, carbon materials are generated from polymers, and the presence of metal salts catalyzes the graphitization of carbon, accelerating the charge-transfer during the charge-discharge process.

To achieve the optimal electrochemical performance of electrospun metal oxide/carbon composite, both the types and ratios of precursors (metal salts, polymer, and other carbon source) and different treatment methods have been widely investigated such as tuning the calcination rates and temperatures during the activation treatments. In a typical example, a RuO_2/activated carbon nanofiber (ACNF) composite with hollow cores has been prepared via electrospinning by using $Ru(acac)_3$/PAN + PMMA as precursors, followed by a calcination procedure (700°C by supplying 30 vol.% steam in a nitrogen carrier gas) [95]. Compared with bare ACNFs, RuO_2/ACNFs have a lower specific surface area but exhibit superior electrochemical properties, which benefit from the introduction of PMMA (presence of micropores and mesopores) and RuO_2 (numerous electrochemically active sites). Similarly, steam activation is an efficient route to produce porous carbon, which can be conducted to enhance the electrochemical behaviors of ZnO/ACNF and V_2O_5/ACNF composites. As results, ZnO/ACNFs ($Zn(Ac)_2$/PAN) reveal a higher specific capacitance of 178.2 in comparison to 163 F g^{-1} for ZnO/ACNFs [96,97]. The supercapacitor cell built by V_2O_5/ACNFs (V_2O_5/PAN) delivered a larger energy density of 68.84 Wh kg^{-1} (organic electrolytic solution) compared to 17.81 Wh kg^{-1} of ACNF electrode (organic electrolytic solution) and the 18.8 Wh kg^{-1} of V_2O_5/CNF (6 M KOH) [98,99]. Without any activation agent or activation process, lignin and pitch were mixed into the spinning solution containing PAN and $Zn(Ac)_2$ to induce porosity on the fiber surface, reducing the cost and increasing the carbon yield and electrical conductivity of the Zn/carbon nanofibers (CNFs) [100]. These characteristics provide it with ideal capacitive performance and good cycling stability (only ~6% decrease after 3000 cycles). Recently, polystyrene as a porogen was added to PAN/VO(acac)$_2$ to produce V_2O_5/CNF with multichannel and good flexibility, serving as binder-free and conductive-free electrode materials for a supercapacitor [101]. Interestingly, there are two other VO$_x$/CNF composites that were used as binder-free and conductive-free electrode materials in the reported literature [102,103]. Their electrochemical results revealed that different types and combinations of precursors (VO_2, V_2O_5, and VO(acac)$_2$) resulted in various compositions and pore structures for VO$_x$/CNF composites through the same calcination process. Furthermore, V/single-walled carbon nanotubes (SWCNTs)/CNFs and graphene oxide (GO)/V_2O_5/CNFs prepared via hybrid electrospinning and carbonizing have also been reported [104,105]. V/SWCNTs/CNFs with hierarchical structure, involving micropores and mesopores, demonstrated a specific surface area of 821 m^2 g^{-1} and a reversible specific capacitance of 479 F g^{-1} at 1 A g^{-1} and retained 94% of its initial capacitance after 5000 cycles; these features are related to the enhanced conductivity and stability (the existence of V−N−C, V−O−C and V = O bonds) caused by the introduction of V and SWCNTs. Such good capacitive behavior is superior to that of SWCNTs/CNFs, V/CNFs, and CNFs. Interestingly, ternary GO/V_2O_5/CNFs displayed a higher specific capacitance than V_2O_5/CNFs.

Electrospun CNFs with encapsulated Co_3O_4 nanoparticles were obtained by using $Co(Ac)_2 \cdot 4H_2O$/PAN as precursors [106]. The formation of onion-like graphitic layers around the Co_3O_4 nanoparticle enhances the electronic conductivity and meanwhile prevents particles from detaching from the carbon material, which contributes the remarkable supercapacitor behaviors, including a capacitance of 586 F g^{-1} at 1 A g^{-1}, good rate performance (~66% capacity retention at

50 A g^{-1}), and excellent cycling properties (74% capacity retention, 2000 cycles). For comparison an electrospun CNF embedded with Co_3O_4 hollow nanoparticles by using $Co(acac)_2$/PAN as a precursor was recently fabricated [107] that could be directly employed as self-supported electrodes. Benefiting from structural characteristics, the integrated binder-free electrode exhibits no decay in its specific capacitive after continuous 2000 cycles at 4 A g^{-1}. In addition, a large number of MnO_x/carbon composites have been deeply investigated by researchers. In addition to the steam activation mentioned above, CO_2 activation was also used for the activation of C/MnO_x [108], leading to a significant increase in the specific surface area of materials (701 $m^2 g^{-1}$) through the interaction between CO_2 and the carbon surface. Its features contribute to the good specific capacitance (213.7 F g^{-1}) and cycling performance ($\sim$97% after 1000 cycles). During the carbonization process of MnO_x/CNF, heating rates play an important role in the electrochemical behavior, which was proved by Shi and coworkers [109]. The better crystallinity of the MnO_x nanoparticles and the graphitic carbon layers on the surface of the composites were achieved for the slow-heated MnO_x/CNF electrode ($MnCl_2$/PAN), corresponding to its better electronic conductivity and good rate capability and cycling performance as free-standing electrodes for a supercapacitor than the fast-heating one. Three other free-standing electrode materials based on MnO_x/CNFs composites have also been reported using $MnCl_2$/PVP, $Mn(OAc)_2 \cdot 4H_2O$/PAN, and $Mn(acac)_2$/PAN as precursors, which yielded the specific capacitances of 578 F g^{-1} at 1 A g^{-1} in 1 M Na_2SO_4, 174.8 F g^{-1} at 2 mV s^{-1} in 0.5 M Na_2SO_4, and 200 F g^{-1} at 2.5 mV s^{-1} in 0.25 M aqueous bis(trifluoromethane)sulfonimide lithium electrolyte, respectively [110−112]. Furthermore, CNFs−MnO_2 derived from PAN/pitch/PMMA delivered a maximum specific capacitance of 183 F g^{-1} (1 mA cm^{-2}) and a superior rate capability in aqueous solution, stemming from the addition of pitch and PMMA [113].

To promote the electrochemical properties, reduced graphene oxide (RGO), acting as a conductive channel, was introduced into MnO_x/CNF (from $Mn(Ac)_2$/PAN + PVP) [114]. The obtained RGO/MnO_x/CNF showed increasing specific capacitances compared with the MCNF-only and RGO-only supercapacitor electrodes, stemming from the synergistic effect of the multiple components. Moreover, ordered and disordered MnO_2/GO hybrids were fabricated by using an electrospinning method and a drop-casting method, respectively, with the same precursor ($Mn(OAc)_2 \cdot 4H_2O$/GO/PVA), confirming that the ordered system is more conducive to the improved electrochemical performance [115].

Two SnO_x combined with CNFs were also served as flexible and free-standing electrode materials for supercapacitors [116]. A typical example is using $Sn(Ac)_2$/PMMA + PAN as precursors to obtain composite materials with tin oxide as the core and carbon as the shell. PMMA as the sacrificial polymer was removed by a simple heating process. A capacitance of 289 F g^{-1} and 88% of its initial capacitance after 5000 cycles were achieved in a two-electrode test system. Another example is using $SnCl_2 \cdot 2H_2O$/PAN as precursors to produce SnO_x/CNF with different valence states of tin (Sn/CNFs, Sn/SnO/CNFs, and SnO_2/CNFs) by tailoring the carbonization processes [117]. Among them, Sn/CNFs exhibited the optimal electrochemical performance as supercapacitor electrodes, and meanwhile a very stable cycle stability was obtained (114% of its initial performance, 10.000 cycles).

Pant and coworkers prepared TiO_2 NPs assembled into CNFs, which exhibited a specific capacitance of 106.57 F g^{-1} [118]. An anatase/rutile mixed-phase TiO_2/hydrogen exfoliated graphene nanofibers with a specific capacitance of 210.5 F g^{-1} in 1 M H_2SO_4 were also fabricated [119]. The improved specific capacitance benefits from the synergistic effect of the pseudocapacitance of the TiO_2 phase and the double-layer capacitance of the graphene sheets. In addition to the above various types of polymers, lignin, GO, and CNT, polypyrrole (PPy) is also a kind of carbon source that could be transformed into an N-doped carbon structure after calcination. As shown in Fig. 15.4, amorphous carbon-coated NiO NFs were obtained via electrospinning, vapor deposition polymerization, and carbonization techniques. Various concentrations of ethanolic $FeCl_3$ solutions as oxidants for the polymerization of PPy led to NiO NFs coated with different thicknesses of carbon. The PPy-coated NiO NFs (NiP NFs) with 5-, 10-, and 15 nm-thick coatings are denoted as NiP_L, NiP_M, and NiP_H, respectively, and the amorphous carbon-coated nickel oxide nanofibers (NiC NFs) with 3-, 7-, and 13 nm-thick coatings are denoted as NiC_L, NiC_M, and NiC_H, respectively [120]. Electrochemical measurements in a three-electrode cell reveal that the increased thickness of the carbon corresponds to a decreasing specific capacitance. The carbon/NiO composite delivered a higher capacitance (288 F g^{-1}) than NiO CNTs, and PPy NTs do.

CNFs embedded with two metal oxides have also been fabricated as the electrode materials for supercapacitors to reveal the synergistic effect between different oxides. For example, an electrospun CNFs$-$Sn$-$ZrO$_2$ composite [121], using ZrO $(NO_3)_3 \cdot 2H_2O$ and $SnCl_2 \cdot 6H_2O$/PAN as precursors, was prepared and exhibited a larger specific capacitance and better stability than the monometallic and pristine electrode materials, owing to the synergistic effect of Sn and ZrO_2 in a carbon nanofibrous matrix. NiO/RuO$_2$/CNF composites with different Ni/Ru weight ratios were prepared and tested in a asymmetrical supercapacitor device (NiO + RuO$_2$/CNF/AC) [122]. Among them, NiO/RuO$_2$/CNF with Ni/Ru = 1:3 delivered a specific capacitance up to about 60 F g^{-1} after 30 cycles. Furthermore, the mixed metal oxides/CNFs could maintain good flexibility and be explored as binder-free and conductive-free electrode materials for supercapacitors, such as Co$_x$O$-$Ag/CNFs and Mn-V-O/CNFs composites. The influences of calcination temperatures (450°C, 500°C, 550°C, and 600°C) and valence states of cobalt (Co^{2+}, Co^{3+}) on the electrochemical behaviors of Co$_x$O-Ag/CNFs composites were investigated [123]. The existence of Ag not only catalyzes the conversion from Co^{2+} to Co^{3+} but also improves the transportation of electrons and ions between the electrode and electrolyte. As a result, the sample that was calcined at 450°C possessed the best electrochemical properties. Meanwhile, the Co$^{2+}-$Ag/CNFs illustrated a higher capacitance than the Co^{2+}@Co$^{3+}-$Ag/CNFs, revealing that the electrochemical performance of Co^{2+} is better than that of Co^{3+}. The fabrication of Mn$-$V$-$O/CNFs with a high surface area (164.7 m^2 g^{-1}) as anodes in asymmetrical supercapacitor devices with high stability was reported by Samir and coworkers [124]. A maximum energy density of 37.77 Wh kg^{-1} at a power density of 900 W kg^{-1} and

Supercapacitors based on electrospun metal oxide nanofibers 377

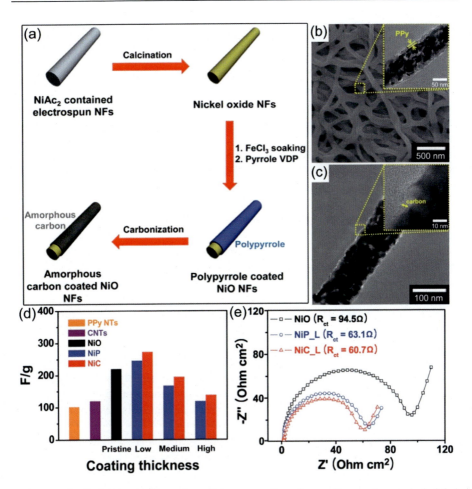

Figure 15.4 (A) Schematic illustration of the preparation of amorphous carbon-coated nickel (II) oxide nanofibers. (B) field emission scanning electron microscope and transmission electron microscope (inset) images of PPy-coated nickel(II) oxide nanofibers. (C) Transmission electron microscope and high resolution-transmission electron microscope (inset) images of NiC_L. (D) Specific capacitance value of the samples at 0.3 A g^{-1}. (E) Electrochemical impedance spectra of three products.
[119] Copyright 2014, Royal Society of Chemistry.

superior Coulombic efficiency were obtained, which are indexed to features of multiple oxidation states of the Mn-V oxide, mesoporous structure, and rough surface.

Three Li$_4$Ti$_5$O$_{12}$/C hybrids produced by different synthetic methods were prepared and utilized as electrode materials for supercapacitors. The first example is the fabrication of hypernetworked Li$_4$Ti$_5$O$_{12}$/carbon hybrid NF sheets using hydroxide/titanium (IV) bis(ammonium-lactato)dihydroxide/ammonium persulfate as precursors through a vapor polymerization of polypyrrole and followed by a calcination process [125]. The assembled cell ($Li_4Ti_5O_{12}$/carbon//AC) delivers a higher energy density of 91 Wh kg^{-1} at the power density of 50 W kg^{-1} than the AC//AC symmetric electrode system. The second example is the preparation of $Li_4Ti_5O_{12}$/activated carbon hybrid nanotubes through electrospinning, in situ TiO_2 sol−gel reaction, hydrothermal reaction, carbonization, and activation treatment [126]. Particularly, during the synthetic process, electrospun PVA NFs have been used as sacrificial templates for the growth of TiO_2 and were involved in the hydrothermal reaction in the $LiEo_x$/LiCl/IPA solution to construct nanotube structure. Compared with the pure $Li_4Ti_5O_{12}$, better charge-transfer kinetics was achieved for the $Li_4Ti_5O_{12}$/activated carbon electrode materials. The third example of the fabrication of porous $Li_4Ti_5O_{12}$/CNFs was reported by Chung and coworkers [127], exhibiting a high capacity, superior rate properties, and outstanding cycle stability when serving as anode materials for hybrid supercapacitors.

Another efficient strategy is the coverage of metal oxides on electrospun CNFs to produce CNF/metal oxide composites with a coaxial heterostructure, which can provide abundant active sites and convenient electron transfer channels. In comparison with metal oxides embedded in carbon materials, the metal oxides grown on the surface of substrates possess more controllable morphologies, which can enlarge the contact areas between the active materials and electrolytes. Furthermore, the morphologies of CNFs can be designed, and a large number of active ingredients can be introduced into CNFs, resulting in better electrochemical properties of the composites.

The modification of RuO_2 on the surface of CNFs was fabricated via an electrospinning and sol−gel process, which can be utilized as binder-free supercapacitor electrodes, illustrating a favorable specific capacitance of 546 F g^{-1} [128]. Similarly, a large number of electrospun carbon nanomaterials have been used as substrates for the growth of MnO_2 such as CNFs, hollow CNFs, porous CNFs, activated CNFs, and hybrid CNFs. For example, the γ-MnO_2/CNFs composite reported by Wang and coworkers displayed various morphologies of γ-MnO_2 involved from nanoparticles to nanoneedles with the increase in reaction time [129]. After a hot pressing process, the flexible CNFs could be obtained for the growth of MnO_2. This composite could serve as free-standing electrodes for supercapacitor, showing a high specific capacitance of 557 F g^{-1} at 1 A g^{-1} in 0.1 M Na_2SO_4 [130]. Three porous CNFs were also achieved via KOH activation, NH_3 activation, and P123 template for the growth of MnO_2, illustrating high specific surface areas and superior capacitive performances [131−133]. An electrospinning technique with a dual nozzle was applied to the preparation of hollow CNFs, which can be hybridized with MnO_2 to show a specific capacitance of 237 F g^{-1} [134]. Recently, MnO_2 on multichannel hollow CNFs was obtained via single-nozzle coelectrospinning of PAN and PMMA precursors. The synthetic routes, corresponding morphology, and electrochemical performance of this material are shown in Fig. 15.5. Both the 1D and hollow structures contribute to the efficient diffusion and contact between active materials and electrolyte, leading to a high specific capacitance of 855 F g^{-1} and good cycling behavior ($\sim$87.3% capacitance retention, 5000 cycles) [135]. For the hybrid CNFs there are not only CNT-CNFs but also some metal oxide−embedded CNFs for the growth of MnO_2. Wang and coworkers have successfully synthesized

Supercapacitors based on electrospun metal oxide nanofibers 379

CNT-CNF/MnO$_2$ and explored the effect of different temperatures on the electrochemical performance of the materials. As a consequence, CNT-CNF/MnO$_2$ showed a higher specific capacitance (374 F g^{-1}) and better rate capability (53.4%) than CNF/MnO$_2$. When temperature increases from 0°C to 75°C, the specific capacitance increases from 365 to 546 F g^{-1} (1 A g^{-1}), while the Coulombic efficiency decreases [112,136]. Furthermore, CNT/CNF composites have also been prepared by using electrospun

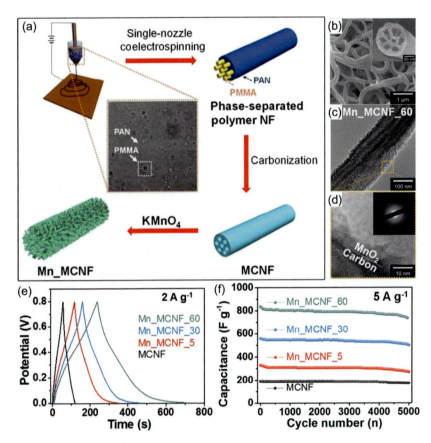

Figure 15.5 (A) Schematic demonstration for the fabrication of multichannel carbon nanofiber and multidimensional MnO$_2$ nanohair-decorated hybrid multichannel carbon nanofiber (inset is optical microscopy image of immiscible polymer solution). (B) Low-resolution and high-resolution cross sectional (inset) field emission scanning electron microscope of multichannel carbon nanofiber. (C) Transmission electron microscopy image of MnO$_2$ nanohair-decorated hybrid multichannel carbon nanofibers after a reaction time of 60 min. (D) High-resolution transmission electron microscope images of MnO$_2$ nanohair-decorated hybrid multichannel carbon nanofibers. (E) Galvanostatic charge discharge curves at 2 A g^{-1} of the multichannel carbon nanofiber and MnO$_2$ nanoair decorated multichannel carbon nanofiber samples. (F) Cycling performance for the samples.
[134] Copyright 2015, Royal Society of Chemistry.

CNFs as skeleton via a vapor deposition process, displaying a high specific surface area and large electrical conductivity (1250 S cm^{-1}). Then MnO_2 was coated on CNT/CNF via an in situ redox deposition, which delivered a higher specific capacitance (517 F g^{-1}, 5 mV s^{-1}), better rate capability, and better cycling stability than the CNF/MnO_2 sample [137]. Our group designed three types of electrospun 1D C/MO$_x$ (M = Mn, Cu, Co) NFs as substrates for the growth of MnO_2 through a redox reaction between low-valence MO$_x$ and $KMnO_4$ [138]. During thermal treatment the existence of low-valence MO$_x$ promotes the conductivity and porosity of the CNF substrates because of the metal catalysis. Compared with CNFs@MnO_2, C/MO$_x$@MnO_2 showed superior electrochemical behaviors, which originates from the desired functions of the individual components and the extra synergistic effect. Recently, FeO$_x$−CNF/MnO_2 was also fabricated and served as electrode materials for supercapacitors [139], resulting in a higher energy density (80.2 Wh kg^{-1}) and power density (57.7 kW kg^{-1}) compared to reported values based on MnO_2, including MnO_2/CNT, MnO_2/graphene, and MnO_2/Zn_2SnO_4. Furthermore, CNFs/Co_3O_4 has been hybridized with MnO_2 though an in situ redox reaction. The self-standing CNFs/Co_3O_4/MnO_2 NF membranes with good flexibility as electrode materials showed a superior capacitive performance of 840 F g^{-1} compared to CNFs/MnO_2 membranes [140], stemming from the increased electrical conductivity and electrochemical activity provided by Co_3O_4-doped CNF backbone. This material could be applied for the flexible supercapacitor device with a maximum energy density of 49.8 Wh kg^{-1}, revealing its bright prospects for use in flexible and lightweight energy storage devices. Additionally, the combination of CNF and MnO_2 was achieved by electrochemical deposition for a metal oxide film supercapacitor in another report, in which α-MnO_2 on superaligned electrospun CNFs yielded a specific capacitance of 141 F g^{-1} and an energy density of 12.5 Wh kg^{-1} [141].

In addition to MnO_2, many other metal oxides have been decorated on the surface of electrospun CNFs for supercapacitor electrodes. For instance, CNFs derived from PAN with polyaniline and graphene sol−gel precursors were used as support for NiO via a chemical deposition and calcination process. Different calcination temperatures led to various electrochemical behaviors of this type of composite. The highest specific capacitance (738 F g^{-1}) was reached at a calcination temperature of 400°C [142]. Compared with Ni-based or Co-based oxides/CNFs, more Ni-based or Co-based hydroxides/CNFs have also been investigated as supercapacitor electrode materials [143−148]. For example, flexible Ni(OH)$_2$/CNFs membrane was prepared with Ni(OH)$_2$ nanoplatelets vertically on CNFs for supercapacitor electrodes, showing 3D macroporous architectures for rapid contact of electrolyte with an active Ni(OH)$_2$ component [146]. The resultant Ni(OH)$_2$/CNF composite membrane displayed a high specific capacitance of 2523 F g^{-1} based on the mass of Ni(OH)$_2$. Furthermore, this membrane exhibited long-term stability with 83% capacitance retention after 1000 cycles, representing a favorable durability performance. The decoration of ZnO on the surface of CNFs is also a strategy to fabricate the supercapacitor electrodes. Typically, CNFs wrapped with ZnO nanoflakes displayed a higher specific capacitance (260 F g^{-1}) than pristine ZnO NFs (118 F g^{-1}) [149]. Interestingly, Mn-doped ZnO/CNFs derived from ZIF-8/PAN were reported by Samuel and coworkers. This composite provides fast electron transfer pathways and high specific surface areas, contributing to superior specific capacitance ($501 \, \text{F g}^{-1}$), excellent cycling stability ($>92\%$ of their initial capacitance after 10,000 cycles), and high energy densities ($72.1 \, \text{Wh kg}^{-1}$) [150]. SnO_2 with different morphologies coated on CNFs was also achieved by solvothermal reaction, chemical deposition, and hydrothermal methods. In detail, the coverage density of SnO_2 nanoparticles on CNFs could be controlled by tailoring the mass ratio of precursors [151]. When the mass ratio of CNFs to stannic chloride pentahydrate ($SnCl_4 \cdot 5H_2O$) reaches 1:7, the electrode material delivered the maximum specific capacitance ($187 \, \text{F g}^{-1}$ at $20 \, \text{mV s}^{-1}$). Copper and its oxides as the hard templates were introduced into CNFs to generate holes, which can be combined with SnO_2 dots as substrates [152]. SnO_2 dots/CNFs were used in the flexible supercapacitor, which exhibited ideal electrochemical performance. In another report, SnO_2 nanoflowers/CNFs were also obtained for batteries [153]. In addition, V_2O_5 films/self-standing CNFs and Fe_3O_4 nanosheets/CNFs have served as electrode materials for supercapacitors, displaying favorable electrochemical behaviors [154,155]. TiO_2 nanoparticles covered on Cu/CuO/porous CNFs were obtained via a course of electrospinning and hydrothermal reaction together with air stabilization and carbonization processes, which were used as electrode materials in a solid-state hybrid supercapacitor [156]. The porous CNFs with a high aspect length-to-volume ratio allow rapid electron transport. Cu with good electrical conductivity contributes to good rate properties. TiO_2 nanoparticles provide both enhanced material mechanical strength and pseudocapacitive activity. These characteristics result in the outstanding rate capability and long cycling stability of $Cu/CuO/PCNF/TiO_2$ and a high energy density ($45.83 \, \text{Wh kg}^{-1}$) for the supercapacitor device.

15.3.3 Metal oxide (metal)/carbon nanofibers/conducting polymer composites

Conducting polymers as well as metal oxides realize the storage of charges on the basis of the fast and reversible faradaic redox reactions. In the meantime the high electrical conductivities of conducting polymers such as polyaniline (PANI), PPy, and poly (ethylenedioxythiophene) (PEDOT) endow them with outstanding electrochemical performance. However, the poor stabilities of conducting polymers become a main obstacle for their further developments. The hybridization of conducting polymer with metal oxide and carbon materials has become a significant attempt to improve the electrochemical performance of electrode materials.

PPy was usually coated on the electrospun CNF via an in situ polymerization strategy. For example, $CNF-MnO_2/PPy$ composite was prepared via such an approach as an electrode material for supercapacitors, showing good electrochemical reversibility at a high sweep rate and yielding a specific capacitance superior to those of $CNF-MnO_2$, CNF-PPy, and CNFs [157]. Similarly, PPy-coated SnO_2/Co_3O_4 NFs (CPSC) were synthesized via a vapor deposition polymerization [158], which can be combined with rGO through freeze drying and thermal reduction to form a CPSC-rGO composite (Fig. 15.6). The specific capacitance of the

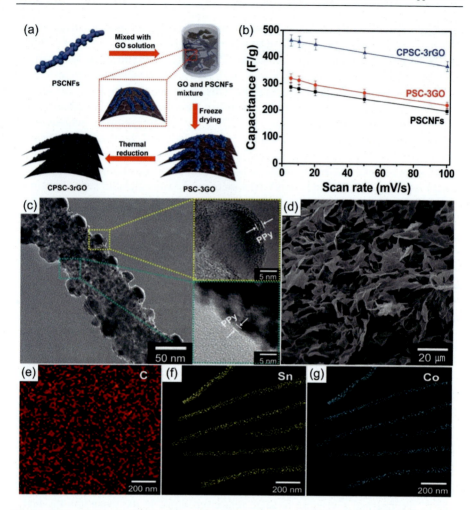

Figure 15.6 (A) Illustrative diagram of the fabrication steps for the CPSC−3rGO nanostructure. (B) Specific capacitance at various scan rates of samples. (C) Transmission electron microscope images of PPy-coated SC nanofibers (inset: enlarged transmission electron microscope). (D) Scanning electron microscope images of CPSC-3rGO nanostructures. Electron energy loss spectroscopy dot mapping of CPSC-3rGO of layer components of (E) C, (F) Sn, and (G) Co atoms.
[159] Copyright 2014, Royal Society of Chemistry.

CPSC−3rGO was 446 F g^{-1}, which is larger than that of PSCNFs (270 F g^{-1}), PSC-3GO (285 F g^{-1}), and 3D rGO (150 F g^{-1}). Furthermore, a flexible symmetric supercapacitor based on the CPSC−rGO composite was assembled, displaying good stability after many bending cycles. Recently, a triaxial electrospinning process was performed to fabricate Mn$_3$O$_4$/thermally exfoliated graphene oxide (TEGO) fibers and Mn$_3$O$_4$/graphene nanoplatelet (GNP) fibers [159], which were used as substrates for the growth of polyaniline by in situ chemical polymerization. The resulting PANI/Mn$_3$O$_4$/GNP-CNFs demonstrated better capacitive behavior and cycling stability as compared to PANI/Mn$_3$O$_4$/TEGO-CNFs, which is attributed not only to the fast ion and electron transport provided by graphene single platelets but also to abundant electroactive sites and shorter diffusion paths supplied by the Mn$_3$O$_4$ and the PANI shell. In addition to the graphene materials, carboxylated graphene quantum dot were also introduced into PVA$-$Co$_3$O$_4$ NFs [160]. PEDOT was coated onto their surface via electropolymerization, which offers a high conductivity and specific surface area, contributing to a specific capacitance of 361.97 F g^{-1}, low equivalent series resistance, and excellent stability (retention of 96% after 1000 cycles).

15.3.4 Other composites

In addition to the aforementioned EMONF$-$based composites, hierarchical MnO$_2$ nanosheets on 1D TiN NFs were fabricated as electrode materials for supercapacitor [161]. The TiN NFs with good electrical conductivity were obtained through nitridation of TiO$_2$ NFs. A high specific capacitance of 386 F g^{-1} (1 A g^{-1}) and excellent cycling stability of 111.7% capacity retention after 4000 cycles (6 A g^{-1}) were achieved for this material.

15.4 Conclusion

In summary, a large number of EMONFs and their composites have been designed and used as electrode materials in the field of supercapacitors, such as single metal oxides, bimetallic or polymetallic oxides, metal oxide/metal oxide (metal hydroxide, metal) composites, metal oxide/carbon composites, and metal oxide (metal)/ CNFs/conducting polymer composites. To improve the electrochemical performance of these materials, the manipulation of the architectures of these materials, including solid NFs, hollow NFs, porous NFs and core-sheath NFs, has been achieved by tailoring the spinning parameters, spinning solution composition, and calcination process, leading to their larger specific surface areas, more abundant pore structures, higher electrochemical activities, and better conductivities. In addition, a large number of symmetrical or asymmetrical supercapacitors have been constructed, revealing the brilliant practical prospect of EMONF$-$based electrode materials. A part of these EMONF$-$based materials even exhibit good prospects for use in flexible devices.

Although significant progress has been made in fabricating electrospun metal oxide-based electrode materials in supercapacitors, there are still significant challenges and obstacles to overcome. In terms of architecture, EMONFs and their composites with hollow and hierarchical structures exhibit advantages in ion diffusion and internal diffusion, which need to be further explored. For the compositions, compared with metal oxides, metal oxide-based composites display superior electrochemical behaviors. To achieve a synergistic effect, the types of EMONF—based composites as electrode materials can be extended to a more advanced level, such as the hybridization of the EMONFs with metal sulfide, metal phosphide, MXenes, MOF, COF, and so on. Furthermore, some EMONF—based membranes show great potential for use in wearable flexible devices, but a better balance of mechanical performance and their electrochemical properties is necessary to meet a wider range of practical applications.

References

[1] G.Z. Chen, Supercapacitor and supercapattery as emerging electrochemical energy stores, Int. Mater. Rev. 62 (4) (2016) 173—202.

[2] M. Salanne, B. Rotenberg, K. Naoi, K. Kaneko, P.L. Taberna, C.P. Grey, et al., Efficient storage mechanisms for building better supercapacitors, Nat. Energy 1 (6) (2016) 16070.

[3] A. Eftekhari, The mechanism of ultrafast supercapacitors, J. Mater. Chem. A 6 (7) (2018) 2866—2876.

[4] Z. Yu, L. Tetard, L. Zhai, J. Thomas, Supercapacitor electrode materials: nanostructures from 0 to 3 dimensions, Energy Environ. Sci. 8 (3) (2015) 702—730.

[5] S. Zhai, L. Wei, H.E. Karahan, X. Chen, C. Wang, X. Zhang, et al., 2D materials for 1D electrochemical energy storage devices, Energy Storage Mater. 19 (2019) 102—123.

[6] C. Lethien, J. Le Bideau, T. Brousse, Challenges and prospects of 3D micro-supercapacitors for powering the internet of things, Energy Environ. Sci. 12 (2019) 96—115.

[7] G. Wang, L. Zhang, J. Zhang, A review of electrode materials for electrochemical supercapacitors, Chem. Soc. Rev. 41 (2) (2012) 797—828.

[8] Y. Zhao, J. Liu, M. Horn, N. Motta, M. Hu, Y. Li, Recent advancements in metal organic framework based electrodes for supercapacitors, Sci. China Mater. 61 (2) (2018) 159—184.

[9] M.S. Balogun, Y. Huang, W. Qiu, H. Yang, H. Ji, Y. Tong, Updates on the development of nanostructured transition metal nitrides for electrochemical energy storage and water splitting, Mater. Today 20 (8) (2017) 425—451.

[10] L. Li, J. Wen, X. Zhang, Progress of two-dimensional $Ti_3C_2T_x$ in supercapacitors, ChemSusChem 13 (6) (2019) 1296—1329.

[11] C. Ji, H.Y. Mi, S.C. Yang, Latest advances in supercapacitors: From new electrode materials to novel device designs, Chin. Sci. Bull. 1 (64) (2019) 9—34.

[12] T.H. Murray, Stirring the simmering "designer baby" pot, Science 343 (6176) (2014) 1208—1210.

[13] T. Brousse, D. Bélanger, J.W. Long, To be or not to be pseudocapacitive? J. Electrochem. Soc. 162 (5) (2015) A5185—A5189.

[14] T.S. Hyun, H.L. Tuller, D.Y. Youn, H.G. Kim, I.D. Kim, Facile synthesis and electrochemical properties of RuO_2 nanofibers with ionically conducting hydrous layer, J. Mater. Chem. 20 (41) (2010) 9172.

[15] J.G. Wang, F. Kang, B. Wei, Engineering of MnO_2-based nanocomposites for high-performance supercapacitors, Prog. Mater. Sci. 74 (2015) 51—124.

[16] J. Bhagwan, A. Sahoo, K.L. Yadav, Y. Sharma, Porous, one dimensional and high aspect ratio Mn_3O_4 nanofibers: fabrication and optimization for enhanced supercapacitive properties, Electrochim. Acta 174 (2015) 992−1001.

[17] X. Ma, M. Xue, F. Li, J. Chen, D. Chen, X. Wang, et al., Gradual-order enhanced stability: a frozen section of electrospun nanofibers for energy storage, Nanoscale 7 (19) (2015) 8715−8719.

[18] J.H. Lee, T.Y. Yang, H.Y. Kang, D.H. Nam, N.R. Kim, Y.Y. Lee, et al., Designing thermal and electrochemical oxidation processes for δ-MnO_2 nanofibers for high-performance electrochemical capacitors, J. Mater. Chem. A 2 (20) (2014) 7197−7204.

[19] E. Lee, T. Lee, B.S. Kim, Electrospun nanofiber of hybrid manganese oxides for supercapacitor: relevance to mixed inorganic interfaces, J. Power Sources 255 (2014) 335−340.

[20] G. Wee, H.Z. Soh, Y.L. Cheah, S.G. Mhaisalkar, M. Srinivasan, Synthesis and electrochemical properties of electrospun V_2O_5 nanofibers as supercapacitor electrodes, J. Mater. Chem. 20 (32) (2010) 6720.

[21] W.F. Mak, G. Wee, V. Aravindan, N. Gupta, S.G. Mhaisalkar, S. Madhavi, High-energy density asymmetric supercapacitor based on electrospun vanadium pentoxide and polyaniline nanofibers in aqueous electrolyte, J. Electrochem. Soc. 159 (9) (2012) A1481−A1488.

[22] H. Jiang, H. Niu, X. Yang, Z. Sun, F. Li, Q. Wang, et al., Flexible Fe_2O_3 and V_2O_5 nanofibers as binder-free electrodes for high-performance all-solid-state asymmetric supercapacitors, Chem. A Eur. J. 24 (42) (2018) 10683−10688.

[23] V. Aravindan, Y.L. Cheah, W.F. Mak, G. Wee, B.V.R. Chowdari, S. Madhavi, Fabrication of high energy-density hybrid supercapacitors using electrospun V_2O_5 nanofibers with a self-supported carbon nanotube network, ChemPlusChem 77 (7) (2012) 570−575.

[24] N.L. Lala, R. Jose, M.M. Yusoff, S. Ramakrishna, Continuous tubular nanofibers of vanadium pentoxide by electrospinning for energy storage devices, J. Nanopart. Res. 14 (11) (2012) 1201.

[25] D.S. Kong, J.M. Wang, H.B. Shao, J.Q. Zhang, C. Cao, Electrochemical fabrication of a porous nanostructured nickel hydroxide film electrode with superior pseudocapacitive performance, J. Alloy. Compd. 509 (18) (2011) 5611−5616.

[26] B. Ren, M. Fan, Q. Liu, J. Wang, D. Song, X. Bai, Hollow NiO nanofibers modified by citric acid and the performances as supercapacitor electrode, Electrochim. Acta 92 (2013) 197−204.

[27] B. Vidhyadharan, N.K.M. Zain, I.I. Misnon, R.A. Aziz, J. Ismail, M.M. Yusoff, et al., High performance supercapacitor electrodes from electrospun nickel oxide nanowires, J. Alloy. Compd. 610 (2014) 143−150.

[28] J. Jia, F. Luo, C. Gao, C. Suo, X. Wang, H. Song, et al., Synthesis of La-doped NiO nanofibers and their electrochemical properties as electrode for supercapacitors, Ceram. Int. 40 (5) (2014) 6973−6977.

[29] M.S. Kolathodi, M. Palei, T.S. Natarajan, Electrospun NiO nanofibers as cathode materials for high performance asymmetric supercapacitors, J. Mater. Chem. A 3 (14) (2015) 7513−7522.

[30] M. Kundu, L. Liu, Binder-free electrodes consisting of porous NiO nanofibers directly electrospun on nickel foam for high-rate supercapacitors, Mater. Lett. 144 (2015) 114−118.

[31] M. Kumar, A. Subramania, K. Balakrishnan, Preparation of electrospun Co_3O_4 nanofibers as electrode material for high performance asymmetric supercapacitors, Electrochim. Acta 149 (2014) 152−158.

[32] D. Wang, Q. Wang, T. Wang, Morphology-controllable synthesis of cobalt oxalates and their conversion to mesoporous Co_3O_4 nanostructures for application in supercapacitors, Inorg. Chem. 50 (14) (2011) 6482−6492.

[33] X. Xia, J. Tu, Y. Zhang, Y. Mai, X. Wang, C. Gu, et al., Freestanding Co_3O_4 nanowire array for high performance supercapacitors, RSC Adv. 2 (5) (2012) 1835.

[34] L. Wang, X. Liu, X. Wang, X. Yang, L. Lu, Preparation and electrochemical properties of mesoporous Co_3O_4 crater-like microspheres as supercapacitor electrode materials, Curr. Appl. Phys. 10 (6) (2010) 1422−1426. \.

[35] Y. Asano, T. Komatsu, K. Murashiro, K. Hoshino, Capacitance studies of cobalt compound nanowires prepared *via* electrodeposition, J. Power Sources 196 (11) (2011) 5215−5222.

[36] R. Tummala, R.K. Guduru, P.S. Mohanty, Nanostructured Co_3O_4 electrodes for supercapacitor applications from plasma spray technique, J. Power Sources 209 (2012) 44−51.

[37] B. Ren, M. Fan, J. Wang, X. Jing, X. Bai, The effect of pluronic P123 on the capacitive behavior of Co_3O_4 as a self-assembeled additive, J. Electrochem. Soc. 160 (9) (2013) E79−E83.

[38] C. Niu, J. Meng, X. Wang, C. Han, M. Yan, K. Zhao, et al., General synthesis of complex nanotubes by gradient electrospinning and controlled pyrolysis, Nat. Commun. 6 (1) (2015) 7402.

[39] G. Binitha, M.S.A.A. Madhavan, P. Praveen, A. Balakrishnan, K.R.V. Subramanian, N. Sivakumar, Electrospun α-Fe_2O_3 nanostructures for supercapacitor applications, J. Mater. Chem. A 1 (38) (2013) 11698.

[40] B. Vidhyadharan, I.I. Misnon, R.A. Aziz, K.P. Padmasree, M.M. Yusoff, R. Jose, Superior supercapacitive performance in electrospun copper oxide nanowire electrodes, J. Mater. Chem. A 2 (18) (2014) 6578−6588.

[41] G. Wang, J. Huang, S. Chen, Y. Gao, D. Cao, Preparation and supercapacitance of CuO nanosheet arrays grown on nickel foam, J. Power Sources 196 (13) (2011) 5756−5760.

[42] H. Zhang, J. Feng, M. Zhang, Preparation of flower-like CuO by a simple chemical precipitation method and their application as electrode materials for capacitor, Mater. Res. Bull. 43 (12) (2008) 3221−3226.

[43] Z. Endut, M. Hamdi, W.J. Basirun, Pseudocapacitive performance of vertical copper oxide nanoflakes, Thin Solid. Films 528 (2013) 213−216.

[44] Y.K. Hsu, Y.C. Chen, Y.G. Lin, Characteristics and electrochemical performances of lotus-like $CuO/Cu(OH)_2$ hybrid material electrodes, J. Electroanalytical Chem. 673 (2012) 43−47.

[45] D.P. Dubal, G.S. Gund, C.D. Lokhande, R. Holze, CuO cauliflowers for supercapacitor application: novel potentiodynamic deposition, Mater. Res. Bull. 48 (2) (2013) 923−928.

[46] Y. Li, S. Chang, X. Liu, J. Huang, J. Yin, G. Wang, et al., Nanostructured CuO directly grown on copper foam and their supercapacitance performance, Electrochim. Acta 85 (2012) 393−398.

[47] B. Vidhyadharan, I.I. Misnon, J. Ismail, M.M. Yusoff, R. Jose, High performance asymmetric supercapacitors using electrospun copper oxide nanowires anode, J. Alloy. Compd. 633 (2015) 22−30.

[48] M.S. Kolathodi, T.S. Natarajan, Development of high-performance flexible solid state supercapacitor based on activated carbon and electrospun TiO_2 nanofibers, Scr. Mater 101 (2015) 84−86.

[49] B. Vidyadharan, P.S. Archana, J. Ismail, M.M. Yusoff, R. Jose, Improved supercapacitive charge storage in electrospun niobium doped titania nanowires, RSC Adv. 5 (62) (2015) 50087−50097.

[50] A. Tyagi, N. Singh, Y. Sharma, R.K. Gupta, Improved supercapacitive performance in electrospun TiO_2 nanofibers through Ta-doping for electrochemical capacitor applications, Catal. Today 325 (2019) 33−40.

[51] Y. Zhang, L. Li, H. Su, W. Huang, X. Dong, Binary metal oxide: advanced energy storage materials in supercapacitors, J. Mater. Chem. A 3 (1) (2015) 43−59.

[52] P.M. Wilde, T.J. Guther, R. Oesten, J. Garche, Strontium ruthenate perovskite as the active material for supercapacitors, J. Electroanal. Chem. 461 (1999) 154160.

[53] Y. Cao, B. Lin, Y. Sun, H. Yang, X. Zhang, Sr-doped lanthanum nickelate nanofibers for high energy density supercapacitors, Electrochim. Acta 174 (2015) 41−50.

[54] Y. Cao, B. Lin, Y. Sun, H. Yang, X. Zhang, Symmetric/asymmetric supercapacitor based on the perovskite-type lanthanum cobaltate nanofibers with Sr-substitution, Electrochim. Acta 178 (2015) 398−406.

[55] W. Wang, B. Lin, H. Zhang, Y. Sun, X. Zhang, H. Yang, Synthesis, morphology and electrochemical performances of perovskite-type oxide $La_xSr_{1-x}FeO_3$ nanofibers prepared by electrospinning, J. Phys. Chem. Solids 124 (2019) 144−150.

[56] Y. Cao, B. Lin, Y. Sun, H. Yang, X. Zhang, Structure, morphology and electrochemical properties of $La_xSr_{1-x}Co_{0.1}Mn_{0.9}O_{3-\delta}$ perovskite nanofibers prepared by electrospinning method, J. Alloy. Compd. 624 (2015) 31−39.

[57] Y. Cao, B. Lin, Y. Sun, H. Yang, X. Zhang, Synthesis, structure and electrochemical properties of lanthanum manganese nanofibers doped with Sr and Cu, J. Alloy. Compd. 638 (2015) 204−213.

[58] G. George, S.L. Jackson, C.Q. Luo, D. Fang, D. Luo, D. Hu, et al., Effect of doping on the performance of high-crystalline $SrMnO_3$ perovskite nanofibers as a supercapacitor electrode, Ceram. Int. 44 (2018) 21982−21992.

[59] N. Arjun, G.T. Pan, T.C.K. Yang, The exploration of lanthanum based perovskites and their complementary electrolytes for the supercapacitor applications, Results Phys. 7 (2017) 920−926.

[60] L. Li, S. Peng, Y. Cheah, P. Teh, J. Wang, G. Wee, et al., Electrospun porous $NiCo_2O_4$ nanotubes as advanced electrodes for electrochemical capacitors, Chem. A Eur. J. 19 (19) (2013) 5892−5898.

[61] G.Q. Zhang, H.B. Wu, H.E. Hoster, M.B. Chan-Park, X.W. Lou, Single-crystalline $NiCo_2O_4$ nanoneedle arrays grown on conductive substrates as binder-free electrodes for high-performance supercapacitors, Energy Environ. Sci. 5 (11) (2012) 9453.

[62] H. Jiang, J. Ma, C. Li, Hierarchical porous $NiCo_2O_4$ nanowires for high-rate supercapacitors, Chem. Commun. 48 (37) (2012) 4465.

[63] H. Wang, Q. Gao, L. Jiang, Facile approach to prepare nickel cobaltite nanowire materials for supercapacitors, Small 7 (17) (2011) 2454−2459.

[64] F.L. Zhu, J.X. Zhao, Y.L. Cheng, H.B. Li, X.B. Yan, Magnetic and electrochemical properties of $NiCo_2O_4$ microbelts fabricated by electrospinning, Acta Phys. Chim. Sin. 28 (12) (2012) 2874−2878.

[65] S. Peng, L. Li, Y. Hu, M. Srinivasan, F. Cheng, J. Chen, et al., Fabrication of spinel one-dimensional architectures by single-spinneret electrospinning for energy storage applications, ACS Nano 9 (2) (2015) 1945−1954.

[66] G. Zhou, J. Zhu, Y. Chen, L. Mei, X. Duan, G. Zhang, et al., Simple method for the preparation of highly porous $ZnCo_2O_4$ nanotubes with enhanced electrochemical property for supercapacitor, Electrochim. Acta 123 (2014) 450−455.

[67] J. Bhagwan, N. Kumar, K.L. Yadav, Y. Sharma, Probing the electrical properties and energy storage performance of electrospun $ZnMn_2O_4$ nanofibers, Solid. State Ion. 321 (2018) 75−82.

[68] J. Bhagwan, S. Rani, V. Sivasankaran, K.L. Yadav, Y. Sharma, Improved energy storage, magnetic and electrical properties of aligned, mesoporous and high aspect ratio nanofibers of spinel-$NiMn_2O_4$, Appl. Surf. Sci. 426 (2017) 913−923.

[69] Y. Lu, M. Zhao, R. Luo, Q. Yu, J. Lv, W. Wang, et al., Electrospun porous $MnMoO_4$ nanotubes as high-performance electrodes for asymmetric supercapacitors, J. Solid. State Electrochem. 22 (3) (2017) 657−666.

[70] V. Sudheendra Budhiraju, A. Sharma, S. Sivakumar, Structurally stable mesoporous hierarchical $NiMoO_4$ hollow nanofibers for asymmetric supercapacitors with enhanced capacity and improved cycling stability, ChemElectroChem 4 (12) (2017) 3331−3339.

[71] R. Aswathy, T. Kesavan, K.T. Kumaran, P. Ragupathy, Octahedral high voltage $LiNi_{0.5}Mn_{1.5}O_4$ spinel cathode: enhanced capacity retention of hybrid aqueous capacitors with nitrogen doped grapheme, J. Mater. Chem. A 3 (23) (2015) 12386−12395.

[72] N. Arun, A. Jain, V. Aravindan, S. Jayaraman, W. Chui Ling, M.P. Srinivasan, et al., Nanostructured spinel $LiNi_{0.5}Mn_{1.5}O_4$ as new insertion anode for advanced Li-ion capacitors with high power capability, Nano Energy 12 (2015) 69−75.

[73] G. Bhuvanalogini, N. Muruganantham, V. Shobana, A. Subramania, Preparation, characterization, and evaluation of $LiNi_{0.4}Co_{0.6}O_2$ nanofibers for supercapacitor applications, J. Solid. State Electrochem. 18 (9) (2014) 2387−2392.

[74] V. Aravindan, J. Sundaramurthy, A. Jain, P.S. Kumar, W.C. Ling, S. Ramakrishna, et al., Unveiling $TiNb_2O_7$ as an insertion anode for lithium ion capacitors with high energy and power density, ChemSusChem 7 (7) (2014) 1858−1863.

[75] J.B. Lee, S.Y. Jeong, W.J. Moon, T.Y. Seong, H.J. Ahn, Preparation and characterization of electro-spun RuO_2-Ag_2O composite nanowires for electrochemical capacitors, J. Alloy. Compd. 509 (11) (2011) 4336−4340.

[76] D.Y. Youn, H.L. Tuller, T.S. Hyun, D.K. Choi, I.D. Kim, Facile Synthesis of Highly Conductive $RuO_2-Mn_3O_4$ Composite Nanofibers *via* Electrospinning and Their Electrochemical Properties, J. Electrochem. Soc. 158 (8) (2011) A970.

[77] T.S. Hyun, H.G. Kim, I.D. Kim, Facile synthesis and electrochemical properties of conducting $SrRuO_3-RuO_2$ composite nanofibre mats, J. Power Sources 195 (5) (2010) 1522−1528.

[78] T.S. Hyun, J.E. Kang, H.G. Kim, J.M. Hong, I.D. Kim, Electrochemical properties of MnO_x-RuO_2 nanofiber mats synthesized by co-electrospinning, Electrochem. Solid-State Lett. 12 (12) (2009) A225.

[79] L. Hu, Y. Deng, K. Liang, X. Liu, W. Hu, $LaNiO_3/NiO$ hollow nanofibers with mesoporous wall: a significant improvement in NiO electrodes for supercapacitors, J. Solid. State Electrochem. 19 (3) (2014) 629−637.

[80] B. Ren, M. Fan, B. Zhang, J. Wang, Novel hollow $NiO@Co_3O_4$ nanofibers for high-performance supercapacitors, J. Nanosci. Nanotechnol. 18 (10) (2018) 7004−7010.

[81] G. Nie, X. Lu, J. Lei, Z. Jiang, C. Wang, Electrospun V_2O_5-doped α-Fe_2O_3 composite nanotubes with tunable ferromagnetism for high-performance supercapacitor electrodes, J. Mater. Chem. A 2 (37) (2014) 15495.

[82] A.V. Radhamani, M. Krishna Surendra, M.S. Ramachandra Rao, Tailoring the supercapacitance of Mn_2O_3 nanofibers by nanocompositing with spinel-$ZnMn_2O_4$, Mater. Des. 139 (2018) 162−171.

[83] K. Mondal, C.Y. Tsai, S. Stout, S. Talapatra, Manganese oxide based hybrid nanofibers for supercapacitors, Mater. Lett. 148 (2015) 142−146.

[84] J. Zhu, Z. Xu, B. Lu, Ultrafine Au nanoparticles decorated $NiCo_2O_4$ nanotubes as anode material for high-performance supercapacitor and lithium-ion battery applications, Nano Energy 7 (2014) 114−123.

[85] S.H. Choi, T.S. Hyun, H. Lee, S.Y. Jang, S.G. Oh, I.D. Kim, Facile synthesis of highly conductive platinum nanofiber mats as conducting core for high rate redox supercapacitor, Electrochem. Solid-State Lett. 13 (6) (2010) A65.

[86] Y.R. Ahn, C.R. Park, S.M. Jo, D.Y. Kim, Enhanced charge-discharge characteristics of RuO_2 supercapacitors on heat-treated TiO_2 nanorods, Appl. Phys. Lett. 90 (12) (2007) 122106.

[87] D. Liu, Q. Wang, L. Qiao, F. Li, D. Wang, Z. Yang, et al., Preparation of nanonetworks of MnO_2 shell/Ni current collector core for high-performance supercapacitor electrodes, J. Mater. Chem. 22 (2) (2012) 483−487.

[88] M.S. Kolathodi, M. Palei, T.S. Natarajan, G. Singh, MnO_2 Encapsulated electrospun TiO_2 nanofibers as electrodes for asymmetric supercapacitors, Nanotechnology 31 (2019) 125401.

[89] E.P. Da Silva, A.F. Rubira, O.P. Ferreira, R. Silva, E.C. Muniz, *In situ* growth of manganese oxide nanosheets over titanium dioxide nanofibers and their performance as active material for supercapacitor, J. Colloid Interface Sci. 555 (2019) 373−382.

[90] L. He, Y. Shu, W. Li, M. Liu, Preparation of La0.7Sr0.3CoO3-δ (LSC)@MnO2 core/shell nanorods as high-performance electrode materials for supercapacitors, J. Mater. Sci Mater. Electron. 30 (2019) 17−25.

[91] Q. Li, R. Yan, Y.F. Zhang, L.M. Dong, Facile Synthesis of α-MoO_3/MnO_2 composite electrodes for high performamance supercapacitor, J. Ovonic Res. 14 (1) (2018) 1−7.

[92] H. Niu, D. Zhou, X. Yang, X. Li, Q. Wang, F. Qu, Towards three-dimensional hierarchical ZnO nanofiber@Ni(OH)$_2$ nanoflake core−shell heterostructures for high-performance asymmetric supercapacitors, J. Mater. Chem. A 3 (36) (2015) 18413−18421.

[93] T. Dong, M. Li, P. Wang, P. Yang, Synthesis of hierarchical tube-like yolk-shell Co_3O_4@$NiMoO_4$ for enhanced supercapacitor performance, Int. J. Hydrogen Energy 43 (31) (2018) 14569−14577.

[94] Y. Huang, Y.E. Miao, H. Lu, T. Liu, Hierarchical $ZnCo_2O_4$@$NiCo_2O_4$ core-sheath nanowires: bifunctionality towards high-performance supercapacitors and the oxygen-reduction reaction, Chem. A Eur. J. 21 (28) (2015) 10100−10108.

[95] K.S. Yang, C.H. Kim, B.H. Kim, Preparation and electrochemical properties of RuO_2-containing activated carbon nanofiber composites with hollow cores, Electrochim. Acta 174 (2015) 290−296.

[96] C.H. Kim, B.H. Kim, Electrochemical behavior of zinc oxide-based porous carbon composite nanofibers as an electrode for electrochemical capacitors, J. Electroanalytical Chem. 730 (2014) 1−9.

[97] C.H. Kim, B.H. Kim, Zinc oxide/activated carbon nanofiber composites for high-performance supercapacitor electrodes, J. Power Sources 274 (2015) 512−520.

[98] B.H. Kim, C.H. Kim, K.S. Yang, A. Rahy, D.J. Yang, Electrospun vanadium pentoxide/carbon nanofiber composites for supercapacitor electrodes, Electrochim. Acta 83 (2012) 335−340.

[99] B.H. Kim, K.S. Yang, D.J. Yang, Electrochemical behavior of activated carbon nanofiber-vanadium pentoxide composites for double-layer capacitors, Electrochim. Acta 109 (2013) 859−865.

[100] S.I. Yun, S.H. Kim, D.W. Kim, Y.A. Kim, B.H. Kim, Facile preparation and capacitive properties of low-cost carbon nanofibers with ZnO derived from lignin and pitch as supercapacitor electrodes, Carbon 149 (2019) 637−645.

[101] G. Huang, C. Li, J. Bai, X. Sun, H. Liang, Controllable-multichannel carbon nanofibers-based amorphous vanadium as binder-free and conductive-free electrode materials for supercapacitor, Int. J. Hydrogen Energy 41 (47) (2016) 22144–22154.

[102] K. Tang, Y. Li, Y. Li, H. Cao, Z. Zhang, Y. Zhang, et al., Self-reduced VO/VO$_x$/carbon nanofiber composite as binder-free electrode for supercapacitors, Electrochim. Acta 209 (2016) 709–718.

[103] X. Chen, B. Zhao, Y. Cai, et al., V-O-C composite nanofibers electrospun from solution precursors as binder- and conductive additive-free electrodes for supercapacitors with outstanding performance, Nanoscale 5 (24) (2013) 12589.

[104] K. Tang, Y. Li, H. Cao, F. Duan, J. Zhang, Y. Zhang, et al., Integrated electrospun carbon nanofibers with vanadium and single-walled carbon nanotubes through covalent bonds for high-performance supercapacitors, RSC Adv. 5 (50) (2015) 40163–40172.

[105] R. Thangappan, S. Kalaiselvam, A. Elayaperumal, R. Jayavel, Synthesis of graphene oxide/vanadium pentoxide composite nanofibers by electrospinning for supercapacitor applications, Solid. State Ion. 268 (2014) 321–325.

[106] S. Abouali, M. Akbari Garakani, B. Zhang, Z.L. Xu, E. Kamali Heidari, J. Huang, et al., Electrospun carbon nanofibers with in Situ encapsulated Co$_3$O$_4$ nanoparticles as electrodes for high-performance supercapacitors, ACS Appl. Mater. Interfaces 7 (24) (2015) 13503–13511.

[107] F. Zhang, C. Yuan, J. Zhu, J. Wang, X. Zhang, X.W.D. Lou, Flexible films derived from electrospun carbon nanofibers incorporated with Co$_3$O$_4$ hollow nanoparticles as self-supported electrodes for electrochemical capacitors, Adv. Funct. Mater. 23 (31) (2013) 3909–3915.

[108] O. Pech, S. Maensiri, Electrochemical performances of electrospun carbon nanofibers, interconnected carbon nanofibers, and carbon-manganese oxide composite nanofibers, J. Alloy. Compd. 781 (2019) 541–552.

[109] L. Shi, H. He, Y. Fang, Y. Jia, B. Luo, L. Zhi, Effect of heating rate on the electrochemical performance of MnO$_x$@CNF nanocomposites as supercapacitor electrodes, Chin. Sci. Bull. 59 (16) (2014) 1832–1837.

[110] X. Zhao, Y. Du, Y. Li, Q. Zhang, Encapsulation of manganese oxides nanocrystals in electrospun carbon nanofibers as free-standing electrode for supercapacitors, Ceram. Int. 41 (6) (2015) 7402–7410.

[111] X. Liu, M. Naylor Marlow, S.J. Cooper, B. Song, X. Chen, N.P. Brandon, et al., Flexible all-fiber electrospun supercapacitor, J. Power Sources 384 (2018) 264–269.

[112] J.G. Wang, Y. Yang, Z.H. Huang, F. Kang, Effect of temperature on the pseudo-capacitive behavior of freestanding MnO$_2$@carbon nanofibers composites electrodes in mild electrolyte, J. Power Sources 24 (2013) 86–92.

[113] J.H. Jeong, B.H. Kim, Synergistic effects of pitch and poly(methyl methacrylate) on the morphological and capacitive properties of MnO$_2$/carbon nanofiber composites, J. Electroanalytical Chem. 809 (2018) 130–135.

[114] O.S. Kwon, T. Kim, J.S. Lee, S.J. Park, H.W. Park, M. Kang, et al., Fabrication of graphene sheets intercalated with manganese oxide/carbon nanofibers: toward high-capacity energy storage, Small 9 (2) (2012) 248–254.

[115] S. Saha, P. Maji, D.A. Pethsangave, A. Roy, A. Ray, S. Some, et al., Effect of morphological ordering on the electrochemical performance of MnO$_2$-Graphene oxide composite, Electrochim. Acta 317 (2019) 199–210.

[116] E. Samuel, B. Joshi, H.S. Jo, Y.I. Kim, M.T. Swihart, J.M. Yun, et al., Flexible and freestanding core-shell SnOx/carbon nanofiber mats for high-performance supercapacitors, J. Alloy. Compd. 201 728 1362-1371.

[117] J. Liu, X. Sun, C. Li, J. Bai, Tin-embedded carbon nanofibers as flexible and free-standing electrode materials for high-performance supercapacitors, Ionics 25 (2019) 4875−4890.

[118] B. Pant, M. Park, S.J. Park, TiO_2 NPs assembled into a carbon nanofiber composite electrode by a one-step electrospinning process for supercapacitor applications, Polymers 11 (5) (2019) 899.

[119] L. Thirugnanam, R. Sundara, Few layer graphene wrapped mixed phase TiO_2 nanofiber as a potential electrode material for high performance supercapacitor applications, Appl. Surf. Sci. 444 (2018) 414−422.

[120] D.H. Shin, J.S. Lee, J. Jun, J. Jang, Fabrication of amorphous carbon-coated NiO nanofibers for electrochemical capacitor applications, J. Mater. Chem. A 2 (10) (2014) 3364−3371.

[121] Y.S. Jang, T. Amna, M.S. Hassan, J.L. Gu, I.S. Kim, H.C. Kim, et al., Improved supercapacitor potential and antibacterial activity of bimetallic CNFs-Sn-ZrO_2 nanofibers: fabrication and characterization, RSC Adv. 4 (33) (2014) 17268−17273.

[122] Y. Wu, R. Balakrishna, M.V. Reddy, A.S. Nair, B.V.R. Chowdari, S. Ramakrishna, Functional properties of electrospun NiO/RuO_2 composite carbon nanofibers, J. Alloy. Compd. 517 (2012) 69−74.

[123] X. Sun, C. Li, J. Bai, Mixed-valent Co_xO-Ag/carbon nanofibers as binder-free and conductive-free electrode materials for high supercapacitor, J. Mater. Sci Mater. Electron. 29 (2018) 19382−19392.

[124] M. Samir, N. Ahmed, M. Ramadan, N.K. Allam, Electrospun mesoporous Mn-V-O@C nanofibers for high performance asymmetric supercapacitor devices with high stability, ACS Sustain. Chem. & Eng. 7 (2019) 13471−13480.

[125] H.S. Choi, T. Kim, J.H. Im, C.R. Park, Preparation and electrochemical performance of hyper-networked $Li_4Ti_5O_{12}$/carbon hybrid nanofiber sheets for a battery-supercapacitor hybrid system, Nanotechnology 22 (40) (2011) 405402.

[126] H.S. Choi, J.H. Im, T. Kim, J.H. Park, C.R. Park, Advanced energy storage device: a hybrid BatCap system consisting of battery-supercapacitor hybrid electrodes based on $Li_4Ti_5O_{12}$-activated-carbon hybrid nanotubes, J. Mater. Chem. 22 (33) (2012) 16986.

[127] H. Xu, X. Hu, Y. Sun, W. Luo, C. Chen, Y. Liu, et al., Highly porous $Li_4Ti_5O_{12}$/C nanofibers for ultrafast electrochemical energy storage, Nano Energy 10 (2014) 163−171.

[128] M.Y. Chung, S.M. Huang, C.T. Lo, Surfactants mediating microstructure and electrochemical supercapacitive properties of ruthenium oxide/electrospun carbon nanofiber composites, J. Electrochem. Soc. 166 (13) (2019) A2870−A2878.

[129] J.G. Wang, Y. Yang, Z.H. Huang, F. Kang, Incorporation of nanostructured manganese dioxide into carbon nanofibers and its electrochemical performance, Mater. Lett. 72 (2012) 18−21.

[130] J.G. Wang, Y. Yang, Z.H. Huang, F. Kang, Coaxial carbon nanofibers/MnO_2 nanocomposites as freestanding electrodes for high-performance electrochemical capacitors, Electrochim. Acta 56 (25) (2011) 9240−9247.

[131] S.C. Lin, Y.T. Lu, J.A. Wang, C.C.M. Ma, C.C. Hu, A flexible supercapacitor consisting of activated carbon nanofiber and carbon nanofiber/potassium-pre-intercalated manganese oxide, J. Power Sources 400 (2018) 415−425.

[132] L. Zhao, J. Yu, W. Li, S. Wang, C. Dai, J. Wu, et al., Honeycomb porous MnO_2 nanofibers assembled from radially grown nanosheets for aqueous supercapacitors with high working voltage and energy density, Nano Energy 4 (2014) 39−48.

[133] D. Zhou, H. Lin, F. Zhang, H. Niu, L. Cui, Q. Wang, et al., Freestanding MnO_2 nanoflakes/porous carbon nanofibers for high-performance flexible supercapacitor electrodes, Electrochim. Acta 161 (2015) 427–435.

[134] S. Hong, S. Lee, U. Paik, Core-shell tubular nanostructured electrode of hollow carbon nanofiber/manganese oxide for electrochemical capacitors, Electrochim. Acta 141 (2014) 39–44.

[135] J. Jun, J.S. Lee, D.H. Shin, S.G. Kim, J. Jang, Multidimensional MnO_2 nanohair-decorated hybrid multichannel carbon nanofiber as an electrode material for high-performance supercapacitors, Nanoscale 7 (38) (2015) 16026–16033.

[136] J.G. Wang, Y. Yang, Z.H. Huang, F. Kang, Synthesis and electrochemical performance of MnO_2/CNTs-embedded carbon nanofibers nanocomposites for supercapacitors, Electrochim. Acta 75 (2012) 213–219.

[137] T. Wang, D. Song, H. Zhao, J. Chen, C. Zhao, L. Chen, et al., Facilitated transport channels in carbon nanotube/carbon nanofiber hierarchical composites decorated with manganese dioxide for flexible supercapacitors, J. Power Sources 274 (2015) 709–717.

[138] G. Nie, X. Lu, M. Chi, M. Gao, C. Wang, General synthesis of hierarchical C/MO_x@MnO_2 (M = Mn, Cu, Co) composite nanofibers for high-performance supercapacitor electrodes, J. Colloid Interface Sci. 509 (2018) 235–244.

[139] M. Zhi, A. Manivannan, F. Meng, N. Wu, Highly conductive electrospun carbon nanofiber/MnO_2 coaxial nano-cables for high energy and power density supercapacitors, J. Power Sources 208 (2012) 345–353.

[140] Y. Huang, Y.E. Miao, W.W. Tjiu, T. Liu, High-performance flexible supercapacitors based on mesoporous carbon nanofibers/Co3O4/MnO2 hybrid electrodes, RSC Adv. 5 (24) (2015) 18952–18959.

[141] Y. Liu, Z. Zeng, R.K. Sharma, S. Gbewonyo, K. Allado, L. Zhang, et al., A bifunctional configuration for a metal-oxide film supercapacitor, J. Power Sources 409 (2019) 1–5.

[142] A.M. Al-Enizi, A.A. Elzatahry, A.M. Abdullah, M.A. Al-Maadeed, J. Wang, D. Zhao, et al., Synthesis and electrochemical properties of nickel oxide/carbon nanofiber composites, Carbon 71 (2014) 276–283.

[143] F. Miao, C. Shao, X. Li, Y. Zhang, N. Lu, K. Wang, et al., One-dimensional heterostructures of beta-nickel hydroxide nanoplates/electrospun carbon nanofibers: Controlled fabrication and high capacitive property, Int. J. Hydrog. Energy 39 (28) (2014) 16162–16170.

[144] C.C. Lai, C.T. Lo, Effect of temperature on morphology and electrochemical capacitive properties of electrospun carbon nanofibers and nickel hydroxide composites, Electrochim. Acta 174 (2015) 806–814.

[145] J. Cai, H. Niu, Z. Li, Y. Du, P. Cizek, Z. Xie, et al., High-performance supercapacitor electrode materials from cellulose-derived carbon nanofibers, ACS Appl. Mater. Interfaces 7 (27) (2015) 14946–14953.

[146] L. Zhang, Q. Ding, Y. Huang, H. Gu, Y.E. Miao, T. Liu, Flexible hybrid membranes with Ni(OH)2 nanoplatelets vertically grown on electrospun carbon nanofibers for high-performance supercapacitors, ACS Appl. Mater. Interfaces 7 (40) (2015) 22669–22677.

[147] T. Mukhiya, B. Dahal, G.P. Ojha, D. Kang, T. Kim, S.H. Chae, et al., Engineering nanohaired 3D cobalt hydroxide wheels in electrospun carbon nanofibers for high-performance supercapacitors, Chem. Eng. J. 361 (2019) 1225–1234.

[148] Z. Tai, J. Lang, X. Yan, Q. Xue, Mutually enhanced capacitances in carbon nanofiber/cobalt hydroxide composite paper for supercapacitor, J. Electrochem. Soc. 159 (4) (2012) A485–A491.

[149] B. Pant, M. Park, G.P. Ojha, J. Park, Y.S. Kuk, E.J. Lee, et al., Carbon nanofibers wrapped with zinc oxide nano-flakes as promising electrode material for supercapacitors, J. Colloid Interface Sci. 522 (2018) 40–47.

[150] E. Samuel, B. Joshi, M.W. Kim, Y.I. Kim, M.T. Swihart, S.S. Yoon, Hierarchical zeolitic imidazolate framework-derived manganese-doped zinc oxide decorated carbon nanofiber electrodes for high performance flexible supercapacitors, Chem. Eng. J. 371 (2019) 657–665.

[151] J. Mu, B. Chen, Z. Guo, M. Zhang, Z. Zhang, C. Shao, et al., Tin oxide (SnO_2) nanoparticles/electrospun carbon nanofibers (CNFs) heterostructures: controlled fabrication and high capacitive behavior, J. Colloid Interface Sci. 356 (2) (2011) 706–712.

[152] Y. Luan, G. Nie, X. Zhao, N. Qiao, X. Liu, H. Wang, et al., The integration of SnO_2 dots and porous carbon nanofibers for flexible supercapacitors, Electrochim. Acta 308 (2019) 121–130.

[153] J. Liang, C. Yuan, H. Li, K. Fan, Z. Wei, H. Sun, et al., Growth of SnO_2 nanoflowers on N-doped carbon nanofibers as anode for Li- and Na-ion, Batteries, Nano-Micro Lett. 10 (2) (2017) 21.

[154] A. Ghosh, E.J. Ra, M. Jin, H.K. Jeong, T.H. Kim, C. Biswas, et al., High pseudocapacitance from ultrathin V_2O_5 films electrodeposited on self-standing carbon-nanofiber paper, Adv. Funct. Mater. 21 (13) (2011) 2541–2547.

[155] J. Mu, B. Chen, Z. Guo, M. Zhang, Z. Zhang, P. Zhang, et al., Highly dispersed Fe_3O_4 nanosheets on one-dimensional carbon nanofibers: synthesis, formation mechanism, and electrochemical performance as supercapacitor electrode materials, Nanoscale 3 (12) (2011) 5034.

[156] 130 M.S. Lal, T. Lavanya, S. Ramaprabhu, An efficient electrode material for high performance solid-state hybrid supercapacitors based on a Cu/CuO/porous carbon nanofiber/TiO_2 hybrid composite, Beilstein J. Nanotechnol. 10 (2019) 781–793.

[157] M.A.A. Mohd Abdah, N. Mohammed Modawe Aldris Edris, S. Kulandaivalu, N. Abdul Rahman, et al., Supercapacitor with superior electrochemical properties derived from symmetrical manganese oxide-carbon fiber coated with polypyrrole, Int. J. Hydrogen Energy 43 (36) (2018) 17328–17337.

[158] J.S. Lee, C. Lee, J. Jun, D.H. Shin, J. Jang, A metal-oxide nanofiber-decorated three-dimensional graphene hybrid nanostructured flexible electrode for high-capacity electrochemical capacitors, J. Mater. Chem. A 2 (30) (2014) 11922.

[159] L.H. Poudeh, I.L. Papst, F.Ç. Cebeci, Y. Menceloglu, M. Yildiz, B.S. Okan, Facile synthesis of single- and multi-layer graphene/Mn_3O_4 integrated 3D urchin-shaped hybrid composite electrodes by core-shell electrospinning, ChemNanoMat 5 (2019) 792–801.

[160] S.N.J. Syed Zainol Abidin, M.S. Mamat, S.A. Rasyid, Z. Zainal, Y. Sulaiman, Electropolymerization of poly(3,4-ethylenedioxythiophene) onto polyvinyl alcohol-graphene quantum dot-cobalt oxide nanofiber composite for high-performance supercapacitor, Electrochim. Acta 261 (2018) 548–556.

[161] K. Xu, Y. Shen, K. Zhang, F. Yang, S. Li, J. Hu, Hierarchical assembly of manganese dioxide nanosheets on one-dimensional titanium nitride nanofibers for high-performance supercapacitors, J. Colloid Interface Sci. 552 (2019) 712–718.

Thermoelectrics based on metal oxide nanofibers

16

Yong X. Gan
Department of Mechanical Engineering, College of Engineering, California State Polytechnic University Pomona, Pomona, CA, United States

16.1 Introduction

The terminology of thermoelectrics comes from the study of thermoelectricity or the thermoelectric effect. The thermoelectric effect refers to the phenomena by which either a temperature gradient results in an electric potential or an electric potential generates a temperature gradient using a thermoelectric module. For practical applications the module may be used as either an electrical power generator or a cooler. The thermoelectric module is typically a solid-state device containing no moving parts. Therefore no noise generates associated with the thermoelectric energy conversion. Maintenance needed is minimum as well. The module could be built very small in volume and light in weight, which allows it to be used for cooling an instrument with compact space. One of the application examples is for integrated circuit board cooling. Flexible thermoelectric modules were made and incorporated into wearable garments [1]. Such garments demonstrate an active cooling function for body thermoregulation, which plays an important role in human comfort and health.

The efficiency of a thermoelectric module depends on the physical properties of the thermoelectric leg material that is used. The value of the figure of merit of a thermoelectric material, z, can be expressed as follows [2]:

$$z = \sigma S^2 / \kappa \tag{16.1}$$

where S is the Seebeck coefficient, σ is the electrical conductivity, and κ is the thermal conductivity.

The dimensionless figure of merit, zT, is more commonly used for the thermoelectric energy conversion efficiency estimation. T is the absolute temperature. If we consider the thermoelectric energy conversion efficiency of a unit instead of a single material, the capital letter Z and letters ZT are used to replace z and zT [3]. A zT or ZT value of 1.0 is approximately equivalent to a Carnot efficiency (η) of 10% when the thermoelectric energy conversion system is considered as a

---

Metal Oxide-Based Nanofibers and Their Applications. DOI: https://doi.org/10.1016/B978-0-12-820629-4.00017-5
© 2022 Elsevier Inc. All rights reserved.

thermodynamic heating or cooling engine. This is because the maximum efficiency of the device, η_{max}, can be estimated by the following formula [4]:

$$\eta_{max} = \frac{T_H - T_C}{T_H} \frac{\sqrt{1 + ZT} - 1}{\sqrt{1 + ZT} + \frac{T_C}{T_H}} \tag{16.2}$$

where T_H is the hot end temperature and T_C is the cold end temperature.

Among various materials, metal oxides have low thermal conductivities and tunable electrical conductivities. They are stable at high temperature in ambient atmosphere. Most metal oxides are almost inert in various corrosion media. In addition, they are relatively easy to process and inexpensive because of their abundance. Such advantages provide the motivation to search for good thermoelectric oxides. Several groups of oxide materials including binary oxide, perovskite oxide, and oxide cobaltite have been investigated as promising thermoelectric materials, as described by Lin et al. [5]. Some of the layered oxides have many advantages because of their peculiar transport behavior [6]. A review of various layered oxides for thermoelectric applications can be found in the work of Yin et al. [7].

This chapter deals with the advances in the thermoelectricity of oxide nanofibers (NFs). The emphasis will be placed on various new processing techniques for making thermoelectric oxide NFs. Several case studies on continuous and discontinuous metal oxide nanofibers (MONFs) will be cited and discussed in detail. First, case studies on the processing and characterization of continuous fibers containing titanium dioxide (TiO_2), cobalt oxide, strontium titanate ($SrTiO_3$), and so on via electrospinning followed by high-temperature reaction and/or heat treatment will be discussed. Then synthesis of discontinuous TiO_2-cobalt(II) oxide (CoO) coaxial NFs or nanowires (NWs) through the template-assisted liquid phase deposition method as described by Hsu et al. [8] will be presented. Device fabrication and characterization of the thermoelectric properties of oxide NFs will be illustrated following the processing and manufacturing technologies. Moreover, the perspectives to future research directions will be discussed, and conclusions will be made.

16.2 Thermoelectric metal oxide nanofiber processing technology

For thermoelectric MONF processing, electrospinning followed by the high-temperature treatment, spray pyrolysis deposition, template-assisted deposition, and electrochemical oxidation are commonly used techniques. Other techniques—such as glass annealing and nanolithography coupled with vapor deposition, are introduced as well. In the following subsections, each of these techniques will be described separately.

16.2.1 Electrospinning

Among various processing and manufacturing techniques, electrospinning is one of the most convenient approaches to long or continuous NF formation. Although electrospinning is considered a conventional fiber-processing technology, it has become one of the most important methods for making thermoelectric oxide NFs. During the process, a polymer solution or melt is electrified by being passed through an intensive electric field. Under the action of the electric force, the polymer is drawn into fibers with diameters of the order of several microns to as fine as several nanometers. To obtain continuous fibers, a sufficiently high direct current (DC) voltage must be applied to the liquid droplet. Once the body of the liquid becomes charged, the electrostatic repulsive force counteracts the surface tension. This counterforce stretches the liquid droplet. At the critical point of stability a stream of liquid erupts from the surface. This point of eruption is known as the Taylor cone. Because of the simultaneous electrification, the solvent is atomized and vaporized from the solution. The molecular cohesion of the liquid is increased sufficiently high. Therefore the stream breakup does not occur, and a stable charged liquid jet at the nozzle is formed. Eventually, the polymer is spun into continuous fibers.

To make oxide NFs, inorganic salts or organometallic compounds should be added into polymer solutions to form the so-called precursors. After the heat treatment at high temperatures in air, the added salts or organometallic compounds are converted into metal oxides. For example, tin sulfate was added into a polyacrylonitrile (PAN) polymer solution for tin dioxide generation [9]. The PAN-based polymer NFs were generated first by using the electrospinning technique. Following electrospinning, a series of heat treatments were performed, which resulted in the formation of composite carbon NFs containing tin (IV) oxide (SnO_2) nanoparticles. The tin dioxide was generated by the following decomposition reaction during the heat treatment [9]:

$$SnSO_4 \rightarrow SnO_2 + SO_2 \uparrow \qquad (16.3)$$

Electrospinning shares the characteristics of both electrospraying droplets [10] and traditional solution dry spinning fibers [11]. The electrospraying occurs when a low-viscosity fluid is electrohydrodynamically jetted. Owing to the low viscosity, the molecular cohesion of the liquid is not high enough. The stream breaks up, forming liquid droplets. The charged liquid droplets fly toward the collecting screen so that electrosprayed products can be obtained [10]. The dry spinning process may require the use of coagulation chemistry or high-pressure gas—such as compressed nitrogen gas [11] to produce solid fibers. The posttreatment following the electrospinning typically causes the change in polymer structures. At the same time, oxide formation within the NFs can be realized, as shown by Lee et al. [10]. Electrospinning is particularly suitable for the production of fibers using large and complex molecules—such as metallo-organic molecules [11]. Electrospinning from molten precursors has also been studied with increasing interest. There is a major advantage in that the process ensures that no solvent is carried over into the final product [12].

Among various oxides, the p-type sodium cobalt oxide with the general formula of $Na_xCo_2O_4$ demonstrates relatively high thermoelectric power and low resistivity. There is increasing interest in using sodium cobalt oxide NFs in thermoelectric energy converters and electronic sensing devices. Maensiri and Nuansing [13] prepared sodium cobalt oxide ($NaCo_2O_4$) continuous NFs with diameters ranging from 20 to 200 nm by electrospinning a sodium acetate/cobalt acetate/PAN solution. The electrospun composite nanofibers underwent calcination treatment at different temperatures of 300°C, 400°C, and 800°C in air for 5 h. The structures of the sodium cobalt oxide NFs were also characterized. The oxide NFs calcined at 300°C and 400°C were polycrystalline. The Seebeck effect coefficient of the polycrystalline $Na_xCo_2O_4$ ($x = 1$), S, is about 100 μ V/K, and the electrical resistivity of the oxide, ρ, is about 200 μΩ•cm at 300 K. Such thermoelectric properties are comparable to those of the $Na_xCo_2O_4$ single crystal. The $NaCo_2O_4$ NFs obtained by calcination at 800°C showed the single-crystalline structure.

The thermoelectric performance of the La_2CuO_4 NFs made via electrospinning were investigated [14]. The electrospun NFs were laid on a Pd-Ag thin-film electrode for thermoelectric effect evaluation. For processing the oxide fibers, several procedures were adopted. First, aqueous solutions of lanthanum nitrate hexahydrate, $La(NO_3)_3 \cdot 6H_2O$, and copper nitrite hemi(pentahydrate), $Cu(NO_3)_2 \cdot 2.5H_2O$, were prepared separately. Then the hydrophilic polymer polyvinyl pyrrolidone (PVP), with a molecular weight of 1,300,000, was dissolved into ethanol. By mixing all three solutions together, the precursor was made for electrospinning the La- and Cu-containing PVP composite NFs. The air-dried composite NFs were calcined at 600°C for 5 h in the ambient atmosphere to generate La_2CuO_4 NFs. Earlier work performed by Liu et al. [15] showed that earth and rare earth metal−doped La_2CuO_4 bulk materials demonstrated very high values of Seebeck coefficient up to 370 μ V/K at 330 K. Hayat et al. [14] believe that the NF specimens could possibly achieve an even higher Seebeck coefficient than the La_2CuO_4 bulk material.

The coaxial electrospinning technique was used to prepare core-shell oxide nanofibers (CNFs) consisting of an n-type thermoelectric Nb-doped $SrTiO_3$ (Nb/STO) core and a thin sheath layer of TiO_2 [16]. The coaxial electrospinning can simultaneously eject multiple fluids (an inner one and an outer one). During coaxial electrospinning, a stainless steel coaxial spinneret was used as the key component for ejecting two solutions. In reference [16], the inner solution was made by adding the as-prepared Nb/STO nanoparticles into a solution comprising distilled water, absolute ethanol, and PVP. This solution in the form of slurry was filled into the inner channel of the coaxial needle. The outer fluid, as the precursor for the TiO_2 shell layer, was made from titanium (IV) isopropoxide, acetylacetone, absolute ethanol, and PVP. The coaxial electrospinning setup consists of a dual-solution feeding unit. This unit accounts for the ejection of two solutions simultaneously through the tip of the spinneret. The feeding rates of the outer and inner solutions were 18 and 3 mL min^{-1}, respectively. The outer fluid, with a higher feeding rate, typically acts as a carrier that draws in the inner fluid at the Taylor cone of the electrospinning jet. If the solutions were immiscible, then a core-shell structure is usually generated. In this case, however, the inner and outer fluids are miscible solutions. This allows for the partial mixing of one solution into another at the tip of the spinneret. In some cases, miscible solutions may result in porosity within the fiber or a fiber with multiple phases from the phase separation during the formation of the fiber or in the postprocessing stage. As is well known, more complex coaxial electrospinning systems may be built by using triaxial and quadaxial (tetraaxial) spinnerets, which allow the tip of the needles to eject more than two solutions at the same time. The coaxially electrospun NFs of the Nb-doped $SrTiO_3$ (Nb/STO) covered by a thin layer of TiO_2 were eventually obtained by calcinating the as-spun fibers in air at the elevated temperature of 500°C for 2 h.

Electrospinning a composite NF web consisting of multiwalled carbon nanotubes (MWCNTs) and copper oxide (CuO) nanoparticles in PVP polymer was conducted [17]. The solution containing 8 wt.% PVP in ethanol was made first. Then MWCNTs were dispersed into the solution by ultrasonic stirring. Next, CuO nanoparticles were added together. Before electrospinning, the mixture was stirred vigorously to ensure the uniform distribution of the CuO nanoparticles in the slurry. Electrospinning was carried out to eject NFs onto a carbon fiber fabric. The electrospun composite NF web had a Seebeck coefficient of 39.21 $\mu V\ K^{-1}$.

The electrospinning technique can be combined with the conventional sol−gel process [18−20]. In situ sol−gel chemical reactions can be triggered during the combined process when metallo-organic compounds are used as the starting materials. For example, barium acetate in acetic acid and titanium isopropoxide in the mixture of 2-methoxyethanol and acetylacetone were used to synthesize barium titanate $(BaTiO_3)$ NFs [18]. At the same time, PVP was dissolved in ethanol and added to the former solution to get the 0.3 M $BaTiO_3$ precursor, which was loaded into a plastic syringe for electrospinning. Barium acetate and titanium isopropoxide are hydrolyzed simultaneously through the sol−gel reactions to initiate the precipitation of $BaTiO_3$ in the polymer fibers. The final product (pure $BaTiO_3NF$) was generated by completely removal of PVP through annealing the as-spun NF at 750°C for 2 h in air.

In addition to making continuous or long fibers, electrospinning can be used as an effective method for refining crystalline grains. Ma et al. [21] studied the structure-property relationship of electrospun thermoelectric $NaCo_2O_4$ NFs. Two precursor solutions for $NaCo_2O_4$ were prepared by dissolving cobalt(II) acetate tetrahydrate and sodium acetate trihydrate into methanol and distilled water, respectively. The mole ratio of Na to Co was controlled in the range from 1.1 to 2. The concentrations of $NaCo_2O_4$ in methanol and water were kept as 0.15 M and 0.4 M, respectively. PVP was added into the two precursor solutions to reach 0.1 $g−mL^{-1}$ concentration for electrospinning. After electrospinning, the as-spun fibers were airdried, and then annealed in air at 750°C for 2 h. The images of the NFs before and after high temperature annealing are shown in Fig. 16.1. The as-electrospun fibers are shown in Fig. 16.1A and C. The high temperature annealed fibers are revealed in Fig. 16.1B and D.

High-resolution transmission electron microscopic (TEM) analysis of the annealed NFs was also performed. It is revealed that the sol−gel-based

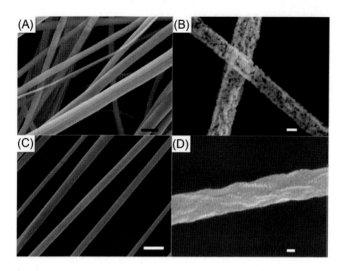

Figure 16.1 Scanning electron microscopic images of NaCo$_2$O$_4$ nanofibers before and after annealing, synthesized with two different solvents: Methanol based nanofibers (A) before and (B) after annealing. Water-based nanofibers (C) before and (D) after annealing. The scale bars in parts A and C represent 1 μm, and those in parts B and D represent 100 nm.
Source: Reprinted with permission from F. Ma, Y. Ou, Y. Yang, Y. Liu, S. Xie, J.F. Li, et al., Nanocrystalline structure and thermoelectric properties of electrospun NaCo2O4 nanofibers, J. Phys. Chem. C. 114 (2010) 22038–22043. Copyright ©2010 American Chemical Society.

electrospinning technique can synthesize thermoelectric NaCo$_2$O$_4$ NFs with a grain size as small as 10 nm. Such a grain size is much finer than that of NaCo$_2$O$_4$ powders made by the conventional sol–gel method.

Kocyigit et al. [22] processed boron-doped strontium-stabilized bismuth cobalt oxide via electrospinning a polyvinyl alcohol (PVA)–based precursor solution. Bismuth(III) acetate, strontium acetate, cobalt(II) acetate, and boric acid were added into a 10% PVA aqueous solution to make the precursor for electrospinning NFs. After electrospinning, calcination of the electrospun NFs at 850°C for 2 h in air atmosphere was conducted to obtain oxide ceramic composite NFs. Then the ceramic NFs were pulverized into powders which were pressed into pellets. The pellets underwent sintering at 850°C to generate consolidated specimens for thermoelectric property measurement. The effect of boron doping on the thermoelectric properties of the nanocrystalline ceramic specimens was examined. It was found that boron doping increased the electrical conductivity and thermal conductivity of the oxide but slightly reduced the Seebeck coefficient. The thermoelectric figure of merit, the zT value, did not change appreciably as a result of the boron doping.

In the study carried out by Lee et al. [23], an oxide-based lateral thermoelectric p-n couple was fabricated through electrospinning. The thermoelectric performance of the p-n couple was evaluated. The electrospun zinc oxide (ZnO) NF was the n-leg, and the lathanium cobaltite (LaSrCoO$_3$) NF was the p-leg of the couple. It was found that the Seebeck coefficients of the n-type ZnO NF was −98.1 μV/K. The p-type NF showed a Seebeck coefficient of 118.8 μV/K. It was demonstrated that the oxide NF-based thermoelectric couple generated an output voltage of 484.7 μV at a small temperature difference of 4.1 K. The procedures for how to fabricate the p-n couple using the two types of electrospun NFs were also described in the paper.

The thermal and electrical responses of the electrospun cobalt oxide particle−containing carbon NF were tested by using a hot air heating gun [24]. The time-dependent response to the thermal wave generated by the hot air heating gun was observed. The hot air gun was pointed at the specimen intermittently. Each time, a strong electrical potential peak was generated. This is an indication of the significant thermoelectric energy conversion effect generated by the cobalt oxide−containing carbon composite NF. Xu et al. [25] provided a more detailed study on making oxide−containing composite carbon NFs. The nickel cobaltite ($NiCo_2O_4$) oxide compound was introduced into carbon NFs by electrospinning followed by calcination. After that, a chemical bath deposition process was used to coat the surface of the carbon NFs with arrays of $Ni(OH)_2$ nanosheets. In Fig. 16.2 the procedures for making the composite NFs are shown. To process the $NiCo_2O_4$ oxide−containing carbon NF, 0.35 g of PAN was dissolved into 10 mL of N,N-dimethylformamide to form a transparent solution. Then 0.25 g of PVP was added into the solution. After that, cobalt acetate ($CoC_4H_6O_4·4H_2O$) and nickel acetate ($NiC_4H_6O_4·6H_2O$) were added into the solution at a mass ratio of 2:1 to generate a reddish-brown homogeneous precursor solution by continuous stirring for 24 h. The electrospinning was performed at 20 kV, 15% relative humidity, and a flow

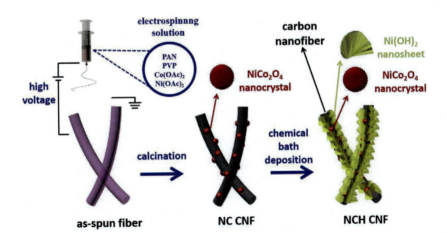

Figure 16.2 Schematic of synthesizing composite nanofibers. NC CNF stands for the $NiCo_2O_4$ oxide−containing carbon nanofiber. NCH CNF represents the $NiCo_2O_4$ oxide and $Ni(OH)_2$ nanosheet decorated core-shell carbon nanofiber.
Source: Reprinted with permission from L. Xu, L. Zhang, B. Chen, J. Yu, Rationally designed hierarchical NiCo2O4-C@Ni(OH)2 core-shell nanofibers for high performance supercapacitors, Carbon 152 (2019) 652−660. Copyright ©2019 Elsevier Ltd.

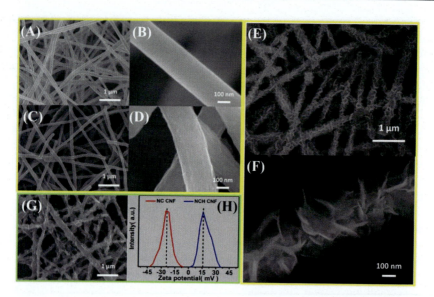

Figure 16.3 Field emission scanning electron microscopic images and zeta potentials of composite nanofibers. (A, B) As-spun nanofibers. (C, D) NC CNF. (E, F) NCH CNF. (G) nickel hydroxide–coated carbon nanofiber. (H) Zeta potentials of NC CNF and NCH CNF. NC CNF is negatively charged, which makes it possible to adsorb positively charged nickel ions. NCH CNF is positively charged.
Source: Reprinted with permission from L. Xu, L. Zhang, B. Chen, J. Yu, Rationally designed hierarchical NiCo2O4-C@Ni(OH)2 core-shell nanofibers for high performance supercapacitors, Carbon 152 (2019) 652–660. Copyright ©2019 Elsevier Ltd.

rate of 1.5 mL h^{-1}. The as-spun fiber was calcined in air at 500°C for 3 h to yield the NiCo$_2$O$_4$ oxide–containing carbon nanofiber (NC CNF).

Next, deposition of Ni(OH)$_2$ nanosheet arrays on the NC CNF was implemented by using the chemical bath deposition approach. 0.02 g of NC CNF was dispersed in a 20-mL mixed aqueous solution that contained a certain amount of Ni(NO$_3$)$_2$ and urea. After being stirred continuously for 20 min, the mixed solution was heated up to 70°C and held for 2 h. Then the solution was aged overnight. The product, nickel hydroxide–coated composite nanofiber (NCH CNF), was collected by filtration and air drying. Fig. 16.3 presents the images of the composite NFs and the zeta potentials of different NFs. Although the work is based on making nickel hydroxide–covered composite NFs for supercapacitor application [25], it provides new insight into multiple functional composite NF-processing technology. In fact, the prepared composite NFs may find application for thermoelectric energy conversion, owing to the existence of the cobalt-based oxide within the NFs and at the surface of the fibers.

16.2.2 Chemical bath deposition

The chemical bath deposition technique is suitable for short or discontinuous thermoelectric MONF preparation. Typically, nanocrystal seeds were prepared first on smooth substrates including silicon wafer, indium tin oxide (ITO)-coated glass, and alumina plate. The nanocrystal seeds were the initiation sites for the growth of NFs in a chemical solution bath. ZnO NW or NF-CuO film heterojunctions were made through the chemical bath deposition, as shown in the work performed by Saltana et al. [26]. ZnO is an n-type semiconductor, while CuO is a p-type semiconductor. Both materials have been considered as candidates for energy conversion applications, including thermoelectric energy conversion. In particular, ZnO has found increasing importance in building low-cost and environmentally friendly thermoelectric devices.

Discontinuous ZnO NFs (NWs) were grown on CuO film via the following steps [26]. First, Si wafers were cleaned by 20% hydrofluoric solution to remove the surface oxide layer. Then, CuO thin film was prepared by using a 0.1 M copper chloride dihydrate ($CuCl_2 \cdot 2H_2O$) solution. The cleaned wafers were dipped vertically into the solution under constant stirring and heating. When the bath temperature reached 60°C, ammonia solution ($NH_3 \cdot H_2O$) was added dropwise to generate the complexants, NH_4^+ and OH^-, for CuO nanoparticle formation. At this stage, the pH value of the solution reached 10. Following that, the solution was heated up to 85°C with continuous stirring. Under such conditions, copper hydroxide $Cu(OH)_2$ and ammonium chloride (NH_4Cl) were generated from the reaction between the complexants and the $CuCl_2 \cdot 2H_2O$. With the increase in bath temperature and reaction time, the intermediate product $Cu(OH)_2$ changed into solid phase CuO as a result of the dehydration. The color of the solution changed from brownish black to completely black, indicating the formation of CuO. The deposited CuO film on the Si wafers was cleaned and dried with flowing N_2 gas. Next, growing ZnO NFs on the CuO film was conducted by the chemical bath deposition technique. Briefly, 0.1 M zinc nitrate hexahydrate $Zn(NO_3)_2 \cdot 6H_2O$ and 0.1 M hexamethylenetetramine ($C_6H_{12}N_4$) were mixed together in a bath to produce a transparent solution. The ZnO NWs or short NFs grew on the CuO film and established good contact between the CuO thin film and the ZnO NFs to form heterojunctions. In Fig. 16.4, the field emission scanning electron microscopic (FESEM) images are shown. Fig. 16.4A is the top view of the ZnO NFs, revealing the formation of the hexagonal top facet of the ZnO NFs. After the ITO coating was applied on the ZnO NFs, the continuous thin layers of ITO can be seen from the FESEM image in Fig. 16.4B. The red circles in the figure are the clear marks for the ITO layers or sections on the top of the ZnO NWs (or short NFs). In Fig. 16.4C the cross section of the ZnO/CuO heterojunction is shown. There are three distinct layers (the silicon wafer substrate, CuO film, and ZnO NW or NF), as shown in the lower part of the FESEM image.

Chu et al. [27] prepared Ni-doped ZnO short NFs (nanorods) via chemical bath deposition. The codeposition concept is introduced in their work. Both zinc oxide and nickel oxide were formed simultaneously from a single bath containing zinc acetate and nickel acetate as the oxide sources. The substrate was a 20×20 mm^2 piece of Corning glass. It was ultrasonically cleaned by using acetone, deionized (DI) water, and isopropyl alcohol for 5 min each. Then, the substrate was air-dried in an oven for about 10 min at 45°C. A 100 nm thick ZnO film served as the seed

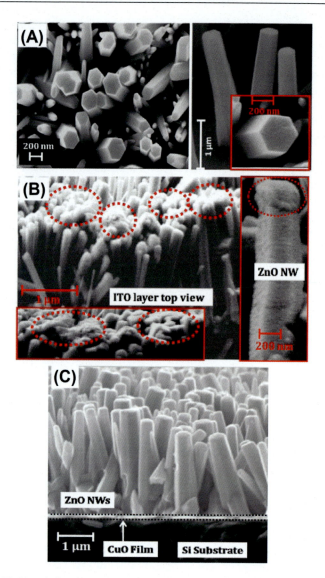

Figure 16.4 Field emission scanning electron microscopic images showing (A) top view of ZnO nanofiber, (B) top view of ZnO nanowire or nanofiber upon ITO deposition (continuous small sections of ITO are shown by red circles). (C) Cross section of the ZnO/CuO heterojunction.
Source: Reprinted with permission from J. Sultana, S. Paul, R. Saha, S. Sikdar, A. Karmakar, S. Chattopadhyay, Optical and electronic properties of chemical bath deposited p-CuO and n-ZnO nanowires on silicon substrates: p-CuO/n-ZnO nanowires solar cells with high open-circuit voltage and short-circuit current, Thin Solid. Films 699 (2020) 137861. Copyright © 2020 Elsevier B.V.

layer, which grew on the Corning glass substrate by using the radio frequency (RF) magnetron sputtering technique. The parameters of work pressure, gas flow, and power of chamber were given as $\sim 5 \times 10^{-6}$ Torr, Ar/O$_2$ = 12/1, and 30 W, respectively. After that, photolithography was carried out to define a micropattern on the substrate using a shadow mask. Then, a 100-nm thick titanium (Ti) film was evaporated onto the seed layer as the contact electrode by electron beam evaporation. The lift-off procedure was applied to form the micro-patterned interdigital transducer (IDT) electrode structure. To grow Ni-doped ZnO NFs, zinc acetate dihydrate [Zn (CH$_3$COO)$_2$•2H$_2$O], hexamethylenetetramine (C$_6$H$_{12}$N$_4$, HMTA), and nickel(II) acetate [Ni(CH$_3$COO)$_2$•6H$_2$O] were used as the starting materials to make the chemical bath deposition solution. The ratio of zinc acetate dihydrate (0.025 M) to HMTA (0.05 M) was kept at 1 to 2. The concentrations of Ni solutions were 4 and 8 mM. The aqueous zinc acetate, nickel acetate, and HMTA solutions were mixed under regular stirring for 20 min at the ambient temperature. After mixing, the resulting solution was immediately transferred to a Teflon-lined autoclave for chemical bath deposition of NF arrays. The autoclave was maintained at 90°C for 3 h in an oven. The as-grown NFs were annealed at 500°C for 15 min in air.

In addition to zinc nitrate and zinc acetate, zinc sulfate was used to synthesis ZnO NFs. A solution was prepared by mixing 1 mL of 1.0 M zinc sulfate (ZnSO$_4$) as a precursor with 1.5 mL of 4.0 M ammonium hydroxide (NH$_4$OH) [28]. During the mixing, the solution was observed to become cloudy. The solution was shaken vigorously until it turned in clear. Then 2 mL of ethanolamine (C$_2$H$_7$NO), a complex agent, was dropped into the solution and sonicated for 10 min to ensure that the solution became homogeneous. Subsequently, 15.5 mL of DI water was added to the solution, which was then stirred well. Following that, an optical fiber as the substrate was immersed in the solution for ZnO NF or nanorod deposition. After the deposition process was completed, the samples were annealed at 200°C for 1 h. The reaction mechanism for ZnO NF formation as described by Rathore et al. [29] can be presented as follows:

$$Zn^{2+} + 4NH_3 \rightarrow Zn(NH_3)_4^{2+} \tag{16.4}$$

$$Zn^{2+} \text{ or } Zn(NH_3)_4^{2+} \text{ or } Zn(OH)_2(NH_3)_4 + 2OH^- \rightarrow Zn(OH)_2 \tag{16.5}$$

$$Zn^{2+} + 2OH^- \leftrightarrow Zn(OH)_2 \leftrightarrow ZnO + H_2O \tag{16.6}$$

Following the similar chemical bath deposition procedures as described by Sultana et al. [26], Kaphle et al. [30] used zinc and copper nitrate compounds to build ZnO NF/CuO heterojunctions. Up to 20% doped cobalt in ZnO was achieved by chemical bath codeposition. It was found that the cobalt doping significantly increased the electrical conductivity of the ZnO NFs or nanorods. The electrical conductivity reached the peak value of 21 S cm^{-1} for the 10% Co-doped ZnO. Without cobalt doping, the pristine ZnO NF showed an electrical conductivity of only 14 S cm^{-1}.

16.2.3 Template-assisted deposition

For discontinuous thermoelectric MONFs synthesis, template-assisted deposition allows the creation of various oxides with tailored structures and expected properties. Su et al. [31] used the anodic aluminum oxide (AAO) nanoporous hard template to deposit TiO_2–CoO core-shell NFs for the thermoelectric property investigation. Electron microscopic analysis results as shown in Fig. 16.5 reveal the structure information. The Seebeck coefficients of the discontinuous NFs encapsulated in the AAO were measured. The highest absolute value of Seebeck coefficient was 393 μ V K^{-1} for the TiO_2 NF-filled AAO. The TiO_2–CoO core-shell NF-filled AAO has a slightly lower Seebeck coefficient absolute value of 300 μ V K^{-1}. Both composites showed n-type behavior. The effect of Ag nanoparticle addition on the thermoelectric behavior was also examined. Although the Ag nanoparticle contributed to increasing the electrical conductivity, the enhancement on the Seebeck coefficient was not obvious.

The TiO_2 hollow NFs as shown in Fig. 16.5A were prepared at first in the nanopores of the AAO template. The initial pore diameter of the AAO was around 20 nm.

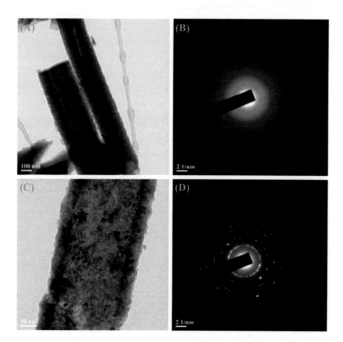

Figure 16.5 Transmission electron microscopic images and X-ray diffraction patterns. (A) TiO_2 discontinuous nanofibers. (B) The diffraction pattern of the TiO_2 nanofibers. (C) A single TiO_2–CoO core-shell nanofiber. (D) The diffraction pattern of the TiO_2–CoO nanofiber.
Source: Reprinted with permission from L. Su, Y.X. Gan, L. Zhang, Thermoelectricity of nanocomposites containing TiO2-CoO coaxial nanocables, Scr. Materialia 64 (2011) 745–748. Copyright © 2010 Acta Materialia Inc. Published by Elsevier Ltd.

The hydrolysis reaction associated with the initiation and growth of TiO_2 nanotubes can be expressed by Eq. (16.7) [8,31]:

$$TiF_6^{2-} + 2H_2O \rightarrow TiO_2 + 6F^- + 4H^+ \tag{16.7}$$

Before the TiO_2 deposition onto the inner wall of the AAO by the reaction as shown in Eq. (16.7), both top and bottom sides of the template were covered with self-assembled monolayers to avoid any TiO_2 deposition on the top and bottom surfaces of the AAO. Consequently, the hollow TiO_2 NFs started growing on the inner wall of the nanopores rather than on the two flat surfaces of the AAO template. The hollow NFs could become solid ones if a sufficiently long time of TiO_2 deposition is allowed so that the TiO_2 can completely fill the nanopores. The selected area X-ray diffraction (XRD) experiment reveals that the as-deposited TiO_2 NF has a noncrystalline or amorphous structure because there is no clear diffraction ring or pattern shown in Fig. 16.5B.

It is noted that the self-assembled monolayers on the front and back sides of the AAO were obtained by smearing an octadecyltetrachlorosilane hexane solution (10 mM concentration) on the two surfaces. After hexane evaporation the controlled TiO_2 deposition on the inner wall of the nanonpores was simply done by immersing the AAO template with the selectively covered monolayers into 10 mL of a mixed aqueous solution containing 0.05 M $(NH_4)_2TiF_6$ and 0.1 M H_3BO_3 for about 30 min. Such carefully selected deposition parameters allowed the obtained TiO_2 NFs in the AAO nanopores to have the hollow structure for hosting another oxide as the core in the core-shell NFs.

The purpose of adding extra H_3BO_3 was to accelerate the TiO_2 deposition by the formation of BF_4^- and H_3O^+ complexants through the reaction illustrated by Eq. (16.8). To shift the hydrolysis reaction toward the right-hand side of Eq. (16.7), the F^- ions should be consumed to promote the formation of the TiO_2. This can be achieved by adding boric acid because of the following reaction:

$$H_3BO_3 + 4H^+ + 4F^- = BF_4^- + H_3O^+ + 2H_2O \tag{16.8}$$

The deposited NFs in AAO were rinsed with the template together in DI water several times and air-dried. Postheat treatment such as annealing at 550°C for 2 h may be carried out to convert the amorphous TiO_2 into one with a crystalline structure.

To improve the thermoelectric performance of the composite material, Ag nanoparticles were produced and embedded into the TiO_2 NFs to increase the electrical conductivity. To achieve the uniform distribution of ultrafine Ag nanoparticles within the TiO_2 NFs, a piece of the prepared TiO_2-AAO was immersed into 10 mL of $AgNO_3$ solution with a concentration of 50 mM. After being soaked in $AgNO_3$ solution and air-dried, the nanofiber-containing specimen was annealed at 500°C for 1 h in air. Since $AgNO_3$ decomposed quite easily following the reaction given by Eq. (16.9), ultrafine Ag precipitations were produced from the decomposition reaction and set uniformly within the hollow TiO_2 NFs.

$$2AgNO_3 \rightarrow 2Ag + 2NO_2\uparrow + O_2\uparrow \tag{16.9}$$

The formation of TiO_2-CoO coaxial NF was done as follows. The self-assembled monolayers were coated again to prevent uneven cobalt oxide deposition. The self-assembled monolayers contain the same substance: octadecyltetrachlorosilane. The AAO template with TiO_2 nanotubes was immersed into a 10 mL solution of 0.05 M Co $(NO_3)_2$ for 20 min to synthesize the core-shell TiO_2-CoO discontinuous or short NFs or nanocables. After the TiO_2-AAO was soaked with Co^{2+} ions and NO_3^- ions, the postheat treatment at 450°C for 30 min was performed to obtain the CoO-TiO_2 core-shell NF. The heat treatment allowed $Co(NO_3)_2$ to decompose into CoO following the reaction in Eq. (16.10). The morphology of such a core-shell fiber is shown in Fig. 16.5C. Fig. 16.5D shows the selected area XRD pattern of the TiO_2-CoO NFs. The crystalline feature formation can be observed clearly through the clearly observed diffraction rings, providing the evidence that there exists nano-scale crystalline structure within the core-shell NFs. To convert the Ti-based hydroxide compound into anatase type crystalline TiO_2, high-temperature calcination at 550°C is typically needed [32]. Since the amorphous titanium dioxide cannot be crystallized in such a short period of time at a temperature as low as 450°C, it is concluded that only the CoO was crystallized, owing to the thermal annealing at 450°C.

$$Co(NO_3)_2 \rightarrow 2CoO + 4NO_2\uparrow + O_2\uparrow \tag{16.10}$$

It should also be indicated that through dissolving the AAO template into the solution by H^+ and F^- ions, the TiO_2-CoO coaxial NFs can be released. The dissolution of the AAO follows the reaction described by Eq. (16.11):

$$Al_2O_3 + 12H^+ + 12F^- = 2H_3AlF_6 + 3H_2O \tag{16.11}$$

The original average size of the nanopores in AAO template is 20 nm; however, after growing the TiO_2 NFs or nanorods, the pore size was expanded to larger than 200 nm. This can be seen from the measured diameters of the prepared NFs as shown by the images in Fig. 16.5A and C. This phenomenon is due to the corrosion of AAO in the solution containing H_3BO_3 and HF. The corrosion caused thinning of the AAO nanopore walls and allowed the pore to merge into bigger-sized ones. Thus the adjacent thicker NFs caused the further expansion of nanopores and grew into even bigger-sized NFs. It has also been reported that both sodium hydroxide (NaOH) and H_3PO_4 can expand the size of the nanopores in AAO, and NaOH has an even faster etching rate [31]. Both HF and H_3BO_3 have pore-expanding mechanisms similar to that of H_3PO_4 because of the dissolution of AAO in the acidic environment. The reactions of the AAO pore expansion in NaOH and H_3PO_4 solutions can be expressed as shown in Eqs. (16.12) and (16.13), respectively:

$$Al_2O_3 + 12OH^- + 12Na^+ \rightarrow 2Na_3AlO_3 + 3H_2O \tag{16.12}$$

$$Al_2O_3 + 9H^+ + 3PO_4^{3-} \rightarrow Al_2(HPO_4)_3 + 3H_2O \tag{16.13}$$

Mbulanga et al. [32] used ZnO nanorods as the sacrificial templates for making hollow titanium dioxide NFs. The ZnO rods were made on a fluorine-doped tin oxide−coated glass substrate by using a zinc nitrate salt solution based chemical bath deposition as described by Sultana et al. [26]. The ZnO surface condition effect was investigated as well. The prepared ZnO nanorods were treated in an aqueous solution consisting of ammonium hexafluorotitanate and boric acid to generate different surface conditions, which allowed investigation of the role of the neutral and polar (zinc/oxygen terminated) surfaces of ZnO rods on the formation of TiO_2 hollow NFs. Annealing at 550°C in air caused the conversion of the Ti-based complexes on the ZnO rods to crystalline anatase phase. TiO_2 hollow NFs were obtained by etching the ZnO nanorod templates in potassium hydroxide (KOH). Morphology observation revealed that a 10 min deposition period led to the development of a mixture of end-caped, open-ended, and perforated TiO_2 nanofibers with dimensions close to those of the ZnO templates. A further step of chemical modification on the surfaces of ZnO rod templates led to the development of anatase phase TiO_2 hollow fibers or tubes with open ends. The surfaces of ZnO rods served as the reaction sites for the deposition of Ti-based complexes. The formation of these complexes on the ZnO surface took place on the neutral side facets of the ZnO rods, while etching of the ZnO rods occurred preferentially along the c-axis, that is, on the ZnO-terminated faces. Therefore the deposition of TiO_2 can be considered to happen in two stages. The first stage featured the development of Ti-based complexes (titanium hydroxide monomers) on the ZnO surface, while the second stage was characterized by the polymerization of these monomers into titanium hydroxide polymers. During the first stage, the deposition starting at the lateral surfaces laid down the foundation for more titanium hydroxide formation.

16.2.4 Chemical spray pyrolysis

As is well known, chemical spray pyrolysis deposition is a fast process that is scalable in manufacturing. It is especially suitable for large-area coating preparation and thin-film deposition. By carefully controlling the processing parameters, it is possible to make efficient thermoelectric NFs, for example, n-type TiO_2 NFs as shown in the work of Hussian et al. [33]. Briefly, the precursor solution for generating titanium dioxide was sprayed onto a glass substrate preheated at 350°C. The solution for the chemical spray pyrolysis consists of titanium chloride ($TiCl_3$), ethanol alcohol (C_2H_5OH), and distilled water. The water has the role of promoting the hydrolysis of $TiCl_3$ to get TiO_2 NFs. The thickness of the TiO_2 NF layer was controlled at 350 nm. The structure of the prepared TiO_2 NF by the spray pyrolysis method at a substrate temperature of 350°C was analyzed by scanning electron microscopy (SEM). From the SEM image shown in Fig. 16.6A, it can be seen that the NFs are piled up together. The size is about 30 nm in diameter. Fig. 16.6B shows the cross section of the piled-up TiO_2 NFs. The NFs demonstrated good continuity and formed the three-dimensional (3D) interlaced dense structure.

In addition, to TiO_2 NFs, ZnO NFs were deposited on glass substrates by the spray pyrolysis deposition technique [34−37]. In the work of Kumar et al. [34], the

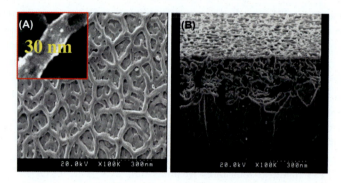

Figure 16.6 Scanning electron microscopic images showing the (A) top view and (B) side view of spray pyrolysis–deposited continuous TiO$_2$ nanofibers at a substrate temperature of 350°C.
Source: Reprinted with permission from H.A.R.A. Hussian, M.A.M. Hassan, I.R. Agool, Synthesis of titanium dioxide (TiO$_2$) nanofiber and nanotube using different chemical method, Optik 127 (2016) 2996–2999. Copyright ©2015 Elsevier, GmbH.

spraying pyrolysis solution was made by dissolving zinc acetate anhydrous Zn(CH$_3$COO)$_2$ into a mixture solvent containing methanol and water. The ratio of methanol to water was 3:1. A small amount of acetic acid as a stabilizer was added to avoid the formation of Zn(OH)$_2$ precipitate. The concentration of the zinc acetate in the solution was about 0.05 M. This solution was forced to pass through a nozzle and was sprayed onto the preheated glass substrate at 450°C under a constant air pressure of 0.2 Torr. Fig. 16.7 shows the SEM images of the as-grown and annealed ZnO NFs. The average diameter of the NFs in the as-grown state was found to be approximately 300 nm, as can be seen from the image shown in Fig. 16.7A. After 1 h annealing at 450°C, the diameter of the NFs remained the same as that of the as-sprayed state, which can be observed from Fig. 16.7B. However, after 4 h of annealing, the diameter of the fibers increased to about 800 nm, as shown in Fig. 16.7C. A further increase in the annealing time would not cause much change in the size of the nanofibers. Therefore the average diameter of the NFs after the 6 h of annealing was still approximately 800 nm, as shown in Fig. 16.7D. The reason for the fiber coarsening with the increase in annealing time was explored. It is probably due to the merging of the finer fibers during the annealing process, resulting in the formation of thicker fibers. It has been observed that the annealing temperature has a direct influence on the thickness of the fibers. The structure of ZnO fibers has also been found to be in better order with a longer annealing period.

Sharmin and Bhuiyan [35] prepared boron-doped zinc oxide NFs by spray pyrolysis of a zinc nitrate solution with the addition of boric acid. It was found that boron (B) increased the conductivity of the NFs. Decreases in the NF diameter due to B doping was also found. Without B doping, the average diameter of the ZnO NF was about 500 nm. After doping B with the concentrations of 0.5, 0.75, 1.0, and 1.5 at%, the average diameters were 320, 240, 180, and 170 nm, respectively. It is evident that the higher the concentration of B, the smaller the diameter of the NFs.

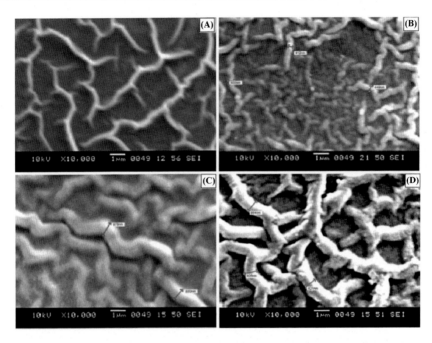

Figure 16.7 Scanning electron microscopic images of spray pyrolysis–deposited ZnO nanofibers (A) as grown, (B) after 1 h of annealing, (C) after 4 h of annealing, and (D) after 6 h of annealing.
Source: Reproduced under the terms and conditions of the liberal Creative Commons Attribution 4.0 International license from N.S. Kumar, K.V. Bangera, G.K. Shivakumar, Effect of annealing on the properties of zinc oxide nanofiber thin films grown by spray pyrolysis technique, Appl. Nanosci. 4 (2014) 209–216. Copyright ©2013 The Authors.

Maity et al. [36] prepared ZnO NFs with an average diameter of 500 nm, which is consistent with the results reported by Sharmin and Bhuiyan [35]. Indium doping can also reduce the fiber size, as was reported by Ilican et al. [37]. The indium-doped zinc oxide NFs have a uniform diameter of 200 nm.

The spray pyrolysis technique not only is suitable for processing TiO_2 and ZnO NFs, but also can make other types of oxide NFs. For example, Zahan and Podder [38] prepared Co_3O_4 via the spray pyrolysis technique. Cobalt oxide (Co_3O_4) NFs were deposited onto glass substrates by spraying cobalt acetate $Co(CH_3COO)_2 \cdot 4H_2O$ precursor solution. Field FESEM images show uniform and well-aligned Co_3O_4 NFs.

16.2.5 Microlithography and nanolithography

Microlithography and nanolithography are lithographic patterning methods for creating features or building architectures at very small scales [39]. Microlithography and nanolithography technologies have been utilized for decades in the semiconducting industry to fabricate integrated circuits. They can generate patterns with a tiny size in the range from micrometers to nanometers. They are usually combined with deposition and etching technology to yield high-resolution topography. Generally, microlithography deals with features smaller than 10 μm, while nanolithography produces features smaller than 100 nm. Microlithography and nanolithography comprise several kinds of methods. Photolithography [40] is one of these methods using ultraviolet (UV) light or laser as the energy beam to develop the photoresists, which are generally made from epoxy or polymethyl methacrylate (PMMA). A prefabricated photomask serves as a master from which the final pattern is made. This technology has found wide applications in fabricating devices including sensors, actuators, and transducers in microelectromechanical systems (MEMS). The advantages of photolithographic technology include fast, scalable, relatively low cost of facilities, and easy to operate [41].

Although photolithographic technology is the most commonly used method, other techniques are also developed. For example, electron beam lithography, as a newer nanolithography method, can achieve much finer patterning resolution than the photolithography. This is because the electron beam with a much shorter wavelength than UV light is used for pattern development in electron beam lithography. The resolution of electron beam lithography has reached as small as a few nanometers. Electron beam lithography is commercially important as a primary route for manufacturing photomasks. Demonstrations can be found in [42] for metal nano-cone fabrication. It is pointed out that electron beam lithography is a maskless lithography technique. During the process, a mask is not needed to generate the final pattern. Instead, the final pattern is created directly from a digital representation on a computer, by controlling the electron beam as it scans across a photoresist-coated substrate. As compared to photolithography, the major advantage of electron beam lithography is the better resolution as mentioned. One of the shortcomings is much slower in speed. In addition, the facilities for electron beam generating are more complicated. The initial investment, operating and maintenance costs are also much higher.

In addition to the above mentioned well-established and commercialized techniques, a large number of promising microlithographic and nanolithographic technologies have been developed. A list of these examples may include direct laser writing [43], extreme ultraviolet (EUV) lithography [44], phase interference lithography [45], X-ray lithography [46], soft lithography or nanoimprinting lithography also called micro/nanoinjection molding [47], scanning probe lithography [48], and magnetolithography [49]. Some of these new techniques have been used successfully for small-scale commercial and/or important research applications. Surface-charge lithography [50], thermal lithography [51], and other new technologies are under research and development.

Preparation of doped, complex oxide NFs was performed using nanolithography combined with the laser vapor deposition technique [52]. The fabricated Nb-doped $SrTiO_3$ NFs were investigated for potential use in high temperature thermoelectric energy conversion devices. The key steps include pulsed laser deposition of the complex oxide at room temperature onto electron beam lithography defined templates of the PMMA photoresist. Following a photoresist lift-off in organic solvents, a post heat treatment procedure was used to crystallize the NFs. Fig. 16.8 schematically shows the experimental procedures. The templates were made from a 4 wt.% polymethyl methacrylate photoresist dissolved in anisole. Pattern writing was done using an electron beam lithography machine. First, the photoresist was spin-coated onto single-crystalline STO substrates with the [100] crystallographic orientation as sketched in Fig. 16.8A. The thickness of the PMMA photoresist reached about 175 nm. After the photoresist was baked, a conductive polymer was spin-coated on it. The conductive layer was baked as well. The function of the conductive layer is preventing charge buildup on the surface of the insulating substrates during the subsequent electron beam lithography. The patterns used for nanofiber fabrication were trenches with horizontal widths ranging from 10 nm to 1 μm and lengths of 500 μm. After the e-beam patterning, the photoresist was developed in a solution consisting of methyl isobutyl ketone in isopropanol for 1 min and was then rinsed in isopropanol. Since the photoresist in the electron-irradiated regions is soluble in the developer solution, some areas of the substrate were exposed. The exposed and developed substrate with a trench style pattern is schematically shown in Fig. 16.8B. Filling the template was performed by using the pulsed laser deposition technique in high vacuum at room temperature. A 248 nm wavelength laser was used to ablate the $SrTi_{0.8}Nb_{0.2}O_3$ (5 at.% Nb) sintered target. The filled template is schematically shown in Fig. 16.8C. Following the pulsed laser deposition, the substrate was soaked in acetone. Fig. 16.8D illustrates the oxide NFs on the substrate after the template is dissolved. Following that, the substrate and NFs were annealed at 650°C for 1 h and then cooled to room temperature. Finally, the NFs were coated with a thin layer of Pt/Au for SEM analysis.

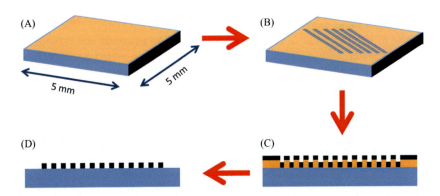

Figure 16.8 Schematic of nanolithography and pulsed laser deposited Nb-STOnanofibers. (A) The polymethyl methacrylate-coated substrate before electron beam lithography. (B) The trenches in the polymethyl methacrylate after electron beam lithography exposure and photoresist development. (C) The template filled with Nb-STOafter pulsed laser deposition. (D) Nanofibers on the substrate after lift-off of the polymethyl methacrylate photoresist.
Source: Reprinted with permission from G. Waller, A. Stein, J.T. Abiade, Nanofabrication of doped, complex oxides, J. Vac. Sci. & Technol. B 30 (2012) 011804. Copyright ©2012 American Vacuum Society.

The SEM images presented by Waller et al. [52] show the actual dimensions of the oxide nanofibers using 10 and 30 nm nominal trench widths. The templates fabricated by using electron beam lithography (EBL) on oxide substrates produced a substantial increase in final dimensions compared to the expected feature width. For a template with a nominal dimension of 10 nm as defined in the pattern generation software, a final dimension of 110 nm was obtained. These dimensions were accurately conveyed to the deposited oxide. For a template with a nominal dimension of 30 nm as defined in the pattern generation software, a final dimension of 150 nm was achieved. The expansion in template dimension and the high dose required for completing the exposure can be explained by a phenomenon known as proximity dosing [52]. On the STO, a portion of the electron beam is dispersed, most likely by backscattering or localized charge concentration, exposing the photoresist in the near proximity to the designated area of irradiation. Consequently, the high dose required to completely expose the resist causes an undesirable pattern broadening. It is believed that a careful design of the irradiation pattern or the use of lower accelerating voltages may alleviate this problem [52]. The electrical behavior of the prepared nanofibers was studied. Nearly 200 times difference in the conductive property between the unannealed Nb/STO and annealed nanofiber was found.

16.2.6 Electrochemical oxidation

Electrochemical oxidation (ECO) is also known as anodic oxidation or anodization. It is commonly used for making Cu-, Zn-, and Fe-based oxide nanostructures. In addition, almost all the so-called valve metals, such as Ti, Al, Mg, Zr, Nb, and Ta [53], show the unique electrochemical oxidation behavior so that oxide NFs or nanotubes from these metals can be generated via the ECO method. The most general layout of ECO consists of two electrodes, operating as anode and cathode, connected to a power source. Sometimes a third electrode is added as the reference electrode. When the energy input and sufficient supporting electrolyte are provided to the system, strong oxidizing species are formed. ECO has recently grown in popularity, thanks to its ease of setup and effectiveness in generating in situ electrochemical oxidation to form the required reactive species at the anode surface. Electrochemical oxidation has been applied to make oxide nanostructures based on the direct oxidation working principle. When voltage is applied crossing the anodic and cathodic electrodes, intermediates of oxygen evolution are formed at the anode. The surface of an "active" anode produces higher state oxides or superoxides. The higher oxide then acts as an intermediate product for NF or nanotube formation. In the literature [54], a piece of titanium foil was used as the anode and a platinum foil was used as the cathode. The electrolyte for anodization was made by mixing 7.5 wt.% water, 90 wt.% glycerol, and 1.5 wt.% ammonium fluoride (NH_4F). After the electrochemical oxidation, titanium dioxide hollow NFs were obtained.

The p-type copper oxide/hydroxide NFs were made by self-organized anodization of a high-purity copper in a 0.1 M Na_2CO_3 electrolyte [55]. The pure copper in the form of foil was cut into 25×10 (mm^2) coupons as the specimens for electrochemical oxidation study. The specimens were degreased by using acetone and ethanol, respectively. Under the ambient temperature condition, the cleaned specimens were electrochemically polished in a 10.0 M H$_3$PO$_4$ at 7.5 V for 60 s. Then the back side and edges of the specimens were coated with an acid-resistant paint to keep an exposed working surface area of 1 cm^2. Self-organized anodization of the copper samples in the 0.1 M Na$_2$CO$_3$ was conducted at 20°C for 1 h in the voltage range from 3.0 to 31.0 V with a 4.0 V increment. Microscopic observation of the sample processed at 31.0 V reveals well-established and faceted NFs, as shown in Fig. 16.9A. At higher magnification it can be seen that in some regions the oxide NFs were selectively etched, as shown in the upper left side of Fig. 16.9B. Raman spectroscopic examination of the samples processed at various voltages reveals the presence of both CuO and Cu$_2$O oxides. As shown in Fig. 16.9C, the peaks in the range of 274−289 cm^{-1} Raman shift come from CuO. This confirms that CuO exists in all the specimens processed at different voltages. For the sample anodized at 30.0 V, a distinct peak at 441 cm^{-1} was found. This peak indicates the presence of Cu$_2$O in the NFs. In the range of 613−635 cm^{-1} a

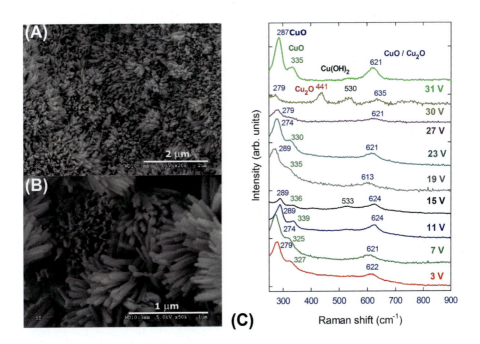

Figure 16.9 Field emission scanning electron microscopic images and Raman spectra of copper oxide nanofibers processed by electrochemical anodization. (A) A low-magnification image. (B) An image with higher magnification. (C) Raman spectra.
Source: Reprinted with permission from W.J. Stępniowski, D. Paliwoda, S.T. Abrahami, M. Michalska-Domańska, K. Landskron, J.G. Buijnsters, et al., Nanorods grown by copper anodizing in sodium carbonate, J. Electroanalytical Chem. 857 (2020) 113628. Copyright ©2019 Elsevier B.V.

Figure 16.10 Field emission scanning electron microscopic images of oxide nanofibers on copper specimens processed by electrochemical anodization in 1.0 M NaOH. (A) A low-magnification image. (B) An image at higher magnification.
Source: Reprinted with permission from W. Jiang, J. He, F. Xiao, S. Yuan, H. Lu, B. Liang, Preparation and antiscaling application of superhydrophobic anodized CuO nanowire surfaces, Ind. & Eng. Chem. Res. 54 (2015) 6874−6883. Copyright ©2015 American Chemical Society.

broad band was found, which is due to the overlap of Cu_2O and CuO signals. It was concluded that anodization of copper in the sodium carbonate aqueous solution generated short NFs made of CuO, Cu_2O, and $Cu(OH)_2$.

Experiments on electrochemical oxidation of copper were also performed in solutions containing strong bases, such as NaOH [56−58] and KOH [59−63]. Jiang at al. [56] used a deaerated 1.0 M NaOH to oxidize copper at a current density of 0.06 mA cm^{-2}. The total anodization time was 5 min at a temperature of 25°C. Following that, the NFs were annealed in a tubular furnace under nitrogen protection. The obtained CNFs were surface modified by a 1.0 wt.% of 1H,1H,2H,2H-perfluorodecyltriethoxysilane (FAS-17) ethanol solution to achieve the superhydrophobic property. The SEM images of the surface modified NFs are shown in Fig. 16.10. From Fig. 16.10A it can be seen that there are a lot of NF clusters. The surface of the specimen was covered completely by the grasslike NFs. At an even higher magnification the image in Fig. 16.10B illustrates the 3D configuration of the NFs. The entangled feature of NFs can still be seen in this figure. The electrochemical anodization of copper in potassium oxalate [64] and potassium carbonate [65] was also reported as typical examples in the literature.

16.2.7 Glass-annealing method

Finally, the glass-annealing technique is discussed here. The method works by heating glassy melt-quenched plates at high temperatures. A long-time slow annealing process allows the formation of tiny whiskers from the surface of the plate specimens. Typically, the sizes of the whiskers are in the submicron and micrometer ranges. By the use of this glass-annealing technique, Funahashi et al. [66] prepared Ca−Co−O-based single-crystalline oxide whiskers with good thermoelectric properties at temperatures higher than 600 K in air. The composition of the whiskers is $Ca_2Co_2O_5$. $Ca_2Co_2O_5$ has a single-phase structure, as determined by the XRD measurement and TEM analysis. The whiskers have a layered structure in which Co-O layers of two different kinds alternate in the direction of the c-axis.

The measured Seebeck effect coefficient of the $Ca_2Co_2O_5$ whiskers is found to be higher than $100\,\mu\,V\,K^{-1}$ at 100 K and increases close to $230\,\mu V\,K^{-1}$ with the increase in temperature. The temperature dependence of electrical resistivity for the whiskers reveals the typical semiconducting-like behavior. The thermoelectric figure of merit (zT) of the $Ca_2Co_2O_5$ whiskers is estimated at a value between 1.2 and 2.7 at temperatures above 873 K, demonstrating high thermoelectric performance at high service temperatures [66].

In the literature [67], Funahashi and Matsubara described the glass-annealing method in synthesizing Ca- and Pb-doped oxide single-crystalline whiskers. To prepare $[(Bi, Pb)_2(Sr, Ca)_2O_4]_xCoO_2$, the starting materials—Bi_2O_3, PbO, $CaCO_3$, $SrCO_3$, and Co_3O_4 powders—were mixed with a cationic composition of Bi:Pb:Ca:Sr:Co = 1:1:1:1:2. The mixture was placed into an alumina crucible to melt at 1573 K for 30 min in air. The melt was quenched by rapid cooling between two copper plates to obtain glassy oxide plates. These glassy plates were heated in flowing O_2 gas at 1193 K for 600 h to grow Ca- and Pb-doped $(Bi_2Sr_2O_4)_xCoO_2$ whiskers at their surfaces. The SEM image of the ribbon-like whiskers was shown by Funahashi and Matsubara [67]. The whiskers showed the following measured dimensions: $1.0-3.0\,\mu m$ in thickness, $20-100\,\mu m$ in width, and about 1.0 mm in length. TEM analysis revealed the layered structure of the whiskers [67]. The CoO_2 layers and rock salt $(Bi, Pb)_2(Sr, Ca)_2O_4$ layers alternate in the c-axis direction. Such a structure allows the whiskers to display superior thermoelectric properties at high temperature in air because the edge-sharing CoO_2 layers act as the conducting and thermoelectric units. Each rock salt layer consists of four ordered sublayers: Sr(Ca)O$-$Bi(Pb)O$-$Bi(Pb)O$-$Sr(Ca)O. These sublayers could slow down the thermal transport in the whiskers. The average composition of the whiskers was found to be $(Bi, Pb)_{2.2}(Sr, Ca)_{2.8}Co_2O_y$ (named as BC-232).

The Seebeck coefficient, S, of the whiskers is around $100\,\mu\,V\,K^{-1}$ at 100 K. The S value increases monotonically with temperature up to 773 K and reaches an asymptotic value of $190\,\mu\,V\,K^{-1}$ at temperatures higher than 773 K. The temperature dependence of electric resistivity was shown to follow a semiconducting-like behavior. But its electrical conductivity value is much lower than those of ordinary semiconductors. The power factor of the BC-232 whiskers increases with the increase in test temperature. The estimated value is above $0.5\,mW\,(m^{-1}\cdot K^{-2})$ at temperatures higher than 650 K and reaches $0.9\,mW\,(m^{-1}\cdot K^{-2})$ at 973 K [67].

16.3 Thermoelectric metal oxide nanofiber device concept and characterization

In this section the MONF device concept and characterization methods will be briefly discussed. For exploring new thermoelectric materials, it is highly desirable to have accurate and fast measurement techniques. To meet such requirements, thermoelectric devices must be built with capabilities to access the local fluctuations or gradients in material properties. Hayat et al. [14] made the La$_2$CuO$_4$ NF-based thermoelectric device by setting the NF mat on Pd-Ag plate electrodes as shown in Fig. 16.11(A). The electrodes were printed on a piece of alumina sheet. The temperature gradient across the electrodes on the thermoelectric device was generated by gradually increasing the temperature of the hot end while keeping the cold end at a fixed temperature of 298 K. The thermoelectric output voltage (V) crossing the two selected electrodes on the device was measured as the temperature increment (ΔT) varied. The data were plotted as shown in Fig. 16.11(B). Fig. 16.11 (C) shows the current versus voltage curve for La$_2$CuO$_4$ NFs at the room temperature. The nonlinear behavior of current as a function of applied voltage indicated a

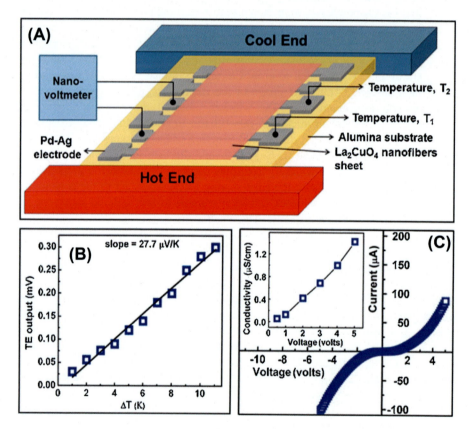

Figure 16.11 Drawing of an oxide nanofiber thermoelectric device and the performance. (A) La$_2$CuO$_4$ nanofibers on Pd-Ag conductive electrodes. (B) Thermoelectric output voltage versus temperature difference (ΔT). (C) Current versus voltage measured at room temperature; inset shows the calculated electrical conductivity versus applied voltage.
Source: Reprinted with permission from K. Hayat, F. Niaz, S. Ali, M. Javid Iqbal, M. Ajmal, M. Ali, et al., Thermoelectric performance and humidity sensing characteristics of La2CuO4 nanofibers, Sens. Actuators B 231 (2016) 102–109. Copyright ©2016 Elsevier B.V.

semiconductor-like behavior of La_2CuO_4 NFs. Electrical conductivity was calculated and shown as an inset in the same sub-figure.

Traditional measurement of electrical conductivity provides a global value for a given material. By contrast, Ma et al. [21] successfully characterized the local conductive behavior of a single $NaCo_2O_4$ NF by the conductive atomic force microscopy (cAFM) using a conductive cantilever tip as the top electrode. Their work shows that it is possible to map the current distribution over the NF surface. The NF was deposited on a Si/SiO_2 substrate with a prepatterned layer of Pt as the bottom electrode. The single $NaCo_2O_4$ NF was placed on both Pt and SiO_2. Building such a nano-scale device allows the shortest conductive path when the cAFM tip scans the NF right above the Pt bottom electrode. The conductive path gradually increases when the tip moves away from the Pt electrode. Therefore a higher current was measured when the tip was closer to the Pt bottom electrode. During the current mapping experiment, it was observed that the current reached the highest for the portion of the NF on top of the Pt electrode. The current decreased gradually in the part of the NF segment away from the Pt. This is because the increased current path resulted in larger electrical resistance. From the current mappings for fiber segments over the Pt and SiO_2 areas, it is found that the current is in the range of nanoamperes for the part of the NF just above the Pt electrode. The current value drops to 10 pA range for the portion of the NF above the SiO_2 (away from the Pt electrode). The locations with defects in the NF can also be revealed by catching the erratic distribution of the current with some local high and low extreme values. This is based on the fact that defects degrade the overall electrically conductive performance of the NF. The local $I-V$ curve was established through the cAFM measurement. From the slope of the $I-V$ curve, the nominal electrical conductivity of the $NaCo_2O_4$ NF was estimated, and the results were shown [21].

The traditional Seebeck coefficient measurement method requires electrical contacts, and the samples have to be heated. An optical contactless measurement technique for determining the thermoelectric transport properties of semiconducting materials has been reported [68]. The carrier populations are not in thermal equilibrium with the lattice of the material during the measurement. Such a contactless method enables access to the local gradients of the two fundamental thermodynamic properties, namely, the temperature and the electrochemical potential. Therefore it is possible to determine the Seebeck coefficient related to the light-induced thermoelectric properties of the material at a micrometer scale. The generalized Planck's law of radiation for fitting the photoluminescence spectra lays the foundation for the technique [69]. It can be used to access the quasi-Fermi-level splitting and the temperature of the carriers in a semiconductor. Measurement of the two parameters allows calculation of the Seebeck coefficient of the material. In addition, from the calibrated photoluminescence intensity profile with the spatial coordinates combined with Callen coupled transport equations and with the kinetic expression for the transport parameters under the relaxation time approximation, the electrical conductivity, thermal electron and hole conductivity, mobility, diffusion coefficients, and heat transferred from the carriers to the lattice can be estimated. All these parameters can be obtained either for electrons or for holes. The approach is also considered applicable for intrinsic semiconductors in the ambipolar regime. The method was applied to a multiquantum well structure of InGaAsP. The luminescence comes from the quantum wells. The transport properties in the plane of the wells inside the whole structure were investigated. Since photoluminescence does not require p-n junction or high electrical conductivities for the measurement, this optical contactless measurement technique of thermoelectric transport parameters involving quasiequilibrium carriers is suitable for determining the properties inside a given layer of the whole structure or in materials with very low conductivities [70].

16.4 Perspectives and conclusions

In conclusion, thermoelectric metallic oxide NFs are promising materials for high-temperature thermoelectric energy conversion applications, owing to their chemical stability. Continuous metallic oxide NFs are especially suitable for building flexible energy converters and sensors. Several kinds of oxide NFs have attracted attention and have been studied intensively, owing to their excellent thermoelectric performances. These include the n-type $SrTiO_3$, $LaAlO_3$, La_2CuO_4, TiO_2, and ZnO and the p-type $NaCo_2O_4$, CuO/Cu_2O, and $Ca_3Co_4O_9$. In view of the processing and manufacturing technologies for thermoelectric oxide NF production, electrospinning, chemical spray pyrolysis, vapor deposition, microlithography and nanolithography, and glass annealing are scalable methods for cost-effective, large-scale applications. There are still many challenges and new fields to explore for thermoelectric oxide nanofibers. One of the biggest challenges is how to improve the thermoelectric performance further. The figure of merit of most currently available for oxide nanofibers is lower than 1.0. New composite oxide NFs and porous oxide NFs are possible new candidates for enhancing the thermoelectric properties. Another issue is that in comparison with other thermoelectric materials such metals and metallic compounds, MONFs have been much less studied. Thermoelectric data are not readily available for oxide NFs. More understanding of the structure-property relationships of composite and porous oxide NFs will have a profound impact on the development of novel thermoelectric materials. In addition, to exploring new oxide NFs and structures, there is a strong need to develop new measurement techniques or improve the current ones for determining the thermal, electrical, and thermoelectric parameters of low-dimension materials, especially MONFs.

References

[1] S. Hong, Y. Gu, J.K. Seo, J. Wang, P. Liu, Y.S. Meng, et al., Wearable thermoelectrics for personalized thermoregulation, Sci. Adv. 5 (2019) eaaw0536.

[2] C.M. Bhandari, D.M. Rowe, CRC Handbook of Thermoelectrics, 5, CRC Press, Boca Raton, FL, 1995, pp. 43–45.

[3] Y.X. Gan, Nanomaterials for Thermoelectric Devices, Jenny Stanford Publishing, Singapore, 2018, pp. 4–70.

[4] Y.X. Gan, Advanced Materials and Systems for Energy Conversion: Fundamentals and Applications, Nova Science Pub Inc., Hauppauge, NY, 2010, pp. 16−91.

[5] Y.H. Lin, J. Lan, C. Nan, Oxide Thermoelectric Materials: Basic Principles and Applications, Wiley-VCH Verlag GmbH & Co. KGaA., Weinheim, Germany, 2019, pp. 77−154.

[6] D. Kenfaui, B. Lenoir, D. Chateigner, B. Ouladdiaf, M. Gomina, J.G. Noudem, Development of multilayer textured $Ca_3Co_4O_9$ materials for thermoelectric generators: influence of the anisotropy on the transport properties, J. Eur. Ceram. Soc. 32 (2012) 2405−2414.

[7] Y. Yin, B. Tudu, A. Tiwari, Recent advances in oxide thermoelectric materials and modules, Vacuum 146 (2017) 356−374.

[8] M.-C. Hsu, I.-C. Leu, Y.-M. Sun, M.-H. Hon, Fabrication of CdS@TiO_2 coaxial composite nanocables arrays by liquid-phase deposition, J. Cryst. Growth 285 (2005) 642−648.

[9] C.A. Bonino, L. Ji, Z. Lin, O. Toprakci, X. Zhang, S.A. Khan, Electrospun carbon-tin oxide composite nanofibers for use as lithium ion battery anodes, ACS Appl. Mater. Interfaces 3 (2011) 2534−2542.

[10] S.J. Lee, S.M. Park, S.J. Han, D.,S. Kim, Electrolyte solution-assisted electrospray deposition for direct coating and patterning of polymeric nanoparticles on non-conductive surfaces, Chem. Eng. J. 379 (2020) 1223182.

[11] D. Ma, B. Liu, X. Jin, L. Zhu, X. Wang, G. Zhang, et al., Rheologic behaviors and continuously dry spinning of polyacetylacetonatozirconium fibers, Mater. Lett. 258 (2020) 126824.

[12] T.M. Robinson, D.W. Hutmacher, P.D. Dalton, The next frontier in melt electrospinning: Taming the jet, Adv. Funct. Mater. 29 (2019) 1904664.

[13] S. Maensiri, W. Nuansing, Thermoelectric oxide $NaCo_2O_4$ nanofibers fabricated by electrospinning, Mater. Chem. Phys. 99 (2006) 104−108.

[14] K. Hayat, F. Niaz, S. Ali, M. Javid Iqbal, M. Ajmal, M. Ali, et al., Thermoelectric performance and humidity sensing characteristics of La_2CuO_4 nanofibers, Sens. Actuators B 231 (2016) 102−109.

[15] Y. Liu, Y.H. Lin, B.P. Zhang, C.W. Nan, J.F. Li, High-temperature electrical transport behavior observed in the $La_{1.96}M_{0.04}CuO_4$ (M: Mg Ca, Sr) polycrystalline ceramics, J. Am. Ceram. Soc. 91 (2008) 2055−2058.

[16] T. Liu, Z. Liu, J. Ren, Q. Zhao, H. He, N. Wang, et al., Operating temperature and temperature gradient effects on the photovoltaic properties of dye sensitized solar cells assembled with thermoelectric-photoelectric coaxial nanofibers, Electrochim. Acta 279 (2018) 177−185.

[17] A. Mohammadi, A. Valipouri, S. Salimian, Nanoparticle-loaded highly flexible fibrous structures exhibiting desirable thermoelectric properties, Diam. Relat. Mater. 86 (2018) 54−62.

[18] X. Xu, Z. Wu, L. Xiao, Y. Jia, J. Ma, F. Wang, et al., Strong piezo-electro-chemical effect of piezoelectric $BaTiO_3$ nanofibers for vibration-catalysis, J. Alloy. Compd. 762 (2018) 915−921.

[19] K. Mondal, Recent advances in the synthesis of metal oxide nanofibers and their environmental remediation applications, Inventions 2 (2017) 9.

[20] Z. Li, S. Liu, S. Song, W. Xu, Y. Sun, Y. Dai, Porous ceramic nanofibers as new catalysts toward heterogeneous reactions, Comp. Commun. 15 (2019) 168−178.

[21] F. Ma, Y. Ou, Y. Yang, Y. Liu, S. Xie, J.F. Li, et al., Nanocrystalline structure and thermoelectric properties of electrospun $NaCo_2O_4$ nanofibers, J. Phys. Chem. C. 114 (2010) 22038−22043.

[22] S. Kocyigit, A. Aytimur, E. Cinar, I. Uslu, A. Akdemir, Boron-doped strontium-stabilized bismuth cobalt oxide thermoelectric nanocrystalline ceramic powders synthesized via electrospinning, JOM 66 (2014) 30−36.

[23] D. Lee, K. Cho, J. Choi, S. Kim, Fabrication and characterization of a thermoelectric pn couple made of electrospun oxide nanofibers, J. Korean Inst. Electr. Electron. Mater. Eng. 28 (2015) 252−256.

[24] S. Patel, S. Kansara, Y.X. Gan, Y.T. Zhao, J.B. Gan, Hydrothermally coated oxide nanoparticle-containing composite fibers. In: Proceedings of the 4th thermal and fluid engineering conference, TFEC2019, Paper No. TFEC-2019-28031, Las Vegas, NV, April 14−17, 2019.

[25] L. Xu, L. Zhang, B. Chen, J. Yu, Rationally designed hierarchical $NiCo_2O_4$-C@Ni $(OH)_2$ core-shell nanofibers for high performance supercapacitors, Carbon 152 (2019) 652−660.

[26] J. Sultana, S. Paul, R. Saha, S. Sikdar, A. Karmakar, S. Chattopadhyay, Optical and electronic properties of chemical bath deposited p-CuO and n-ZnO nanowires on silicon substrates: p-CuO/n-ZnO nanowires solar cells with high open-circuit voltage and short-circuit current, Thin Solid. Films 699 (2020) 137861.

[27] Y.L. Chu, L.W. Ji, Y.J. Hsiao, H.Y. Lu, S.J. Young, I.T. Tang, et al., Fabrication and characterization of Ni-doped ZnO nanorod arrays for UV photodetector application, J. Electrochem. Soc. 167 (2020) 067506.

[28] N.A.M. Yahya, M.R.Y. Hamid, B.H. Ong, N.A. Rahman, M.A. Mahdi, M.H. Yaacob, H-2 gas sensor based on Pd/ZnO nanostructures deposited on tapered optical fiber, IEEE Sens. J. 20 (2020) 2982−2990.

[29] N. Rathore, D.V.S. Rao, S.K. Sarkar, Growth of a polarity controlled ZnO nanorod array on a glass/FTO substrate by chemical bath deposition, RSC Adv. 5 (2015) 28251−28257.

[30] A. Kaphle, E. Echeverria, D.N. Mcllroy, P. Hari, Enhancement in the performance of nanostructured CuO-ZnO solar cells by band alignment, RSC Adv. 10 (2020) 7839−7854.

[31] L. Su, Y.X. Gan, L. Zhang, Thermoelectricity of nanocomposites containing TiO_2-CoO coaxial nanocables, Scr. Materialia 64 (2011) 745−748.

[32] C.M. Mbulanga, W.E. Goosen, R. Betz, J.R. Botha, Effect of surface properties of ZnO rods on the formation of anatase-phase TiO_2 tubes prepared by liquid deposition method, Appl. Phys. A: Mater. Sci. Process. 126 (2020) 180.

[33] H.A.R.A. Hussian, M.A.M. Hassan, I.R. Agool, Synthesis of titanium dioxide (TiO_2) nanofiber and nanotube using different chemical method, Optik 127 (2016) 2996−2999.

[34] N.S. Kumar, K.V. Bangera, G.K. Shivakumar, Effect of annealing on the properties of zinc oxide nanofiber thin films grown by spray pyrolysis technique, Appl. Nanosci. 4 (2014) 209−216.

[35] M. Sharmin, A.H. Bhuiyan, Modifications in structure, surface morphology, optical and electrical properties of ZnO thin films with low boron doping, J. Mater. Sci-Mater Electron. 30 (2019) 4867−4879.

[36] R. Maity, S. Das, M.K. Mitra, K.K. Chattopadhyaya, Synthesis and characterization of ZnO nano/microfibers thin films by catalyst free solution route, Phys. E 25 (2005) 605−612.

[37] S. Ilican, Y. Caglar, M. Caglar, F. Yakuphanoglu, Electrical conductivity, optical and structural properties of indium-doped ZnO nanofiber thin film deposited by spray pyrolysis method, Phys. E 35 (2006) 131−138.

[38] M. Zahan, J. Podder, Surface morphology, optical properties and Urbach tail of spray deposited Co_3O_4 thin films, J. Mater. Sci.: Mater. Electron. 30 (2019) 4259−4269.

[39] Y. Li, Yang, M.H. Hong, Parallel laser micro/nano-processing for functional device fabrication, Laser Photon. Rev. 14 (2020) 1900062.

[40] S.H. Lee, S.E. Seo, K.H. Kim, J. Lee, C.S. Park, B.H. Jun, et al., Single photomask lithography for shape modulation of micropatterns, J. Ind. Eng. Chem. 84 (2020) 196−201.

[41] U.S. Kim, S.Y. Baek, T.W. Kim, J.W. Park, Cold tribo-nanolithography on metallic thin-film surfaces, J. Nanosci. Nanotechnol. 20 (2020) 4318−4321.

[42] D. Eschimese, F. Vaurette, T. Melin, S. Arscott, Precise tailoring of evaporated gold nano-cones using electron beam lithography and lift-off, Nanotechnology 31 (2020) 225302.

[43] X.H. Yu, Q.W. Zhang, D.F. Qi, S.W. Tang, S.X. Dai, P.Q. Zhang, et al., Femtosecond laser-induced large area of periodic structures on chalcogenide glass via twice laser direct-writing scanning process, Opt. Laser Technol. 124 (2020) 105977.

[44] F. Rahman, D.J. Carbaugh, J.T. Wright, P. Rajan, S.G. Pandya, S. Kaya, A review of polymethyl methacrylate (PMMA) as a versatile lithographic resist - With emphasis on UV exposure, Microelectron. Eng. 224 (2020) 111238.

[45] D.B. You, J.H. Park, B.S. Kang, D.H. Yun, B.S. Shin, A fundamental study of a surface modification on silicon wafer using direct laser interference patterning with 355-nm UV laser, Sci. Adv. Mater. 12 (2020) 516−519.

[46] K.H. Jang, J.J. Choi, J.H. Kim, X-ray LIGA microfabricated circuits for a sub-THz wave folded waveguide traveling-wave-tube amplifier, J. Korean Phys. Soc. 75 (2019) 716−723.

[47] F.D. Arisoy, I. Czolkos, A. Johansson, T. Nielsen, J.J. Watkins, Low-cost, durable master molds for thermal-NIL, UV-NIL, and injection molding, Nanotechnology 31 (2020) 015302.

[48] K. Xu, J. Chen, High-resolution scanning probe lithography technology: a review, Appl. Nanosci. 10 (2020) 1013−1022.

[49] A. Bardea, A. Yoffe, Magneto-lithography, a simple and inexpensive method for high throughput, surface patterning, IEEE Trans. Nanotechnol. 16 (2017) 439−444.

[50] J. Pablo-Navarro, S. Sangiao, C. Magen, J.M. Maria de Teresa, Diameter modulation of 3D nanostructures in focused electron beam induced deposition using local electric fields and beam defocus, Nanotechnology 30 (2019) 505302.

[51] Y. Meng, J.K. Behera, Z.W. Wang, J.L. Zheng, J.S. Wei, L.C. Wu, et al., Nanostructure patterning of $C-Sb_2Te_3$ by maskless thermal lithography using femtosecond laser pulses, Appl. Surf. Sci. 508 (2020) 145228.

[52] G. Waller, A. Stein, J.T. Abiade, Nanofabrication of doped, complex oxides, J. Vac. Sci. Technol. B 30 (2012) 011804.

[53] Y.L. Cheng, Z.A. Zhu, Q.H. Zhang, X.J. Zhuang, Y.L. Cheng, Plasma electrolytic oxidation of brass, Surf. Coat. Technol. 385 (2020) 125366.

[54] Y.X. Gan, B.J. Gan, E. Clark, L.S. Su, L.H. Zhang, Converting environmentally hazardous materials into clean energy using a novel nanostructured photoelectrochemical fuel cell, Mater. Res. Bull. 47 (2012) 2380−2388.

[55] W.J. Stępniowski, D. Paliwoda, S.T. Abrahami, M. Michalska-Domańska, K. Landskron, J.G. Buijnsters, et al., Nanorods grown by copper anodizing in sodium carbonate, J. Electroanal Chem 857 (2020) 113628.

[56] W. Jiang, J. He, F. Xiao, S. Yuan, H. Lu, B. Liang, Preparation and antiscaling application of superhydrophobic anodized CuO nanowire surfaces, Ind. Eng. Chem. Res 54 (2015) 6874−6883.

[57] J. Wu, X. Li, B. Yadian, H. Liu, S. Chun, B. Zhang, et al., Nano-scale oxidation of copper in aqueous solution, Electrochem. Commun. 26 (2013) 21−24.

[58] F. Caballero-Briones, A. Palacios-Padros, O. Calzadilla, F. Sanz, Evidence and analysis of parallel growth mechanisms in Cu_2O films prepared by Cu anodization, Electrochim. Acta 55 (2010) 4353−4358.

[59] X. Wu, H. Bai, J. Zhang, F. Chen, G. Shi, Copper hydroxide nanoneedle and nanotube arrays fabricated by anodization of copper, J. Phys. Chem. B 109 (2005) 22836−22842.

[60] Z. Cheng, D. Ming, K. Fu, N. Zhang, K. Sun, pH-controllable water permeation through a nanostructured copper mesh film, ACS Appl. Mater. Interfaces 4 (2012) 5826−5832.

[61] F. Xiao, S. Yuan, B. Liang, G. Li, S.O. Pehkonen, T.J. Zhang, Superhydrophobic CuO nanoneedle-covered copper surfaces for anticorrosion, J. Mater. Chem 3 (2015) 4374−4388.

[62] L. Shooshtari, R. Mohammadpour, A.I. Zad, Enhanced photoelectrochemical processes by interface engineering, using Cu_2O nanorods, Mater. Lett 163 (2016) 81−84.

[63] W.J. Stępniowski, S. Stojadinovic, R. Vasilic, N. Tadic, K. Karczewski, S.T. Abrahami, et al., Morphology and photoluminescence of nanostructured oxides grown by copper passivation in aqueous potassium hydroxide solution, Mater. Lett 198 (2017) 89−92.

[64] T.G. Satheesh Babu, T. Ramachandran, Development of highly sensitive nonenzymatic sensor for the selective determination of glucose and fabrication of a working model, Electrochim. Acta 55 (2010) 1612−1618.

[65] W.J. Stępniowski, D. Paliwoda, Z. Chen, K. Landskron, W.Z. Misiolek, Hard anodization of copper in potassium carbonate aqueous solution, Mater. Lett 252 (2019) 182−185.

[66] R. Funahashi, I. Matsubara, H. Ikuta, T. Takeuchi, U. Mizutani, S. Sodeoka, An oxide single crystal with high thermoelectric performance in air, Jpn. J. Appl. Phys. Part. 2-Lett 39 (2000) L1127−L1129.

[67] R. Funahashi, I. Matsubara, Thermoelectric properties of Pb- and Ca-doped $(Bi_2Sr_2O_4)$ $(_x)CoO_2$ whiskers, Appl Phys Lett 79 (2001) 362−364.

[68] F. Gibelli, L. Lombez, J. Rodiere, J.F. Guillemoles, Optical imaging of light-induced thermopower in semiconductors, Phys Rev Appl 5 (2016) 024005.

[69] F. Gibelli, L. Lombez, J.F. Guillemoles, Two carrier temperatures non-equilibrium generalized Planck law for semiconductors, Phys. B-Conden Matter 498 (2016) 7−14.

[70] F. Gibelli, L. Lombez, J.F. Guillemoles, Accurate radiation temperature and chemical potential from quantitative photoluminescence analysis of hot carrier populations, J. Phys-Condens Matter 29 (2017) 06LT02.

Index

Note: Page numbers followed by "*f*" and "*t*" refer to figures and tables, respectively.

A

Active fiber composites (AFCs), 236–237
Additive micro/nano manufacturing, 120–121
Ag@ZnO/TiO$_2$ nanofibrous membrane, 182f
Air filter
  characterization of, 194–198
    permeability, 196–197
    pore size, and size distribution, 195–196
    porosity, 194–195
    quality factor, 197–198
    single-fiber efficiency, 197
    surface area, 196
  mechanism, 192–194
  nanofibrous particulate matter, 198–200
Air pollutants, 191–192
  filtration for. *See* Air filter
Air pollution, 191–192
Al$_2$O$_3$ fibers, 334f
Al-SnO$_2$/PANI nanofibers, 92–93, 93f
Aluminum oxide (Al$_2$O$_3$), 334–336
  in water treatment, 180
Anode materials, 306–307
Anodes, 319
Anodic aluminum oxide (AAO), 406–409
Assembled cell model, 352f
As-spun MONFs, heating of, 118–119
As-spun nanofibers, 39, 47–48
Atomic force microscopy, 230
Au/La$_2$O$_3$-doped SnO$_2$ NFs, 153

B

Barium titanate (BaTiO$_3$) nanofibers, 72–73, 236
BaTiO$_3$ nanotubes, TEM analysis of, 225, 226f
Bernoulli's principle, 66

Bimetallic/polymetallic oxides, 368–370
Biodegradable *vs.* bio-based polymer, 98t
Biopolymers, 97–105
  defined, 97–99
  examples of, 97–99
Biosensor, 114, 121, 121f
Bipolar resistive switching (BRS), 251, 257, 265
Bode Z_{Im} plots, 323
Bohr exciton radius, 284–286
Boltzmann's constant, 281–282
Bottom-up approaches, 65
British Standards Institution, 65
Brunauer–Emmett–Teller (BET) method, 196, 369–370
Butler–Volmer equation, 312

C

Calcinated hematite nanofibers, 166f
Calcination, 8, 35–36
  in air, 35–36
Calcined TiO$_2$ nanofibers, 119f
Carbon monoxide (CO), 191–192
Carbon nanofibers (CNFs), 373–383
Carbon tubes bundles (CTBs), 340–343
Cathode, 302, 318–319, 318f
Cathode materials, 305–306
CA/TiO$_2$/Ag NP composite nanofibers, 103–104, 104f
Cellulose, 99
Centrifugal jet spinning, 3
Centrifugal spinning process, 69–70, 73f, 340
CeO$_2$ composite nanofiber, 77–78, 79f
CeO$_2$ nanofibers, 36, 37f
Ceramics, 219
Charge extraction and transport, 283
Chemical bath deposition, 402–405

Chemical spray pyrolysis, 409–411
Chemical vapor deposition (CVD), for nanofiber synthesis, 116
Chemiresistive-based metal oxide gas sensors, 143–144
ChEt/ChOx dual enzyme loaded nanofibers, 125f
Chlorine, 178
Citric acid, in spinning solution, 11
Citric acid-based sol–gel process, 12f
CO_2 laser supersonic drawing method, 69f
Co_3O_4, 366, 374–375
Coaxial electrospinning technique, 42f, 142, 143f, 225f, 398–399
Cobalt tetraoxide (Co_3O_4)–ZnO C-S NFs, 149–150
Cobalt-based metal oxides, 315–317, 315f, 317f
Cobalt-free metal oxides, 318–319, 318f
Cold plasma. *See* Dielectric barrier discharge plasma
Colloid electrospinning, 12–13
Complementary metal oxide semiconductor (CMOS), 257
Complex oxide nanofibers, 261–262
Composite fibers, 293–294
Conduction band minimum (CBM), 279–280, 282–283
Conductive atomic force microscopy (cAFM), 419
Conventional water treatment processes, 174f
Copper oxide (CuO), 258, 336
 as antimicrobial agent, 178
 for electronic devices, 161
Copper oxide (CuO) nanofibers, 73–74, 74f
Core-sheath nanofibers, 41
Core-shell fiber, 408
Core–shell nanowires, 262–270, 263t, 267f, 268f, 269f
Core-shell structure, 398–399
Crystal class, classification of, 216f
C–S SnO_2/Au-doped In_2O_3 NF sensor, 150

D

Depth filtration, 192–193, 193f
Dielectric barrier discharge plasma, 73–74
Dielectric elastomers (DEs), 90–91

Drawing method, for nanofibers, 67–68, 69f
Dye-sensitized solar cell (DSSC), 279–280

E

Elastomers, 93–96
Electroactive polymers (EAPs), 90–93
Electrochemical biosensing, 122, 122f
Electrochemical deposition, for nanofiber synthesis, 116
Electrochemical impedance spectroscopy (EIS), 319–324, 320f, 320t, 322f
Electrochemical oxidation (ECO), 414–416
Electrode, 302
 architectures, 304–305
 preparation, 307–310, 308t
Electron beam lithography (EBL), 412, 414
Electronic state, 286–294, 286f
Electrons, excitation of, 280–281, 281f
Electrospinning (ES) process, 3–6, 139, 141, 307–311, 309t, 371, 397–402, 402f
 of colloids. *See* Colloid electrospinning
 components, 4f
 disadvantages of, 65–66
 for fabrication of MONFs, 117–120
  horizontal, 117–118, 117f
  limitations, 120
  vertical, 117–118
 features of, 89–90
 large-scale production, 16–19
 limitations of, 90
 mechanism of, 159–160
 metal oxide and composite nanofibers obtained from, 7t
 for metal oxide nanofibers, 6–16
 parameters, 13–16, 362
 of polysaccharides, 99
 principle of, 31–32
 procedure for MONFs, 160–161
 for producing metal oxide nanofibers, 33–40, 33f
  converting amorphous to crystalline structure, 38–39
  physical and chemical modifications, 39–40
  precursor solution, directly electrospinning of, 34–35
  removing polymeric component in composite nanofibers, 35–37

Index 427

for randomly oriented nanofibers, 50*f*
safety, 20–21
setup, 139–140, 140*f*, 160*f*
sol–gel electrospinning process, 8–11
solution, viscosity of, 141
Electrospun $BaTiO_3$ NFs, 236
Electrospun metal oxide nanofiber
(EMONF), 32, 118*f*, 163–164,
362–383
morphology of, 39–40
Electrospun nanofiber cathodes,
morphologies of, 310*f*
Electrospun nanofibers, 21, 51–52
Enzyme-linked immunosorbent assay
(ELISA) standard, 125–126
Energy storage materials, synthesis of one-
dimensional metal oxide–based
crystals
aluminum oxide, 334–336
copper oxide, 336
iron oxide, 336–338
manganese oxide, 338–340
nickel oxide, 340–343
silicon oxide and silicates, 343–344
tin oxide, 344
titanium oxides and titanates, 344–347
tungsten oxide and tungstates, 348–349
vanadium oxide, 349–351
zinc oxide, 352–354
zirconate fibers, 354–356
Equivalent circuit (EC) model, 319–323
Escherichia coli, 178
photocatalytic antibacterial efficiency
against, 182*f*
European Union (EU), 219–220
Exhaust gas pollutants, 200–201
Experimental symmetrical cell, 309*f*

F

Fabrication metal oxide nanofibers, oxides,
254*f*
Fe_2O_3/PVP composite nanofiber, 11, 12*f*
Fermi energy level, 281–286, 288–289,
291–293
Ferratrane, 10–11, 11*f*
Ferric oxide (Fe_2O_3) nanofibers, 50–51,
75–77, 76*f*
Ferroelectric materials, 215–216
Ferroelectricity, 217

Field emission scanning electron microscopy
(FESEM), 337*f*, 349*f*, 403, 404*f*,
415*f*, 416*f*
Forcespinning process. *See* Centrifugal
spinning process

G

Gadolinia-doped ceria (GDC), 302–303,
311, 316, 319
Gas filter, 200–205
MONFs as catalyst support matrix and
catalyst, 200–203
metal oxide nanofibrous photocatalysts,
203–205, 204*f*
Gas-sensing applications, of metal oxide
nanofibers, 143–153
composite NFs, 145–150
importance of, 143–144
loaded/doped, 150–153
pristine, 144–145
Glass-annealing method, 416–417
Graphene/vanadium ions, 351*f*
Gravitational fiber drawing technique,
120–121

H

Hammer's method, 348–349
High electrical conductivity, 287
High resistance state (HRS), 250–251,
269–270
High-energy-density storage devices, 236
Highest occupied molecular orbital
(HOMO), 279–280, 282–283
Hollow nanofibers, 41–43
Homovalent ion substitution, 292–293
Horizontal electrospinning, 117–118, 117*f*
Hydrocarbons (HCs), 191–192
Hydrogen oxidation reaction (HOR), 302
Hydrothermal method, for nanofiber
synthesis, 116
Hydrothermal synthesis, 220–222
of ferroelectric one-dimensional
nanostructures, 222*t*

I

Ideal gas sensor, 143
In situ formation of Ag nanoparticles, 184*f*
In situ polymerization method, 180
In_2O_3 nanofiber, 52–53

428   Index

Indium tin oxide (ITO), 51
  for electronic devices, 161
In-fiber porosity, 43–46
Inorganic nanofiber nanogenerators, 234t
Inorganic nanoparticles, 183
Inorganic piezoelectric materials, 216–220
Interfiber porosity, 43–46
Intermediate temperature–solid oxide fuel
      cells (IT-SOFCs), 302–303, 306,
      312, 315–318
International Commission on Non-Ionizing
      Radiation Protection, 20
Iron nitrate composite fibers, 77f
Iron oxide (Fe_2O_3), 336–338
  for electronic devices, 161

K

Knudsen number, 198–199

L

Lanthanum strontium cobalt ferrite (LSCF),
      305–306, 309–311, 315–317
La-substituted $SrTiO_3$ (LST), 319
Lead zirconate titanate (PZT), 215–216,
      219–220
  electrospinning of, 223–224
  hydrothermal synthesis, 221
  nanofiber generator, 232f
Lead-free KNN fibers, 225–226
Lithium ion hybrid supercapacitors, 370
Low resistance state (LRS), 250–251,
      269–270
Lowest unoccupied molecular orbital
      (LUMO), 279–280, 282–283

M

Magnesium oxide (MgO), in wastewater
      treatment, 178
Magnetic field–assisted method, 50–51
Magnetospinning technique, 120
Manganese oxide (MnO_2), 248, 338–340,
      338f
Melt mixing method, 180
Memristors switching, 250–253, 250t, 253f,
      255t
Metal oxide composite nanofibers, 145–150
Metal oxide nanofibers (MONFs), 3, 6, 32,
      36–37, 113, 247f
  advanced structures of, 40–55

alignment and patterns, control of,
    49–52
core-sheath, 40–43
hierarchical surface structures, control
    of, 46–49
hollow, 40–43
in-fiber and interfiber porosity,
    43–46
side-by-side, 40–43
three-dimensional fibrous aerogels,
    53–54
welding at cross points, 52–53
antibacterial activity of, 179f
applications of, 113, 114f
biosensing applications of, 121–128, 129t
  cobalt oxides, 126–128
  hematite, 126
  magnetite, 126
  titanium dioxide, 123
  ZnO, 123–126
as catalyst support matrix and catalyst,
    200–203
challenges, and future scope, 130–131
conductive and transparent networks,
    161–163
diameter of, 35–36
for electronic devices, 161, 165t
electrospinning process for, 6–16, 33–40
electrospun metal oxide nanofibers,
    362–370
  bimetallic or polymetallic oxides,
    368–370
  single metal oxides, 362–368
electrospun metal oxide nanofiber–based
    composites, 370–383
  metal oxide/carbon nanofibers/
    conducting polymer composites,
    381–383
  metal oxide/carbon-based composites,
    372–381
  metal oxide/metal oxide composites,
    371–372
fabrication of, 38f
gas-sensing, 143–153, 162
  composite NFs, 145–150
  importance of, 143–144
  loaded/doped, 150–153
  pristine, 144–145
mass production of, 54–55

Index

memristors and resistive switching, 250–253
photoelectrochemical properties of, 176–178
polymer matrix for, 39
recent computational advances, 128–130, 130f
recent trends, 249
resistive switching in, 254–262
  complex oxide nanofibers, 261–262
  core–shell nanowires, 262–270
  CuO, 258
  Nb_2O_5, 261
  NiO, 254–257
  TiO_2, 257–258
  VO_2, 261
  WO_3, 261
  zinc oxide, 258–261
synthesis of, 71–79, 114–121, 115f
  advanced microfabrication and nanofabrication strategies, 120–121
  barium titanate, 72–73
  copper oxide, 73–74, 74f
  ferric oxide, 75–77, 76f
  NiO, CeO_2
  physicochemical route, 115–116
  spinning technique, 116–120
  and NiO-CeO_2 composite nanofibers, 77–78, 79f
  silica, 71–72
  tin oxide, 71
  titanium dioxide, 78–79
  tungsten oxide, 74–75, 75f
  zinc oxide, 78–79
thermoelectric MONFs device concept and characterization, 417–420, 418f
thermoelectric MONFs processing technology, 396–417
  chemical bath deposition, 402–405
  chemical spray pyrolysis, 409–411
  electrochemical oxidation, 414–416
  electrospinning, 397–402
  glass-annealing method, 416–417
  microlithography and nanolithography, 411–414
  template-assisted deposition, 406–409
as water purifiers, 175–178, 177f
Metal oxide nanofibrous photocatalysts, 203–205, 204f

Metal oxide/carbon nanofibers/conducting polymer composites, 381–383
Metal oxide/carbon-based composites, 372–381, 377f, 379f
Metal oxide/metal oxide composites, 371–372, 373f
Metal oxide–based biosensors, 122
Metal-doped NFs, 150
Metal–insulator–metal (MIM), 250, 266–267
Metallic nanofibers, preparation of, 15–16
Methane (CH_4), 191–192
Microelectromechanical systems (MEMS), 411–412
Microemulsion, for nanofiber synthesis, 116
Microlithography and nanolithography, 411–414, 413f
Mixed ionic-electronic conductor (MIEC), 304–306, 324–327, 324f, 326f
Model development, 325–326
Model results, 326–327, 326f
Molten salt synthesis, 228
Morphotropic phase boundary, (MPB), 219
Multineedle electrospinning technology, 16–18, 18f
  modifications of, 18–19

N
$NaCo_2O_4$ nanofibers, 398–400, 400f
Nanocarving, 3
Nanofiber *vs.* conventional solid oxide fuel cell electrodes, 311–319
Nanofibers, 3, 113
  characteristics of, 31
  defined, 65
  in electronic devices, 159–161
  features, 114
  in-fiber pores, 43–46
  inherent properties of, 159
  interfiber pores, 43–46
  and nanowire synthesis, 355f
  nonelectrospinning techniques, 65–70
  of polylactide blends, 104–105
  production techniques, 65
  solvent vapor treatment of, 52
  welding of, 52–53
Nanofibrous particulate matter filter, 198–200
Nanofillers, 90

Nanogenerators, 231–235, 231*f*
  fabrication procedure of, 233*f*
Nanomaterial, defined, 65
Nanotechnology, 173
National Science Foundation, 65
Nb_2O_5, 261
Needleless spinning technology, 16–18
Nickel(II)oxide (NiO), 254–257, 340–343, 365–366
NiO composite nanofiber, 77–78, 79*f*
$NiO-CeO_2$ composite nanofiber, 77–78, 79*f*
Nitrogen oxides (NOx), 191–192
Nitrogen-doped carbon-coated Ni nanofibers, 15–16, 17*f*
Nonelectrospinning techniques, for nanofibers, 65–70
  centrifugal spinning, 69–70
  drawing method, 67–68, 69*f*
  plasma-induced, 67, 67*f*
  solution blow spinning, 66–67
  template synthesis, 68–69
Nonelectrospinning, for fabrication of MONFs, 120
Nonvolatile random access memory (NV-RAM), 257

O
Ohm's law, 325
One dimenstional nanostructured ferroelectric, and piezoelectric materials, 227*t*
One-dimensional (1D) material, 254, 278–279, 284–286
  nanostructures, synthesis of, 220
Organic polymers, 7–8
Oxide one-pot synthesis (OOPS) process, 10–11
Oxygen reduction reaction (ORR), 301–303, 323
Oxygen storage components (OSC), 201–202
Ozone (O_3), 191–192

P
Palladium (Pd)-loaded SnO_2 NFs, 150
$PANI/TiO_2$ nanofibers, 92
PANI/zinc oxide (ZnO) nanofibers, 91–92, 92*f*
Particulate matter (PM), 191–192

Patent ductus arteriosus (PDA) coating, 184–185
Percolation threshold, 304
Perovskite, 305
Perovskite-type metal oxides (ABO_3), 368
Phase separation, for nanofiber synthesis, 116
Photoanode, 284–286, 285*f*, 291*f*
Photocatalysis, 203–205, 204*f*
Photocatalytic oxidation (PCO), 203–204
Photoconversion efficiency (PCE), 277–279, 284–288, 293–294
Photocurrent, generation of, 284
Photolithographic technology, 412
Photovoltage, generation of, 281–283
Photovoltaic (PV) cells, 277–284, 286–287, 289–290, 292–294
  one-dimensional nanomaterials in, 278*f*
  sensitized photovoltaic cells, n-type doped nanofiber photoanode for, 290*t*, 292*t*
Physicochemical route, for MONF fabrication, 115–116
Piezoelectric materials, 215–216
  inorganic, 216–220
  material and structural characterizations, 229
  potential applications, 231–237
    high-energy-density storage devices, 236
    nanogenerators, 231–235, 231*f*
    structural health monitoring, 236–237
Piezoelectric nanogenerator, 231–235, 231*f*
Piezoelectric NFs, material and structural characterizations, 229
Piezoelectricity, 216
Piezoforce microscopy (PFM), 230
PLA/TiO_2 nanofibers, 101–102, 102*f*
Planck's law, 419–420
Plasma-enhanced chemical vapor deposition (PECVD), for nanofiber synthesis, 116
Plasma-induced technique, 3, 67, 67*f*
Platinum (Pt)–chromium oxide (Cr_2O_3)–tungsten oxide (WO_3) composite NFs, 149
Point-of-care immunosensor device, 125–126, 126*f*
Poly (methyl methacrylate) (PMMA), 340
Polyacrylic acid, 34

Index

Polyacrylonitrile (PAN), 7–8, 34, 46, 362, 373–376, 378–381, 397
Polyaniline (PANI), 90–92, 92f
Polyaniline composite, electrospun fibers of, 181
Polydimethylsiloxane (PDMS), 232
Polyethylene oxide (PEO), 7–8
Polyhydroxyalkanoates (PHAs), 97–101
Polylactic acid (PLA), 97–100
Polymer matrix polyvinylidene fluoride (PVDF), 236
Polymeric nanocomposites, 89
Polymer–metal oxide composite fibers, for water treatment, 179–186
Polymers
    biodegradable $vs.$ bio-based, 98t
    biopolymers, 97–105
    category of, 90
    elastomers, 93–96
    electroactive, 90–93
Polymetallic oxides, 368–370
Polymethyl methacrylate (PMMA), 34, 411–413
Polypyrrole (PPy), 376, 381–383
Polystyrene (PS), 7–8
Polyvinyl acetate (PVAc), 34, 362, 366–368
Polyvinyl alcohol (PVA), 7–8, 34, 97–100, 336–338, 366, 400
Polyvinyl butyral (PVB), 7–8
Polyvinyl pyrrolidone (PVP), 6–8, 34, 140–141, 336–338, 340–344, 348–349, 354–355, 362–363, 365–368, 398–399
Polyvinylidene fluoride (PVDF) nanofiber, 52–53, 343–344
Porosity, of filter medium, 194–195
Porous tin oxide (SnO_2) NFs, 144
Precursor solution, 309
$PrFeO_3$ hollow nanofibers, 144–145, 147f
Pristine and Rh-doped SnO_2 NFs, 152–153
Pristine metal oxide nanofibers, 144–145
Pseudomonas aeruginosa, 178
Pt-doped In_2O_3 porous NFs, 151–152
PU nanofibers, 94–96
PU/CuO composite nanofibers, 94–95, 95f
    XRD patterns of, 96f

Q

Quartz (SiO_2), as piezoelectric material, 218

R

R_{act}, 312
Radio frequency (RF), 403–405
Raman spectroscopy, 229–230
    atomic force microscopy, 230
Recent trends, 249
Reduced graphene oxide (rGO), 145–147
Remember every situation encourages transmission (RESET), 250–251
Resistive random access memory (RRAM), 257
Resistive switching, 250–253, 250t, 253f, 255t
    operation modes of, 252f
rGO/SnO_2 nanofibers, gas-sensing mechanism of, 148–149, 148f
Ruddesdlen–Popper (RP), 306, 318–319
Ruthenium oxide (RuO_2), 248, 362–363, 371–374, 378–380

S

Scanning electron microscopy (SEM), 260f, 335f, 343f, 366, 409–410, 410f, 411f, 414, 416
Seebeck coefficient, 398–401, 406, 417, 419–420
Selected area electron diffraction (SAED) pattern, 123, 125f
Side-by-side electrospinning method, 43
Silica (SiO_2) nanofibers, 71–72
Silica microfibers and nanofibers, 186f
Silicon oxide and silicates, 339f, 343–344
Silver (Ag)-doped $LaFeO_3$ NFs, 153
Single electron transistor (SET), 250–251
Single metal oxides, 362–368, 364f, 367f
Single-fiber mechanisms, 193, 193f
Single-needle electrospinning method, 16–18
Sintering. *See* Calcination
SiO_2 nanofibers, 53–54
SiO_2 nanoparticle aerogels, 53–54
SnO_2 NF, 150
SnO_2–indium oxide (In_2O_3) composite NFs, 149
Sodium chloride (NaCl), 74
Soft dielectric EAPs, 90–91
Solar cells, metal oxide nanofibers in photoanode in, 284–286, 285f
    reducing energy trap states, 286–294

432 Index

Solar cells, metal oxide nanofibers in
(*Continued*)
  composite fibers, 293–294
  Fermi energy level, 288–289
  homovalent ion substitution, 292–293
  improving crystallinity through high
    sintering, 287–288
  n-type doping induced diffusion
    coefficient improvement, 289–290
  p-type doping induced schottky-barrier,
    291–292
  role of nanofibers in, 277–279
  sensitized photovoltaic cells,
      photoconversion mechanism in,
      279–284
    charge extraction and transport, 283
    excitation of electrons, 280–281
    generation of photocurrent, 284
    generation of photovoltage, 281–283
Solar energy conversion, 279f
Sol–gel electrospinning process, 15
  hydrolysis and condensation reaction, 9f
  precursor solution for, 8–11
Sol–gel method, 399–400
  for nanofiber synthesis, 116
Sol–gel template synthesis, 228
Solid oxide fuel cells, metal oxide
      nanofiber-based electrodes in, 301f,
      302f
  electrospinning, 307–311
    electrode preparation, 307–310
    typical electrode structures, 310–311
  nanofiber solid oxide fuel cell electrodes,
      structure-performance relationship in,
      319–327
    electrochemical impedance
      spectroscopy, 319–324
    infiltrated mixed ionic-electronic
      conductor nanofiber electrodes, one-
      dimensional pseudohomogeneous
      model of, 324–327
  nanofiber *vs.* conventional solid oxide fuel
      cell electrodes, electrochemical
      performance of, 311–319
    cobalt-based metal oxides, 315–317
    cobalt-free metal oxides, 318–319
    strontium-doped lanthanum manganite,
      313–315

state-of-the-art architectures and materials
    for, 303–307
  anode materials, 306–307
  cathode materials, 305–306
  electrode architectures, 304–305
Solution blow spinning (SBS), 66–67, 73f
Solution mixing method, 180
Solvent vapor treatment, 52
Solvents, 140–141
Specific operating temperature, 52–53
Spinel structure (AB_2O_4), 369–370
Spinning technique, for fabrication of
    MONFs, 116–120
  electrospinning, 117–120
  nonelectrospinning, 120
Spray pyrolysis technique, 411
Staphylococcus aureus, 178
Strontium-doped lanthanum manganite,
    313–315, 313f, 314t
Structural health monitoring, 236–237
Sulfur dioxide (SO_2), 191–192
Supercapacitors, electrochemical
    performance of, 361
Surface filtration, 192–193, 193f
Synthesizing composite nanofibers, 401f

T
Taylor cone, 5–6, 5f, 159–160
Template synthesis, 68–69
Template-assisted deposition, 406–409
Thermal expansion coefficient (TEC),
    306–307
Thermoelectric energy conversion system,
    395–396
Thermoelectric module, 395
3-aminopropyltriethoxysilane (APS), 236
Three-dimensional fibrous aerogels, 53–54,
    54f
Three-dimensional (3D) printing technique,
    120–121
Three-phase boundary (TPB), 302–305,
    311–314, 326–327
Three-way catalyst (TWC), 201–202
Tin oxide (SnO_2) nanofibers, 71, 344
Titanates, 344–347
Titanium dioxide (Ti_2O), 248–249,
    257–260, 265, 287–290, 293–294,
    368, 371–372, 376

Index

effect of heat treatment temperature on, 142*f*
for electronic devices, 161
nanofibers, 34–35, 48*f*
photocatalysis process of, 175–176
Titanium dioxide (TiO_2)-SnO_2 C–S composite NFs, 149
Titanium dioxide nanofibers, 78–79, 125*f*
for biosensing, 123
for esterified cholesterol sensing, 124*f*
Titanium oxide (TiO_2), 344–347, 345*f*
Transmission electron microscope (TEM), 258–260, 347*f*, 348–349, 366, 399–400, 416–417
x-ray diffraction patterns, 406*f*
Transparent conductive electrodes (TCEs), 161–162
Triethanolamine (TEA), 9–10, 10*f*
Triisopropanolamine (TIS), 9–10, 10*f*
Tungstates, 348–349
Tungsten oxide (WO_3) nanofibers, 74–75, 75*f*, 348–349
24-needle electrospinning system, 18–19, 19*f*
Typical electrode structures, 310–311

U

Ultraviolet (UV), 411–412
Ultraviolet light-activated gas-sensing, 147*f*
Unipolar resistive switching (URS), 251, 257
Uric acid biosensor, 124–125

V

V_2O_5, 363–365
Valence band maximum (VBM), 282–283
Vanadium oxide, 349–351
Vapor–liquid–solid (VLS), 3, 257
Vertical electrospinning, 117–118
V_{Nernst}, 312
VO_2, 261
Volatile organic solvents, 140–141

W

Water pollution, 173
Water treatment
metal oxides in, 174–175
polymer–metal oxide composite fibers for, 179–186
WO_3, 261
World Health Organization (WHO), 191

X

X-ray diffraction (XRD), 229, 407–408

Y

Yttria-stabilized zirconia (YSZ), 302–303, 313–315

Z

Zinc oxide (ZnO), 248–249, 258–261, 287–288, 293–294, 352–354
as antimicrobial agent, 178
for electronic devices, 161
nanofibers, 78–79
for biosensing, 123–126
FE-SEM images of, 127*f*
as piezoelectric material, 218
as sensing material, 144, 146*f*
Zirconate fibers, 353*f*, 354–356
Zirconatrane, 10–11, 11*f*
ZnO-SnO_2 NFs, 150, 151*f*
ZrO_2/PVP composite nanofiber, 11, 12*f*

Printed in the United States
by Baker & Taylor Publisher Services